Photosynthesis Bibliography

volume 8 1977

References no. 28496-32554 / AAS-ZUT

Editors Z. Šesták & J. Čatský

Springer-Science+Business Media, B.V. 1981

Contributors:
Z. Šesták
J. Čatský
I. Tichá
J. Pospíšilová
J. Solárová
D. Hodáňová

ISBN 978-90-6193-047-1   ISBN 978-94-017-2638-2 (eBook)
DOI 10.1007/978-94-017-2638-2

PREFACE

The bibliography includes papers in all fields of photosynthesis research - from studies of model biochemical and biophysical systems of the photosynthesis mechanism to primary production studied by the so-called growth analysis. In addition to papers devoted entirely to photosynthesis, papers on other topics are included if they contain data on photosynthetic activity, photorespiration, chloroplast structure, chlorophyll and carotenoid synthesis and destruction, *etc.*, or if they contain valuable methodological information (measurement of selected environmental factors, leaf area, *etc.*). In many branches it has been very difficult to define the limits of interest for photosynthesis researchers. This problem has arisen *e.g.* in topics dealing with the transport of gases, where - in addition to the papers on CO_2 transfer - some papers on water vapour transfer are included, these being of general application. On the other hand, many papers dealing with the anatomy and physiology of stomata have been omitted, if the aspect of carbon dioxide or water vapour exchange has not been discussed.

This volume contains references to papers published in the year 1977, and, similarly to Vol. 7, also addenda including references published in the preceding period (*i.e.* 1966 - 1976). The numbers of these additional references are labelled with an asterisk in the list of references.

To maximize the value of the bibliography the references are arranged alphabetically by authors' names, and each volume is provided with three indexes. The Authors' Index contains all names of authors, co-authors and editors. The Subject Index covers only primary items chosen according to their interest for photosynthesis researchers. Starting with Volume 6, the Subject Index has been newly arranged and enlarged. It contains more details on the electron transport chain, carbon fixation pathways, gas exchange on leaf and canopy level, *etc.* In the Plant Index, only important crop plants and selected plant types and groups are indexed.

Cumulative indexes accompany Volumes 1, 5, and then every fifth volume, *i.e.* Volumes 10, 15, *etc.*

We have tried to cover fully the relevant papers which have appeared in the most important scientific periodicals and books. Articles published in local journals, mimeographed booklets, *etc.*, were chosen mostly from reprints and lists of publications received directly from the authors. Only abstracts published in regular journals were included.

Since some 3000 relevant papers are currently published every year and included in this bibliography, and since the majority of citations have been checked with the originals, collecting and preparing for publication of such a large amount of material would have been impossible without the collaboration of the authors of the relevant publications. The courtesy of those authors who have already supplied us with reprints is highly appreciated.

We acknowledge with thanks the cooperation of our colleagues from the Institute of Experimental Botany of the Czechoslovak Academy of Sciences in Prague, especially Mrs. DRAHOMÍRA TĚŽKÁ, Mrs. LUDMILA HÁVOVÁ, Mrs. LENKA KOLČABOVÁ, Mrs. MARIE MANDLOVÁ and Mrs. MARTA ŠMÍDOVÁ who helped in preparing the card material. Mr. PETR ZÁZVORKA and the librarian of the Institute, Mrs. ZORA ZAWOYSKA helped us with checking the references.

Dr. Z. ŠESTÁK and Dr. J. ČATSKÝ

Institute of Experimental Botany
Czechoslovak Academy of Sciences

Flemingovo n. 2
CS-160 00 PRAHA 6
Czechoslovakia

INSTRUCTIONS FOR USE

All references are arranged alphabetically according to the authors' names and the year of publication. They are numbered and these numbers are used in the indexes. In case of a book title, the number is preceded by B. An asterisk preceding the number denotes the reference published in the preceding period (1966 - 1976).

The references contain the original unshortened title of the paper (book). English, French and German titles are cited in the original language. Titles in other languages are supplemented with a translation in English (sometimes using the title of the respective English abstract or a shortened title with omitted deadweight words). Titles of Japanese, Chinese *etc.* papers are given in English translation only. The journals' names are abbreviated mainly according to the "Style Manual for Biological Journals" (Second Edition, Amer. Institute of Biological Sciences, Washington, D.C. 1964), *e.g.* :

Abhandlungen	chinese	Industry	Publishers
Abstract	Chromatography	inorganic	quantitative
Abteilung	Commission	Institute	Quarterly
Academy	Communication	international	Radiation
Acta	comparative	Investigation	Radiobiology
Africa	Comptes rendus	italian	Rastenii
agricultural	Conference	Izvestiya	Recherche
Agriculture	Congress	Jahrbuch	Report
Agronomy	Contribution	japanese	Research
Akademie (-emiya)	Cytochemistry	Japan	Review
Algology	Cytology	Journal	royal
allgemeine	czechoslovak	Klasse	russian
american	Dendrology	Laboratory	russkii
America	Department	Landwirtschaft	scandinavicus
analytical	Deutschland	Letters	Science
Anatomy	Disease	Limnology	Section
angewandte	Dissertation	Magazin	Series (-iya)
Annals	Doklady	marine	Society
annual	Dopovidi	Mathematics	sovetskii
anorganisch (-nic)	Ecology	Microbiology	soviet
applied	Education	miscellaneous	special
Arbeit	Embryology	molecular	SSSR
Archiv	Encyclopedia	Monograph	Station
Atmosphere	Engineer	moskovskii	Supplement
atomic	Enzymology	Mycology	Survey
Australia	european	national	Symposium
Beiheft	experimental	natural	technical
Belgique	Experiment	Naturforschung	Technology
Bericht	Faculty	neerlandicus	Tijdschrift
biochemical	Federation	Netherland	Transaction
Biochemistry	Fizika	New Zealand	Travail (-aux)
biokhimicheskii	Fiziologiya	nuclear	tropical
Biokhimiya	Forestry	Oceanography	Trudy
biological(-ogicheskii)	Forschung	Optics	ukrainian
Biology (-ogiya)	Foundation	organic	UK
biophysical	France	original	US, USA
Biophysics	Gazette	Otdelenie	USSR
Bodenkunde	general	Pathology	University
bolgarskii	genetical	Pflanzen-	végétal
botanical (-anicheskii)	Genetics	Philosophy	Virology
Botany	Gesellschaft	physical	Virusforschung
british	Giornale	Physics	Volume
Bulletin	helveticus	physiological	Weekblad
Canada	Histochemistry	Physiology	Wetenschappen
cellular (-ulaire)	Histology	Phytopathology	Wissenschaft
central	Horticulture	Plant (-arum)	Zeitschrift
chemical	hungaricus	polish	Zeitung
Chemistry	Husbandry	Proceedings	Zentralblatt
chimicus	imperial	Publication	Zhurnal

The numbers at the end of each reference of a journal article denote : volume (issue) : first page - last page, year of publication. The number of issue is given only in the journal where each issue is paginated separately.

Book titles are cited according to the title page, not to the book jacket or cover (if the names of the editors are not given on the title page, they are not cited in the reference). The publishing house, place and year of publication are included.

Brackets at the end of the reference give bibliographic details and explanations to the contents, not given in the original. The following abbreviations are used most often :

ab	abstract	Jap.	Japanese
Arm.	Armenian	Latv.	Latvian
Belorus.	Belorussian	Lithu.	Lithuanian
Bulg.	Bulgarian	Norweg.	Norwegian
Car	carotenoids	PC	paper chromatography
CC	column chromatography	PhAR	photosynthetically active
Chin.	Chinese		radiation
Chl	chlorophyll	Pol.	Polish
Croat.	Croatian	Ps	photosynthesis
E	English	R	Russian
F	French	Roum.	Roumanian
G	German	Span.	Spanish
GC	gas chromatography	Swed.	Swedish
Georg.	Georgian	TLC	thin-layer chromatography
Hung.	Hungarian	Tr	transpiration
IRGA	infra-red gas analyser	Ukr.	Ukrainian
Ital.	Italian	Uz.	Uzbeg

The transliteration of Cyrillic characters is in accordance with the BSI-ASA//SC-Z39 draft table, *i.e.* :

Translit.	Cyrill.	Translit.	Cyrill.
a	а	p	п
b	б	r	р
ch	ч	s	с
d	д	sh	ш
e	е	shch	щ
ė	з	t	т
f	ф	ts	ц
g	г	u	у
i	и	v	в
ĭ	й	y	ы
k	к	ya	я
kh	х	yu	ю
l	л	z	з
m	м	zh	ж
n	н	"	ъ
o	о	'	ь

Several exceptions apply for Ukrainian, Belorussian and Serbian :

Translit.	Cyrill.		Translit.	Cyrill.	Translit.	Cyrill.
Ukr. y	и	Serbian :	ć	ħ	č	ч
i	і		dj	ђ	š	ш
ï	ї		dž	џ	c	ц
Beloruss.			h	х		
ŭ	ў		j	ј		
			lj	љ		
			nj	њ		

Authors' names are presented in spelling used in the original paper. If this spelling does not correspond to the original spelling used by the author (*e.g.* Russian papers of English authors), one spelling is referred to the other in the Authors' Index.

Printers' errors in the original papers are marked by underlining the respective words (letters).

E R R A T A

Ref. no.	For	Read
Volume 2		
10125	celss	cells
Volume 5, Part 1		
19426 last line		- 295, 1974.

*28496 - AASE, J.K., BROWN, J.H., BENCI, J.F. : Radiation penetration in canopies of
         barley cultivars isogenic for color. - Can. J. Plant Sci. *54* : 611 - 615,
         1974.

28497 - ABDULLAEV, Kh.A. : Polyploidy and the genome control of chloroplast number
         and function. - In : COOMBS, J. (ed.) : 4th International Congress on Photo-
         synthesis. P. 1. UKISES, London 1977.

28498 - ABDULRAHMAN, F.S., WINSTEAD, J.E. : Chlorophyll levels and leaf ultrastruct-
         ure as ecotypic characters in *Xanthium strumarium* L. - Amer. J. Bot. *64* :
         1177 - 1181, 1977.

28499 - ACKER, S., DURANTON, J. : Protochlorophyllide holochromes and structure of
         the plastids. - In : COOMBS, J. (ed.) : 4th International Congress on Photo-
         synthesis. Pp. 1 - 2. UKISES, London 1977.

28500 - ACKERSON, R.C., KRIEG, D.R., HARING, C.L., CHANG, N. : Effects of plant water
         status on stomatal activity, photosynthesis, and nitrate reductase activity of
         field grown cotton. - Crop Sci. *17* : 81 - 84, 1977.

28501 - ACKERSON, R.C., KRIEG, D.R., MILLER, T.D., STEVENS, R.G. : Water relations
         and physiological activity of potatoes. - J. amer. Soc. hort. Sci. *102* : 572-
         - 575, 1977. [Ps.]

28502 - ACOCK, B., CHARLES-EDWARDS, D.A., HEARN, A.R. : Growth response of a *Chrysan-
         themum* crop to the environment. I. Experimental techniques. - Ann. Bot. *41* :
         41 - 48, 1977. [Assimilation chamber.]

28503 - ADAMS, C.J., HILLS, F.J. : A power parabola for an assymetrical response. -
         Agron. J. *69* : 124 - 125, 1977. [Production.]

28504 - ADAMS, J.A., JOHNSON, H.B., BINGHAM, F.T., YERMANOS, D.M. : Gaseous exchange
         of *Simmondsia chinensis* (Jojoba) measured with a double isotope porometer and
         related to water stress, salt stress, and nitrogen deficiency. - Crop Sci.
         17 : 11 - 15, 1977. [Ps.]

28505 - ADAMS, J.E., ARKIN, G.F. : A light interception method for measuring row crop
         ground cover. - Soil Sci. Soc.Amer. J. *41* : 789 - 792, 1977.

28506 - ADAMS, M.W.W., HALL, D.O. : Purification and properties of the membrane-bound
         hydrogenase from the photosynthetic bacterium *Rhodospirillum rubrum*. - In :
         COOMBS, J. (ed.) : 4th International Congress on Photosynthesis. Pp. 2 - 3.
         UKISES, London 1977.

28507 - ADMIRAAL, W. : Salinity tolerance of benthic estuarine diatoms as tested with
         a rapid polarographic measurement of photosynthesis. - Mar. Biol. *39* : 11 -
         - 18, 1977.

28508 - AGEEVA, O.G. : Effects of light on thermostability of Hill reaction in pea
         and spinach chloroplasts. - Photosynthetica *11* : 1 - 4, 1977.

28509 - AHARONI, N., BLUMENFELD, A., RICHMOND, A.E. : Hormonal activity in detached
         lettuce leaves as affected by leaf water content. - Plant Physiol. *59* : 1169-
         - 1173, 1977. [Resistances.]

28510 - AHMAD, I., WAINWRIGHT, S.J. : Tolerance to salt, partial anaerobiosis, and
         osmotic stress in *Agrostis stolonifera*. - New Phytol. *79* : 605 - 612, 1977.
         [Growth analysis.]

28511 - AHMED, S.I., KENNER, R.A. : A study of *in vitro* electron transport activity
         in marine phytoplankton as a function of temperature. - J. Phycol. *13* : 116 -
         - 121, 1977.

28512 - AKANOV, È.N. : Avtomaticheskaya sistema dlya nepreryvnogo izmereniya intensiv-
         nosti fotosinteza rasteniĭ v germeticheskikh fitotronakh. [Automatic system
         for continuous recording the rate of photosynthesis of plants in hermetical-
         ly sealed air-conditioned cabinets.] - Fiziol. Rast. *24* : 653 - 659, 1977.
         [In R, ab : E.]

28513 - AKAZAWA, T. : Structure, enzymic activities and biosynthesis of ribulose bis-
         phosphate carboxylase. - In : COOMBS, J.(ed.) : 4th International Congress
         on Photosynthesis. P. 3. UKISES, London 1977.

28514 - ÅKERLUND, H.-E., ANDERSSON, B., ALBERTSSON, P.-Å. :Isolation of Photosystem II vesicles possessing inside-out properties. - In : COOMBS, J. (ed.) : 4th International Congress on Photosynthesis. P. 4. UKISES, London 1977.

28515 - AKHMANOV, S.A., BORISOV, A.Yu., DANELYUS, R.V., PISKARSKAS, A.S., RAZZHIVIN, A.P., SAMUILOV, V.D. : Rezonansnaya pikosekundnaya spektroskopiya fizicheskikh stadiĭ fotosinteza s pomoshch'yu perestraivaemykh generatorov. [Resonance picosecond spectroscopy of physical stages of photosynthesis using transformable generators.] - Pis'ma ZhĒTF 26 : 655 - 658, 1977. [In R.]

28516 - AKIYAMA, T., OKUBO, T., TAKAHASHI, S. :[Ecological efficiencies of energy conversion in pasture. III. Dry matter production in Sasa nipponica community.] - J. jap. Soc. Grassland Sci. 23 : 43 - 51, 1977. [Growth analysis; in Jap., ab : E.]

28517 - AKIYAMA, T., OKUBO, T., TAKAHASHI, S. : [Ecological efficiencies of energy conversion in pasture. IV. Dynamics of nitrogen and energy in Sasa nipponica community.] - J. jap. Soc. Grassland Sci. 23 : 52 - 59, 1977. [In Jap., ab : E.]

28518 - AKOYUNOGLOU, G. : Biogenesis of the photosynthetic membranes in higher plants. - In : PACKER, L., PAPAGEORGIOU, G.C., TREBST, A. (ed.) : Bioenergetics of Membranes. Pp. 71 - 84. Elsevier/North-Holland Biomedical Press, Amsterdam - - Oxford - New York 1977.

28519 - AKOYUNOGLOU, G. : Development of the Photosystem II unit. - In : COOMBS, J. (ed.) : 4th International Congress on Photosynthesis. Pp. 4 - 5. UKISES, London 1977.

28520 - AKOYUNOGLOU, G., TSIMILLI-MICHAEL, M. : Effect of plastid development on the 518 nm absorbance change in bean leaves. - Plant Sci. Lett. 10 : 129 - 139, 1977.

28521 - AKSENOV, S.M., KOZHEMYAKOV, A.P. : Vliyanie rasteniĭ na produtsirovanie uglekislogo gaza pochvami vyrabotannykh torfyanikov. [Effect of plants on the production of CO_2 by soils of worked peat bogs.] - Vestn. leningrad. Univ., Biol. 1977 (3) : 123 - 129, 1977. [In R, ab : E.]

B28522 - AKULOVA, E.A. (ed.) : Bibliograficheskiĭ Ukazatel' Nauchnykh Rabot Sotrudnikov Instituta Fotosinteza Akademii Nauk SSSR (1965 - 1975). [Bibliography of Scientific Papers of Co-workers of the Institute of Photosynthesis of the Academy of Sciences of the U.S.S.R. (1965 - 1975).] - Pushchino 1977. [In R.]

28523 - AKULOVA, E.A. : Flavonoly - endogennye regulyatory energeticheskogo obmena khloroplastov. [Flavonols - endogenous regulators of energy exchange in chloroplasts.] - In : AKULOVA, E.A., MUZAFAROV, E.N. (ed.) : Regulyatsiya Energeticheskogo Obmena Khloroplastov i Mitokhondriĭ Endogennymi Fenol'nymi Ingibitorami. Pp. 100 - 125, 129, 134. Pushchino 1977. [In R, ab : E.]

28524 - AKULOVA, E.A., MUZAFAROV, E.N., IVANOV, B.N. : Ingibiruyushchiĭ effekt kvertsetina na fotokhimicheskie reaktsii khloroplastov. [Inhibiting effect of quercetin on photochemical reactions of chloroplasts.] - Dokl. Akad. Nauk SSSR 233 : 958 - 961, 1977. [In R, ab : E.]

28525 - AKULOVA, E.A., ROSHCHINA, V.V. : Fotosinteticheskiĭ perenos elektronov na u-rovne tsitokhroma f i plastotsianina. [Photosynthetic electron transfer at the level of cytochrome f and plastocyanin.] - Biokhimiya 42 : 2140 - 2148, 1977. [In R, ab : E.]

28526 - AKULOVICH, N.K., PARSHYKAVA, T.A., ARLOŬSKAYA, K.I. : Dasledavanne karatkakhvalevaga peratvarėnnya spektral'nykh form khlarafilidu postėtyyaliravanykh listsyaŭ pry admoŭnaĭ tėmperatury. [Short-wave transformation of spectral forms of chlorophyllide from postetiolated leaves at a negative temperature.] - Vestsi Akad. Navuk belarus. SSR, Ser. biyal. Navuk 1977 (2) : 34 - 41, 1977. [In Belorus., ab : R.]

28527 - ALBERDA, T., LOUWERSE, W., Van LAAR, H.H., KREMER, D., De WIT, C.T., De VOS, N.M., BODLAENDER, K.B.A., TOXOPEUS, H. : Crop photosynthesis : methods and compilation of data obtained with a mobile field equipment. - Agr. Res. Rep. (Wageningen) 865 : 1 - 46, 1977.

28528 - **ALBERTE, R.S., TEL-OR, E.** : Characteristics of the photosynthetic apparatus of heterocyst and vegetative cells of *Nostoc* and *Anabaena*. - Plant Physiol. *59* (6, Suppl.) : 129, 1977.

28529 - **ALBERTE, R.S., THORNBER, J.P., FISCUS, E.L.** : Water stress effects on the content and organization of chlorophyll in mesophyll and bundle sheath chloroplasts of maize. - Plant Physiol. *59* : 351 - 353, 1977.

28530 - **ALEF, K., KLEMME, J.-H.** : Characterization of a soluble NADH-independent nitrate reductase from the photosynthetic bacterium *Rhodopseudomonas capsulata*. - Z. Naturforsch. *32C* : 954 - 956, 1977.

B28531 - **ALEXANDROV, V.Ya.** : Cells, Molecules and Temperature. Conformational Flexibility of Macromolecules and Ecological Adaptation. (Ecological Studies Vol. 21.) - Springer-Verlag, Berlin - Heidelberg - New York 1977. [Ps.]

28532 - **ALI, H.C.** : Comparison of chlorophyll content and stomata size of inbred lines and their hybrids of corn (*Zea mays* L.). - Z. Acker- Pflanzenbau *145* : 166 - 170, 1977.

28533 - **ALIEV, D.A., KAZIBEKOVA, È.G.** : Ob arkhitektonike i fotosinteticheskoĭ funktsii vysokourozhaĭnoĭ pshenitsy. [Architectonics and photosynthetic function of high-yielding wheat.] - Fiziol. Rast. *24* : 962 - 968, 1977. [In R, ab : E.]

28534 - **AL-KHAFAF, S., WIERENGA, P.J., WILLIAMS, B.C.** : A flotation method for determining root mass in soil. - Agron. J. *69* : 1025 - 1026, 1977.

28535 - **ALLEN, C.F., BRENDLER, S.** : Quinones of *Phormidium luridum*. - In : COOMBS, J. (ed.) : 4th International Congress on Photosynthesis. Pp. 5 - 6. UKISES, London 1977.

*28536 - **ALLEN, H.L., OCEVSKI, B.T.** : Limnological studies in a large, deep, oligotrophic lake (Lake Ohrid, Yugoslavia). Evaluation of nutrient availability and control of phytoplankton production through *in situ* radiobioassay procedures. - Arch. Hydrobiol. *77* : 1 - 21, 1976.

28537 - **ALLEN, H.L., OCEVSKI, B.T.** : Limnological studies in a large, deep, oligotrophic lake (Lake Ohrid, Yugoslavia). A summary of nutritional radiobioassay responses of the pelagial phytoplankton. - Hydrobiologia *53* : 49 - 54, 1977. [Production.]

28538 - **ALLEN, J.F.** : Effects of washing and osmotic shock on catalase activity of intact chloroplast preparations. - FEBS Lett. *84* : 221 - 224, 1977. [Ps.]

28539 - **ALLEN, J.F.** : Superoxide and photosynthetic reduction of oxygen. - In : MICHELSON, A.M., McCORD, J.M., FRIDOVICH, I. (ed.) : Superoxide and Superoxide Dismutases. Pp. 417 - 436. Academic Press, London - New York 1977.

28540 - **ALLEN, M.J.** : Direct conversion of radiant into electrical energy using plant systems. - In : BUVET, R., ALLEN, M.J., MASSUÉ, J.-P. (ed.) : Living Systems as Energy Converters. Pp. 271 - 274. North-Holland Publ. Co., Amsterdam - New York - Oxford 1977.

28541 - **AL'-MUSAVI, R.A.** : Zavisimost' razvitiya tsianofagov *Anabaena variabilis* ot fotosinteza i dykhaniya. [Relation of the development of cyanophage in *Anabaena variabilis* to photosynthesis and respiration.] - Mikrobiologiya *46* : 725 - - 729, 1977. [In R, ab : E.]

28542 - **ALSCHER, R., JAGENDORF, A.T., FONDA, S.A.** : The polyribosome-membrane association in chloroplasts. - Plant Physiol. *59* (6, Suppl.) : 9, 1977.

28543 - **ALTOSAAR, I., BOHM, B.A., TAYLOR, I.E.P.** : Isolation and properties of a ferredoxin from leaves of *Sambucus racemosa* L. - Can. J. Biochem. *55* : 159 - - 164, 1977.

28544 - **ALVIM, P. de T.** : Cacao. - In : ALVIM, P. de T., KOZLOWSKI, T.T. (ed.) : Ecophysiology of Tropical Crops. Pp. 279 - 313. Academic Press, New York - San Francisco - London 1977. [Growth analysis.]

B28545 - **ALVIM, P. de T., KOZLOWSKI, T.T.** (ed.) : Ecophysiology of Tropical Crops. - Academic Press, New York - San Francisco - London 1977. [Ps.]

*28546 - AMANO, S., HINO, A., DAITO, H., KURAOKA, T. : [Photosynthetic activity in several kinds of fruit trees. I. Effect of some environmental factors on the rate of photosynthesis.] - Engei Gakkai Zasshi *41* : 144 - 150, 1972. [In Jap., ab : E.]

28547 - AMESZ, J. : Low temperature reactions in photosynthesis. - In : CASTELLANI, A. (ed.) : Research in Photobiology. Pp. 121 - 128. Plenum Press, New York - London 1977.

28548 - AMESZ, J. : Photosynthesis. Biophysical aspects. - Progr. Bot. *39* : 48 - 61, 1977.

28549 - AMESZ, J. : Plastoquinone. - In : TREBST, A., AVRON, M. (ed.) : Photosynthesis I. (Encycl. Plant Physiol. N.S. Vol. 5.) Pp. 238 - 246. Springer-Verlag, Berlin - Heidelberg - New York 1977.

28550 - AMESZ, J., DUYSENS, L.N.M. : Primary and associated reactions of system II. - - In : BARBER, J. (ed.) : Primary Processes of Photosynthesis. Pp. 149 - 185. Elsevier, Amsterdam - New York - Oxford 1977.

28551 - AMESZ, J., GLASBERGEN, J.M., DE GROOTH, B.G. : Secondary electron transport at low temperatures in chloroplasts and chromatophores. - In : COOMBS, J. (ed.) : 4th International Congress on Photosynthesis. Pp. 6 - 7. UKISES, London 1977.

28552 - AMIRDZHANOV, A.G. : Radiatsionnye faktory i transpiratsionnyĭ raskhod vinogradnika. [Radiation factors and transpiration expenditure of a vineyard.] - Fiziol. Rast. *24* : 790 - 796, 1977. [Ps; in R, ab : E.]

*28553 .- ANDERSEN, J.L., HERGENRADER, G.L. : Comparative primary productivity of two flood control reservoirs in the Salt Valley watershed of eastern Nebraska. - Trans. Nebr. Acad. Sci. *2* : 134 - 143, 1973.

28554 - ANDERSON, J.M. : The molecular organization of chloroplast thylakoids. - In : BRINKLEY, B.R., PORTER, K.R. (ed.) : International Cell Biology, 1976 - 1977. Pp. 183 - 192. Rockefeller Univ. Press, New York 1977.

28555 - ANDERSON, J.W., DONE, J. : A polarographic study of glutamate synthase activity in isolated chloroplasts. - Plant Physiol. *60* : 354 - 359, 1977. [Ps.]

28556 - ANDERSON, J.W., DONE, J. : Polarographic study of ammonia assimilation by isolated chloroplasts. - Plant Physiol. *60* : 504 - 508, 1977. [Ps.]

28557 - ANDERSON, L.E. : Isolation of pea leaf chloroplast glucose-6-phosphate dehydrogenase by affinity gel chromatography. - Plant Physiol. *59* (6, Suppl.): 6, 1977.

28558 - ANDERSON, L.E., DUGGAN, J.X. : Inhibition of light modulation of chloroplast enzyme activity by sulfite. One of the lethal effects of SO_2. - Oecologia *28* : 147 - 151, 1977.

28559 - ANDERSON, L.E., NEHRLICH, S.C. : Isolation of pea leaf ribulose-5-phosphate kinase by affinity gel chromatography. - In : COOMBS, J. (ed.) : 4th International Congress on Photosynthesis. P. 7. UKISES, London 1977.

28560 - ANDERSON, L.E., NEHRLICH, S.C. : Photosynthetic electron-transport system controls cytoplasmic glucose-6-phosphate dehydrogenase activity in pea leaves. - FEBS Lett. *76* : 64 - 66, 1977.

28561 - ANDERSON, R.S., DOKULIL, M. : Assessments of primary and bacterial production in three large mountain lakes in Alberta, Western Canada. - Int. Rev. ges. Hydrobiol. *62* : 97 - 108, 1977.

28562 - ANDERSSON, B., ÅKERLUND, H.-E., ALBERTSSON, P.-Å. : Light-induced reversible proton extrusion by spinach-chloroplast photosystem II vesicles isolated by phase partition. - FEBS Lett. *77* : 141 - 145, 1977.

28563 - ANDREEVA, N.E., BARASHKOV, B.I., ZAKHAROVA, G.V., SHUBIN, V.V., CHIBISOV, A. K. : Priroda élektronno-vozbuzhdennogo sostoyaniya v okislitel'no-vosstanovitel'nykh reaktsiyakh pigmentov. I. Fotookislenie khlorofilla *a* p-benzokhinonom. [Nature of electron-excited state in pigment redox reactions. I. Photooxidation of chlorophyll *a* with *p*-benzoquinone.] - Biofizika *22* : 761 - - 765, 1977. [In R, ab : E.]

28564 - **ANDREO, C.S., VALLEJOS, R.H.** : An essential arginyl residue in the soluble chloroplast ATPase. - FEBS Lett. *78* : 207 - 210, 1977.

28565 - **ANDREOLI, M.G., OLTOLINI, A., PAPINI, C., TORRICELLI, P., VEZZANI, S.** : Analisi critica sui metodi della determinazione della clorofilla "*a*" nel Plancton. [Critical analysis of the methods for determining chlorophyll *a* in plankton.] - Ateneo parmense, Acta nat. *13* (1) : 45 - 80, 1977. [In Ital., ab : E.]

28567 - **ANDREWS, P., COLLINS, W.J., STERN, W.R.** : The effect of withholding water during flowering on seed production in *Trifolium subterraneum* L. - Aust. J. agr. Res. *28* : 301 - 307, 1977. [Growth analysis.]

28566 - **ANDREWS, T.J., ABEL, K.M.** : Photosynthetic carbon metabolism in tropical seagrasses. - In : COOMBS, J. (ed.) : 4th International Congress on Photosynthesis. P. 8. UKISES, London 1977.

28568 - **ANKAR, S., HOBRO, R., LARSSON, U.** : Report from the biological groups of phytoplankton counting, chlorophyll measurements and benthic macrofauna sampling. - Ambio *1977* (spec. Rep. 5) : 245 - 248, 1977.

28569 - **ANTIA, N.J., CHENG, J.Y.** : Reexamination of the carotenoid pigments of the unicellular blue-green alga *Agmenellum quadruplicatum*. - J. Fish. Res. Board Can. *34* : 659 - 668, 1977.

28570 - **ANTON, J.A., KWONG, J., LOACH, P.A.** : Synthesis of covalently linked porphyrin dimers and trimers as models for the primary electron-donor unit in photosynthesis. - In : OLSON, J.M., HIND, G. (ed.) : Chlorophyll-Proteins, Reaction Centers, and Photosynthetic Membranes. Pp. 370 - 371. Brookhaven nat. Lab., Upton 1977.

28571 - **AOKI, M., YABUKI, K.** : Studies on the carbon dioxide enrichment for plant growth, VII. Changes in dry matter production and photosynthetic rate of cucumber during carbon dioxide enrichment. - Agr. Meteorol. *18* : 475 - 485, 1977.

28572 - **AOSHIMA, H., SEKIYA, J., KAJIWARA, T., HATANAKA, A.** : Application of a spin trapping method to the free radical produced in the chloroplasts of *Thea sinensis* leaves in the presence of linoleic acid. - Agr. biol. Chem. *41* : 1787- - 1788, 1977.

*28573 - **D'AOUST, B.G., WHITE, R., WELLS, J.M., OLSEN, D.A.** : Coral-algal associations: capacity for producing and sustaining elevated oxygen tensions *in situ*. - Undersea biomed. Res. *3* : 35 - 40, 1976.

28574 - **APEL, K.** : *Acetabularia* : A model system for studying the biosynthesis of chloroplast membranes. - In : WOODCOCK, C.L.F. (ed.) : Progress in *Acetabularia* Research. Pp. 137 - 151. Academic Press, New York - San Francisco - London 1977.

28575 - **APEL, K.** : Chlorophyll-proteins.from *Acetabularia mediterranea*. - In : OLSON, J.M., HIND, G. (ed.) : Chlorophyll-Proteins, Reaction Centers, and Photosynthetic Membranes. Pp. 149 - 161. Brookhaven nat. Lab., Upton 1977.

28576 - **APEL, K.** : The light-harvesting chlorophyll *a/b* · protein complex of the green alga *Acetabularia mediterranea*. Isolation and characterization of two subunits. - Biochim. biophys. Acta *462* : 390 - 402, 1977.

28577 - **APEL, K.** : Thylakoid membrane proteins of the green alga *Acetabularia mediterranea*. - In : BOGORAD, L., WEIL, J.H. (ed.) : Acides Nucléiques et Synthèse des Protéines chez les Végétaux. Coll. Int. C.N.R.S. No. 261. Pp. 403 - - 412. Édit. C.N.R.S., Paris 1977.

28578 - **APEL, P.** : Kornwachstum und Assimilatzwischenspeicherung bei Winterweizen und Sommergerste. - In : Produkce Biomasy a Tvorba Výnosů Polních Plodin. Vol. 1. Pp. 69 - 73. Česká vědeckotechnická Společnost zemědělská, Praha 1977.

28579 - **APOSTOLOVA, R.D., KSENZHEK, O.S.** : Vliyanie ionnogo sostava vneshneĭ sredy na kislorodnyĭ obmen vodorosli *Acetabularia*. [The effect of ionic composition of the environment on oxygen exchange of *Acetabularia*.] - Fiziol. Rast. *24* : 51 - 56, 1977. [Ps; in R, ab : E.]

28580 - APPELT, N., KNOBLOCH, K. : A cytochrome c similar to "c_2" as the electron acceptor during thiosulfate oxidation in *Rhodopseudomonas palustris*. - Hoppe-Seyler Z. physiol. Chem. *358* : 1174 - 1175, 1977.

28581 - APPLEBY, A.J. : Some irreversible thermodynamic aspects of photosynthesis. - In : COOMBS, J. (ed.) : 4th International Congress on Photosynthesis. Pp. 8 - - 9. UKISES, London 1977.

28582 - ARATA, H., TAKAMIYA, K.-I., NISHIMURA, M. : Delayed fluorescence from bacteriochlorophyll in *Chromatium vinosum* chromatophores. - Biochim. biophys. Acta *459* : 36 - 46, 1977.

28583 - ARATA, H., TAKAMIYA, K., NISHIMURA, M. : Delayed fluorescence from bacteriochlorophyll in *Chromatium vinosum* chromatophores : characteristics in the presence of o-phenanthroline. - J. Biochem. (Tokyo) *81* : 1133 - 1139, 1977.

*28584 - ARGYROUDI-AKOYUNOGLOU, J.H. : Effect of cations on the reconstitution of heavy subchloroplast fractions (grana) in disorganized low-salt agranal chloroplasts. - Arch. Biochem. Biophys. *176* : 267 - 274, 1976.

28585 - ARGYROUDI-AKOYUNOGLOU, J.H. : Development of the cation-induced stacking capacity in higher plant thylakoids during their biogenesis. - In : PACKER, L., PAPAGEORGIOU, G.C., TREBST, A. (ed.) : Bioenergetics of Membranes. Pp. 85 - - 98. Elsevier/North-Holland Biomedical Press, Amsterdam - Oxford - New York 1977.

28586 - ARGYROUDI-AKOYUNOGLOU, J.H., TSAKIRIS, S. : Onset of the cation-induced stacking property in thylakoids of greening *Pisum sativum*. - In : COOMBS, J. (ed.): 4th International Congress on Photosynthesis. Pp. 9 - 10. UKISES, London 1977.

28587 - ARMOND, P.A., ARNTZEN, C.J. : Localization and characterization of photosystem II in grana and stroma lamellae. - Plant Physiol. *59* : 398 - 404, 1977.

28588 - ARMOND, P.A., SCHREIBER, U. : Effect of high temperatures on photosynthetic electron transport activity and chlorophyll fluorescence characteristics. - In : COOMBS, J. (ed.) : 4th International Congress on Photosynthesis. Pp. 10- - 11. UKISES, London 1977.

28589 - ARMOND, P.A., SCHREIBER, U., BJÖRKMAN, O. : Photosynthetic acclimation to temperature in *Larrea divaricata* : Light harvesting efficiency and capacity of photosynthetic electron transport reactions. - Carnegie Inst. Year Book *76* : 335 - 341, 1977.

28590 - ARMOND, P.A., STAEHELIN, L.A., ARNTZEN, C.J. : Spatial relationship of photosystem I, photosystem II, and the light-harvesting complex in chloroplast membranes. - J. Cell Biol. *73* : 400 - 418, 1977.

28591 - ARNOLD, W.A., AZZI, J.R. : Two effects of electrical fields on chloroplasts. - Plant Physiol. *60* : 449 - 451, 1977.

28592 - ARNON, D.I. : Photosynthesis 1950-75 : Changing concepts and perspectives. - In : TREBST, A., AVRON, M. (ed.) : Photosynthesis I. (Encycl. Plant Physiol. N.S. Vol. 5.) Pp. 7 - 56. Springer-Verlag, Berlin - Heidelberg - New York 1977.

28593 - ARNON, D.I., CHAIN, R.K. : Ferredoxin-catalyzed photophosphorylations : Concurrence, stoichiometry, regulation, and quantum efficiency. - Plant Cell Physiol. *1977* (spec. Issue 3 - Photosynthetic Organelles. Structure and Function) : 129 - 147, 1977.

28594 - ARNON, D.I., CHAIN, R.K. : Role of oxygen in ferredoxin-catalyzed cyclic photophosphorylations. - FEBS Lett. *82* : 297 - 302, 1977,

28595 - ARNTZEN, C.J., ARMOND, P.A., BRIANTAIS, J.-M., BURKE, J.J., NOVITZKY, W.P. : Dynamic interactions among structural components of the chloroplast membrane. - In : OLSON, J.M., HIND, G. (ed.) : Chlorophyll-Proteins, Reaction Centers, and Photosynthetic Membranes. Pp. 316 - 337. Brookhaven nat. Lab., Upton 1977.

28596 - ARO, E.-M., VALANNE, N. : The effect of magnesium on the proportions of chlorophyll-protein complexes in polyacrylamide gels. - In : COOMBS, J. (ed.) : 4th International Congress on Photosynthesis. P. 11. UKISES, London 1977.

28597 - ARONOFF, S., KWOK, E. : Biosynthesis of chlorophyll *b*. - Can. J. Biochem.
55 : 1091 - 1095, 1977.

28598 - ARUGA, Y., MIURA, A. : *In vivo* absorption spectra and pigment content of the
two types of color mutants of *Porphyra*. - J. Phycol. *13* (Suppl.) : 5, 1977.

28599 - ASADA, K., TAKAHASHI, M., TANAKA, K., NAKANO, Y. : Formation of active oxygen
and its fate in chloroplasts. - In : HAYAISHI, O., ASADA, K. (ed.) : Biochemical and Medical Aspects of Active Oxygen. Pp. 45 - 63. Jap. sci. Soc. Press,
Tokyo 1977.

28600 - ASAMI, S., TAKABE, T., AKAZAWA, T. : Biosynthetic mechanism of glycolate in
Chromatium. IV. Glycolate-glycine transformation. - Plant Cell Physiol. *18* :
149 - 159, 1977.

28601 - ASENSI, A., DELEPINE, R., GUGLIELMI, G. : Nouvelles observations sur l'ultrastructure du plastidome des Phéophycées. - Bull. Soc. phycol. France *22* :
192 - 205, 1977.

28602 - ASHENDEN, T.W., MANSFIELD, T.A. : Influence of wind speed on the sensitivity
of ryegrass to SO_2. - J. exp. Bot. *28* : 729 - 735, 1977. [Boundary layer resistance.]

28603 - ASHIDA, K., KIKUCHI, R. : Phycocyanins of *Porphyra yezoensis*. - J. Phycol.
13 (Suppl.) : 5, 1977.

28604 - ASHTON, F.M., VILLIERS, O.T. de, GLENN, R.K., DUKE, W.B. : Localization of
metabolic sites of action of herbicides. - Pest. Biochem. Physiol. *7* : 122 -
- 141, 1977.

28605 - ASLAM, M., HUNT, L.A. : Effect of ear and tiller removal on flag leaf photosynthesis in spring wheat. - Plant Physiol. *59* (6, Suppl.) : 100, 1977.

28606 - ASLAM, M., LOWE, S.B., HUNT, L.A. : Effect of leaf age on photosynthesis and
transpiration of cassava (*Manihot esculenta*). - Can. J. Bot. *55* : 2288 - 2295,
1977.

28607 - ASPERGES, M., CAUBERGS, R. : Study of different pigments in *Cladonia pleuro-
ta* (FLOERKE) SCHAER. - Bull. Soc. roy. bot. Belg. *110* : 260 - 268, 1977.

28608 - ASSELIN, A., TRUDEL, M.J. : Incorporation de polyribonucléotides synthétiques
dans des chloroplastes isolés de *Triticum aestivum* cv. Opal. - Can. J. Plant
Sci. *57* : 865 - 872, 1977.

28609 - ASTAUROVA, O.B., AFANASOVA, L.A., SALAMAKHA, O.V. : Izozimnyĭ sostav malat-
degidrogenazy u dvukh vidov atsetabulyarii. [Isoenzyme composition of malate
dehydrogenase in two species of *Acetabularia*.] - Ontogenez *8* : 423 - 429,
1977. [In R, ab : E.]

28610 - ASTAUROVA, O.B., AFANASOVA, L.A., YAZYKOV, A.A. : Induktsiya izozimov malat-
degidrogenazy v khloroplastnoĭ fraktsii atsetabularii pri yadernykh trans-
plantatsiyakh. [Induction of malate dehydrogenase isoenzymes in the chloroplast fraction of *Acetabularia* following nuclear transplantation.] - Ontogenez *8* : 429 - 432, 1977. [In R, ab : E.]

28611 - ATALLAH, K., RAUSCHENBACH, P., SIMON, H., BERTHOLD, F., KOLBE, W. : Determination of the liquid scintillation counting efficiency of ^3H and/or ^{14}C labelled samples independently of the degree of colour and/or chemical quenching. - Z. Naturforsch. *32C* : 143 - 149, 1977.

28612 - ATANASIU, L., FABIAN-GALAN, G. : Progrese recente în cunoașterea căilor de
asimilare fotosintetică a carbonului. [Recent progress in investigation of
the photosynthetic carbon assimilation.] - In : Progrese și Perspective în
Biologie. Pp. 41 - 65. București 1977. [In Roum.]

28613 - ATKINS, C.A., KUO, J., PATE, J.S., FLINN, A.M., STEELE, T.W. : Photosynthetic
pod wall of pea (*Pisum sativum* L.). Distribution of carbon dioxide-fixing enzymes in relation to pod structure. - Plant Physiol. *60* : 779 - 786, 1977.

28614 - ATKINS, C.A., PATE, J.S. : An IRGA technique to measure CO_2 content of small
volumes of gas from internal atmospheres of plant organs. - Photosynthetica
11 : 214 - 216, 1977.

28615 - ATWOOD, D.K., KINARD, W.F., BARCELONA, M.J., JOHNSON, E.C. : Comparison of
polarographic electrode and Winkler titration determinations of dissolved
oxygen in oceanographic samples. - Deep-Sea Res. 24 : 311 - 313, 1977.

28616 - AUCLAIR, D. : Effets des poussières sur la photosynthèse II. Influence des
polluants particulaires sur la photosynthèse du Pin sylvestre et du Peuplier.
- Ann. Sci. forest. 34 : 47 - 57, 1977.

28617 - AUCLAIR, D., CAPUT, C. : Modifications de la photosynthèse de Pinus pinea L.
lors d'une pollution artificielle sub-nécrotique par le dioxyde de soufre in
situ. - Compt. rend. Séances Acad. Agr. France 63 : 563 - 569, 1977.

*28618 - AUFHAMMER, W., SOLANSKY, S. : Ein Beitrag zur Kontrolle des Assimilatetrans-
portes unter dem Einfluß von Wirkstoffen. - Z. Acker- Pflanzenbau 135 : 143 -
- 155, 1972.

28619 - AUNE, J., SMITH, B.J., ABOUL-ELA, M.M. : The effect of lysosomal enzymes on
chloroplasts in vitro. - Tex. J. Sci. 28 : 85 - 90, 1977.

*28620 - AUNG, L.H., WRIGHT, R.D., De HERTOGH, A.A. : Carbohydrate and dry matter chan-
ges in organs of Tulipa gesneriana L. during low temperature treatment. -
HortScience 11 : 37 - 39, 1976.

28621 - AUSLÄNDER, W., JUNGE, W. : Proton release into the internal phase of thylakoids linked to transitions between oxidation states of the oxygen evolving
enzyme system. - In : COOMBS, J. (ed.) : 4th International Congress on Photosynthesis. P. 12. UKISES, London 1977.

28622 - AUSSENAC, G., DUCREY, M. : Etude bioclimatique d'une futaie feuillue (Fagus
silvatica L. et Quercus sessiliflora SALISB.) de l'Est de la France I. Analyse des profils microclimatiques et des caractéristiques anatomiques et mor-
phologiques de l'appareil foliaire. - Ann. Sci. forest. 34 : 265 - 284, 1977.

28623 - AUST, H.J., DOMES, W., KRANZ, J. : Influence of CO_2 uptake of barley leaves
on incubation period of powdery mildew under different light intensities.
- Phytopathology 67 : 1469 - 1472, 1977.

28624 - AUSTIN, R.B., EDRICH, J.A., FORD, M.A., BLACKWELL, R.D. : The fate of the
dry matter, carbohydrates and ^{14}C lost from the leaves and stems of wheat
during filling. - Ann. Bot. 41 : 1309 - 1321, 1977.

28625 - AVADHANI, P.N., HIGGS, R.E.A. : C4 photosynthesiş among the "Adinandra Belu-
kar" species. - Plant Physiol. 59 (6, Suppl.) : 66, 1977.

28626 - AVARMAA, R., SOOVIK, T., TAMKIVI, R., TÕNISSOO, V. : Fluorescence life-times
of chlorophyll A and some related compounds at low temperatures. - Stud. biophys. 65 : 213 - 218, 1977.

*28627 - AVARMAA, R.A., MAURING, K.Kh. : Singlet-tripletnyĭ opticheskiĭ rezonans v
tverdom rastvore khlorofilla. [Singlet-triplet optical resonance in a solid
solution of chlorophyll.] - Opt. Spektroskop. 41 : 670 - 671, 1976. [In R.]

28628 - AVERINA, N.G., POLIKARPOVA, N.N., SHLYK, A.A. : Vliyanie khloramfenikola i
tsiklogeksimida na sintez δ-aminolevulinatdegidratazy v zelenykh i zeleneyushchikh prorostkakh yachmenya. [Effect of chloramphenicol and cycloheximide on
the synthesis of δ-aminolevulinate dehydratase in green and greening barley
shoots.] - Biokhimiya 42 : 2064 - 2070, 1977. [In R, ab : E.]

28629 - AVERY, D.J. : Maximum photosynthetic rate - a case study in apple. - New Phytol. 78 : 55 - 63, 1977.

28630 - AVERY, D.J. : Photosynthesis and limitations to productivity in apple. - In :
COOMBS, J. (ed.) : 4th International Congress on Photosynthesis. P. 13. UKISES,
London 1977.

28631 - AVILOV, I.A., MASLOV, Yu.I. : Spektroskopicheskaya kharakteristika nekotorykh
shtammov sinezelenykh vodorosleĭ. [Spectroscopic characteristics of some
strains of blue-green algae.] - In : NASYROV, Yu.S. (ed.) : Genetika Fotosinteza. Pp. 227 - 233. Donish, Dushanbe 1977. [In R.]

28632 - AVRATOVŠČUKOVÁ, N. : Genetika fotosyntézy. [Genetics of photosynthesis.] -
Stud. Inform. ÚVTIZ, zákl. Vědy Zeměděl. 1977 (1) : 1 - 96, 1977. [In Czech,
ab : E, R.]

28633 - **AVRON, M.** : Energy transduction in chloroplasts. - Annu. Rev. Biochem. *46* : 143 - 155, 1977.

28634 - **AVRON, M., FORK, D.C.** : The rate of proton gradient decay in chloroplasts as an indicator of membrane parameters. - Carnegie Inst. Year Book *76* : 231 - - 235, 1977.

28635 - **AVRON, M., SCHREIBER, U.** : Proton gradients as possible intermediary energy transducers during ATP-driven reverse electron flow in chloroplasts. - Carnegie Inst. Year Book *76* : 236, 1977.

28636 - **AVRON, M., SCHREIBER, U.** : Proton gradients as possible intermediary energy transducers during ATP-driven reverse electron flow in chloroplasts. - FEBS Lett. *77* : 1 - 6, 1977. [Device for simultaneous observation of fluorescence changes.]

28637 - **AXELSSON, L.** : The photostability of different chlorophyll forms in dark grown leaves of wheat. III. Dependence on age of the plants. - Physiol. Plant. *41* : 217 - 222, 1977.

28638 - **AXELSSON, L.** : The stabilization of the chlorophyll pigments against high intensity irradiation during the greening of dark grown leaves of wheat - physiological aspects on the low photostability of chlorophyll. - In : COOMBS, J. (ed.) : 4th International Congress on Photosynthesis. Pp. 13 - 14. UKISES, London 1977.

28639 - **BABAYAN, R.S., SAAKYAN, M.A., AĬRAPETYAN, R.B.** : Ob ugnetenii rostovykh protsessov u defektivnykh po sintezu khlorofilla prorostkov yachmenya. [Inhibition of growth processes in barley seedlings with defective chlorophyll synthesis.] - Fiziol. Rast. *24* : 637 - 639, 1977. [In R.]

28640 - **BACCARINI MELANDRI, A., CASADIO, R., MELANDRI, B.A.** : Thermodynamics and kinetics of photophosphorylation in bacterial chromatophores and their relation with the transmembrane electrochemical potential difference of protons. - Europe. J. Biochem. *78* : 389 - 402, 1977.

28641 - **BACCARINI-MELANDRI, A., MELANDRI, B.A.** : A role for ubiquinone-10 in the b-c_2 segment of the photosynthetic bacterial electron transport chain. - FEBS Lett. *80* : 459 - 464, 1977.

28642 - **BACCARINI MELANDRI, A., MELANDRI, B.A.** : Proton translocation in facultative photosynthetic bacteria. - In : MARRÈ, E., CIFERRI, O. (ed.) : Regulation of Cell Membrane Activities in Plants. Pp. 19 - 24. North-Holland Publ. Comp., Amsterdam - Oxford - New York 1977.

28643 - **BACCARINI MELANDRI, A., MELANDRI, B.A.** : The role of ubiquinone-10 in the b-c_2 segment of the cyclic photosynthetic chain of bacterial chromatophores. - In : PACKER, L., PAPAGEORGIOU, G.C., TREBST, A. (ed.) : Bioenergetics of Membranes. Pp. 199 - 204. Elsevier/North-Holland Biomedical Press, Amsterdam - Oxford - New York 1977.

28644 - **BACHOFEN, R., HANSELMANN, K.W., SNOZZI, M., ZÜRRER, H., CUENDET, P.A., ZUBER, H.** : Isolation of light-harvesting and reaction-center complexes from *Rhodospirillum rubrum*. - In : OLSON, J.M., HIND, G. (ed.) : Chlorophyll-Proteins, Reaction Centers, and Photosynthetic Membranes. Pp. 365 - 366. Brookhaven nat. Lab., Upton 1977.

28645 - **BADGER, M.R., COLLATZ, G.J.** : Studies on the kinetic mechanism of ribulose-1,5-bisphosphate carboxylase and oxygenase reactions, with particular reference to the effect of temperature on kinetic parameters. - Carnegie Inst. Year Book *76* : 355 - 361, 1977.

28646 - **BADGER, M.R., KAPLAN, A., BERRY, J.A.** : The internal CO_2 pool of *Chlamydomonas reinhardtii* : response to external CO_2. - Carnegie Inst. Year Book *76* : 362 - - 366, 1977.

28647 - **BADOUR, S.S., TAN, C.K., WAYGOOD, E.R.** : Observations on cell development in *Chlamydomonas segnis (Chlorophyceae)* at low and high carbon dioxide tension. - J. Phycol. *13* : 80 - 86, 1977.

28648 - BAES, C.F., Jr., GOELLER, H.E., OLSON, J.S., ROTTY, R.M. : Carbon dioxide and climate : The uncontrolled experiment. - Amer. Sci. 65 : 310 - 320, 1977.

*28649 - BAGAUTDINOVA, R.I., MOKRONOSOV, A.T., GRISHCHUK, G.I. : Izmenenie fotosinteticheskogo apparata pri narushenii morfofiziologicheskikh korrelyatsiĭ u rasteniĭ. [Change of the photosynthetic apparatus during disturbances of the morphological correlations in plants.] - Nauch. Dokl. vyssh. Shkoly, biol. Nauki 19 (11) : 87 - 92, 1976. [In R.]

28650 - BAHL, J. : Chlorophyll, carotenoid, and lipid content in Triticum sativum L. plastid envelopes, prolamellar-bodies, stroma lamellae, and grana. - Planta 136 : 21 - 24, 1977.

28651 - BAHL, J., FRANCKE, B., MONÉGER, R. : Effets comparés d'éclairements continus et intermittents sur l'évolution ultrastructurale et pigmentaire des plastes de feuilles étiolées de blé. - Biol. cellulaire 30 : 283 - 292, 1977.

28652 - BAHR, J.T., BOURQUE, P.D., SMITH, H.J. : Solubility properties of fraction I proteins of maize, cotton, spinach, and tobacco. - J. agr. Food Chem. 25 : 783 - 789, 1977.

28653 - BAIJOT, E., DeCALLONNE, J.R., MEYER, J.A. : Modifications induced by benomyl and related compounds into chloroplasts spectral patterns, photosynthetic rates and chlorophyll contents of Spinacia oleracea and Cucumis melo. - Bull. environ. Contam. Toxicol. 17 : 431 - 437, 1977.

28654 - BAILISS, K.W. : Gibberellins, abscisic acid and virus-induced stunting. - In: KIRALY, Z. (ed.) : Current Topics in Plant Pathology. Pp. 361 - 373. Akad. Kiadó, Budapest 1977. [Chl.]

28655 - BAKER, N.R., LEECH, R.M. : Development of Photosystem I and Photosystem II activities in leaves of light-grown maize (Zea mays). - Plant Physiol. 60 : 640 - 644, 1977.

28656 - BAKER, N.R., LEECH, R.M. : Development of the photosynthetic apparatus in light-grown maize. - In : COOMBS, J. (ed.) : 4th International Congress on Photosynthesis. Pp. 14 - 15. UKISES, London 1977.

28657 - BAKER, N.R., STRASSER, R.J., BUTLER, W.L. : Development of energy transfer between PSII and PSI in greening bean leaves. - In : COOMBS, J. (ed.) : 4th International Congress on Photosynthesis. P. 15. UKISES, London 1977.

28658 - BAKER, T.S., EISENBERG, D., EISERLING, F. : Ribulose bisphosphate carboxylase : A two-layered, square-shaped molecule of symmetry 422. - Science 196 : 293 - 295, 1977.

28659 - BAKER, T.S., SUH, S.W., EISENBERG, D. : Structure of ribulose-1,5-bisphosphate carboxylase-oxygenase: form III crystals. - Proc. nat. Acad. Sci. USA 74 : 1037 - 1041, 1977.

28660 - BAKKER-GRUNWALD, T. : ATPase. - In : TREBST, A., AVRON, M. (ed.) : Photosynthesis I. (Encycl. Plant Physiol. N.S. Vol. 5.) Pp. 369 - 373. Springer-Verlag, Berlin - Heidelberg - New York 1977.

*28661 - BALAN, V.V. : Dinamika soderzhaniya khlorofilla v list'yakh yabloni s pal'mettnoĭ formoĭ krony v zavisimosti ot podvoya, dozy predplantazhnogo udobreniya i ploshchadi pitaniya. [Dynamics of chlorophyll formation in leaves of apple tree with palmette crown form in dependence on rootstock, dose of presowing fertilizer and area of feeding.] - Tr. kishinev. sel'skokhoz. Inst. 154 : 55 - 59, 1976. [In R.]

28662 - BALASIMHA, D., RAM, G., TEWARI, M.N. : Cytokinin-coumarin interaction in relation to growth, sulfhydryl, chlorophylls, and peroxidase activity in Phaseolus radiatus L. - Biochem. Physiol. Pflanzen 171 : 49 - 54, 1977.

28663 - BALDY, P., VERIN, C., CAVALIÉ, G. : 2-phosphoglycollate phosphohydrolase from French bean leaves : purification and some catalytic properties of two isoenzymes. - In : COOMBS, J. (ed.) : 4th International Congress on Photosynthesis. Pp. 15 - 16. UKISES, London 1977.

28664 - BALNOKIN, Yu.V., FOKHT, A.S. : Amperometricheskiĭ metod opredeleniya skorosti kislorodnogo obmena fotosinteziruyushchikh organizmov pri razlichnom soderzhanii kisloroda v srede. [Amperometric technique for determination of the rate of oxygen exchange in photosynthetic organisms at different oxygen content in the medium.] - Fiziol. Rast. *24* : 207 - 214, 1977. [In R, ab : E.]

28665 - **BALNOKIN, Yu.V., FOKHT, A.S.** : Metod odnovremennoĭ registratsii skorosti vydeleniya kisloroda i skorosti ėlektronnogo transporta v reaktsii Khilla. [A method for simultaneous registration of the rate of oxygen evolution and the rate of electron transport in Hill reaction.] - Fiziol. Rast. *24* : 431 - 436, 1977. [In R, ab : E.]

28666 - **BALOCH, A.K., BUCKLE, K.A., EDWARDS, R.A.** : Separation of carrot carotenoids on Hyflo Super-Cel-magnesium oxide-calcium sulfate thin layers. - J. Chromatogr. *139* : 149 - 155, 1977.

28667 - **BALTIMORE, B.G., MALKIN, R.** : Appearance of membrane-bound iron-sulfar centers and photosystem 1 reaction center during greening of barley leaves. - Plant Physiol. *60* : 76 - 80, 1977.

28668 - **BALTIMORE, B.G., MALKIN, R.** : The appearance of membrane-bound iron-sulfur centers and the Photosystem I reaction center during greening of barley leaves. - In : COOMBS, J. (ed.) : 4th International Congress on Photosynthesis. Pp. 16 - 17. UKISES, London 1977.

*28669 - **BALTSCHEFFSKY, H.** : Protein structure and the molecular evolution of biological energy conversion. - BioSystems *6* : 217 - 223, 1975.

28670 - **BALTSCHEFFSKY, H.** : Protein β-structure and the molecular evolution of biological energy conversion. - In : BUVET, R., ALLEN, M.J., MASSUE, J.-P. (ed.) : Living Systems as Energy Converters. Pp. 81 - 88. North-Holland Publ. Co., Amsterdam - New York - Oxford 1977. [Ps.]

28671 - **BALTSCHEFFSKY, M.** : Conversion of solar energy into energy-rich phosphate compounds. - In : BUVET, R., ALLEN, M.J., MASSUE, J.-P. (ed.) : Living Systems as Energy Converters. Pp. 199 - 207. North-Holland Publ. Co., Amsterdam - New York - Oxford 1977.

28672 - **BALTSCHEFFSKY, M.** : Flash induced ATP formation and stimulation of ATPase in *R. rubrum* chromatophores. - In : COOMBS, J. (ed.) : 4th International Congress on Photosynthesis. P. 17. UKISES, London 1977.

28673 - **BANSE, K.** : Determining the carbon-to-chlorophyll ratio of natural phytoplankton. - Mar. Biol. *41* : 199 - 212, 1977.

28674 - **BARANNIKOVA, Z.D., BUREN', V.M., VITKOVSKAYA, V.V., PODVALKOVA, I.A., KRUZHILIN, A.S.** : Ustoĭchivost' i ontogenez kul'turnykh rasteniĭ v issledovaniyakh V.A. Novikova. [Resistance and ontogeny of cultivated plants in the research of V.A. Novikov.] - Fiziol. Rast. *24* : 180 - 184, 1977. [Ps, Chl; in R.]

28675 - **BARBER, D.J.W., RICHARDS, J.T.** : Energy transfer in the accessory pigments *R*-phycoerythrin and *C*-phycocyanin. - Photochem. Photobiol. *25* : 565 - 569, 1977.

28676 - **BARBER, J.** : Energy conversion and ion fluxes in chloroplasts. - In : Fertilizer Use and Production of Carbohydrate + Lipids. Pp. 83 - 93. Int. Potash Inst., Worblaufen - Bern 1977.

B28677 - **BARBER, J.** (ed.) : Primary Processes of Photosynthesis. - Elsevier, Amsterdam - New York - Oxford 1977.

28678 - **BARBER, J.** : Thylakoid membrane surface charges in relation to prompt and delayed chlorophyll fluorescence. - In : PACKER, L., PAPAGEORGIOU, G.C., TREBST, A. (ed.) : Bioenergetics of Membranes. Pp. 459 - 469. Elsevier/North-Holland Biomedical Press, Amsterdam - Oxford - New York 1977.

28679 - **BARBER, J., HALLIWELL, B.** : Photosynthesis at Reading. - Nature *270* : 104 - - 105, 1977.

28680 - **BARBER, J., MAURO, S., LANNOYE, R.** : The relationship between the yield factors for prompt and delayed fluorescence. - FEBS Lett. *80* : 449 - 454, 1977.

28681 - **BARBER, J., MILLS, J.** : Cation interaction with chloroplast membranes : The diffuse double layer and chlorophyll fluorescence. - In : THELLIER, M., MON-

NIER, A., DEMARTY, M., DAINTY, J. (ed.) : Echanges Ioniques Transmembranaires chez les Végétaux.   Pp. 553 - 558. Édit. CNRS, Paris 1977.

28682 - BARBER, J., MILLS, J., LOVE, A. : Electrical diffuse layers and their influence on photosynthetic processes. - FEBS Lett. *74* : 174 - 181, 1977.

28683 - BARBER, J., SEARLE, G.F.W. : 9-aminoacridine as a probe of the electrical double layer associated with the thylakoid membrane. - In : COOMBS, J. (ed.): 4th International Congress on Photosynthesis. P. 18. UKISES, London 1977.

28684 - BARCLAY, G.F., OPARKA, K.J., JOHNSTON, R.P.C. : Induced disruption of sieve element plastids in *Heracleum mantegazzianum* L. - J. exp. Bot. *28* : 709 - 717, 1977.

28685 - BARDEN, J.A. : Apple tree growth, net photosynthesis, dark respiration, and specific leaf weight as affected by continuous and intermittent shade. - J. amer. Soc. hort. Sci. *102* : 391 - 394, 1977.

28686 - BARG, R., UMIEL, N. : Effects of sugar concentration on growth greening and shoot formation in callus cultures from four genetic lines of tobacco. - Z. Pflanzenphysiol. *81* : 161 - 166, 1977.

28687 - BARKO, J.W., MURPHY, P.G., WETZEL, R.G. : An investigation of primary production and ecosystem metabolism in a Lake Michigan dune pond. - Arch. Hydrobiol. *81* : 155 - 187, 1977.

28688 - BARLOW, E.W.R., BOERSMA, L., YOUNG, J.L. : Photosynthesis, transpiration, and leaf elongation in corn seedlings at suboptimal soil temperatures. - Agron. J. *69* : 95 - 100, 1977.

28689 - BARNES, A. : The influence of the length of the growth period and planting density on total crop yield. - Ann. Bot. *41* : 883 - 895, 1977. [Growth analysis.]

28690 - BAR-NUN, S., OHAD, I. : Presence of polypeptides of cytoplasmic and chloroplastic origin in isolated photoactive preparations of photosystems I and II in *Chlamydomonas reinhardi* y-1. - Plant Physiol. *59* : 161 - 166, 1977.

28691 - BAR-NUN, S., SCHANTZ, R., OHAD, I. : Appearance and composition of chlorophyll-protein complex I and II during chloroplasts membrane biogenesis in *Chlamydomonas reinhardi* y-1. - Biochim. biophys. Acta *459* : 451 - 467, 1977.

28692 - BAROUCH, Y., CLAYTON, R.K. : Ubiquinone reduction and proton uptake by chromatophores of *Rhodopseudomonas sphaeroides* R-26. Periodicity of two in consecutive light flashes. - Biochim. biophys. Acta *462* : 785 - 788, 1977.

28693 - BARR, R., CRANE, F.L. : The effect of prostaglandins on photosynthesis. - Proc. Indiana Acad. Sci. *86* : 117 - 122, 1977.

28694 - BARR, R., CRANE, F.L. : The effect of various plastoquinone analogs on electron transport in spinach chloroplasts. - Plant Physiol. *59* (6, Suppl.) : 23, 1977.

28695 - BARR, R., CRANE, F.L. : Evidence for α-tocopherol function in the electron transport chain of chloroplasts. - Plant Physiol. *59* : 433 - 436, 1977.

28696 - BARR, R., CRANE, F.L., BEYER, G., MAXWELL, L.A., FOLKERS, K. : The effect of various substituted p-benzoquinones on electron transport in spinach chloroplasts. - Europe. J. Biochem. *80* : 51 - 54, 1977.

28697 - BARRATT, D.H.P., WOOLHOUSE, H.W. : The measurement of chloroplast protein turnover in *Phaseolus vulgaris*. - In : COOMBS, J. (ed.) : 4th International Congress on Photosynthesis. P. 19. UKISES, London 1977.

28698 - BARRETT, J., ANDERSON, J.M. : Subchloroplast membrane fragments of *Ecklonia radiata*. - In : COOMBS, J. (ed.) : 4th International Congress on Photosynthesis. Pp. 19 - 20. UKISES, London 1977.

28699 - BARRETT, J., ANDERSON, J.M. : Thylakoid membrane fragments with different chlorophyll *a*, chlorophyll *c* and fucoxanthin compositions isolated from the brown seaweed *Ecklonia radiata*. - Plant Sci. Lett. *9* : 275 - 283, 1977.

*28700 - BARSKIĬ, E.L., BORISOV, A.Yu., SAMUILOV, V.D. : Kompleksy reaktsionnykh tsen-
trov fotosinteziruyushchikh bakteriĭ. [Reaction centre complexes of photosynthesizing bacteria.] - Uspekhi sovrem. Biol. *82* : 222 - 235, 1976. [In R.]

28701 - BARSKY, E.L., SAMUILOV, V.D. : Bacteriochlorophyll photobleaching centered at
890 nm in *Rhodospirillum rubrum* subchromatophore particles. - In : COOMBS, J.
(ed.) : 4th International Congress on Photosynthesis. Pp. 20 - 21. UKISES,
London 1977.

28702 - BARSTOW, J.M., ERBISCH, F.H. : Effects of acute *gamma* radiation and winter tem-
perature-light conditions on photosynthesis of *Cladonia mitis*. - Bryologist
80 : 83 - 87, 1977.

28703 - BARTHOLOMEW, D.P., KADZIMIN, S.B. : Pineapple. - In : ALVIM, P. de T., KOZ-
LOWSKI, T.T. (ed.) : Ecophysiology of Tropical Crops. Pp. 113 - 156. Academic Press, New York - San Francisco - London 1977. [Ps.]

28704 - BARTHOLOMEW, P.W., CHESTNUTT, D.M.B. : The effect of a wide range of fertilizer nitrogen application rates and defoliation intervals on the dry-matter
production, seasonal response to nitrogen, persistence and aspects of chemical composition of perennial ryegrass (*Lolium perenne* cv. S.24). - J. agr. Sci.
88 : 711 - 721, 1977.

28705 - BASHI, E., ROTEM, J. : The effects of photosynthesis and exogenous glucose
on infection of tomato by *Stemphylium botryosum* f.sp. *lycopersici*. - Phytopathol. Z. *88* : 69 - 77, 1977.

28706 - BASSHAM, J.A. : Increasing crop production through more controlled photosynthesis. - Science *197* : 630 - 638, 1977.

28707 - BASSHAM, J.A. : Synthesis of organic compounds from carbon dioxide in land
plants. - In : MITSUI, A., MIYACHI, S., SAN PIETRO, A., TAMURA, S. (ed.) :
Biological Solar Energy Conversion. Pp. 151 - 166. Academic Press, New York
- San Francisco - London 1977.

28708 - BASZYŃSKI, T., TUKENDORF, A. : Restoration of Photosystem I in heptane-extracted spinach chloroplasts by β-carotene. - In : COOMBS, J. (ed.) : 4th
International Congress on Photosynthesis. P. 21. UKISES, London 1977.

25709 - BAUMANN, I., BAUMANN, G., GÜNTHER, G. : Effect of 3-amino-1,2,4-triazole
(amitrole) on cytoplasmic and chloroplast rRNA of oat seedlings. - Biochem.
Physiol. Pflanzen *171* : 157 - 163, 1977. [Chl.]

28710 - BAUR, J.R., MILLER, F.R., BOVEY, R.W. : Effects of preharvest desiccation
with glyphosate on grain sorghum seed. - Agron. J. *69* : 1015 - 1018, 1977.
[Chl.]

28711 - BAYLY, I.L., FREEMAN, E.A. : Seasonal variation of selected cations in *Acorus
calamus* L. - Aquatic Bot. *3* : 65 - 84, 1977.

*28712 - BAZHANOVA, N.V., ALTUNYAN, M.G. : K voprosu o razdelenii pigmentov plastid
metodom khromatografii na bumage i tonkom sloe. [Separation of plastid pigments by paper and thin layer chromatography.] - Biol. Zh. Arm. *29* (9) : 49 -
- 52, 1976. [In R, ab : Arm.]

*28713 - BAZHANOVA, N.V., GASPARYAN, O.B. : Nakoplenie pigmentov i vitaminov v zelenom
korme, vyrashchennom v kamere iskusstvennogo klimata v zavisimosti ot osveshcheniya. [Accumulation of pigments and vitamins in green fodder grown in an
artificial climate chamber in dependence on irradiance.] - Soobshch. Inst.
agrokhim. Probl. Gidroponiki Akad. Nauk arm. SSR *15* : 126 - 135, 1976. [In
R, ab : Arm., E.]

*28714 - BAZHANOVA, N.V., GEVORKYAN, A.G. : O razlichii deĭstviya temperaturnogo fak-
tora i ul'trafioletovogo oblucheniya na vzaimoprevrashcheniya ksantofillov
u vysokogornykh rasteniĭ. [Different action of temperature and ultraviolet
irradiation on interconversion of xanthophylls in high-mountain plants.] -
In : Vtoroĭ Vsesoyuznyĭ Biokhimicheskiĭ S"ezd. Tezisy Sektsionnykh Soobshche-
niĭ. 19 Sektsiya. Pp. 59 - 60. FAN, Tashkent 1969. [In R.]

*28715 - BAZHANOVA, N.V., GEVORKYAN, A.G., OGANESYAN, D.A. : Dinamika nakopleniya i
sootnoshenie pigmentov v rasteniyakh al'piĭskoi i subal'piĭskoĭ zon vysoko-
goriĭ Armenii. [Dynamics of accumulation and relation of pigments in plants

of the alpine and sub-alpine zones of Armenian high mountains.] - Biol. Zh. Arm. *21* (10) : 78 - 82, 1968. [In R, ab : Arm.]

*28716 - BAZHANOVA, N.V., GEVORKYAN, A.G., OGANESYAN, D.A. : Reaktsii vzaimoprevrash-cheniya ksantofillov u vysokogornykh rasteniĭ. [Reactions of interconversion of xanthophylls in high-mountain plants.] - Biol. Zh. Arm. *22* (8) : 58 - 64, 1969. [In R, ab : Arm.]

28717 - BEADLE, C.L. : The response of calculated mesophyll conductance and some photosynthetic partial processes to plant water status. - In : COOMBS, J. (ed.): 4th International Congress on Photosynthesis. P. 22. UKISES, London 1977.

28718 - BEADLE, C.L., JARVIS, P.G. : The effects of shoot water status on some photosynthetic partial processes in Sitka spruce. - Physiol. Plant. *41* : 7 - 13, 1977.

28719 - BEALE, S.I. : Biosynthesis of photosynthetic pigments - pathways and regulation. - In : COOMBS, J. (ed.) : 4th International Congress on Photosynthesis. Pp. 22 - 23. UKISES, London 1977.

28720 - BEARDEN, A.J., MALKIN, R. : Chloroplast photosynthesis : the reaction center of Photosystem I. - In : OLSON, J.M., HIND, G. (ed.) : Chlorophyll-Proteins, Reaction Centers, and Photosynthetic Membranes. Pp. 247 - 266. Brookhaven nat. Lab., Upton 1977.

28721 - BECACOS-KONTOS, T. : Primary production and environmental factors in an oligotrophic biome in Aegean Sea. - Mar. Biol. *42* : 93 - 98, 1977.

28722 - BECHER, B., EBREY, T.G. : The quantum efficiency for the photochemical conversion of the purple membrane protein. - Biophys. J. *17* : 185 - 191, 1977.

28723 - BECK, J.C., LEVINE, R.P. : Synthesis of chloroplast membrane lipids and chlorophyll in synchronous cultures of *Chlamydomonas reinhardi*. - Biochim. biophys. Acta *489* : 360 - 369, 1977.

28724 - BEDDARD, G., PORTER, G. : Energy Transfer in a model of the photosynthetic unit. - In : COOMBS, J. (ed.) : 4th International Congress on Photosynthesis. P. 23. UKISES, London 1977.

28725 - BEDDARD, G.S., DAVIDSON, R.S., TRETHEWEY, K.R. : Quenching of chlorophyll fluorescence by β-carotene. - Nature *267* : 373 - 374, 1977.

28726 - BEDDARD, G.S., PORTER, G. : Excited state annihilation in the photosynthetic unit. - Biochim. biophys. Acta *462* : 63 - 72, 1977.

28727 - BEDU, S. : Existence d'un système phosphohydrolasique complexe sur les thylakoïdes de chloroplastes de feuilles de Lupin (*Lupinus luteus* L.). - Physiol. vég. *15* : 641 - 656, 1977.

28728 - BEER, S., ESHEL, A., WAISEL, Y. : Carbon metabolism in seagrasses. I. The utilization of exogenous inorganic carbon species in photosynthesis. - J. exp. Bot. *28* : 1180 - 1189, 1977.

28729 - BEESON, K.W., POTASEK, M.J. : Mechanism of electron transfer in solar energy conversion. - In : COOMBS, J. (ed.) : 4th International Congress on Photosynthesis. P. 24. UKISES, London 1977.

28730 - BEFFAGNA, N., COCUCCI, S., MARRÈ, E. : Stimulating effect of fusicoccin on K-activated ATPase in plasmalemma preparations from higher plant tissues. - Plant Sci. Lett. *8* : 91 - 98, 1977.

28731 - BEHBOUDIAN, M.H. : Water relations of eggplant. - Plant Physiol. *59* (Suppl.): 55, 1977. [Ps.]

28732 - BEHBOUDIAN, M.H. : Water relations of cucumber, tomato, and sweet pepper. - Med. Landbouwhogesch. Wageningen *77* (6) : 1 - 84, 1977. [Ps, resistances.]

28733 - BEHNKE, H.-D. : Regular occurring massive deposits of phytoferritin in the phloem of succulent *Centrospermae*. - Z. Pflanzenphysiol. *85* : 89 - 92, 1977. [Chloroplasts.]

*28734 - BEHNKE, H.-D. : Sieve-element plastids of *Fouquieria, Frankenia (Tamaricales)*, and *Rhabdodendron (Rutaceae)*, taxa sometimes allied with *Centrospermae (Caryophyllales)*. - Taxon *25* : 265 - 268, 1976.

28735 - **BEINHAUER, R.** : Photosynthetisch aktive Strahlung im Gewächshaus. - Gartenbauwissenschaft *42* (3) : 109 - 113, 1977.

28736 - **BEISENHERZ, W.W.** : Zur Synthese der Plastidenribosomen während der Leukoplas-
ten-Chloroplasten-Transformation in Callusgewebe von *Nicotiana tabacum*. - Z.
Pflanzenphysiol. *83* : 449 - 457, 1977.

28737 - **BEKASOVA, O.D., EVSTIGNEEV, V.B.** : O roli fikobilinovykh pigmentov v fotosinteze. [Role of phycobilin pigments in photosynthesis.] - Biofizika *22* : 429 -
- 435, 1977. [In R, ab : E.]

28738 - **BEKINA, R.M.** : Reduction of O_2 in Photosystem II. - In : **COOMBS, J.** (ed.) :
4th International Congress on Photosynthesis. Pp. 25 - 26. UKISES, London
1977.

28739 - **BEKINA, R.M., LEBEDEVA, A.F.** : Izuchenie reaktsiĭ pogloshcheniya kisloroda
(reaktsiĭ Melera) v khloroplastakh v prisutstvii kremnievomolibdenovoĭ kisloty. [Oxygen uptake (Mehler reaction) in chloroplasts in the presence of molyb-
dosilicic acid.] - Biokhimiya *42* : 693 - 699, 1977. [In R, ab : E.]

28740 - **BELIN, C.W.** : Benthic microalgal primary productivity in the Pettaquamscutt
River, RI. USA. - J. Phycol. *13* (Suppl.) : 6, 1977.

28741 - **BELL, D., HAUG, A., GOOD, N.** : Stimulation of a microsecond-delayed fluorescence in chloroplasts by uncouplers. - Plant Physiol. *59* (6, Suppl.) : 23,
1977.

28742 - **BELL, L., SHUVALOVA, N., KRUPENKO, A., GANAGO, I.** : Conservation of energy of
light-stimulated oxidative processes in *Chlorella* cells. - In : **COOMBS, J.**
(ed.) : 4th International Congress on Photosynthesis. Pp. 24 - 25. UKISES,
London 1977.

28743 - **BELLOTTI, A., BRANCA, C., COGHI, E.** : Proprietà biologiche di composti 2,1-
-benzisotiazolici. Nota III. Attività fitotossica e inibitrice della reazione
di Hill di mono- e di-acilderivati del 3-amino-2,1-benzisotiazolo. [Biologic-
al properties of 2,1-benzisothiazoles. III. Phytotoxicity and ability to inhibit the Hill reaction of some mono- and di-acylderivatives of 3-amino-2,1-
-benzisothiazole.] - Ateneo parmense, Acta nat. *13* : 601 - 612, 1977. [In
Ital., ab : E.]

28744 - **BENDALL, D.S.** : Electron and proton transfer in chloroplasts. - In : **NORTH-
COTE, D.H.** (ed.) : Plant Biochemistry II. Int. Rev. Biochem. Vol. 13. Pp.
41 - 78. Univ. Park Press, Baltimore 1977.

28745 - **BENDALL, D.S.** : Development of photosystems during greening of etiolated barley. - In : **COOMBS, J.** (ed.) : 4th International Congress on Photosynthesis.
Pp. 26 - 27. UKISES, London 1977.

28746 - **BENDALL, D.S., WOOD, P.M.** : Kinetics of electron transfer through higher-plant
plastocyanin. - In : **COOMBS, J.** (ed.) : 4th International Congress on Photosynthesis. Pp. 27 - 28. UKISES, London 1977.

28747 - **BENEDETTI, E. de, GARLASCHI, F.M.** : On the estimation of proton gradient and
osmotic volume in chloroplast membranes. - J. Bioenerg. Biomembranes *9* : 195-
- 201, 1977.

28748 - **BENEDICT, C.R.** : The $\delta^{13}C$ values of the marine angiosperm, *Thalassia testudinum*. - Plant Physiol. *59* (6, Suppl.) : 65, 1977.

*28749 - **BENEDICT, C.R., KOHEL, R.J., SCHUBERT, A.M.** : Transport of ^{14}C-assimilates to
cottonseed : Integrity of funiculus during seed filling stage. - Crop Sci.
16 : 23 - 27, 1976.

28750 - **BENEMANN, J.** : Hydrogen and methane production through microbial photosynthe-
sis. - In : **BUVET, R., ALLEN, M.J., MASSUÉ, J.-P.** (ed.) : Living Systems as
Energy Converters. Pp. 285 - 297. North-Holland Publ. Co., Amsterdam - New
York - Oxford 1977.

28751 - **BENGIS, C., NELSON, N.** : Subunit structure of chloroplast photosystem I reaction center. - J. biol. Chem. *252* : 4564 - 4569, 1977.

28752 - BENGTSON, C., KLOCKARE, B., LARSSON, S., SUNDQVIST, C. : The effect of phytohormones on chlorophyll(ide), protochlorophyll(ide) and carotenoid formation
in greening dark grown wheat leaves. - Physiol. Plant. *40* : 198 - 204, 1977.

28753 - BEN-HAYYIM, G. : The role of Mg^{2+} in light-induced reactions : its transport
across the thylakoid membrane and its effect on quantum yield. - In : COOMBS,
J. (ed.) : 4th International Congress on Photosynthesis. Pp. 28 - 29. UKISES,
London 1977.

28754 - BEN-HAYYIM, G., NEUMANN, J. : Proton translocation and ATP formation coupled
to electron transport from H_2O to the primary acceptor of photosystem 2. -
Europe. J. Biochem. *72* : 57 - 61, 1977.

28755 - BENNETT, J. : Phosphorylation of chloroplast membrane proteins. - In : COOMBS,
J. (ed.) : 4th International Congress on Photosynthesis. P. 29. UKISES, London 1977.

28756 - BENNETT, J. : Phosphorylation of chloroplast membrane polypeptides. - Nature
269 : 344 - 346, 1977.

28757 - BENNETT, J.P., RUNECKLES, V.C. : Effects of low levels of ozone on growth of
crimson clover and annual ryegrass. - Crop Sci. *17* : 443 - 445, 1977. [Growth
analysis.]

28758 - BENNOUN, P., GIRARD, J., CHUA, N.-H. : A uniparental mutant of *Chlamydomonas
reinhardtii* deficient in the chlorophyll-protein complex CP1. - Mol. gen. Genet. *153* : 343 - 348, 1977.

28759 - BENNOUN, P., SCHMIDT, G., CHUA, N.H. : A mutant of *Chlamydomonas reinhardtii*
inactive on the donor side of Photosystem II. - In : COOMBS, J. (ed.) : 4th
International Congress on Photosynthesis. P. 29. UKISES, London 1977.

28760 - BENSASSON, R. : Pigments involved in the photomotion of microorganisms. - In:
CASTELLANI, A. (ed.) : Research in Photobiology. Pp. 85 - 94. Plenum Press,
New York - London 1977. [Chl, Car, biliproteins.]

28761 - BERAYA, L.G., SHOSHITAISHVILI, Ts.M. : K ustanovleniyu optimal'nogo sootnoshe-
niya mezhdu nadzemnymi i podzemnymi chastyami vinogradnoĭ lozy. [Determining
optimal relationship between aerial and underground parts of grape vine.] -
Tr. gruz. nauch.-issled. inst. SVIV *24* : 290 - 296, 1977. [Photosynthates;
in R.]

28762 - BERCHTOLD, M., BACHOFEN, R. : Possible role of cyclic AMP in the synthesis
of chlorophyll in *Chlorella fusca*. - Arch. Microbiol. *112* : 173 - 177, 1977.

*28763 - BERDYKULOV, Kh.A., AGZAMOV, A. : Usvoenie uglekisloty suspenzieĭ khlorelly,
kul'tiviruemoĭ v poluproizvodstvennykh usloviyakh. [CO_2 assimilation by a
Chlorella suspension cultivated under pilot plant conditions.] - Uzb. biol.
Zh. *1975* (3) : 66 - 67, 1976. [in R.]

28764 - BERDYKULOV, Kh.A., AGZAMOV, A., NURIEVA, D. : Podbor usloviĭ kul'tivirovaniya
i nekotorye fiziologicheskie osobennosti *Ankistrodesmus angustus*. [Choice of
cultivation conditions and some physiological peculiarities of *Ankistrodesmus
angustus*.] - Uzb. biol. Zh. *1977* (3) : 16 - 18, 1977. [Ps; in R, ab : Uzb.]

28765 - BERENSON, J.A., BENEMANN, J.R. : Immobilization of hydrogenase and ferredoxins
on glass beads. - FEBS Lett. *76* : 105 - 107, 1977.

28766 - BERG, S.P., IZAWA, S. : Pathways of silicomolybdate photoreduction and the
associated photophosphorylation in tobacco chloroplasts. - Biochim. biophys.
Acta *460* : 206 - 219, 1977.

28767 - BERGER, R., LIAAEN-JENSEN, S., McALISTER, V., GUILLARD, R.L. : Carotenoids of
Prymnesiophyceae (Haptophyceae). - Biochem. Syst. Ecol. *5* : 71 - 75, 1977.

28768 - BERGMANN, H. : Möglichkeiten der Transpirationseffektuierung und Stabilisie-
rung des Pflanzenwasserhaltes unter besonderer Berücksichtigung einer Agro-
chemikalienanwendung. - Arch. Acker- Pflanzenbau Bodenk. *21* : 767 - 788, 1977.
[Ps.]

28769 - BERING, C.L., LOACH, P.A. : Effects of 2,5-dibromo-3-methyl-6-isopropyl ben-
zoquinone (DBMIB) on photochemical events in *Rhodospirillum rubrum*. - Photochem. Photobiol. *26* : 607 - 615, 1977.

28770 - BERKALOFF, C. : Carotenoproteins in extrachloroplastic structures of the green alga *Protosiphon botryoides*. - Plant Sci. Lett. *10* : 45 - 48, 1977.

28771 - BERKALOFF, C., DUVAL, J.C. : Cation induced fluorescence changes in brown algae isolated chloroplasts. - In : COOMBS, J. (ed.) : 4th International Congress on Photosynthesis. Pp. 29 - 30. UKISES, London 1977.

28772 - BERLYN, M.B., ZELITCH, I. : Photosynthetic characteristics of photoautotrophically—grown tobacco callus. - Plant Physiol. *59* (6, Suppl.) : 62, 1977.

28773 - BERMAN, T., STILLER, M. : Simultaneous measurement of phosphorus and carbon uptake in Lake Kinneret by multiple isotopic labelling and differential filtration. - Microbial Ecol. *3* : 279 - 288, 1977.

28774 - BERRY, J.A., OSMOND, C.B., LORIMER, G.H. : Kinetic and steady-state studies of ^{18}O fixation into photorespiratory intermediates by intact leaves of C_3 species. - Carnegie Inst. Year Book *76* : 307 - 313, 1977.

28775 - BERS, É.P. : Vliyanie usloviĭ kul'tivirovaniya na produktivnost' i nekotorye fiziologicheskie kharakteristiki kletok *Chlamydomonas reinhardii*. [Effect of growing conditions on the productivity and some physiological characteristics of *Chlamydomonas reinhardii* cells.] - Tr. petergof. biol. Inst. *25* (Eksperimental'naya Al'gologiya) : 34 - 46, 1977. [Chl; in R, ab : E.]

*28776 - BERSENEVA, G.P., KRUPATKINA, D.K. : Soderzhanie khlorofilla *a* v tropicheskoĭ chasti Atlanticheskogo okeana i ego svyaz' so skorost'yu fotosinteza. [Chlorophyll *a* content in the tropical part of Atlantic Ocean and its relation to photosynthetic rate.] - Biol. Morya *37* (Produktsiya i Metabolicheskie Protsessy u Morskikh Organizmov) : 3 - 9, 1976. [In R.]

*28777 - BERÜTER, J. : Untersuchungen über das Verhalten von Spinat und Gerste gegenüber dem herbiziden Wirkstoff Ioxynil. - Vierteljahresschr. naturforsch. Ges. Zürich *116* (2) : 213 - 252, 1971. [Ps.]

28778 - BERZBORN, R.J., LOCKAU, W. : Antibodies. - In : TREBST, A., AVRON, M. (ed.): Photosynthesis I. (Encycl. Plant Physiol. N.S. Vol. 5.) Pp. 283 - 296. Springer-Verlag, Berlin - Heidelberg - New York 1977. [Ps.]

28779 - BERZBORN, R.J., MÜLLER, D. : Correlation of grana in chloroplasts with the variability in the size of "photophosphorylation unit". - In : COOMBS, J. (ed.) : 4th International Congress on Photosynthesis. Pp. 30 - 31. UKISES, London 1977.

28780 - BETHLENFALVAY, G., NORRIS, R.F. : Desmedipham phytotoxixity to sugarbeets (*Beta vulgaris*) under constant *versus* variable light, temperature, and moisture conditions. - Weed Sci. *25* : 407 - 411, 1977. [Ps.]

28781 - BETHLENFALVAY, G.J., PHILLIPS, D.A. : Interaction between CO_2 and N_2 reduction in peas. - Plant Physiol. *59* (6, Suppl.) : 129, 1977.

28782 - BETHLENFALVAY, G.J., PHILLIPS, D.A. : Ontogenetic interactions between photosynthesis and symbiotic nitrogen fixation in legumes. - Plant Physiol. *60* : 419 - 421, 1977.

28783 - BETHLENFALVAY, G.J., PHILLIPS, D.A. : Effect of light intensity on efficiency of carbon dioxide and nitrogen reduction in *Pisum sativum* L. - Plant Physiol. *60* : 868 - 871, 1977.

28784 - BEYELER, W., BACHOFEN, R. : Initial events of light-dependent ATP-synthesis in spinach subchloroplast particles. - In : COOMBS, J. (ed.) : 4th International Congress on Photosynthesis. Pp. 31 - 32. UKISES, London 1977.

28785 - BHAGSARI, A.S., ASHLEY, D.A., BROWN, R.H., BOERMA, H.R. : Leaf photosynthetic characteristics of determinate soybean cultivars. - Crop Sci. *17* : 929 - 932, 1977.

28786 - BHAGWAT, A.S., MITRA, J., SANE, P.V. : Studies on enzymes of C-4 pathway : Part III - Regulation of malic enzyme of *Zea mays* by fructose-1,6-diphosphate & other metabolites. - Indian J. exp. Biol. *15* : 1008 - 1012, 1977.

28787 - BHAGWAT, A.S., SANE, P.V. : Studies on enzymes C-4 pathway : Part IV - Comparative study of inhibitors of malic enzyme isolated from CAM, C-4 & C-3 plants. - Indian J. exp. Biol. *15* : 1013 - 1015, 1977.

28788 - BHARATI, M.P. : Density and environment interaction in soybean: a review. - Nepalese J. Agr. *12* : 239 - 247, 1977. [Growth analysis.]

28789 - BHARDWAJ, R., SINGHAL, G.S. : Effect of water stress on photochemical activity and membrane organisation of chloroplast during greening of barley seedlings. - In : COOMBS, J. (ed.) : 4th International Congress on Photosynthesis. Pp. 32 - 33. UKISES, London 1977.

*28790 - BHARGAVA, R.M.S., DWIVEDI, S.N. : Seasonal distribution of phytoplankton pigments in the estuarine system of Goa. - Indian J. mar. Sci. *5* : 87 - 90, 1976.

28791 - BHARGAVI, K., RAO, I.M., SWAMY, P.M. : Promotion of rooting and delay of senescence in detached leaves of *Gomphrena globosa* by growth regulators and Ca^{2+}. - Indian J. exp. Biol. *15* : 1069 - 1070, 1977. [Chl.]

28792 - BIALEK, G.E., HORVÁTH, G., GARAB, G.I., MUSTÁRDY, L.A., FALUDI-DÁNIEL, Á. : Selective scattering spectra as an approach to internal structure of granal and agranal chloroplasts. - Proc. nat. Acad. Sci. USA *74* : 1455 - 1457, 1977.

28793 - BIANCHI, A., LAUDI, G. : Azione esercitata da alcuni inibitori respiratori sulla crescita e sull'inverdimento di plantule di *Picea abies* e di *Larix decidua*. [Action of some respiratory inhibitors on growth and greening of seedlings of *Picea abies* and *Larix decidua*.] - G. bot. ital. *111* : 219 - 226, 1977. [In Ital., ab : E.]

28794 - BICKEL, H., SCHULTZ, G. : On the synthesis of aromatic amino acids and prenylquinones in isolated chloroplasts. - In : COOMBS, J. (ed.) : 4th International Congress on Photosynthesis. Pp. 33 - 34. UKISES, London 1977.

28795 - BIDWELL, R.G.S. : Photosynthesis and light and dark respiration in freshwater algae. - Can. J. Bot. *55* : 809 - 818, 1977.

28796 - BIELESKI, R.L., REDGWELL, R.J. : Synthesis of sorbitol in apricot leaves. - Aust. J. Plant Physiol. *4* : 1 - 10, 1977. [Photosynthates.]

28797 - BIENFANG, P., GUNDERSEN, K. : Light effects on nutrient-limited, oceanic primary production. - Mar. Biol. *43* : 187 - 199, 1977.

28798 - BIGGINS, J. : Regulation and stoichiometry of energy coupling during PSI cyclic electron flow : P/e_2 ratio. - In : COOMBS, J. (ed.) : 4th International Congress on Photosynthesis. P. 34. UKISES, London 1977.

28799 - BILDERBACK, T.E., MATTSON, R.H. : Whitefly host preference associated with selected biochemical and phenotypic characteristics of poinsettias. - J. amer. Soc. hort. Sci. *102* : 327 - 331, 1977. [Chl.]

28800 - BILLORE, S.K., MALL, L.P. : Dry matter structure and its dynamics in *Sehima* community. II. Dry matter dynamics. - Trop. Ecol. *18* : 29 - 35, 1977.

28801 - BILLORE, S.K., MEHTA, S.C., MALL, L.P. : Changes in chlorophyll and carotenoid in summer leaves of a tropical deciduous tree *Buchanania lanzan* SPRENG. - Sci. Cult. *43* : 324 - 325, 1977.

28802 - BINDER, A., JAGENDORF, A.T. : Isolation and characterization of the subunits of spinach chloroplast coupling factor (CF_1). - In : COOMBS, J. (ed.) : 4th International Congress on Photosynthesis. P. 35. UKISES, London 1977.

28803 - BINGHAM, G.E., COYNE, P.I. : A portable, temperature-controlled, steady-state porometer for field measurements of transpiration and photosynthesis. - Photosynthetica *11* : 148 - 160, 1977.

28804 - BINGHAM, S., SCHIFF, J.A. : Developmental relationships of plastid membrane polypeptides in *Euglena gracilis* var. *bacillaris*. - In : OLSON, J.M., HIND, G. (ed.) : Chlorophyll-Proteins, Reaction Centers, and Photosynthetic Membranes. P. 360. Brookhaven nat. Lab., Upton 1977.

28805 - BINGHAM, S., STILLER, J., SCHIFF, J.A. : Thylakoid membrane polypeptides from mutants of *Euglena gracilis* var. *bacillaris*. - Plant Physiol. *59* (6, Suppl.): 9, 1977.

28806 - BINKLEY, S.F., MERRITT, C. : Sampling light intensity in a young pine plantation. - Can. J. Forest Res. *7* : 700 - 702, 1977.

28807 - BIRD, I.F., CORNELIUS, M.J., KEYS, A.J. : Effects of temperature on photosynthesis by maize and wheat. - J. exp. Bot. *28* : 519 - 524, 1977.

28808 - BIRMINGHAM, B.C., COLMAN, B. : Measurement of HCO_3^- compensation points of freshwater algae. - Plant Physiol. *59* (6, Suppl.) : 65, 1977.

28809 - BISCOE, P.V., GALLAGHER, J.N. : Weather, dry matter production and yield. - In : LANDSBERG, J.J., CUTTING, C.V. (ed.) : Environmental Effects on Crop Physiology. Pp. 75 - 100. Academic Press, London - New York - San Francisco 1977. [Ps.]

28810 - BISCOE, P.V., INCOLL, L.D., LITTLETON, E.J., OLLERENSHAW, J.H. : Barley and its environment. VII. Relationships between irradiance, leaf photosynthetic rate and stomatal conductance. - J. appl. Ecol. *14* : 293 - 302, 1977.

28811 - BISHOP, D.F., WOOD, W.A. : An assay for δ-aminolevulinic acid synthetase based on a specific semiautomatic determination of picomole quantities of δ- -[^{14}C]aminolevulinate. - Anal. Biochem. *80* : 466 - 482, 1977.

28812 - BISHOP, D.G., NOLAN, W.G. : Release of plastocyanin from chloroplast membranes by the polyene antibiotic amphotericin B. - In : COOMBS, J. (ed.) : 4th International Congress on Photosynthesis. Pp. 35 - 36. UKISES, London 1977.

28813 - BISHOP, H.G., GRAMSHAW, D. : Effect of sowing rate, grass competition and cutting frequency on persistence and productivity of two lucerne (*Medicago sativa*) cultivars at Biloela, Queensland. - Aust. J. exp. Agr. anim. Husb. *17* : 105 - 111, 1977.

28814 - BISHOP, N.I., FRICK, M., JONES, L.W. : Photohydrogen production in green algae : Water serves as the primary substrate for hydrogen and oxygen production. - In : MITSUI, A., MIYACHI, S., SAN PIETRO, A., TAMURA, S. (ed.) : Biological Solar Energy Conversion. Pp. 3 - 22. Academic Press, New York - San Francisco - London 1977.

28815 - BISHOP, N.I., FRICK, M., SENGER, H. : On the mechanism of photohydrogen evolution by green algae : the involvement of PS-I and PS-II. - Plant Physiol. *59* (6, Suppl.) : 130, 1977.

28816 - BISHOP, N.I., WONG, J. : On the identity of chloroplast membrane peptides, as seen with SDS-PAGE, with known components of the photosynthetic electron transport system. - In : COOMBS, J. (ed.) : 4th International Congress on Photosynthesis. Pp. 36 - 37. UKISES, London 1977.

28817 - BISWAL, U.C., SINGHAL, G., MOHANTY, P. : Alteration of membrane characteristics and photochemical activities of chloroplasts isolated from dark stress induced barley leaves. - In : COOMBS, J. (ed.) : 4th International Congress on Photosynthesis. P. 37. UKISES, London 1977.

28818 - BJÖRKMAN, O., BADGER, M. : Thermal stability of photosynthetic enzymes in heat- and cool-adapted C_4 species. - Carnegie Inst. Year Book *76* : 346 - 354, 1977.

28819 - BLACKWOOD, G.C., LEAVER, C.J. : The effect of light on protein synthesis in green leaves. - In : BOGORAD, L., WEIL, J.H. (ed.) : Acides Nucléiques et Synthèse des Protéines chez les Végétaux. Coll. Int. C.N.R.S. No. 261. Pp. 611 - 615. Édit. CNRS, Paris 1977. [Fraction 1 protein.]

28820 - BLAIR, B.O., BAUMGARDNER, M.F. : Detection of the green and brown wave in hardwood canopy covers using multidate, multispectral data from LANDSAT-1. - Agron. J. *69* : 808 - 811, 1977.

28821 - BLAKE, J., FERRELL, W.K. : The association between soil and xylem water potential, leaf resistance, and abscisic acid content in droughted seedlings of Douglas-fir (*Pseudotsuga menziesii*). - Physiol. Plant. *39* : 106 - 109, 1977.

28822 - BLANKENSHIP, R., PARSON, W. : The role of Fe in primary and secondary reactions of bacterial reaction centers. - In : COOMBS, J. (ed.) : 4th International Congress on Photosynthesis. Pp. 37 - 38. UKISES, London 1977.

28823 - BLANKENSHIP, R.E., MCGUIRE, A., SAUER, K. : Rise time of EPR signal II$_{vf}$ in chloroplast photosystem II. - Biochim. biophys. Acta *459* : 617 - 619, 1977.

28824 - BLANKENSHIP, R.E., SCHAAFSMA, T.J., PARSON, W.W. : Magnetic field effects on radical pair intermediates in bacterial photosynthesis. - Biochim. biophys. Acta 461 : 297 - 305, 1977.

28825 - BLATT, M.R., BRIGGS, W.R. : A recording microphotometer for measurement of chloroplast orientation movements in single algal filaments. - Carnegie Inst. Year Book 76 : 278 - 281, 1977.

28826 - BLAUROCK, A.E., KING, G.I. : Asymmetric structure of the purple membrane. - Science 196 : 1101 - 1104, 1977.

28827 - BLAZHEVA, N., RADEVA, V. : Vliyanie na preparata CCC v"rkhu dobiva na sukho veshchestvo i semena i s"d"rzhanieto na pigmenti v lyutsernata. [Chlorcholine chloride effect on the dry matter, seed yield and pigments content of alfalfa.] - Rasteniev"dni Nauki 14 (1) : 110 - 118, 1977. [In Bulg., ab : E, R.]

28828 - BLEE, E., SCHANTZ, R. : Biosynthesis of galactolipids in Euglena gracilis Z. - In : COOMBS, J. (ed.) : 4th International Congress on Photosynthesis. Pp. 38 - 39. UKISES, London 1977. [Chl.]

28829 - BLINN, D.W., TOMPKINS, T., ZALESKI, L. : Mercury inhibition on primary productivity using large volume plastic chambers in situ. - J. Phycol. 13 : 58 - - 61, 1977.

28830 - BLOESCH, J., STADELMANN, P., BÜHRER, H. : Primary production, mineralization, and sedimentation in the euphotic zone of two Swiss lakes. - Limnol. Oceanogr. 22 : 511 - 526, 1977.

28831 - BLOK, M.C., HELLINGWERF, K.J., VAN DAM, K. : Reconstitution of bacteriorhodopsin in a millipore filter system. - FEBS Lett. 76 : 45 - 50, 1977.

28832 - BLUM, H., SALERNO, J.C., PRINCE, R.C., LEIGH, J.S., Jr., OHNISHI, T. : Electron paramagnetic resonance determination of a low-lying excited state in Chromatium vinosum high-potential iron protein. - Biophys. J. 20 : 23 - 31, 1977.

28833 - BLUMENFELD, L.A., GOLDFELD, M.G., DMITROVSKY, L.G., HANGULOV, S.V. : Non-cyclic photophosphorylation and electron transport in chloroplasts under flash excitation. - In : COOMBS, J. (ed.) : 4th International Congress on Photosynthesis. Pp. 39 - 40. UKISES, London 1977.

28834 - BOARDMAN, N.K. : Chloroplasts-structure and photosynthesis. - Bot. Monogr. (Oxford) 14 : 85 - 104, 442 - 483, 1977.

28835 - BOARDMAN, N.K. : Comparative photosynthesis of sun and shade plants. - Annu. Rev. Plant Physiol. 28 : 355 - 377, 1977.

28836 - BOARDMAN, N.K. : Development of chloroplast structure and function. - In : TREBST, A., AVRON, M. (ed.) : Photosynthesis I. (Encycl. Plant Physiol. N.S. Vol. 5.) Pp. 583 - 600. Springer-Verlag, Berlin - Heidelberg - New York 1977.

28837 - BOARDMAN, N.K. : Solar energy conversion in photosynthesis and its potential contribution to world demand for liquid and gaseous fuels. - In : COOMBS, J. (ed.) : 4th International Congress on Photosynthesis. Pp. 40 - 41. UKISES, London 1977.

28838 - BOARDMAN, N.K. : The energy budget in solar energy conversion in ecological and agricultural systems. - In : BUVET, R., ALLEN, M.J., MASSUE, J.-P. (ed.): Living Systems as Energy Converters. Pp. 307 - 317. North-Holland Publ. Co., Amsterdam - New York - Oxford 1977.

28839 - BOARDMAN, N.K., THORNE, S.W. : Effect of a low concentration of glutaraldehyde on proton uptake, phosphorylation and fluorescence quenching in chloroplasts. - Plant Cell Physiol. 1977 (spec. Issue 3 - Photosynthetic Organelles. Structure and Function) : 157 - 163, 1977.

28840 - BOCHAROV, E.A., DZHANUMOV, D.A., KLIMOV, S.V.: Sostav fosfolipidov i galaktolipidov v khloroplastakh kontrastnykh po morozoustoĭchivosti sortov ozimykh kul'tur. [Composition of phospholipids and galactolipids in chloroplasts of winter crop cultivars of contrasting hardiness.] - Nauch. Tr. nauch.-issled. Inst. sel'. Khoz. tsentr. Raĭonov nechernozem. Zony 41 (Sozdanie Sortov Zernovykh Kul'tur Intensivnogo Tipa) : 114 - 121, 1977. [In R.]

28841 - BÖCHER, M., KLUGE, M. : Der C_4-Weg der C-Fixierung bei *Spinacea oleracea*. I.
^{14}C-Markierungsmuster suspendierter Blattstreifen unter dem Einfluß des Sus-
pensionsmedium. - Z. Pflanzenphysiol. *83* : 347 - 361, 1977.

28842 - BODSON, M., KING, R.W., EVANS, L.T., BERNIER, G. : The role of photosynthesis in flowering of the long-day plant *Sinapis alba*. - Aust. J. Plant Physiol. *4* : 467 - 478, 1977.

*28843 - BOGENRIEDER, A. : Vergleichende physiologisch-ökologische Untersuchungen an
Populationen subalpiner Pflanzen aus Schwarzwald und Alpen. - Oecol. Plant.
9 : 131 - 156, 1974. [Ps, Chl.]

28844 - BOGENRIEDER, A., KLEIN, R. : Die Rolle des UV-Lichtes beim sog. Auspflanzungs-
schock von Gewächshaussetzlingen. - Angew. Bot. *51* : 99 - 107, 1977. [Ps.]

28845 - BÖGER, P. : Some properties and functions of plastocyanin. - In : COOMBS, J.
(ed.) : 4th International Congress on Photosynthesis. Pp. 41 - 42. UKISES,
London 1977.

28846 - BÖGER, P., BEESE, B., MILLER, R. : Long-term effects of herbicides on the
photosynthetic apparatus. II. Investigations on bentazone inhibition. - Weed
Res. *17* : 61 - 67, 1977.

28847 - BÖGER, P., VETTER, H. : Herbizide im modernen Pflanzenbau. Der Photosynthese-
apparat als Angriffsort für neue Wirkstoffe. - Naturwiss. Rundschau *30* : 322-
- 331, 1977.

28848 - BOGOMOLNI, R.A. : Light energy conservation processes in *Halobacterium halobium* cells. - Fed. Proc. *36* : 1833 - 1839, 1977.

28849 - BOGORAD, L. : Genes for chloroplast ribosomal RNAs and ribosomal proteins :
Gene dispersal in eukaryotic genomes. - In : BRINKLEY, B.R., PORTER, K.R.
(ed.) : International Cell Biology 1976-1977. Pp. 175 - 182. Rockefeller
Univ. Press, New York 1977.

28850 - BOGORAD, L., DAVIDSON, J.N., HANSON, M.R. : The genetics of the chloroplast
ribosome in *Chlamydomonas reinhardi*. - In : BOGORAD, L., WEIL, J.H. (ed.) :
Nucleic Acids and Protein Synthesis in Plants. Pp. 135 - 154. Plenum Press,
New York 1977.

28851 - BÖHME, H. : On the role of ferredoxin and ferredoxin-NADP$^+$ reductase in cyclic electron transport of spinach chloroplasts. - Europe. J. Biochem. *72* :
283 - 289, 1977.

28852 - BÖHME, H. : Structural and quantitative analysis of membrane-bound components of the photosystem I complex of spinach chloroplasts by immunological
methods. - In : PACKER, L., PAPAGEORGIOU, G.C., TREBST, A. (ed.) : Bioenergetics of Membranes. Pp. 329 - 337. Elsevier/North-Holland Biomedical Press,
Amsterdam - Oxford - New York 1977.

*28853 - BOĬCHENKO, E.A. : Soedineniya metallov v évolyutsii rasteniĭ v biosfere.
[Metal compounds in the evolution of plants in the biosphere.] - Izv. Akad.
Nauk SSSR, Ser. biol. *1976* : 378 - 385, 1976. [Ps; in R, ab : E.]

28854 - BOĬCHENKO, E.A., UDEL'NOVA, T.M. : Znachenie metallov v okislitel'no-vosstanovitel'nykh funktsiyakh rasteniĭ. [Importance of metals in the oxidation-reduction functions of plants.] - Izv. Akad. Nauk SSSR, Ser. biol. *1977* : 194 -
- 200, 1977. [Ps; in R, ab : E.]

28855 - BOKHARI, U.G. : Regrowth of western wheatgrass utilizing ^{14}C-labeled assimilates stored in belowground parts. - Plant Soil *48* : 115 - 127, 1977.

28856 - BOLHÀR-NORDENKAMPF, H.R. : The efficiency of net leaf photosynthesis as related to the parameters light, CO_2 tension, temperature and nutrient concentration. - In : COOMBS, J. (ed.) : 4th International Congress on Photosynthesis. P. 42. UKISES, London 1977.

28857 - BOLIN, B. : Changes of land biota and their importance for the carbon cycle.
- Science *196* : 613 - 615, 1977.

28858 - BOLTON, J.R. : Photochemical conversion and storage of solar energy. - J.
solid State Chem. *22* : 3 - 8, 1977. [Ps.]

28859 - **BOLTON, J.R.** : Photosystem I photoreactions. - In : **BARBER, J.** (ed.) : Primary Processes of Photosynthesis. Pp. 187 - 202. Elsevier, Amsterdam - New York - Oxford 1977.

28860 - **BOLTON, J.R.** : Solar energy conversion efficiency in photosynthesis - or - why two photosystems ? - In : **COOMBS, J.** (ed.) : 4th International Congress on Photosynthesis. P. 43. UKISES, London 1977.

28861 - **BOMBÓWNA, M.** : Biocenoza potoku wysokogorskiego pozostajacego pod wpływem turystyki. 1. Chemizm wody Rybiego Potoku i zawartość chlorofilu w glonach osiadlych oraz sestonie a zanieczyszczenie. Biocenosis of a high mountain stream under the influence of tourism. 1. Chemistry of the Rybi Potok waters and the chlorophyll content in attached algae and seston in relation to the pollution. - Acta hydrobiol. *19* : 243 - 255, 1977. [Chl.]

28862 - **BOMSEL, J.L.**, **SELLAMI, A.** : *In vivo* study of exchangeability of intact molecules of adenine nucleotides between chloroplastic and non-chloroplastic compartments of wheat leaf (*Triticum vulgare*, VILL.). - In : **COOMBS, J.** (ed.) : 4th International Congress on Photosynthesis. Pp. 43 - 44. UKISES, London 1977.

28863 - **BONATTI, P.M.**, **BIANCHI, A.**, **ALESSANDRI, M.** : Plastidi in *Libocedrus decurrens "aurea"* TERR. - [Plastids in *Libocedrus decurrens "aurea"* TERR.] - Caryologia *30* : 177 - 187, 1977. [In Ital., ab : E.]

28864 - **BONDARENKO, V.I.**, **TKALICH, I.D.** : Vliyanie usloviĭ vegetatsii na formirovanie rasteniĭ, fotosintez i produktivnost' ozimoĭ pshenitsy. [Effect of vegetation conditions on formation of plants, photosynthesis and productivity of winter wheat.] - Fiziol. Biokhim. kul't. Rast. *6* : 576 - 581, 1977. [In R, ab : E.]

28865 - **BONHOMME, R.**, **VARLET-GRANCHER, C.** : Application aux couverts végétaux des lois de rayonnement en milieu diffusant I. - Establissement des lois et vérifications expérimentales. - Ann. agron. *28* : 567 - 582, 1977.

28866 - **BONHOMME, R.**, **VARLET GRANCHER, C.**, **CHARTIER, M.**, **ARTIS, P.** : Utilisation de l'énergie solaire par une culture de *Vigna sinensis* IV. - Influence de l'âge et des éclairements passés sur le potentiel photosynthétique des feuilles cotylédonaires. - Ann. agron. *28* : 159 - 169, 1977.

28867 - **BONNET, F.**, **VERNOTTE, C.**, **BRIANTAIS, J.-M.**, **ETIENNE, A.-L.** : Kinetics of chlorophyll fluorescence at 77 K in *Chlorella* and chloroplasts. Effects of CCCP, ferricyanide and DCMU. - Biochim. biophys. Acta *461* : 151 - 158, 1977.

28868 - **BONOMI, F.**, **PAGANI, S.**, **CERLETTI, P.** : Insertion of sulfide into ferredoxins catalyzed by rhodanese. - FEBS Lett. *84* : 149 - 152, 1977.

28869 - **BONUGLI, K.J.**, **DAVIES, D.D.** : The regulation of potato phosphoenolpyruvate carboxylase in relation to a metabolic pH-stat. - Planta *133* : 281 - 287, 1977.

28870 - **BONZI, L.M.**, **FABBRI, F.** : The separation of mitochondrion-like structures from chloroplasts in sieve parenchyma cells of *Arum italicum* MILL. - Caryologia *30* : 493 - 494, 1977.

28871 - **BONZON, M.**, **GREPPIN, H.** : Migration of adult spinach chloroplasts in S-Rho space, before and after photoperiodic floral induction. - Z. Pflanzenphysiol. *81* : 260 - 268, 1977.

28872 - **BOOTE, K.J.** : Effect of fruit and vegetative apices on photosynthesis of peanut leaves. - Plant Physiol. *59* (Suppl.) : 98, 1977.

28873 - **BOOTE, K.J.** : Root : shoot relationships. - Proc. Soil Crop Sci. Soc. Florida *36* : 15 - 23, 1977. [Ps production.]

28874 - **BORCHERS, C.**, **SWANSON, C.A.** : The kinetics of compensated translocation in *Phaseolus vulgaris* L. - Plant Physiol. *59* (6, Suppl.) : 125, 1977. [Photosynthates.]

*28875 - **BORGNA, P.**, **VICARINI, L.**, **MAZZA, M.** : Diacilaminodifenildisolfuri ad azione inibente la reazione di Hill. [Diacylamino diphenyl disulfides showing inhibition of the Hill reaction.] - Farmaco, Ed. sci. *29* (2) : 120 - 128, 1974. [In Ital., ab : E.]

28876 - BORISEVICH, G.P., KONONENKO, A.A., RUBIN, A.B. : Electrochromism of pigments in chromatophore films from photosynthesizing bacteria. - Photosynthetica *11*: 81 - 87, 1977.

28877 - BORISOV, A.Yu. : Photosynthesis as the prototype of new solar energetics. - In : COOMBS, J. (ed.) : 4th International Congress on Photosynthesis. Pp. 44- - 45. UKISES, London 1977.

28878 - BORISOV, A.Yu., FETISOVA, Z.G., GODIK, V.I. : Energy transfer in photoactive complexes obtained from green bacterium *Chlorobium limicola*. - Biochim. biophys. Acta *461* : 500 - 509, 1977.

28879 - BORNEFELD, T. : The quantum requirement of photophosphorylation and its stoichiometric relation to reduction in *Anacystis in vivo*. - Z. Pflanzenphysiol. *85* : 393 - 401, 1977.

28880 - BOROWITZKA, M.A., LARKUM, A.W.D., DAY, R. : Seasonal aspects of the productivity of coral reef algal turf communities. - J. Phycol. *13* (Suppl.) : 8, 1977. [Ps.]

28881 - BORRISS, H., SCHMERDER, B. : Die Wirkung von Äthylen und Benzylaminopurin auf die Chlorophyllbildung von dormanten und nachgereiften *Agrostemma*-Embryonen. - Biol. Rundschau *15* : 183 - 186, 1977.

28882 - BORTMAN, S.J., TRELEASE, R.N. : Isolation and characterization of glyoxysomes from cotton cotyledons. - Plant Physiol. *59* (6, Suppl.) : 80, 1977.

28883 - BORZENKOVA, R.A., D'YACHENKO, O.Z., BORTNIKOVA, I.F. : Fitokhromnyĭ kontrol' replikatsii i rosta plastid v ĕtiolirovannykh prorostkakh fasoli. [Phytochrome control of plastid replication and growth in etiolated seedlings of bean.] - Fiziol. Biokhim. kul't. Rast. *9* : 492 - 496, 1977. [In R.]

28884 - BOSE, S., ARNTZEN, C.J. : Reversible inactivation of photosystem II reaction centers by cation depletion of chloroplast membranes. - In : COOMBS, J. (ed.): 4th International Congress on Photosynthesis. Pp. 45 - 46. UKISES, London 1977.

28885 - BOSE, S., ARNTZEN, C.J. : Salt-induced activation of photosystem II units in isolated chloroplasts. - Plant Physiol. *59* (6, Suppl.) : 24, 1977.

28886 - BOSE, S., BURKE, J.J., ARNTZEN, C.J. : Cation-induced microstructural changes in chloroplast membranes : Effects on photosystem II activity. - In : PACKER, L., PAPAGEORGIOU, G.C., TREBST, A. (ed.) : Bioenergetics of Membranes. Pp. 245 - 256. Elsevier/North-Holland Biomedical Press, Amsterdam - Oxford - New York 1977.

28887 - BOSSELAERS, J., CLIJSTERS, H., van POUCKE, M. : Partial reactions of photosynthesis in *Phaseolus vulgaris* chloroplasts *in vitro* after application of kinetin. - In : COOMBS, J. (ed.) : 4th International Congress on Photosynthesis. P. 46. UKISES, London 1977.

28888 - BOTHE, H. : Flavodoxin. - In : TREBST, A., AVRON, M. (ed.) : Photosynthesis I. (Encycl. Plant Physiol. N.S. Vol. 5.) Pp. 216 - 221. Springer-Verlag, Berlin - Heidelberg - New York 1977.

28889 - BOTHE, H. : The relationship between H_2-metabolism, nitrogen fixation and photosynthesis in *Anabaena*. - In : COOMBS, J. (ed.) : 4th International Congress on Photosynthesis. P. 47. UKISES, London 1977.

28890 - BOTKIN, D.B. : Forests, lakes, and the anthropogenic production of carbon dioxide. - BioScience *27* : 325 - 331, 1977. [Ps.]

28891 - BOTTOMLEY, W., HIGGINS, T.J.V., WHITFIELD, P.R., LEAVER, C.J. : The products of *in vitro* protein synthesizing systems programmed by chloroplast and cytoplasmic RNA and chloroplast DNA. - In : BOGORAD, L., WEIL, J.H. (ed.) : Acides Nucléiques et Synthèse des Protéines chez les Végétaux. Coll. Int. C.N.R.S. No. 261. Pp. 413 - 418. Édit. CNRS, Paris 1977.

28892 - BOUCHER, F., REST, M. van der, GINGRAS, G. : Structure and function of carotenoids in the photoreaction center from *Rhodospirillum rubrum*. - Biochim. biophys. Acta *461* : 339 - 357, 1977.

28893 - **BOUGES-BOCQUET, B.** : Cytochrome f and plastocyanin kinetics in *Chlorella pyrenoidosa*. - COOMBS, J. (ed.) : 4th International Congress on Photosynthesis. P. 48. UKISES, London 1977.

28894 - **BOUGES-BOCQUET, B.** : Cytochrome f and plastocyanin kinetics in *Chlorella pyrenoidosa*. I. Oxidation kinetics after a flash. - Biochim. biophys. Acta *462* : 362 - 370, 1977.

28895 - **BOUGES-BOCQUET, B.** : Cytochrome f and plastocyanin kinetics in *Chlorella pyrenoidosa*. II. Reduction kinetics and electric field increase in the 10 ms range. - Biochim. biophys. Acta *462* : 371 - 379, 1977.

28896 - **BOURDU, R.** : Les structures cellulaires en relation avec la productivité. - In : MOYSE, A. (ed.) : Les Processus de la Production Végétale Primaire. Pp. 203 - 234. Gauthier-Villars, Paris 1977.

28897 - **BOURQUE, D.P., HORN, N.A., CAPEL, M.S.** : Altered chloroplast ribosomes of a streptomycin resistant mutant of *Nicotiana tabacum*. - Plant Physiol. *59* (6, Suppl.) : 110, 1977.

28898 - **BOUSQUET, J.F., SKAJENNIKOFF, M., BETHENOD, O., CHARTIER, P.** : Action dépressive de l'ochracine, phytotoxine synthétisée par le *Septoria nodorum* (BERK.) BERK., sur l'assimilation du CO_2 par des plantules de Blé. - Ann. Phytopathol. *9* : 503 - 510, 1977.

28899 - **BOWEN, G.D., CARTWRIGHT, B.** : Mechanisms and models of plant nutrition. - In: RUSSELL, J.S., GREACEN, E.L. (ed.) : Soil Factors in Crop Production in a Semi-Arid Environment. Pp. 197 - 223. Univ. Queensland Press, St. Lucia 1977. [Ps.]

28900 - **BOWEN, M.S., WARD, H.B.** : Growth and photosynthesis in laboratory cultures of two cryptomonad algae. - J. Phycol. *13* (Suppl.) : 8, 1977.

28901 - **BOWES, G., HOLADAY, A.S., VAN, T.K., HALLER, W.T.** : Photosynthesis and photorespiratory carbon metabolism in aquatic plants. - In : COOMBS, J. (ed.) : 4th International Congress on Photosynthesis. P. 49. UKISES, London 1977.

28902 - **BOWES, G., VAN, T.K., GARRARD, L.A., HALLER, W.T.** : Adaptation to low light levels by hydrilla. - J. aquat. Plant Management *15* : 32 - 35, 1977. [Ps.]

28903 - **BOWES, J.M., ITOH, S., CROFTS, A.R.** : Involvement of protons in the reactions of photosystem 2 as studied by effects of pH of microsecond fluorescence yield changes and delayed fluorescence. - In : COOMBS, J. (ed.) : 4th International Congress on Photosynthesis. P. 50. UKISES, London 1977.

28904 - **BOWYER, J.R., CROFTS, A.R.** : The cytochromes and ubiquinone involved in photosynthetic electron transport in *Rhodopseudomonas capsulata*. - In : COOMBS, J. (ed.) : 4th International Congress on Photosynthesis. P. 51. UKISES, London 1977.

28905 - **BOYCE, C.O.L., OYEWOLE, S.H., FULLER, R.C.** : Localization of photosynthetic reaction center in *Chlorobium limicola*. - In : OLSON, J.M., HIND, G. (ed.) : Chlorophyll-Proteins, Reaction Centers, and Photosynthetic Membranes. P. 365. Brookhaven nat. Lab., Upton 1977.

28906 - **BOYER, P.D.** : Conformational coupling in oxidative phosphorylation and photophosphorylation. - Trends biochem. Sci. *2* (2) : 38 - 41, 1977.

28907 - **BOYER, P.D., CHANCE, B., ERNSTER, L., MITCHELL, P., RACKER, E., SLATER, E.C.:** Oxidative phosphorylation and photophosphorylation. - Annu. Rev. Biochem. *46*: 955 - 1026, 1977.

28908 - **BOYER, P.D., GRESSER, M., VINKLER, C., HACKNEY, D., CHOATE, G.** : Nucleotide binding and ATPase subunit cooperativity in energy transduction by mitochondria and chloroplasts. - In : VAN DAM, K., VAN GELDER, B.F. (ed.) : Structure and Function of Energy-Transducing Membranes. Pp. 261 - 274. Elsevier/North Holland Biomedical Press, Amsterdam 1977.

28909 - **BRADBEER, J.W.** : Chloroplasts - structure and development. - In : SMITH, H. (ed.) : The Molecular Biology of Plant Cells. Vol. 14. Pp. 64 - 84. Blackwell Scientific Publications, Oxford 1977.

28910 - BRADBEER, J.W., ARRON, G.P., HERRERA, A., KEMBLE, R.J., MONTES, G., SHERRATT, D., WARA-ASWAPATI, O. : The greening of leaves. - In : JENNINGS, D.H. (ed.): Integration of Activity in the Higher Plant. Pp. 195 - 219. Cambridge Univ. Press, New York 1977.

28911 - BRADBEER, J.W., ARRON, G.P., KEMBLE, R., WARA-ASWAPATI, O. : The synthesis of some enzymes of the photosynthetic carbon cycle in bean leaves. - In : BOGORAD, L., WEIL, J.H. (ed.) : Acides Nucléiques et Synthèse des Protéines chez les Végétaux. Coll. Int. C.N.R.S. No. 261. Pp. 453 - 456. Édit. CNRS, Paris 1977.

28912 - BRADBEEER, J.W., SHERRATT, D. : The light-induced binding of plastid ribosomes to thylakoids in developing leaves of *Phaseolus vulgaris* L. - In : COOMBS, J. (ed.) : 4th International Congress on Photosynthesis. P. 52. UKISES, London 1977.

28913 - BRADBURY, I.K., HOFSTRA, G. : Assimilate distribution patterns and carbohydrate concentration changes in organs of *Solidago canadensis* during an annual developmental cycle. - Can. J. Bot. 55 : 1121 - 1127, 1977.

28914 - BRADBURY, I.K., MALCOLM, D.C. : The effect of phosphorus and potassium on transpiration, leaf diffusive resistance and water-use efficiency in Sitka spruce (*Picea sitchensis*) seedlings. - J. appl. Ecol. 14 : 631 - 641, 1977.

28915 - BRADY, C.J., SCOTT, N.S. : Chloroplast polyribosomes and synthesis of fraction 1 protein in the developing wheat leaf. - Aust. J. Plant Physiol. 4 : 327 - 335, 1977.

28916 - BRADY, C.J., SCOTT, N.S. : The persistence of plastid polyribosomes and Fraction 1 protein synthesis in ageing wheat leaves. - In : BOGORAD, L., WEIL, J.H. (ed.) : Acides Nucléiques et Synthèse des Protéines chez les Végétaux. Coll. Int. C.N.R.S. No. 261. Pp. 387 - 393. Édit. CNRS, Paris 1977.

28917 - BRAND, J.J. : Spectral changes in *Anacystis nidulans* induced by chilling. - Plant Physiol. 59 : 970 - 973, 1977.

28918 - BRANDLE, J.R., CAMPBELL, W.F., SISSON, W.B., CALDWELL, M.M. : Net photosynthesis, electron transport capacity, and ultrastructure of *Pisum sativum* L. exposed to ultraviolet-B radiation. - Plant Physiol. 60 : 165 - 169, 1977.

28919 - BRANDLE, J.R., HINCKLEY, T.M., BROWN, G.N. : The effects of dehydration-rehydration cycles on protein synthesis of black locust seedlings. - Physiol. Plant. 40 : 1 - 5, 1977. [Ps.]

28920 - BRANDON, P.C., BOEKEL-MOL, G.N. van : β-bromo-β-nitrostyrene as a facile and photosystem I-specific electron acceptor. - FEBS Lett. 80 : 201 - 204, 1977.

28921 - BRANDT, A.B., KISELEVA, M.I. : Izmenenie udel'nogo fotosinteza kletok khlorelly i soderzhashchegosya v nikh khlorofilla v khode tsikla razvitiya ětoĭ vodorosli. [Changes in specific photosynthesis and chlorophyll content of *Chlorella* cells in the course of their development.] - Fiziol. Rast. 24 : 818 - 823, 1977. [In R, ab : E.]

28922 - BRANDT, A.B., KISELEVA, M.I. : K voprosu o fotosinteticheskoĭ ěffektivnosti zelenoĭ oblasti spektra. [Photosynthetic efficiency of green spectral region.] - Biofizika 22 : 668 - 670, 1977. [In R, ab : E.]

*28923 - BRANDT, P. : Regulation der Synthese von Plastidenproteinen in *Euglena gracilis*, Stamm Z. - Planta 130 : 81 - 83, 1976.

28924 - BRANDT, P. : The significance of the inactivation of the plastidial DNA-dependent RNA-polymerase by higher temperature for the development of the chloroplasts in *Euglena gracilis*, strain Z. - In : COOMBS, J. (ed.) : 4th International Congress on Photosynthesis. P. 53. UKISES, London 1977.

28925 - BRANGEON, J., PRIOUL, J.L., REYSS, A. : Reversibility of light-induced adaptive responses in the rye-grass *Lolium multiflorum*. - In : COOMBS, J. (ed.) : 4th International Congress on Photosynthesis. Pp. 53 - 54. UKISES, London 1977.

28926 - **BRAR, G., THIES, W.** : Contribution of leaves, stem, siliques and seeds to dry matter accumulation in ripening seeds of rapeseed, *Brassica napus* L. - Z. Pflanzenphysiol. *82* : 1 - 13, 1977.

28927 - **BRAUN, G.** : Über die Ursachen und Kriterien der Immissionsresistenz bei Fichte, *Picea abies* (L.) KARST. I. Morphologisch-anatomische Immissionsresistenz. - Europe. J. Forest Pathol. *7* : 23 - 43, 1977. [Chl.]

28928 - **BRAVDO, B., PALGI, A., LURIE, S.** : Changes in ribulose diphosphate carboxylase/oxygenase during tomato fruit ripening. - Israel J. Bot. *26* : 47 - 48, 1977.

28929 - **BRAVDO, B., PALGI, A., LURIE, S.** : The activity of some glycolate pathway enzymes at various stages of tomato fruit ripening. - In : COOMBS, J. (ed.) : 4th International Congress on Photosynthesis. P. 54. UKISES, London 1977.

28930 - **BRAVDO, B.-A.** : Oscillatory transpiration and CO_2 exchange of citrus leaves at the CO_2 compensation concentration. - Physiol. Plant. *41* : 36 - 41, 1977.

28931 - **BRAVDO, B.-A., PALGI, A., LURIE, S., FRENKEL, C.** : Changing ribulose diphosphate carboxylase/oxygenase activity in ripening tomato fruit. - Plant Physiol. *60* : 309 - 312, 1977.

28932 - **BRETON, J.** : Dichroism of transient absorbance changes in the red spectral region using oriented chloroplasts. II. *P*-700 absorbance changes. - Biochim. biophys. Acta *459* : 66 - 75, 1977.

28933 - **BRETON, J., PAILLOTIN, G.** : Dichroism of transient absorbance changes in the red spectral region using oriented chloroplasts. I. Field indicating absorbance changes. - Biochim. biophys. Acta *459* : 58 - 65, 1977.

28934 - **BRIANTAIS, J.M., VERNOTTE, C.** : Regulation of excitons transfer between the two photosystems by cations concentration changes in chloroplasts. - In : THELLIER, M., MONNIER, A., DEMARTY, M., DAINTY, J. (ed.) : Transmembrane Ionic Exchange in Plants. Pp. 559 - 567. Édit. C.N.R.S., Paris 1977.

28935 - **BRIANTAIS, J.-M., VERNOTTE, C., LAVERGNE, J., ARNTZEN, C.J.** : Identification of S_2 as the sensitive state to alkaline photoinactivation of photosystem II in chloroplasts. - Biochim. biophys. Acta *461* : 61 - 74, 1977.

28936 - **BRIANTAIS, J.-M., VERNOTTE, C., LAVERGNE, J., ARNTZEN, C.J., PICAUD, M.** : Inactivation of S_2 by alkaline pH inside the thylakoid. - In : PACKER, L., PAPAGEORGIOU, G.C., TREBST, A. (ed.) : Bioenergetics of Membranes. Pp. 287 - - 296.Elsevier/North-Holland Biomedical Press, Amsterdam - Oxford - New York 1977.

28937 - **BRIGGS, S.V.** : Estimates of biomass in a temperate mangrove community. - Aust. J. Ecol. *2* : 369 - 373, 1977.

28938 - **BRIN, G.P., ALIEV, Z.Sh.** : Deĭstvie rastvoriteleĭ na fotokhimicheskuyu aktivnost' i spektral'nye svoĭstva khloroplastov. [Effect of solvents on photochemical activity and spectral properties of chloroplasts.] - In : NASYROV, Yu.S. (ed.) : Genetika Fotosinteza. Pp. 209 - 215. Donish, Dushanbe 1977. [In R.]

28939 - **BRINKHUIS, B.H.** : Seasonal variations in salt-marsh macroalgae photosynthesis. I. *Ascophyllum nodosum* ecad *scorpioides*. - Mar. Biol. *44* : 165 - 175, 1977.

28940 - **BRINKHUIS, B.H.** : Seasonal variations in salt-marsh macroalgae photosynthesis. II. *Fucus vesiculosus* and *Ulva lactuca*. - Mar. Biol. *44* : 177 - 186, 1977.

28941 - **BRINKMAN, M.A., FREY, K.J.** : Growth analysis of isoline-recurrent parent grain yield differences in oats. - Crop Sci. *17* : 426 - 430, 1977.

28942 - **BRINKMANN, G., SENGER, H.** : Light-dependent formation of thylacoid membranes during the development of the photosynthetic apparatus in pigment mutant C-2A' of *Scenedesmus*. - In : COOMBS, J. (ed.) : 4th International Congress on Photosynthesis. P. 55. UKISES, London 1977.

28943 - **BRITH-LINDNER, M., ROSENHECK, K.** : The circular dichroism of bacteriorhodopsin : asymmetry and light-scattering distortions. - FEBS Lett. *76* : 41 - 44, 1977.

28944 - **BRITTON, G., LOCKLEY, W.J.S., POWLS, R., GOODWIN, T.W., HEYES, L.M.** : Carotenoid transformations during chloroplast development in *Scenedesmus obliquus* PG1 demonstrated by deuterium labelling. - Nature *288* : 81 - 82, 1977.

28945 - **BRITTON, G., POWLS, R.** : Phytoene, phytofluene and ζ-carotene isomers from a *Scenedesmus obliquus* mutant. - Phytochemistry *16* : 1253 - 1255, 1977.

28946 - **BRITTON, G., POWLS, R., LOCKLEY, W.J.S., GOODWIN, T.W.** : Carotenoid transformations during chloroplast development in a ζ-carotenic mutant strain of *Scenedesmus obliquus* greening in D_2O. - In : **COOMBS, J.** (ed.) : 4th International Congress on Photosynthesis. Pp. 55 - 56. UKISES, London 1977.

28947 - **BRITTON, G., POWLS, R., SCHULZE, R.M.** : The effect of illumination on the pigment composition of the ζ-carotenic mutant, PG1, of *Scenedesmus obliquus*. - Arch. Microbiol. *113* : 281 - 284, 1977.

28948 - **BRITTON, G., SINGH, R.K., GOODWIN, T.W.** : Carotenoid biosynthesis in *Rhodomicrobium vannielii*. Experiments with nicotine and 2-(4-chlorophenylthio)triethylammonium chloride (CPTA). - Biochim. biophys. Acta *488* : 475 - 483, 1977.

28949 - **BRITTON, G., SINGH, R.K., MALHOTRA, H.C., GOODWIN, T.W., BEN-AZIZ, A.** : Biosynthesis of 1,2-dihydrocarotenoids in *Rhodopseudomonas viridis* : experiments with inhibitors. - Phytochemistry *16* : 1561 - 1566, 1977.

28950 - **BRITZ, S.J.** : Inhibitor studies on the mechanism of rhythmic chloroplast movement in *Ulva*. - J. Phycol. *13* (Suppl.) : 9, 1977.

28951 - **BRODA, E.** : The evolution of photosynthesis. - Precambryan Res. *4* (2) : 117 - - 132, 1977.

28952 - **BRODA, E.** : Two kinds of lithotrophs missing in nature. - Z. allgem. Mikrobiol. *17* : 491 - 493, 1977. [Ps.]

*28953 - **BRODY, M.** : Chloroplast membranes. Lipid, protein, and chlorophyll interactions. - In : **PACKER, L.** (ed.) : Biomembranes. Architecture, Biogenesis, Bioenergetics and Differentiation. Pp. 331 - 351. Academic Press, New York - London 1974.

28954 - **BRODY, S.S., BRODY, M.** : Asymmetric and symmetric bimolecular membranes containing photosynthetic pigments. - Photochem. Photobiol. *26* : 57 - 58, 1977.

28955 - **BROOKS, A.S., TORKE, B.G.** : Vertical and seasonal distribution of chlorophyll *a* in Lake Michigan. - J. Fish. Res. Board Can. *34* : 2280 - 2287, 1977.

28956 - **BROUERS, M.** : Electron spin resonance of protochlorophyllide aggregates. - Plant Sci. Lett. *10* : 13 - 17, 1977.

28957 - **BROVCHENKO, M.I.** : Énergozavisimost' évakuatsii assimilyatov v apoplast i zagruzki okonchaniĭ provodyashcheĭ sistemy lista. [Energy dependent evacuation of assimilates into the apoplast and loading of endings of the conducting system of the leaf.] - Fiziol. Rast. *24* : 327 - 334, 1977. [In R, ab : E.]

28958 - **BROWN, A.P., FAUSET, C.R.** : Spectral dependence of light scatter by chloroplast suspensions. - In : **COOMBS, J.** (ed.) : 4th International Congress on Photosynthesis. Pp. 56 - 57. UKISES, London 1977.

28959 - **BROWN, A.S., TROXLER, R.F.** : Bilin-apoprotein linkage in rhodophytan phycobiliproteins : the role of cysteine. - FEBS Lett. *82* : 206 - 210, 1977.

28960 - **BROWN, A.S., TROXLER, R.F.** : Properties and *N*-terminal sequence of allophycocyanin from the unicellular rhodophyte *Cyanidium caldarium*. - Biochem. J. *163* : 571 - 581, 1977.

*28961 - **BROWN, D.H.** : Toxicity studies on the components of an oil-spill emulsifier using *Lichina pygmaea* and *Xanthoria parietina*. - Mar. Biol. *18* : 291 - 297, 1973. [Ps.]

28962 - **BROWN, D.H., HOOKER, T.N.** : The significance of acidic lichen substances in the estimation of chlorophyll and phaeophytin in lichens. - New Phytol. *78* : 617 - 624, 1977.

28963 - **BROWN, G.N., BIXBY, J.A., MELCAREK, P.K., HINCKLEY, T.M., ROGERS, R.** : Xylem pressure potential and chlorophyll fluorescence as indicators of freezing survival in black locust and western hemlock seedlings. - Cryobiology *14* : 94 - 99, 1977.

28964 - **BROWN, J.** : Fluorescence emission from the oxidized Photosystem I reaction center, $P700^+$. - In : OLSON, J.M., HIND, G. (ed.) : Chlorophyll-Proteins, Reaction Centers, and Photosynthetic Membranes. P. 362. Brookhaven nat. Lab., Upton 1977.

28965 - **BROWN, J.S.** : Fluorescence spectroscopy of a $P700$-chlorophyll-protein complex. - In : COOMBS, J. (ed.) : 4th International Congress on Photosynthesis. P. 57. UKISES, London 1977.

28966 - **BROWN, J.S.** : Fluorescence spectroscopy of chlorophyll-protein complexes. - In : PACKER, L., PAPAGEORGIOU, G.C., TREBST, A. (ed.) : Bioenergetics of Membranes. Pp. 297 - 304. Elsevier/North-Holland Biomedical Press, Amsterdam - Oxford - New York 1977.

28967 - **BROWN, J.S.** : Spectral studies of $P700$-chlorophyll a-protein complexes. - Carnegie Inst. Year Book *76* : 209 - 212, 1977.

28968 - **BROWN, J.S.** : Spectroscopy of chlorophyll in biological and synthetic systems. - Photochem. Photobiol. *26* : 319 - 326, 1977.

28969 - **BROWN, J.S.** : Fluorescence spectroscopy of a $P700$-chlorophyll-protein complex. - Photochem. Photobiol. *26* : 519 - 525, 1977.

28970 - **BROWN, L.F., TRLICA, M.J.** : Interacting effects of soil water, temperature and irradiance on CO_2 exchange rates of two dominant grasses of the shortgrass prairie. - J. appl. Ecol. *14* : 197 - 204, 1977.

28971 - **BROWN, L.F., TRLICA, M.J.** : Carbon dioxide exchange of blue grama swards as influenced by several ecological variables in the field. - J. appl. Ecol. *14*: 205 - 213, 1977.

28972 - **BROWN, L.F., TRLICA, M.J.** : Simulated dynamics of blue grama production. - J. appl. Ecol. *14* : 215 - 224, 1977. [Ps.]

28973 - **BROWN, T.H.** : Rate of loss of dry matter and change in chemical composition of nine pasture species over summer. - Aust. J. exp. Agr. anim. Husb. *17* : 75 - 79, 1977.

28974 - **BROWNING, G., SAUNDERS, P.F.** : Membrane localised gibberellins A_9 and A_4 in wheat chloroplasts. - Nature *265* : 375 - 377, 1977.

28975 - **BROWSE, J.A., DROMGOOLE, F.I., BROWN, J.M.A.** : Photosynthesis in the aquatic macrophyte *Egeria densa*. I. $^{14}CO_2$ fixation at natural CO_2 concentrations. - Aust. J. Plant Physiol. *4* : 169 - 176, 1977.

28976 - **BRUNNER, U., ELLER, B.M.** : Spectral properties of juvenile and adult leaves of *Piper betle* and their ecological significance. - Physiol. Plant. *41* : 22 - - 24, 1977. [Chl.]

*28977 - **BRYANT, G.A.** : The auto-ecology of the diatoms *Melosira* and *Cyclotella* in Southern Ontario lakes. - J. Phycol. *10* (Suppl.) : 12 - 13, 1974. [Chl.]

28978 - **BUCHOLTZ, M.L., MAUDINAS, B., PORTER, J.W.** : Effects of *in vivo* inhibitors of carotene biosynthesis on the synthesis of carotenes by a soluble tomato plastid enzyme system. - Chem.-biol. Inter. *17* : 359 - 362, 1977.

28979 - **BUESA, R.J.** : Photosynthesis and respiration of some tropical marine plants. - Aquat. Bot. *3* : 203 - 216, 1977.

28980 - **BUETOW, D.E., KISSIL, M.S., ZABLEN, L.** : Evolution of chloroplast ribosomal RNA. - In : BOGORAD, L., WEIL, J.H. (ed.) : Acides Nucléiques et Synthèse des Protéines chez les Végétaux. Coll. Int. C.N.R.S. No. 261. Pp. 227 - 233. Édit. CNRS, Paris 1977. [Chl.]

28981 - **BUGGELN, R.G., BAL, A.K.** : Effects of auxins and chemically related non-auxins on photosynthesis and chloroplast ultrastructure in *Alaria esculenta (Laminariales)*. - Can. J. Bot. *55* : 2098 - 2105, 1977.

28982 - **BUGLEWICZ, E.G., HERGENRADER, G.L.** : The impact of artificial reduction of light on a eutrophic farm pond. - Trans. Nebr. Acad. Sci. *4* : 23 - 33, 1977. [Production.]

28983 - **BULLEID, N.C.** : Adenosine triphosphate analysis in marine ecology : a review and manual. - Div. Fisheries Oceanogr. CSIRO Rep. *75* : 1 - 18, 1977.

28984 - **BUL'ON, V.V.** : Vzaimosvyaz' mezhdu soderzhaniem khlorofilla *a* v planktone i proizrachnost'yu vody po disku Sekki. [Correlation between chlorophyll *a* content in plankton and water transparency according to the Secchi disc.] - Dokl. Akad. Nauk SSSR *236* : 505 - 508, 1977. [In R.]

28985 - **BUL'ON, V.V.** : Vnekletochnaya produktsiya fitoplanktona. [Extracellular production of phytoplankton.] - Uspekhi sovrem. Biol. *84* : 294 - 304, 1977. [In R.]

28986 - **BUNCE, J.A.** : Leaf elongation in relation to leaf water potential in soybean. - J. exp. Bot. *28* : 156 - 161, 1977.

28987 - **BUNCE, J.A.** : Nonstomatal inhibition of photosynthesis at low water potentials in intact leaves of species from a variety of habitats. - Plant Physiol. *59* : 348 - 350, 1977.

28988 - **BUNCE, J.A., MILLER, L.N., CHABOT, B.F.** : Competitive exploitation of soil water by five eastern North American tree species. - Bot. Gaz. *138* : 168 - - 173, 1977. [Ps.]

28989 - **BUNCE, J.A., PATTERSON, D.T., PEET, M.M., ALBERTE, R.S.** : Light acclimation during and after leaf expansion in soybean. - Plant Physiol. *60* : 255 - 258, 1977.

28990 - **BURIEL, J.F., ALBERTE, R.S.** : Photosynthetic potential of loblolly pine seedlings *vs* sugar maple developed under three light intensities. - Plant Physiol. *59* (6, Suppl.) : 99, 1977.

28991 - **BURKE, J.J., DITTO, C.L., ARNTZEN, C.J.** : Cation-mediated organizational changes of the light-harvesting complex of pea chloroplasts. - Plant Physiol. *59* (6, Suppl.) : 24, 1977.

28992 - **BURNS, D.D., MIDGLEY, M.** : Localization and possible role of an adenosine triphosphatase in *Chlorobium thiosulfatophilum*. - Europe. J. Biochem. *67* : 323 - - 333, 1976.

28993 - **BURRIS, J.E.** : Photorespiration in marine algae. - Plant Physiol. *59* (6, Suppl.) : 66, 1977.

28994 - **BURRIS, J.E.** : Photosynthesis, photorespiration, and dark respiration in eight species of algae. - Mar. Biol. *39* : 371 - 379, 1977.

28995 - **BUSCHMANN, C., LICHTENTHALER, H.K.** : Hill-activity and P700 concentration of chloroplasts isolated from radish seedlings treated with -indoleacetic acid, kinetin or gibberellic acid. - Z. Naturforsch. *32C* : 798 - 802, 1977.

28996 - **BUTCHER, J.E., BOYER, M.G., FOWLE, C.D.** : Some changes in pond chemistry and photosynthetic activity following treatment with increasing concentrations of chlorpyrifos. - Bull. environm. Contam. Toxicol. *17* : 752 - 758, 1977.

28997 - **BUTLER, W.L.** : Energy distribution in the photosynthetic apparatus of plants. - In : OLSON, J.M., HIND, G. (ed.) : Chlorophyll-Proteins, Reaction Centers, and Photosynthetic Membranes. Pp. 338 - 346. Brookhaven nat. Lab., Upton 1977.

28998 - **BUTLER, W.L.** : Chlorophyll fluorescence : A probe for electron transfer and energy transfer. - In : TREBST, A., AVRON, M. (ed.) : Photosynthesis I. (Encycl. Plant Physiol. N.S. Vol. 5.) Pp. 149 - 167. Springer Verlag, Berlin - Heidelberg - New York 1977.

28999 - **BUTLER, W.L., STRASSER, R.J.** : A tripartite model for the photochemical apparatus of photosynthesis. - In : COOMBS, J. (ed.) : 4th International Congress on Photosynthesis. P. 58. UKISES, London 1977.

29000 - **BUTLER, W.L., STRASSER, R.J.** : Does the rate of cooling affect fluorescence properties of chloroplasts at -196 °C ? - Biochim. biophys. Acta *462* : 283 - - 289, 1977.

29001 - **BUTLER, W.L., STRASSER, R.J.** : Tripartite model for the photochemical apparatus of green plant photosynthesis. - Proc. nat. Acad. Sci. USA *74* : 3382 - - 3385, 1977.

29002 - BUTT, V.S., HARRISON, S.J., WALTON, N.J. : The fate of glyoxylate in leaf peroxisomes. - In : COOMBS, J. (ed.) : 4th International Congress on Photosynthesis. P. 59. UKISES, London 1977.

*29003 - BUTTERY, B.R., BUZZELL, R.I. : Flavonol glycoside genes and photosynthesis in soybeans. - Crop Sci. *16* : 547 - 550, 1976.

29004 - BUTTERY, B.R., BUZZELL, R.I. : The relationship between chlorophyll content and rate of photosynthesis in soybeans. - Can. J. Plant Sci. *57* : 1 - 5, 1977.

29005 - BÜTTNER, R., SALZER, J. : Untersuchungen über die Nettophotosynthese bei Apfelsorten. 1.Mitt. Vergleich der blattflächenbezogenen Nettophotosynthese vom 4 Apfelsorten während der Vegetationsperioden 1971 und 1972. - Arch. Züchtungsforsch. *7* : 95 - 101, 1977.

B29006 - BUVET, R., ALLEN, M.J., MASSUÉ, J.-P. (ed.) : Living Systems as Energy Converters. - North-Holland Publ. Co., Amsterdam - New York - Oxford 1977. [Ps.]

29007 - BUXTON, D.R., BRIGGS, R.E., PATTERSON, L.L., WATKINS, S.D. : Canopy characteristics of narrow-row cotton as influenced by plant density. - Agron. J. *69* : 929 - 933, 1977.

29008 - BYKOV, O.D., GALKIN, V.I., ZHITLOVA, N.A., KOSHKIN, V.A. : Vliyanie temperatury na izmenenie intensivnosti fotosinteza i dykhaniya diploidov i poliploidov kartofelya. [Effect of temperature on changes in photosynthetic and respiration rates of potato diploids and polyploids.] - In : NASYROV, Yu.S. (ed.) : Genetika Fotosinteza. Pp. 241 - 249. Donish, Dushanbe 1977. [In R.]

29009 - BYSZEWSKI, W. : Zusammenhang zwischen der Produktivität und der Ertragsfähigkeit von Zuckerrüben. - In : Produkce Biomasy a Tvorba Výnosů Polních Plodin. Vol. 1. Pp. 115 - 135. Česká vědeckotechnická Společnost zemědělská, Praha 1977. [Ps.]

29010 - BYTEVA, I.M., STOPOLYANSKAYA, L.V. : Izuchenie roli kisloroda v protsesse regeneratsii fotovosstanovlennogo khlorofilla i ego analogov v rastvore. [Role of oxygen during the regeneration of photoreduced chlorophyll and its analogs in solution.] - Biofizika *22* : 715 - 716, 1977. [In R, ab : E.]

29011 - CADÉE, G.C., HEGEMAN, J. : Distribution of primary production of the benthic microflora and accumulation of organic matter on a tidal flat area, Balgzand, Dutch Wadden Sea. - Neth. J. Sea Res. *11* : 24 - 41, 1977.

29012 - CAHEN, D., MALKIN, S., OHAD, I. : Development of photosystem II activity in *Chlamydomonas reinhardi* mutants. Insertion of photosystem II units into inactive preexisting membranes *versus* continuous formation of new photosynthetic membranes. - Plant Physiol. *60* : 845 - 849, 1977.

29013 - CAIRNS, J., Jr. : Quantification of biological integrity. - In : BALLENTINE, R.K., GUARRAIA, L.J. (ed.) : The Integrity of Water. Pp. ¡71 - 187. U.S. Environ. Protect. Agency, Washington 1977. [Aquatic production.]

29014 - CALDWELL, D.E. : Accessory pigment fluorescence for quantitation of photosynthetic microbial populations. - Can. J. Microbiol. *23* : 1594 - 1597, 1977.

29015 - CALDWELL, M.M., OSMOND, C.B., NOTT, D.L. : C_4-pathway photosynthesis at low temperature in cold-tolerant *Atriplex* species. - Plant Physiol. *60* : 157 - - 164, 1977.

29016 - CALDWELL, M.M., WHITE, R.S., MOORE, R.T., CAMP, L.B. : Carbon balance, productivity, and water use of cold-winter desert shrub communities dominated by C_3 and C_4 species. - Oecologia *29* : 275 - 300, 1977.

29017 - CALVAYRAC, R., DUBERTRET, G., LAVAL-MARTIN, D. : Evolution of some photosynthetic parameters during photoheterotrophic growth of *Euglena gracilis* Z in the presence of DCMU. - In : COOMBS, J. (ed.) : 4th International Congress on Photosynthesis. Pp. 59 - 60. UKISES, London 1977.

29018 - CALVIN, M. : Energy and materials *via* photosynthesis. - In : BUVET, R., ALLEN, M.J., MASSUÉ, J.-P. (ed.) : Living Systems as Energy Converters. Pp. 231 - - 259. North-Holland Publ. Co., Amsterdam - New York - Oxford 1977.

29019 - **CALVIN, M.** : Hydrocarbons *via* photosynthesis. - Energy Res. *1* : 299 - 327, 1977.

29020 - **CAMACHO-B, S.E.** : Some aspects of stomatal behavior of citrus. - Proc. int. Soc. Citricult. *1* : 66 - 69, 1977. [Ps.]

29021 - **CAMMACK, R., RAO, K.K., BARGERON, C.P., HUTSON, K.G., ANDREW, P.W., ROGERS, L.J.** : Midpoint redox potentials of plant and algal ferredoxins. - Biochem. J. *168* : 205 - 209, 1977.

29022 - **CAMPBELL, C.A., CAMERON, D.R., NICHOLAICHUK, W., DAVIDSON, H.R.** : Effect of fertilizer N and soil moisture on growth, N content, and moisture use by spring wheat. - Can. J. Soil Sci. *57* : 289 - 310, 1977. [Growth analysis.]

B29023 - **CAMPBELL, G.S.** : An Introduction to Environmental Biophysics. - Springer-Verlag, New York - Heidelberg - Berlin 1977. [Ps.]

*29024 - **CAMPBELL, W.F., EVANS, J.O., REED, S.C.** : Effects of glyphosate on chloroplast ultrastructure of quackgrass mesophyll cells. - Weed Sci. *24* : 22 - 25, 1976.

29025 - **CAMPILLO, A.J., HYER, R.C., MONGER, T.G., PARSON, W.W., SHAPIRO, S.L.** : Light collection and harvesting processes in bacterial photosynthesis investigated on a picosecond time scale. - Proc. nat. Acad. Sci. USA *74* : 1997 - 2001, 1977.

29026 - **CAMPILLO, A.J., SHAPIRO, S.L.** : Picosecond relaxation measurements in biology. - In : SHAPIRO, S.L. (ed.) : Topics in Applied Physics. Vol. 18: Ultrashort Light Pulses - Picosecond Techniques and Applications. Pp. 317 - 376. Springer-Verlag, Berlin - Heidelberg 1977. [Ps.]

29027 - **CAMPILLO, A.J., SHAPIRO, S.L., GEACINTOV, N.E., SWENBERG, C.E.** : Single-pulse picosecond determination of 735 nm fluorescence risetime in spinach chloroplasts. - FEBS Lett. *83* : 316 - 320, 1977.

29028 - **CAMPION, A., EL-SAYED, M.A., TERNER, J.** : Resonance Raman kinetic spectroscopy of bacteriorhodopsin on the microsecond time scale. - Biophys. J. *20* : 369 - - 375, 1977.

29029 - **CAMPION, A., TERNER, J., EL-SAYED, M.A.** : Time-resolved resonance Raman spectroscopy of bacteriorhodopsin. - Nature *265* : 659 - 661, 1977.

29030 - **CANAANI, O.D., SAUER, K.** : Analysis of the subunit structure of protochlorophyllide holochrome by sodium dodecyl sulfate-polyacrylamide gel electrophoresis. - Plant Physiol. *60* : 422 - 429, 1977.

29031 - **CANAANI, O.D., SAUER, K.** : The subunit structure of protochlorophyllide-holochrome. - In : OLSON, J.M., HIND, G. (ed.) : Chlorophyll-Proteins, Reaction Centers, and Photosynthetic Membranes. Pp. 360 - 361. Brookhaven nat. Lab., Upton 1977.

29032 - **CANVIN, D.T., FOCK, H.** : Effect of inhibitors and environmental treatments on carbon flow in the glycolate pathway. - In : COOMBS, J. (ed.) : 4th International Congress on Photosynthesis. P. 60. UKISES, London 1977.

29033 - **CAPEL, M.S., BOURQUE, D.P.** : Electrophoretic characterization of chloroplast ribosomal proteins assisted by computer analysis. - Plant Physiol. *59* (6, Suppl.) : 110, 1977.

29034 - **CAPLAN, S.R., EISENBACH, M., COOPER, S., GARTY, H., KLEMPERER, G., BAKKER, E.P.** : Light-driven proton and sodium ion transport in bacteriorhodopsin-containing particles. - In : PACKER, L., PAPAGEORGIOU, G.C., TREBST, A. (ed.) : Bioenergetics of Membranes. Pp. 101 - 114. Elsevier/North-Holland Biomedical Press, Amsterdam - Oxford - New York 1977.

29035 - **CAPLE, M., CHOW, H.-S., BURNS, R.M., STROUSE, C.E.** : Models of pigment aggregation based on crystallographic investigations : Recent results pertinent to the composition of the light-harvesting system of green sulfur bacteria. - In : OLSON, J.M., HIND, G. (ed.) : Chlorophyll-Proteins, Reaction Centers, and Photosynthetic Membranes. Pp. 56 - 63. Brookhaven nat. Lab., Upton 1977.

29036 - **CAPRON, T.M., MANSFIELD, T.A.** : Inhibition of growth in tomato by air polluted with nitrogen oxides. - J. exp. Bot. *28* : 112 - 116, 1977. [Growth analysis.]

29037 - **CARDENAS, J., ORTEGA, T., RIVAS, J., LOSADA, M.** : Regulation by reducing power to ferredoxin-nitrate reductase and NADP reductase of *Nostoc*. - In : COOMBS, J. (ed.) : 4th International Congress on Photosynthesis. P. 61. UKISES, London 1977.

29038 - **CARITHERS, R.P., YOCH, D.C., ARNON, D.I.** : Isolation and characterization of bound iron-sulfur proteins from bacterial photosynthetic membranes. II. Succinate dehydrogenase from *Rhodospirillum rubrum* chromatophores. - J. biol. Chem. *252* : 7461 - 7467, 1977.

29039 - **CARLING, D.E., MILLIKEN, D.F.** : Some physiological changes in *Vinca rosea* L. associated with mycoplasma-like disorders. - Plant Cell Physiol. *18* : 1379 - - 1381, 1977. [Chl.]

29040 - **CARLSEN, B.** : Barley mutants with defects in photosynthetic carbon dioxide fixation. - Carlsberg Res. Commun. *42* : 199 - 209, 1977.

29041 - **CARLSON, R.E.** : A trophic state index for lakes. - Limnol. Oceanogr. *22* : 361 - 369, 1977. [Chl.]

29042 - **CARLSON, R.W., BAZZAZ, F.A.** : Growth reduction in American sycamore (*Plantanus occidentalis* L.) caused by Pb-Cd interaction. - Environ. Pollut. *12* : 243 - 253, 1977. [Ps.]

29043 - **CARLSSON, R., SUNDQVIST, C.** : Stimulated respiration in darkness and during illumination after treatment with δ-aminolevulinic acid. - In : COOMBS, J. (ed.) : 4th International Congress on Photosynthesis. Pp. 61 - 62. UKISES, London 1977.

29044 - **CARMELI, C.** : Exchange reactions. - In : TREBST, A., AVRON, M. (ed.) : Photosynthesis I. (Encycl. Plant Physiol. N.S. Vol. 5.) Pp. 492 - 500. Springer--Verlag, Berlin - Heidelberg - New York 1977. [Ps.]

29045 - **CARMER, S.G.** : Treatment designs to estimate optimum plant density for maximum corn grain yield. - Agron. J. *69* : 803 - 807, 1977. [Models.]

29046 - **CARMI, A., KOLLER, D.** .:Endogenous regulation of photosynthetic rate in primary leaves of bean (*Phaseolus vulgaris* L.). - Ann. Bot. *41* : 59 - 67, 1977.

29047 - **CAROLIN, R.C., JACOBS, S.W.L., VESK, M.** : The ultrastructure of Kranz cells in the family *Cyperaceae*. - Bot. Gaz. *138* : 413 - 419, 1977.

*29048 - **CARPENA, O., NAVARRO, S., COSTA, F., VERDU, J.** : Separacion e identificacion de carotenoides en hojas de citrus. [Separation and identification of carotenoids from *Citrus* leaves.] - An. Edafol. Agrobiol. *31* : 43 - 59, 1972. [In Span., ab : E.]

*29049 - **CARR, J.L.** : The primary productivity and physiology of *Ceratophyllum demersum*. I. Gross macro primary productivity. - Aust. J. mar. Freshwater Res. *20*: 115 - 126, 1969.

29050 - **CARRIER, J.M.** : Inhibition of oxygen release by anoxia in a C_3-plant (*Nicotiana tabacum* cv. Wisconsin 38). Comparison with C_4-plants. - Planta *135* : 39 - - 43, 1977.

29051 - **CARRIER, J.M., NEVE, N.** : Oxidation-reduction states of pyridine nucleotides in maize leaves submitted to anoxia and dark or light treatments. - In : COOMBS, J. (ed.) : 4th International Congress on Photosynthesis. Pp. 62 - 63. UKISES, London 1977.

29052 - **CARTWRIGHT, S.C., LUSH, W.M., CANNY, M.J.** : A comparison of translocation of labelled assimilate by normal and lignified sieve elements in wheat leaves. - Planta *134* : 207 - 208, 1977.

29053 - **CARY, J.W.** : Photosynthesis of sugarbeets under N and P stress : Field measurements and carbon balance. - Agron. J. *69* : 739 - 744, 1977.

29054 - **CARY, J.W.** : Relations between CO_2 exchange rate, CO_2 compensation, and mesophyll resistance from a simple field method. - Crop Sci. *17* : 453 - 456, 1977.

29055 - **CASADORO, G., RASCIO, N.** : Relationship between plastids with a membrane-bound body and cell differentiation gradients in belladonna. - Caryologia *30* : 189 - - 198, 1977.

29056 - CASADORO, G., RASCIO, N., PAGANELLI CAPPELLETTI, E.M. : Membrane-bound plastidial inclusions in belladonna (*Atropa belladonna* L.). - Biol. cell. *29* :
61 - 66, 1977. [Chloroplast.]

29057 - CASPERS, N. : Seasonal variations of caloric values in herbaceous plants. -
Oecologia *26* : 379 - 383, 1977.

29058 - CASTELFRANCO, P.A., SCHWARCZ, S., WEINSTEIN, J.D. : Protoporphyrin IX from
glutamate in isolated chloroplasts. - Plant Physiol. *59* (6, Suppl.) : 8,
1977.

29059 - CASTELFRANCO, P.A., WEINSTEIN, J.D., SCHWARCZ, S. : 5-aminolevulinic acid
(ALA) and protoporphyrin IX from glutamate in isolated chloroplasts. - In :
COOMBS, J. (ed.) : 4th International Congress on Photosynthesis. P. 63. UKISES,
London 1977.

29060 - CASTRO, P.R.C., LUCCHESI, A.A., ALVES, E., PARANHOS, S.B. : Análise de crescimento de cana-de-açúcar. [Growth analysis of the sugar cane plant.] - Brasil Acucareiro *6* : 358 - 362, 1977. [In Port., ab : E.]

29061 - CAVALIERI, A.J., HUANG, A.H.C. : Effect of NaCl on the *in vitro* activity of
malate dehydrogenase in salt marsh halophytes of the U.S. - Physiol. Plant.
41 : 79 - 84, 1977.

29062 - CERFF, R. : The kinetic mechanism of NADP-glyceraldehyde 3-phosphate dehydrogenase. - In : COOMBS, J. (ed.) : 4th International Congress on Photosynthesis. P. 64. UKISES, London 1977.

B29063 - CERNUSCA, A. : Alpine Grasheide Hohe Tauern. Ergebnisse der Ökosystemstudie
1976. Veröffentl. Österr. MaB-Hochgebirgsprogrammes Hohe Tauern. Vol.1. -
Universitätsverlag Wagner, Innsbruck 1977. [Ps.]

29064 - CERNUSCA, A. : Bestandesstruktur, Mikroklima, Bestandesklima und Energiehaus-
halt von Pflanzenbeständen des alpinen Grasheidegürtels in den Hohen Tauern.
Erste Ergebnisse der Projektstudie 1976. - In : CERNUSCA, A. (ed.) : Alpine
Grasheide Hohe Tauern. Veröffentl. Österr. MaB-Hochgebirgsprogrammes Hohe
Tauern. Vol. 1. Pp. 25 - 45. Universitätsverlag Wagner, Innsbruck 1977.

29065 - CERNUSCA, A., DECKER, P. : Respiratorischer Kohlenstoffverbrauch im alpinen
Grasheidegürtel der Hohen Tauern. Erste Ergebnisse der Projektstudie 1976.
- In : CERNUSCA, A. (ed.) : Alpine Grasheide Hohe Tauern. Veröffentl. Österr.
MaB-Hochgebirgsprogrammes Hohe Tauern. Vol.1. Pp. 123 - 131. Universitäts-
verlag.Wagner, Innsbruck 1977.

29066 - ČERVENKA, K. : Výměna CO_2 listy jabloní. [CO_2 exchange by apple leaves.] - In :
In : HUZULÁK, J., MASAROVIČOVÁ, E. (ed.) : Fotosyntéza a Vodný Režim Drevín.
Pp. 167 - 174. Modra - Piesky 1977. [In Czech, ab : E, R.]

29067 - CERVIGNI, T., TRIOLO, L., MOSCONI, C., GIORGI, B. : Perspectives of a high
yielding yellow mutant of *Triticum durum*. - In : COOMBS, J. (ed.) : 4th International Congress on Photosynthesis. P. 65. UKISES, London 1977. [Chl.]

29068 - CHABOT, B.F., CHABOT, J.F. : Effects of light and temperature on leaf anatomy
and photosynthesis in *Fragaria vesca*. - Oecologia *26* : 363 - 377, 1977.

29069 - CHABOT, J.F., CHABOT, B.F. : Ultrastructure of the epidermis and stomatal complex of balsam fir (*Abies balsamea*). - Can. J. Bot. *55* : 1064 - 1075, 1977.

29070 - CHAIN, R.K., ARNON, D.I. : Quantum efficiency of photosynthetic energy conversion. - Proc. nat. Acad. Sci. USA *74* : 3377 - 3381, 1977.

*29071 - CHALUPA, V. : Fotosyntéza lesních dřevin ve vztahu k vodnímu provozu. [Photo-
synthesis of forest woody plants in relation to water relations.] - Zprávy
les. Výzkumu (Praha) *20* (4) : 14 - 16, 1974. [In Czech.]

29072 - CHALUPA, V. : Vliv přesazování sazenic smrku (*Picea abies* (L.) KARST.) a bo-
rovice (*Pinus silvestris* L.) na jejich vodní provoz a výškový růst. [The influence of the transplanting of Norway spruce and Scots pine plants on their
water regime and height growth.] - Práce VÚLHM *51* : 19 - 39, 1977. [In Czech,
ab : E, R.]

29073 - **CHAMOROVSKIĬ, S.K., LUKASHEV, E.P., KONONENKO, A.A., VENEDIKTOV, P.S., SHIN-KAREV, B.P., RUBIN, A.B.** : Teoreticheskoe rassmotrenie temperaturnykh zavisimosteĭ kinetiki vosstanovleniya bakteriokhlorofilla P_{890}, okislennogo impul's-nym lazernym i postoyannym svetom, v khromatoforakh *Ectothiorhodospira shaposhnikovii*. [Theoretical examination of temperature dependent kinetics of reduction of bacteriochlorophyll P890, oxidized by impulse laser and direct light in chromatophores of *Ectothiorhodospira shaposhnikovii*.] – Nauch. Dokl. vyssh. Shkoly, biol. Nauki *20* (6) : 38 - 43, 1977. [In R.]

29074 - **CHAMOROVSKY, S.K., PYT'EVA, N.F., RUBIN, A.B.** : Light-induced electron transport reactions in *Chromatium minutissimum* chromatophores poised at different redox potentials. - Stud. biophys. *66* : 129 - 143, 1977.

29075 - **CHAMPIGNY, M.L.** : Adenine nucleotides and the control of photosynthetic activities. - In : COOMBS, J. (ed.) : 4th International Congress on Photosynthesis. P. 66. UKISES, London 1977.

29076 - **CHAMPIGNY, M.-L., BISMUTH, A.** : Inorganic carbon transport across the spinach chloroplast envelope. - Plant Cell Physiol. *1977* (spec. Issue 3 - Photosynthetic Organelles. Structure and Function) : 365 - 375, 1977.

29077 - **CHAN, A.T.** : A comparison of growth rates of diatoms and dinoflagellates in relation to light intensity and cell size. - J. Phycol. *13* (Suppl.) : 11, 1977. [Chl.]

*29078 - **CHANCE, B., DUTTON, P.L., LEIGH, J.S.** : Spectroscopy within the cell. - In : SMITH, R.A. (ed.) : Very High Resolution Spectroscopy. Pp. 205 - 226. Academic Press, London - New York 1976. [Ps.]

29079 - **CHANG, C.K.** : Stacked double-macrocyclic ligands III. Spectral properties of cofacial diporphyrins as a function of the inter-chromophore separation | (1). - J. heterocycl. Chem. *14* : 1285 - 1288, 1977. [$P700$.]

29080 - **CHANG, J.C., DAS, T.P.** : Theory of hyperfine interactions in bacteriochlorophyll and related systems. - In : OLSON, J.M., HIND, G. (ed.) : Chlorophyll--Proteins, Reaction Centers, and Photosynthetic Membranes. P. 371. Brookhaven nat. Lab., Upton 1977.

29081 - **CHAN VAN NI, NIKANDROV, V.V., BRIN, G.P., KRASNOVSKIĬ, A.A.** : Vzaimosvyaz' fotovydeleniya i fotopogloshcheniya kisloroda khloroplastami : aktivatsiya i ingibirovanie. [Light-induced production and consumption of oxygen by chloroplasts : activation and inhibition.] - Biokhimiya *42* : 1298 - 1306, 1977. [In R, ab : E.]

29082 - **CHAN VAN NI, NIKANDROV, V.V., KRASNOVSKIĬ, A.A.** : Millisekundnoe poslesvechenie khloroplastov. Aktivatsiya i podavlenie. [Delayed luminescence in millisecond time range of chloroplasts. Activation and inhibition.] - Biofizika *22* : 1056 - 1061, 1977. [In R, ab : E.]

29083 - **CHAPADOS, C., LEBLANC, R.M.** : Aggregation of chlorophylls in monolayers. Infrared study of chlorophyll a in a mono- and multilayer arrays. - Chem. Phys. Lett. *49* : 180 - 182, 1977.

29084 - **CHAPMAN, D.** : The molecular organisation of cell membranes. - In : PACKER, L., PAPAGEORGIOU, G.C., TREBST, A. (ed.) : Bioenergetics of Membranes. Pp. 3 - 10. Elsevier/North-Holland Biomedical Press, Amsterdam - Oxford - New York 1977. [*Halobacterium* membrane.]

29085 - **CHAPMAN, K.S.R., HATCH, M.D.** : Regulation of mitochondrial NAD-malic enzyme involved in C_4 pathway photosynthesis. - Arch. Biochem. Biophys. *184* : 298 - - 306, 1977.

29086 - **CHARLES-EDWARDS, D.A., ACOCK, B.** : Growth response of a *Chrysanthemum* crop to the environment. II. A mathematical analysis relating photosynthesis and growth. - Ann. Bot. *41* : 49 - 58, 1977.

29087 - **CHARLES-EDWARDS, D.A., ACOCK, B.** : Photosynthesis and dry matter production in the U.K. - In : COOMBS, J. (ed.) : 4th International Congress on Photosynthesis. P. 67. UKISES, London 1977.

29088 - CHARLES-EDWARDS, D.A., THORNLEY, J.H.M. : Some aspects of plant modelling. -
In : UNGER, K. (ed.) : Biophysikalische Analyse pflanzlicher Systeme. Pp. 96-
- 104. VEB Gustav Fischer Verlag, Jena 1977.

29089 - CHARLTON, M.N. : Carbon budgets in Lake Ontario. - J. Fish. Res. Board Can.
34 : 1240 - 1241, 1977. [Ps and productivity.]

29090 - CHARTIER, P., BETHENOD, O. : La productivité primaire à l'échelle de la feuil-
le. - In : MOYSE, A. (ed.) : Les Processus de la Production Végétale Primaire.
Pp. 77 - 112. Gauthier-Villars, Paris 1977.

29091 - CHARTIER, P., MOROT-GAUDRY, J.F., BETHENOD, O., THOMAS, D.A. : The net assimilation of C_3 and C_4 plants as influenced by light and carbon dioxide, and
an analysis of the role of the gene *opaque 2* in young maize. - In : LANDS-
BERG, J.J., CUTTING, C.V. (ed.) : Environmental Effects on Crop Physiology.
Pp. 125 - 136. Academic Press, London - New York - San Francisco 1977.

29092 - CHATURVEDI, S.N., ZABKA, G. : Studies on dark fixation of carbon dioxide in
Kalanchoë I. Effect of water stress and growth retardants. - Ann. Bot. *41* :
493 - 500, 1977.

29093 - CHATURVEDI, S.N., ZABKA, G. : Studies on dark fixation of carbon dioxide in
Kalanchoë II. Effect of interaction of photoperiodic induction with water
stress and growth retardants. - Ann. Bot. *41* : 501 - 505, 1977.

29094 - CHEKALIN, N.M. : Tipy indutsirovannykh makromutatsiĭ u chiny posevnoĭ (*Lathyrus sativus* L.) Soobshchenie I. Tipy khlorofil'nykh mutatsiĭ. [Types of induced macromutations in common vetchling (*Lathyrus sativus* L.). Communication
I. Types of chlorophyll mutations.] - Genetika *13* (1) : 23 - 31, 1977. [In
R, ab : E.]

29095 - CHEKULAEVA, L.N. : Nekotorye osobennosti purpurnykh membran, vydelennykh iz
kletok galofilov. [Some peculiarities of purple membranes isolated from halo-
phile cells.] - Biofizika *22* : 728 - 730, 1977. [In R, ab : E.]

29096 - CHEMERIS, Yu.K., KHITROV, Yu.A., UGOLKOVA, N.G., VENEDIKTOV, P.S. : Issledovanie zamedlennoĭ fluorestsentsii khlorofila v kul'ture sinkhronno delyash-
chikhsya kletok khlorelly. [Delayed chlorophyll fluorescence in the culture
of synchronously dividing *Chlorella* cells.] - Fiziol. Rast. *24* : 976 - 980,
1977. [In R, ab : E.]

29097 - CHENIAE, G., MARTIN, I.F. : The inactivation of O_2 evolution *via* attack of
the S_2-state by Tris and the reactivation of compounds that increase the rate
of decay of S_2, S_3. - In : COOMBS, J. (ed.) : 4th International Congress on
Photosynthesis. P. 67. UKISES, London 1977.

29098 - CHERNYAD'EV, I.I., USPENSKAYA, V.È., KONDRAT'EVA, E.N., DOMAN, N.G. : Ob as-
similyatsii ugleroda iskhodnym i mutantnym shtammami *Rhodopseudomonas palustris.* [Carbon assimilation by the wild and mutant strains of *Rhodopseudomonas
palustris.*] - In : NASYROV, Yu.S. (ed.) : Genetika Fotosinteza. Pp. 222 - 226.
Donish, Dushanbe 1977. [In R.]

29099 - CHERRY, R.J., HEYN, M.P., OESTERHELT, D. : Rotational diffusion and exciton
coupling of bacteriorhodopsin in the cell membrane of *Halobacterium halobium*.
- FEBS Lett. *78* : 25 - 30, 1977.

29100 - CHERRY, R.J., MÜLLER, U., SCHNEIDER, G. : Rotational diffusion of bacteriorhodopsin in lipid membranes. - FEBS Lett. *80* : 465 - 469, 1977.

29101 - CHEVALLIER, D., DOUCE, R., NURIT, F. : Interactions entre mitochondries et
chloroplastes dans la cellule. II. Action du DBMIB, de l'antimycine *A* et du
FCCP sur la spore de *Funaria hygrometrica*. - Can. J. Bot. *55* : 1650 - 1659,
1977.

29102 - CHIKOV, V.I., LOZOVAYA, V.V., TARCHEVSKIĬ, I.A. : Dnevnaya dinamika fotosinteza tselogo rasteniya pshenitsy. [Diurnal dynamics of photosynthesis of a
whole wheat plant.] - Fiziol. Rast. *24* : 691 - 698, 1977. [In R, ab : E.]

29103 - CHILDS, S.W., GILLEY, J.R., SPLINTER, W.E. : A simplified model of corn growth
under moisture stress. - Trans. ASAE *20* : 858 - 865, 1977. [Ps.]

29104 - CHIN, H.F., NEALES, T.F., WILSON, J.H. : The effects of cotyledon excision on growth and leaf senescence in soya-bean plants. - Ann. Bot. *41* : 771 - 777, 1977. [Ps, Chl.]

29105 - CHIPMAN, D.M., SHOSHAN, V., SHAVIT, N. : Kinetics of binding of fluorescent nucleotide analogs to CF_1. - In : COOMBS, J. (ed.) : 4th International Congress on Photosynthesis. P. 68. UKISES, London 1977.

29106 - CHIRGADZE, Y.N., GARBER, M.B., NIKONOV, S.V. : Crystallographic study of plastocyanins. - J. mol. Biol. *113* : 443 - 447, 1977.

29107 - CHOE, H.T., THIMANN, K.V. : The retention of photosynthetic activity by senescing chloroplasts of oat leaves. - Planta *135* : 101 - 107, 1977.

29108 - CHOLLET, R. : Evaluation of the light/dark $^{14}CO_2$ efflux assay of photorespiration. - Plant Physiol. *59* (6, Suppl.) : 42, 1977.

29109 - CHOLLET, R. : The biochemistry of photorespiration. - Trends Biochem. Sci. *2* (7) : 155 - 159, 1977.

29110 - CHONAN, N., KAWAHARA, H., MATSUDA, T. : [Effect of nitrogen application on ultrastructure of the chloroplast in rice plants.] - Jap. J. Crop Sci. *46* : 387 - 392, 1977. [In Jap., ab : E.]

29111 - CHONAN, N., KAWAHARA, H., MATSUDA, T. : [Ultrastructural development of mesophyll cells in rice seedlings.] - Jap. J. Crop Sci. *46* : 147 - 156, 1977. [Peroxisome; in Jap., ab : E.]

29112 - CHONAN, N., KAWAHARA, H., MATSUDA, T. : [Changes in chloroplast ultrastructure during the senescence of leaves in rice plants.] - Jap. J. Crop Sci. *46* : 379 - 386, 1977. [In Jap., ab : E.]

29113 - CHOW, P.N.P. : Bleaching of chlorophylls in alcohol extracts with benzoyl peroxide for liquid scintillation counting of ^{14}C-labeled compounds. - Anal. Biochem. *80* : 507 - 512, 1977.

29114 - CHOW, W.S., HOPE, A.B. : Proton translocation, electron transport and photophosphorylation in isolated chloroplasts. - Aust. J. Plant Physiol. *4* : 647 - - 665, 1977.

29115 - CHRISTELLER, J.T., LAING, W.A., SUTTON, W.D. : Carbon dioxide fixation by lupin root nodules. 1. Characterization, association with phosphoenolpyruvate carboxylase, and correlation with nitrogen fixation during nodule development. - Plant Physiol. *60* : 47 - 50, 1977. [Comparison with leaf PEPC.]

29116 - CHRISTENSEN, B., WIUM-ANDERSEN, S. : Seasonal growth of mangrove trees in Southern Thailand. I. The phenology of *Rhizophora apiculata* BL. - Aquatic Bot. *3* : 281 - 286, 1977.

29117 - CHUECA, A., LÁZARO, J.J., LÓPEZ GORGÉ, J. : Fructose-1,6-diphosphatase from spinach leaf chloroplasts : subunit structure. - Plant Sci. Lett. *8* : 71 - - 77, 1977.

29118 - CHUNAEV, A.S. : Svetoustoĭchivost' zelenykh vodorosleĭ : dve podsistemy. [Light resistance of green algae : two subsystems.] - Vestn. leningrad. Univ. *1977* [9, Biol. 2] : 134, 1977. [Chl; in R.]

29119 - CHUNG, H.-H., BARNES, R.L. : Photosynthate allocation in *Pinus taeda*. I. Substrate requirements for synthesis of shoot biomass. - Can. J. Forest Res. *7* : 106 - 111, 1977.

29120 - CHUROVÁ, K. : Analýza produkčného procesu ovsa na zelenú hmotu pri rozličnej vlhkosti pôdneho substrátu. [Analysis of production process of oats green mass at different humidity of soil substrate.] - Ved. Práce výsk. Ústavu rast. Výroby Piešťanoch *14* : 49 - 59, 1977. [In Slovak, ab : E, G.]

*29121 - CIRELI, B., BÜTÜN, G. : Güneş ve gölge yapraklarda pigment maddeleri üzerine bir ön çalişma. [Preliminary observation on the pigments of sun and shade leaves.] - Ege Univ. Fen Fak. Ilmi Raporlar Serisi *189* : 1 - 7, 1975. [In Türk., ab : E.]

29122 - CLARK, B.J., PRIOUL, J.-L., COUDERC, H. : The physiological response to cutting in Italian ryegrass. - J. Brit. Grassland Soc. *32* : 1 - 5, 1977. [Ps.]

29123 - **CLARK, J.F., MORGAN, T.R.** : Effect of rainfall on crop yields. - J. aust. Inst. agr. Sci. *43* : 131 - 132, 1977.

29124 - **CLARKE, R.H., CONNORS, R.E., FRANK, H.A., HOCH, J.C.** : Investigation of the structure of the reaction center in photosynthetic systems by optical detection of triplet state magnetic resonance. - Chem. phys. Lett. *45* : 523 - 528, 1977.

29125 - **CLARKE, R.H., FRANK, H.A., HOBART, D., LEENSTRA, W.A.** : The triplet states of chlorophylls and chlorophyll complexes. - In : COOMBS, J. (ed.) : 4th International Congress on Photosynthesis. P. 69. UKISES, London 1977.

29126 - **CLARKE, R.H., HOBART, D.R.** : Structural aspects of the reaction center of photosynthetic bacteria calculated from triplet state zero-field splittings. - FEBS Lett. *82* : 155 - 158, 1977.

*29127 - **CLASEN, J., BERNHARDT, H.** : The use of algal assays for determining the effect of iron and phosphorus compounds on the growth of various algal species. - Water Res. *8* : 31 - 44, 1974. [Chl.]

29128 - **CLAUHS, R.P., GRUN, P.** : Changes in plastid and mitochondrion content during maturation of generative cells of *Solanum (Solanaceae)*. - Amer. J. Bot. *64* : 377 - 383, 1977.

29129 - **CLAUS, S.** : Zur Modellierung von Teilsystemen bei der Stoffproduktion von Pflanzen. - In : UNGER, K. (ed.) : Biophysikalische Analyse pflanzlicher Systeme. Pp. 56 - 69. VEB Gustav Fischer Verlag, Jena 1977.

29130 - **CLAUS, S., UNGER, K.** : Energie- und Wasserhaushalt eines Blattes unter Berücksichtigung der Stomataregelung. - In : UNGER, K. (ed.) : Biophysikalische Analyse pflanzlicher Systeme. Pp. 131 - 139. VEB Gustav Fischer Verlag, Jena 1977. [Ps.]

29131 - **CLAUSSEN, W.** : Einfluβ der Frucht auf Netto-Assimilationsleistung und Netto--Photosyntheseraten der Aubergine (*Solanum melongena* L.). - Gartenbauwissenschaft *42* : 61 - 65, 1977.

29132 - **CLAUSSEN, W.** : Netto-Photosyntheseraten sowie Saccharose- und Stärkestoffwechsel in unterschiedlich alten Blättern der Aubergine (*Solanum melongena* L.). - Angew. Bot. *51* : 129 - 141, 1977.

29133 - **CLAUSSEN, W., BILLER, E.** : Die Bedeutung der Saccharose- und Stärkegehalte der Blätter für die Regulierung der Netto-Photosyntheseraten. - Z. Pflanzenphysiol. *81* : 189 - 198, 1977.

29134 - **CLAYTON, R.K.** : Fluorescence of photosynthetic reaction centers at low temperatures. - Plant Cell Physiol. *1977* (spec. Issue 3 - Photosynthetic Organelles. Structure and Function) : 87 - 96, 1977.

29135 - **CLAYTON, R.K.** : Photosynthesis and solar energy conversion. - In : OLSON, J. M., HIND, G. (ed.) : Chlorophyll-Proteins, Reaction Centers, and Photosynthetic Membranes. Pp. 1 - 15. Brookhaven nat. Lab., Upton 1977.

29136 - **CLEMENS, J., PEARSON, C.J.** : The effect of waterlogging on the growth and ethylene content of *Eucalyptus robusta* SM. (Swamp Mahogany). - Oecologia *29* : 249 - 255, 1977. [Growth analysis.]

29137 - **CLEMENT, A., GARBAYE, J., TACON, F. LE** : Importance des ectomycorhizes dans la résistance au calcaire du Pin noir (*Pinus nigra* ARN. ssp. *nigricans* HOST). - Ecol. Plant. *12* : 111 - 131, 1977. [Chlorosis.]

29138 - **CLIJSTERS, H., VAN ASSCHE, F., MARCELLE, R.** : Effects of zinc nutrition at sublethal concentrations on photosynthesis of primary bean leaves. - In : COOMBS, J. (ed.) : 4th International Congress on Photosynthesis. Pp. 69 - 70. UKISES, London 1977.

29139 - **CLOERN, J.E.** : Effects of light intensity and temperature on *Cryptomonas ovata (Cryptophyceae)* growth and nutrient uptake rates. - J. Phycol. *13* : 389 - - 395, 1977. [Chl.]

29140 - **COBB, A.H.** : Symbiosis between the chloroplasts of the marine alga *Codium fragile* and the mollusc *Elysia viridis*. - In : COOMBS, J. (ed.) : 4th International Congress on Photosynthesis. P. 70. UKISES, London 1977. [Ps.]

29141 - COBB, A.H. : The relationship of purity to photosynthetic activity in preparations of *Codium fragile* chloroplasts. - Protoplasma *92* : 137 - 146, 1977.

29142 - COCKBURN, W., McAULAY, A. : Changes in metabolite levels in *Kalanchoë daigremontiana* and the regulation of malic acid accumulation in Crassulacean Acid
Metabolism. - Plant Physiol. *59* : 455 - 458, 1977.

29143 - CODD, G.A., SALLAL, A.K.J. : Comparative studies on the thylakoids from
light- and dark-grown cultures of the blue-green alga *Chlorogloea fritschii*.
- In : COOMBS, J. (ed.) : 4th International Congress on Photosynthesis. P.
71. UKISES, London 1977.

29144 - CODD, G.A., STEWART, W.D.P. : Quaternary structure of the D-ribulose 1,5-di-
phosphate carboxylase from the cyanelles of *Cyanophora paradoxa*. - FEMS Lett.
1 : 35 - 38, 1977.

29145 - CODD, G.A., STEWART, W.D.P. : Ribulose-1,5-diphosphate carboxylase in heterocysts and vegetative cells of *Anabaena cylindrica*. - FEMS Microbiol. Lett.
2 : 247 - 249, 1977.

29146 - CODD, G.A., STEWART, W.D.P. : D-ribulose 1,5-diphosphate carboxylase from the
blue-green alga *Aphanocapsa* 6308. - Arch. Microbiol. *113* : 105 - 110, 1977.

29147 - COEN, D.M., BEDBROOK, J.R., BOGORAD, L., RICH, A. : Maize chloroplast DNA
fragment encoding the large subunit of ribulosebisphosphate carboxylase. -
Proc. nat. Acad. Sci. USA *74* : 5487 - 5491, 1977.

29148 - COGDELL, R.J., CELIS, S., CELIS, A., CROFTS, A.R. : Reaction centre carotenoid band shifts. - In : COOMBS, J. (ed.) : 4th International Congress on Photosynthesis. P. 72. UKISES, London 1977.

29149 - COGDELL, R.J., CELIS, S., CELIS, H., CROFTS, A.R. : Reaction centre carotenoid band shifts. - FEBS Lett. *80* : 190 - 194, 1977.

29150 - COHEN, C.E., BAZZAZ, M.B., FULLETT, S.H., REBEIZ, C.A. : Chloroplast biogenesis. XX. Accumulation of porphyrin and phorbin pigments in cucumber cotyledons during photoperiodic greening. - Plant Physiol. *60* : 743 - 746, 1977.

29151 - COHEN, Y., KRUMBEIN, W.E., SHILO, M. : Solar Lake (Sinai). 3. Bacterial distribution and production. - Limnol. Oceanogr. *22* : 621 - 634, 1977. [Ps,
Chl.]

29152 - COHEN-BAZIRE, G., BÉGUIN, S., RIMON, S., GLAZER, A.N., BROWN, D.M. : Physi-
co-chemical and immunological properties of allophycocyanins. - Arch. Microbiol. *111* : 225 - 238, 1977.

*29153 - COÏC, Y., LESAINT, C. : Influence de la modalité de déficience en phosphore
sur l'équilibre photosynthèse-protidosynthèse. - Compt. rend. Séances Acad.
Agr. France *62* : 1251 - 1256, 1976.

29154 - COLES, S.L., JOKIEL, P.L. : Effects of temperature on photosynthesis and respiration in hermatypic corals. - Mar. Biol. *43* : 209 - 216, 1977.

29155 - COLLARD, R.C., JOINER, J.N., CONOVER, C.A., McCONNELL, D.B. : Influence of
shade and fertilizer on light compensation point of *Ficus benjamina* L. - J.
amer. Soc. hort. Sci. *102* : 447 - 449, 1977.

29156 - COLLATZ, G.J. : Influence of certain environmental factors on photosynthesis
and photorespiration in *Simmondsia chinensis*. - Planta *134* : 127 - 132, 1977.

29157 - COLLINS, O.D.G., SUTCLIFFE, J.F. : The relationship between transport of individual elements and dry matter from the cotyledons of *Pisum sativum* L. -
Ann. Bot. *41* : 163 - 171, 1977. [Photosynthates.]

29158 - COLLINS, W.B. : Comparison of growth and tuber development in three potato
cultivars with diverse canopy size. - Can. J. Plant Sci. *57* : 797 - 801,
1977. [Growth analysis.]

29159 - COLMAN, B. : Excretion of glycolate by *Chlorella fusca*. - Plant Physiol. *59*
(6, Suppl.) : 91, 1977. [Ps.]

29160 - COLMAN, B. : Regulation of glycolate excretion in *Chlorella fusca*. - In :
COOMBS, J. (ed.) : 4th International Congress on Photosynthesis. Pp. 72 - 73.
UKISES, London 1977.

29161 - CONJEAUD, H., MATHIS, P. : Sub-microsecond study of triplet states and of the carotenoid absorption shift in chloroplasts. - In : COOMBS, J. (ed.) : 4th International Congress on Photosynthesis. Pp. 73 - 74. UKISES, London 1977.

29162 - CONNOR, D.J., LEGGE, N.J., TURNER, N.C. : Water relations of mountain ash (*Eucalyptus regnans* F. MUELL.) forests. - Aust. J. Plant Physiol. *4* : 753 - 762, 1977. [Stomatal resistance.]

29163 - CONOVER, C.A., POOLE, R.T. : Effects of cultural practices on acclimatization of *Ficus benjamina* L. - J. amer. Soc. hort. Sci. *102* : 529 - 531, 1977. [Chl.]

29164 - CONSTABLE, G.A., GLEESON, A.C. : Growth and distribution of dry matter in cotton (*Gossypium hirsutum* L.). - Aust. J. agr. Res. *28* : 249 - 256, 1977.

29165 - COOMBS, J., BALDRY, C.W., YUILL-HIGGS, A. : PEP carboxylase-assays and artefacts. - In : COOMBS, J. (ed.) : 4th International Congress on Photosynthesis. P. 74. UKISES, London 1977.

29166 - COOPER, J.P. : Photosynthetic efficiency of maize compared with other field crops. - Ann. appl. Biol. *87* : 237 - 242, 1977.

29167 - COPE, F.W. : Cooperative interactions in nerve membrane potential and in photosynthesis, evidenced by non-linear Arrhenius plots and critical exponents. - Physiol. Chem. Phys. *9* : 247 - 258, 1977.

29168 - COPONY, W., PAMFIL, C. : Über Erfahrungen mit biometrischen Modellen bei der Untersuchung der Stoffproduktion von Kartoffeln. - In : UNGER, K. (ed.) : Biophysikalische Analyse pflanzlicher Systeme. Pp. 80 - 84. VEB Gustav Fischer Verlag, Jena 1977.

29169 - CORNIC, G. : Analyse du dégagement de CO_2 dans un air sans CO_2. - In : MOYSE, A. (ed.) : Les Processus de la Production Végétale Primaire. Pp. 137 - 156. Gauthier-Villars, Paris 1977.

29170 - COSSETTE, C., LAW, K.N. : Utilisation du feuillage des arbres. Un procédé pour l'extraction des protéines et d'une pâte de chlorophylle contenues dans les aiguilles des épinettes. - Ann. ACFAS *44* (2) : 117 - 121, 1977. [Chl.]

29171 - COSTE, B., JACQUES, G., MINAS, H.J. : Sels nutritifs et production primaire dans le Golfe du Lion et ses abords. - Ann. Inst. océanogr. (Paris) *53* : 189 - 202, 1977.

29172 - COSTES, C., MONTIES, B. : Spectroscopic effects of reactions between electrophilic reagents and epoxycarotenoids violaxanthin and neoxanthin. - Physiol. vég. *15* : 667 - 678, 1977.

29173 - COTTON, T.M., LOACH, P.A., SHIPMAN, L.L., KATZ, J.J. : Characterization of chlorophyll *a*-bifunctional ligand adducts at low temperature. - In : OLSON, J.M., HIND, G. (ed.) : Chlorophyll-Proteins, Reaction Centers, and Photosynthetic Membranes. P. 370. Brookhaven nat. Lab., Upton 1977.

29174 - COTTON, T.M., VAN DUYNE, R.P. : Resonance Raman spectroelectrochemistry of the primary photochemical species in photosynthetic bacteria. - In : COOMBS, J. (ed.) : 4th International Congress on Photosynthesis. P. 75. UKISES, London 1977.

*29175 - COUCH, R., GANGSTAD, E.O. : Response of waterhyacinth to laser radiation. - Weed Sci. *22* : 450 - 453, 1974. [Ps.]

29176 - COUGHLAN, S. : Glycolate metabolism in *Thalassiosira pseudonana*. - J. exp. Bot. *28* : 78 - 83, 1977.

29177 - COUGHLAN, S. : The effect of organic substrates on the growth, photosynthesis and dark survival of marine algae. - Brit. phycol. J. *12* : 155 - 162, 1977.

29178 - COUGHLAN, S., TATTERSFIELD, D. : Photorespiration in larger littoral algae. - Bot. Mar. *20* : 265 - 266, 1977.

29179 - COWAN, I.R., FARQUHAR, G.D. : Stomatal function in relation to leaf metabolism and environment. - In : JENNINGS, D.H. (ed.) : Integration of Activity in the Higher Plant. Pp. 471 - 505. Cambridge Univ. Press, Cambridge - London - New York - Melbourne 1977. [Ps.]

29180 - **COX, R.P.** : Some properties of chloroplast cytochrome b-563 in an enriched preparation obtained by digitonin treatment. - In : **COOMBS, J.** (ed.) : 4th International Congress on Photosynthesis. P. 76. UKISES, London 1977.

29181 - **CRAIG, S., GOODCHILD, D.J.** : Leaf ultrastructure of *Triodia irritans* : a C_4 grass possessing an unusual arrangement of photosynthetic tissues. - Aust. J. Bot. *25* : 277 - 290, 1977.

29182 - **CRAMER, W.A.** : Cytochromes. - In : **TREBST, A., AVRON, M.** (ed.) : Photosynthesis I. (Encycl. Plant Physiol. N.S. Vol. 5.) Pp. 227 - 237. Springer-Verlag, Berlin - Heidelberg - New York 1977.

29183 - **CRAMER, W.A., WHITMARSH, J.** : Photosynthetic cytochromes. - Annu. Rev. Plant Physiol. *28* : 133 - 172, 1977.

29184 - **CRANE, F.L., BARR, R.** : Stimulation of photosynthesis by carbonyl compounds and chelators. - Biochem. biophys. Res. Commun. *74* : 1362 - 1368, 1977.

29185 - **CRANE, N.L., SOMMERFELD, M.R.** : Phytoplankton ecology of Lynx Lake, Arizona. - Southwest. Nat. *22* : 305 - 320, 1977. [Chl.]

29186 - **CRAUBNER, H., KOENIG, F., SCHMID, G.H.** : Molecular weight and the dodecyl sulphate binding of a thylakoid membrane polypeptide involved in a reaction on the oxygen-evolving side of photosystem II. - Z. Naturforsch. *32C* : 384 - - 391, 1977.

29187 - **CRESSWELL, C.F., LEWIS, O.A.M.** : The regulation of photosynthetic carbon by nitrate nitrogen in C_4 plants. - In : **COOMBS, J.** (ed.) : 4th International Congress on Photosynthesis. Pp. 76 - 77. UKISES, London 1977.

29188 - **CRILL, P.A.** : The photosynthesis-light curve : A simple analog model. - J. theor. Biol. *64* : 503 - 516, 1977.

29189 - **CROMBIE, W.M.L.** : The influence of photosynthesis and SKF inhibitors on cannabinoid production in *Cannabis sativa*. - Phytochemistry *16* : 1369 - 1371, 1977.

29190 - **CROSBIE, T.M., MOCK, J.J., PEARCE, R.B.** : Variability and selection advance for photosynthesis in Iowa stiff stalk synthetic maize population. - Crop Sci. *17* : 511 - 514, 1977.

29191 - **CROSSLAND, C.J., BARNES, D.J.** : Gas-exchange studies with the staghorn coral *Acropora acuminata* and its zooxanthellae. - Mar. Biol. *40* : 185 - 194, 1977. [Ps.]

29192 - **CROWTHER, D., CELIS, H., ALMANZA de CELIS, S., CROFTS, A.R.** : Studies of bacterial reaction centres-reconstitution of proton pumping in phospholipid vesicles. - In : **COOMBS, J.** (ed.) : 4th International Congress on Photosynthesis. Pp. 77 - 78. UKISES, London 1977.

29193 - **CRUZ, A. de la, HACKNEY, C.T.** : Energy value, elemental composition, and productivity of belowground biomass of a *Juncus* tidal marsh. - Ecology *58* : 1165 - 1170, 1977.

29194 - **CSATORDAY, K.** : Detection of pigment forms in *Anacystis nidulans* by fluorescence reabsorption spectroscopy. - In : **COOMBS, J.** (ed.) : 4th International Congress on Photosynthesis. P. 78. UKISES, London 1977.

29195 - **CSATORDAY, K.** : Detection of pigment forms in *Anacystis nidulans* by fluorescence reabsorption spectroscopy. - FEBS Lett. *79* : 379 - 382, 1977.

29196 - **CSATORDAY, K., HORVÁTH, G.** : Synchronization of *Anacystis nidulans*. Oxygen evolution during the cell cycle. - Arch. Microbiol. *111* : 245 - 246, 1977.

29197 - **CUENDET, P.A., ZUBER, H.** : Characterization of a protein associated with bacteriochlorophyll and lipid from *Rhodospirillum rubrum*. - Experientia *33* : 790, 1977.

29198 - **CUENDET, P.A., ZUBER, H.** : Isolation and characterization of a bacteriochlorophyll-associated chromatophore protein from *Rhodospirillum rubrum* G-9. - FEBS Lett. *79* : 96 - 100, 1977.

29199 - **CUENDET, P.A., ZUBER, H.** : Lichtsammlerpigment-Protein Komplexe. - Ber. deut. bot. Ges. *90* : 493 - 496, 1977.

29200 - **CUENDET, P.A., ZUBER, H., ZUERRER, H., SNOZZI, M.** : On the localization of
a bacteriochlorophyll-associated protein in the chromatophore membrane of
Rhodospirillum rubrum. - In : COOMBS, J. (ed.) : 4th International Congress
on Photosynthesis. P. 79. UKISES, London 1977.

29201 - **CUMMINS, W.R., CHIA-LOOI, A., SVOBODA, J.** : A paradox in the measurements of
net photosynthetic and net production maxima in high arctic *Saxifraga cernua.*
- In : COOMBS, J. (ed.) : 4th International Congress on Photosynthesis. Pp.
79 - 80. UKISES, London 1977.

29202 - **CUNNINGHAM, G.L., SYVERTSEN, J.P.** : The effect of nonstructural carbohydrate
levels on dark CO_2 release in creosotebush. - Photosynthetica *11* : 291 - 295,
1977.

29203 - **CUTLER, J.M., RAINS, D.W.** : Effects of irrigation history on responses of
cotton to subsequent water stress. - Crop Sci. *17* : 329 - 335, 1977. [Growth
analysis.]

29204 - **CUTLER, J.M., RAINS, D.W., LOOMIS, R.S.** : Role of changes in solute concentration in maintaining favorable water balance in field-grown cotton. - Agron. J.
69 : 773 - 779, 1977. [Ps.]

29205 - **CYRONAK, M.J., BRITTON, G., SIMPSON, K.L.** : Rhodoxanthin, the red pigment of
Equisetum arvense sporophytes. - Phytochemistry *16* : 612 - 613, 1977.

29206 - **CZARNOWSKI, M.** : Ekofizjologiczna ocena szacunkowa produkcji fotosyntetycznej
liści drzew. [Ecophysiological estimation of photosynthetic production of
tree leaves.] - In : HUZULÁK, J., MASAROVIČOVÁ, E. (ed.) : Fotosyntéza a Vodný Režim Drevín. Pp. 239 - 250. Modra-Piesky 1977. [In Pol., ab : E, R.]

29207 - **CZARNOWSKI, M.** : Seasonal changes of photosynthetic rate in deciduous tree
growing within the region of industrial emissions. - Bull. Acad. pol. Sci.,
Sér. Sci. biol. *25* : 443 - 450, 1977.

29208 - **CZECZUGA, B.** : Adaptive significance of carotenoids in *Chlorophyta* subjected
to different light conditions. - Bull. Acad. pol., Sér. Sci. biol. *26* : 507 -
- 510, 1977.

29209 - **CZECZUGA, B.** : Carotenoids in leaves and their galls. - Marcellia *40* : 177 -
- 180, 1977.

*29210 - **CZICHI, U., KINDL, H.** : A model of closely assembled consecutive enzymes on
membranes : Formation of hydroxycinnamic acids from L-phenylalanine on thylakoids of *Dunaliella marina.* - Hoppe-Seyler's Z. physiol. Chem. *356* : 475 -
- 485, 1975.

29211 - **DADAY, A., PLATZ, R.A., SMITH, G.D.** : Anaerobic and aerobic hydrogen gas
formation by the blue-green alga *Anabaena cylindrica.* - Appl. environm. Microbiol. *34* : 478 - 483, 1977.

29212 - **DADYKIN, V.P., SAMSONOVA, L.P.** : O vliyanii plenochnykh antitranspirantov na
drevesnye rasteniya.[Effect of film antitranspirants on trees.] - Fiziol.
Rast. *24* : 574 - 581, 1977. [Ps; in R, ab : E.]

29213 - **DAKWA, J.T.** : Estimation of fresh and dry weights of cocoa seedlings from
plant height. - Trop. Agr. *54* : 173 - 178, 1977.

29214 - **DALEY, L.S., BIDWELL, R.G.S.** : Phosphoserine and phosphohydroxypyruvic acid.
Evidence for their role as early intermediates in photosynthesis. - Plant
Physiol. *60* : 109 - 114, 1977.

29215 - **DALEY, L.S., RAY, T.B., VINES, H.M., BLACK, C.C., Jr.** : Characterization of
phosphoenolpyruvate carboxykinase from pineapple leaves *Ananas comosus* (L.)
MERR. - Plant Physiol. *59* : 618 - 622, 1977.

29216 - **DALEY, L.S., VINES, H.M.** : ATPase and fluctuations in Pi pool in CAM plants.
- Plant Physiol. *59* (6, Suppl.) : 116, 1977.

29217 - **DALEY, L.S., VINES, H.M.** : Diurnal fluctuations of inorganic orthophosphate in
pineapple (*Ananas comosus* L., MERR.) leaves and a possible role of ATPase. -
Plant Sci. Lett. *10* : 289 - 298, 1977. [C-cycles.]

29218 - **DALGARN, D., HUBER, D., NEWMAN, D.W.** : Lipid labeling in normal and nitrogen--deficient squash (*Cucurbita maxima*) leaves. - Physiol. Plant. *40* : 153 - 156, 1977. [Chl.]

29219 - **DALTON, C.** : Greening, growth and photosynthesis of cultured spinach cells. - In : COOMBS, J. (ed.) : 4th International Congress on Photosynthesis. P. 80. UKISES, London 1977.

29220 - **DALTON, C.C., STREET, H.E.** : The influence of applied carbohydrates on the growth and greening of cultured spinach (*Spinacia oleracea* L.) cells. - Plant Sci. Lett. *10* : 157 - 164, 1977. [Ps, Chl.]

29221 - **DAMISCH, W.** : Einige Aspekte der Ertragsbildung bei Getreide. - In : Produkce Biomasy a Tvorba Výnosů Polních Plodin. Vol. 1. Pp. 40 - 58. Česká vědecko-technická Společnost zemědělská, Praha 1977. [Growth analysis.]

29222 - **DAMISCH, W.** : Quantitative Beziehungen zwischen Energieaufwand und Stoffspeicherung bei Sommergerste. - In : UNGER, K. (ed.) : Biophysikalische Analyse pflanzlicher Systeme. Pp. 183 - 192. VEB Gustav Fischer Verlag, Jena 1977. [Ps.]

29223 - **DAMISCH, W., WIBERG, A.** : Ergebnisse über den Einfluss der Temperatur auf Stoffzuwachs und Ertragsbildung bei Sommergerste. - Arch. Acker- Pflanzenbau Bodenk. *21* : 485 - 494, 1977. [Ps.]

29224 - **DAMON, P.E.** : El carbono 14 y la unidad de las ciencias. [Carbon-14 and unity of science.] - Acta cien. venez. *28* : 249 - 256, 1977. [Ps; in Span.]

29225 - **DANCSHÁZY, Z., ORMOS, P., KARVALY, B.** : The molecular mechanism of the regulation of bacteriorhodopsin-generated photoelectric membrane potential. - In: COOMBS, J. (ed.) : 4th International Congress on Photosynthesis. P. 81. UKISES, London 1977.

29226 - **DANDONNEAU, Y.** : Variations nycthémérales de la profondeur du maximum de chlorophylle dans le Dôme d'Angola. - Cah. ORSTOM, Sér. Océanographie *15* : 27 - - 37, 1977.

29227 - **DANON, A., BRITH-LINDNER, M., CAPLAN, S.R.** : Biogenesis of the purple membrane of *Halobacterium Halobium*. - Biophys. Struct. Mechanism *3* : 1 - 17, 1977.

29228 - **DANON, A., CAPLAN, S.R.** : Light-dependent CO_2 fixation in anaerobic *Halobacterium halobium*. - In : PACKER, L., PAPAGEORGIOU, G.C., TREBST, A. (ed.) : Bioenergetics of Membranes. Pp. 115 - 118. Elsevier/North-Holland Biomedical Press, Amsterdam - Oxford - New York 1977.

29229 - **DANON, A., CAPLAN, S.R.** : CO_2 fixation by *Halobacterium halobium*. - FEBS Lett. *74* : 255 - 258, 1977.

29230 - **DARBYSHIRE, J.** : Large scale production of photosynthetic prokaryotes. - In : COOMBS, J. (ed.) : 4th International Congress on Photosynthesis. P. 82. UKISES, London 1977.

29231 - **DAS, V.S.R., RAGHAVENDRA, A.S.** : Biochemical and biophysical characterisation of the mesophyll and bundle sheath cells in some C_4 plants. - In : COOMBS, J. (ed.) : 4th International Congress on Photosynthesis. P. 83. UKISES, London 1977.

29232 - **DAS, V.S.R., SANTAKUMARI, M.** : Stomatal characteristics of some dicotyledonous plants in relation to the C_4 and C_3 pathways of photosynthesis. - Plant Cell Physiol. *18* : 935 - 938, 1977.

29233 - **DAS GUPTA, D.K., BASUCHAUDHURI, P.** : Molybdenum nutrition of rice under low and high nitrogen level. - Plant Soil *46* : 681 - 685, 1977.

29234 - **DAS GUPTA, D.K., KALER, K.V.I.S.** : Charge transport in $\underline{B}$-carotene. - J. biol. Phys. *5* : 95 - 107, 1977.

29235 - **DAS GUPTA, D.K., KALER, K.V.I.S.** : Charge transport in semiconducting biopolymers. - In : Amorphous and Liquid Semiconductors. Proceedings of the 7th International Conference on Amorphous and Liquid Semiconductors. Pp. 780 - - 784. Edinburgh 1977. [Car.]

*29236 - DAUNICHT, H.J. : Control of plant growth and development by carbon dioxide
          concentration. - Acta Hort. *39* : 167 - 174, 1974. [Ps.]

 29237 - DAUTKULOV, A.D., ZHEKSEMBIEVA, R.O. : Vliyanie kaliĭnykh udobreniĭ na nekoto-
          rye fiziologicheskie pokazateli u lyutserny. [Effect of potassium fertilizers
          on some physiological characteristics in alfalfa.] - Fiziol. Biokhim. kul't.
          Rast. *9* : 68 - 73, 1977. [In R, ab : E.]

 29238 - DAVENPORT, H.E., DUPONT, M.S. : Some effects of seasonal variations in plant
          growth conditions upon the composition and activity of isolated chloroplasts.
          - Plant Cell Physiol. *1977* (spec. Issue 3 - Photosynthetic Organelles. Struc-
          ture and Function) : 275 - 281, 1977.

 29239 - DAVID, K.A.V., FAY, P. : Effects of long-term treatment with acetylene on
          nitrogen-fixing microorganisms. - Appl. environm. Microbiol. *34* : 640 - 646,
          1977. [Ps.]

 29240 - DAVID, K.A.V., THOMAS, J., RAO, N.S. : Quantitative chlorophyll & carotenoid
          pattern in radiation-induced mutants of jute *Corchorus olitorius*. - Indian J.
          exp. Biol. *15* : 85 - 87, 1977.

 29241 - DAVIES, B.H. : Carotenoids in higher plants. - In : TEVINI, M., LICHTENTHALER,
          H.K. (ed.) : Lipids and Lipid Polymers in Higher Plants. Pp. 199 - 217. Sprin-
          ger-Verlag, Berlin - Heidelberg - New York 1977.

 29242 - DAVIES, P.S. : Carbon budgets and vertical zonation of Atlantic reef corals.
          - In :Proceedings, Third International Coral Reef Symposium. Vol. 1. Pp. 391-
          - 396. Univ. Miami, Miami 1977.

 29243 - DAVIES, W.J., KOZLOWSKI, T.T. : Variations among woody plants in stomatal
          conductance and photosynthesis during and after drought. - Plant Soil *46* :
          435 - 444, 1977.

 29244 - DAVIS, D.J., JANOVITZ, E.B., GROSS, E.L. : Regulation of excitation energy
          distribution in subchloroplast particles. Photosystem II. - Arch. Biochem.
          Biophys. *184* : 197 - 203, 1977.

 29245 - DAVIS, D.J., SAN PIETRO, A. : Chemical modification of spinach ferredoxin :
          Evidence for the involvement of a complex between ferredoxin and ferredoxin:
          :NADP oxidoreductase in NADP photoreduction. - Biochem. biophys. Res. Commun.
          74 : 33 - 40, 1977.

 29246 - DAVIS, D.J., SAN PIETRO, A. : Interactions between spinach ferredoxin and
          other electron carriers. The involvement of a ferredoxin:cytochrome *c* complex
          in the ferredoxin-linked cytochrome *c* reductase activity of ferredoxin:NADP+
          oxidoreductase. - Arch. Biochem. Biophys. *182* : 266 - 272, 1977.

 29247 - DAVIS, S.D., van BAVEL, C.H.M., McCREE, K.J. : Effect of leaf aging upon sto-
          matal resistance in bean plants. - Crop Sci. *17* : 640 - 645, 1977.

 29248 - DAVISON, W. : Sampling and handling procedures for the polarographic measure-
          ment of oxygen in hypolimnetic waters. - Freshwater Biol. *7* : 393 - 401, 1977.

 29249 - DAVTYAN, V.A., AVAKYAN, G.S. : O vliyanii osveshchennosti na nekotorye fizio-
          logicheskie pokazateli kulisnykh nasazhdeniĭ. [Effect of irradiance on some
          physiological characteristics of windbreak stands.] - Biol. Zh. Arm. *30* (7) :
          27 - 33, 1977. [Ps, Chl; in R, ab : Arm.]

 29250 - DAVTYAN, V.A., KAZARYAN, V.V. : O sezonnykh izmeneniyakh soderzhaniya khloro-
          filla i prochnosti ego svyazi s lipoproteidnym kompleksom u vechnozelenykh
          rasteniĭ. [Seasonal changes in the content of chlorophyll and its stability
          in relation to the lipoprotein complex in evergreen plants.] - Tr. bot. Inst.
          Akad. Nauk arm. SSR *20* (Voprosy Individual'nogo Razvitiya Vysshikh Rasteniĭ):
          128 - 135, 1977. [In R.]

 29251 - DAVTYAN, V.A., KHURSHUDYAN, A.P. : O vliyanii kol'tsevaniya na fotosintez i
          rost plodov yabloni. [Effect of ringing on photosynthesis and growth of ap-
          ples.] - Biol. Zh. Arm. *30* (9) : 42 - 48, 1977. [In R, ab : Arm.]

 29252 - DAWES, C.J. : A photosynthetic and biochemical comparison of *Eucheuma* from
          the Gulfs of Mexico and California. - J. Phycol. *13* (Suppl.) : 16, 1977.

29253 - DAWES, C.J., STANLEY, N.F., MOON, R.E. : Physiological and biochemical studies on the ι-carrageenan producing red alga *Eucheuma uncinatum* SETCHELL and
GARDNER from the Gulf of California. - Bot. mar. *20* : 437 - 442, 1977.

29254 - DAY, F.P., Jr., MONK, C.D. : Net primary production and phenology on a Southern Appalachian watershed. - Amer. J. Bot. *64* : 1117 - 1125, 1977.

29255 - DAY, W. : Stomatal resistance in different gases. - J. appl. Ecol. *14* : 643 -
- 647, 1977.

29256 - DEASON, T.R., CZYGAN, F.-C., SOEDER, C.J. : Taxonomic significance of secondary carotenoid formation in *Neospongiococcum (Chlorococcales, Chlorophyta)*. -
New Phycol. *13* : 176 - 180, 1977.

29257 - DEASON, T.R., SCHNEPF, E. : Fine structure of *Nautococcus mammilatus (Chlorococcales, Chlorophyceae)*, a coccoid alga with tomentose cell walls. - J. Phycol. *13* : 218 - 224, 1977. [Chloroplast.]

29258 - DECLEIRE, M., DE CAT, W. : Detection rapide de traces d'herbicides inhibiteurs
de photosynthèse par l'inhibition de la reduction des nitrites à la lumière.
- Z. Pflanzenphysiol. *82* : 310 - 314, 1977.

29259 - De GREEF, J.A., De PROFT, M. : Studies on ethylene evolution in greening
plants. - Plant Physiol. *59* (6, Suppl.) : 39, 1977. [Chl.]

29260 - DE GREEF, J.A., VERBELEN, J.P. : Control of plastid growth in greening plants
by phytochrome mediated interorgan co-operation. - In : **COOMBS, J.** (ed.) :
4th International Congress on Photosynthesis. P. 84. UKISES, London 1977.

29261 - De GREEF, J.A., VERBELEN, J.P. : Plastid development in etiolated bean leaves
under uncoupling conditions of oxidative phosphorylation. - Ann. Bot. *41* :
1371 - 1373, 1977.

29262 - DEGROOTE, D., KENNEDY, R.A. : Photosynthesis in *Elodea canadensis* MICHX. Four-
-carbon acid synthesis. - Plant Physiol. *59* : 1133 - 1135, 1977.

29263 - DE GROOTH, B.G., AMESZ, J. : Electrochromic absorbance changes of photosynthetic pigments in *Rhodopseudomonas sphaeroides*. I. Stimulation by secondary e-
lectron transport at low temperature. - Biochim. biophys. Acta   *462* : 237 -
- 246, 1977.

29264 - DE GROOTH, B.G., AMESZ, J. : Electrochromic absorbance changes of photosynthetic pigments in *Rhodopseudomonas sphaeroides*. II. Analysis of the band
shifts of carotenoid and bacteriochlorophyll. - Biochim. biophys. Acta *462* :
247 - 258, 1977.

29265 - DE GROOTH, B.G., AMESZ, J. : Electrochromism in chloroplasts and in chromatophores of the purple bacterium *Rhodopseudomonas sphaeroides*. - In : **PACKER,
L., PAPAGEORGIOU, G.C., TREBST, A.** (ed.) : Bioenergetics of Membranes. Pp.
227 - 232. Elsevier/North-Holland Biomedical Press, Amsterdam - Oxford - New
York 1977.

29266 - de GROOTH, B.G., ROMIJN, J.C., PULLES, M.P.J. : Oscillating absorbance changes in purple bacteria. - In : **COOMBS, J.** (ed.) : 4th International Congress
on Photosynthesis. P. 85. UKISES, London 1977.

29267 - DEITZER, G.F., HOPKINS, D.W., WAGNER, E. : Effect of light on oscillations of
enzyme activity in *Chenopodium rubrum* L. - Plant Physiol. *59* (6, Suppl.) : 92,
1977. [Ps.]

29268 - DE JONG, D.W. : Chloroplast pigments in yellow virescent and precocious to-
bacco mutants and their relationship to photosynthetic efficiency. - In :
COOMBS, J. (ed.) : 4th International Congress on Photosynthesis. Pp. 85 - 86.
UKISES, London 1977.

29269 - DEKOV, D., RADKOV, P., VITKOV, M. : Prouchvane v"rkhu listnata pov"rkhnost na
nyakoi sortove fasul pri polivni usloviya. [Foliage study on some irrigated
bean varieties.] - Rasteniev"dni Nauki *14* (3) : 27 - 33, 1977. [In Bulg., ab :
E, R.]

29270 - DELANEY, M.E., OWEN, W.J., ROGERS, L.J. : Accumulation of sugars and polysac-
charide accompanying an inhibition of the light reaction in photosynthesis. -
J. exp. Bot. *28* : 1153 - 1162, 1977.

29271 - **de la TORRE, A., CHUECA, A., LOPEZ GORGE, J.** : Isolation and properties of crystalline ferredoxin from *Lactuca sativa*. - In : COOMBS, J. (ed.) : 4th International Congress on Photosynthesis. Pp. 86 - 87. UKISES, London 1977.

29272 - **del CAMPO, F.F., PICOREL, R., RAMIREZ, J.M.** : Some properties of a phototrophic *Rhodospirillum rubrum* strain which lacks *P800*. - In : COOMBS, J. (ed.) : 4th International Congress on Photosynthesis. P. 87. UKISES, London 1977.

29273 - **DELEENS, E.** : Progressive $\delta^{13}C$ shift in the leaves of *Kalanchoe blossfeldiana* during growth and photoperiodic change. - In : COOMBS, J. (ed.) : 4th International Congress on Photosynthesis. P. 88. UKISES, London 1977.

29274 - **DELEENS, E., GARNIER-DARDART, J.** : Carbon isotope composition of biochemical fractions isolated from leaves of *Bryophyllum daigremontianum* BERGER, a plant with Crassulacean Acid Metabolism : some physiological aspects related to CO_2 dark fixation. - Planta *135* : 241 - 248, 1977.

29275 - **DELEPELAIRE, P., BENNOUN, P.** : Energy transfer and site of energy trapping in Photosystem I. - In : COOMBS, J. (ed.) : 4th International Congress on Photosynthesis. P. 89. UKISES, London 1977.

29276 - **D'ELIA, C.F., RYTHER, J.H., LOSORDO, T.M.** : Productivity and nitrogen balance in large scale phytoplankton cultures. - Water Res. *11* : 1031 - 1040, 1977.

29277 - **DELIU, C., FABIAN, A., MOLDOVAN, I.** : Cercetări privind fotosinteza unor plante acvatice în condiţii de poluarea mediului cu detergenţi. [Study of aquatic plant photosynthesis in a detergent-polluted medium.] - Contrib. bot. (Cluj-Napoca) *1977* : 191 - 197, 1977. [In Roum., ab : F.]

29278 - **DELOSME, R.** : Increase of the transmembrane electric field between 6 and 60 μs following flash excitation of Photosystem I. - In : COOMBS, J. (ed.) : 4th International Congress on Photosynthesis. Pp. 89 - 90. UKISES, London 1977.

29279 - **DELRIEU, M.J.** : Oscillatory deactivation of the states S_2 and S_3 of photosynthetic system II in various conditions. - In : COOMBS, J. (ed.) : 4th International Congress on Photosynthesis. P. 91. UKISES, London 1977.

*29280 - **DelROSARIO,D.A., PUTNAM, A.R.** : Enhancement of foliar activity of linuron with carbaryl. - WeedScience *21* : 465 - 468, 1973. [Ps.]

29281 - **DE LUCA, P., ALFANI, A., VIRZO DE SANTO, A.** : CAM, transpiration, and adaptive mechanisms to xeric environments in the succulent *Cucurbitaceae*. - Bot. Gaz. *138* : 474 - 478, 1977.

29282 - **DEL VALLE-TASCON, S., GIMENEZ-GALLEGO, G., RAMIREZ, J.M.** : Photooxidase system of *Rhodospirillum rubrum*. I. Photooxidations catalyzed by chromatophores isolated from a mutant deficient in photooxidase activity. - Biochim. biophys. Acta *459* : 76 - 87, 1977.

29283 - **DEMCHENKO, S.I., AVETISOV, V.A., BUTENKO, R.G.** : Analiz prirody khlorofil'-nykh khimer u *Arabidopsis* v M_1 s pomoshch'yu kul'tury tkani. Soobshchenie I. Vyrashchivanie regenerantov iz defektnoĭ po khlorofillu tkani. [Analysis of nature of chlorophyll chimeres in M_1 of *Arabidopsis* using tissue cultures. Communication I. Growing of regenerants from a chlorophyll-deficient tissue.] - In : NASYROV, Yu.S. (ed.) : Genetika Fotosinteza. Pp. 161 - 167. Donish, Dushanbe 1977. [In R.]

29284 - **DEMETER, S., KE, B.** : Electrochemical and spectro-kinetic evidence for an intermediate electron acceptor in Photosystem I. - Biochim. biophys. Acta *462* : 770 - 774, 1977.

29285 - **DEMIDOV, É.D., BELL, L.N.** : O svetovom poroge fotofosforilirovaniya v izolirovannykh khloroplastakh. [Light threshold of photophosphorylation in isolated chloroplasts.] - Fiziol. Rast. *24* : 424 - 426, 1977. [In R.]

29286 - **DENCH, J., JACKSON, C., MOORE, A.L.** : Glycine decarboxylase activity in intact spinach leaf mitochondria. - In : COOMBS, J. (ed.) : 4th International Congress on Photosynthesis. P. 92. UKISES, London 1977. [Photorespiration.]

29287 - **DENMAN, K., OKUBO, A., PLATT, T.** : The chlorophyll fluctuation spectrum in the sea. - Limnol. Oceanogr. *22* : 1033 - 1038, 1977.

29288 - **DENMAN, K.L.** : Short term variability in vertical chlorophyll structure. - Limnol. Oceanogr. *22* : 434 - 441, 1977.

*29289 - **DENMEAD, O.T.** : Temperate cereals. - In : MONTEITH, J.L. (ed.) : Vegetation and the Atmosphere. Vol. 2. Case Studies. Pp. 1 - 31. Academic Press, London - New York - San Francisco 1976. [Ps.]

29290 - **DENNIS, J.G.** : Distribution patterns of belowground standing crop in arctic tundra at Barrow, Alaska. - Arct. alp. Res. *9* : 113 - 127, 1977.

29291 - **DESAI, T.S., TATAKE, V.G., SANE, P.V.** : Characterization of the low temperature thermoluminescence band Z_V in leaf. An explanation for its variable nature. - Biochim. biophys. Acta *462* : 775 - 780, 1977. [Chl.]

29292 - **DESCHAMPS, P.Y., LECOMTE, P., VIOLLIER, M.** : Remote sensing of ocean color and detection of chlorophyll content. - In : Proceedings of the Eleventh International Symposium on Remote Sensing of Environment. Pp. 1021 - 1033. Environ. Res. Inst. Michigan, Ann Arbor 1977.

29293 - **DESORTOVÁ, B., FOTT, J., STRAŠKRABA, M.** : Metodika stanovenia a hodnotenia koncentrácie chlorofylu v povrchových vodách. [Methods of determination and evaluation of chlorophyll concentration in surface waters.] - In : Informačný Bulletin pre Chemickú, Biologickú, Mikrobiologickú a Technologickú Cinnost' na Úseku Podnikov Povodí, Vodovodov a Kanalizácií. Pp.1 - 36. MLVH SSR, Bratislava 1977. [In Slovak.]

29294 - **DEVANATHAN, M.A.V.** : Photosynthetic productivity in natural environments. - In : COOMBS, J. (ed.) : 4th International Congress on Photosynthesis. P. 93. UKISES, London 1977.

*29295 - **DEVIDÉ, Z.** : Biophysical investigations of white-spotted leaves. - Period. Biol. *76* (1) : 41, 1974. [Ps.]

*29296 - **De VILLIERS, O.T., ASHTON, F.M.** : Effects of IAA, GA and ethrel on biochemical processes in isolated mesophyll cells. - Agroplantae *8* : 87 - 90, 1976. [Ps.]

29297 - **deVILLIERS, O.T., ASHTON, F.M.** : Metabolic activity of isolated leaf cells of *Phaseolus vulgaris* in relation to leaf development. - Plant Physiol. *59* : 1072 - 1075, 1977. [Ps.]

29298 - **De VILLIERS, O.T., ASHTON, F.M., GLENN, R.K.** : The effect of ethanol and acetone on metabolic processes in isolated plant cells. - Agroplantae *9* : 67 - - 69, 1977.

29299 - **DeYOE, D.R., BROWN, G.N.** : Adaptive reorganization in the lipid contingent of eastern white pine chloroplast lamellae during the onset of winter. - Plant Physiol. *59* (6, Suppl.) : 5, 1977.

29300 - **DeYOE, D.R., BROWN, G.N.** : Compositional analysis of the photosynthetic membrane system of western hemlock chloroplasts. - Can. J. Bot. *55* : 2399 - - 2407, 1977.

29301 - **DICKSON, R.E.** : EDTA-promoted exudation of ^{14}C-labeled compounds from detached cottonwood and bean leaves as related to translocation. - Can. J. Forest Res. *7* : 277 - 284, 1977.

29302 - **DICKSON, R.E.** : Translocation of ^{14}C-photosynthate from cottonwood leaves : a time course study. - Plant Physiol. *59* (6, Suppl.) : 125, 1977.

29303 - **DIGBY, P.S.B.** : Photosynthesis and respiration in the coralline algae, *Clathromorphum circumscriptum* and *Corallina officinalis* and the metabolic basis of calcification. - J. mar. biol. Ass. UK *57* : 1111 - 1124, 1977.

29304 - **DILLEY, R.A., GIAQUINTA, R.T., PROCHASKA, L.J., ORT, D.R.** : Control of proton translocation in the chloroplast water oxidation system. - In : JUNGREIS, A.M., HODGES, T.K., KLEINZELLER, A., SCHULTZ, S.G. (ed.) : Water Relations in Membrane Transport in Plants and Animals. Pp. 55 - 67. Academic Press, New York - San Francisco - London 1977.

29305 - **DILLEY, R.A., PROCHASKA, L.J.** : Photosystem-specific intramembrane domains of H^+ deposition related to membranes conformational changes. - In : COOMBS, J.

(ed.) : 4th International Congress on Photosynthesis. P. 94. UKISES, London 1977.

*29306 - DIMITROV, Kh., DIMITROVA, M. : Vliyanie na mineralnoto khranene i vodosnab-dyavaneto v"rkhu fotosintezata i rastezha na topolite. [Effect of mineral nutrition and water supply on poplar tree photosynthesis and growth.] - Fiziol. Rast. (Sofia) 2 (3) : 43 - 51, 1976. [In Bulg., ab : E, R.]

29307 - DINER, B. : Double hitting in Photosystem II viewed from the donor side. - In : COOMBS, J. (ed.) : 4th International Congress on Photosynthesis. P. 90. UKISES, London 1977.

29308 - DINER, B.A. : Dependence of the deactivation reactions of photosystem II on the redox state of plastoquinone pool A varied under anaerobic conditions. Equilibria on the acceptor side of photosystem II. - Biochim. biophys. Acta 460 : 247 - 258, 1977.

29309 - DINER, B.A., JOLIOT, P. : Oxygen evolution and manganese. - In : TREBST, A., AVRON, M. (ed.) : Photosynthesis I. (Encycl. Plant Physiol. N.S. Vol. 5.) Pp. 187 - 205. Springer-Verlag, Berlin - Heidelberg - New York 1977.

29310 - DISMUKES, C., McGUIRE, A., FRIESNER, R., SAUER, K. : The mechanism of initial charge separation in photosynthetic plants and algae as revealed by electron spin polarization in $P700$ and EPR studies on oriented chloroplasts. - In : COOMBS, J. (ed.) : 4th International Congress on Photosynthesis. P. 95. UKISES, London 1977.

29311 - DI TORO, D.M., THOMANN, R.V., O'CONNOR, D.J., MANCINI, J.L. : Estuarine phytoplankton biomass models - verification analyses and preliminary applications. - In : GOLDBERG, E.D., McCAVE, I.N., O'BRIEN, J.J., STEELE, J.H. (ed.) : The Sea : Ideas and Observations on Progress in the Study of the Seas. Vol. 6. Pp. 969 - 1020. J. Wiley & Sons, New York 1977.

29312 - DITTRICH, P., RASCHKE, K. : Malate metabolism in isolated epidermis of Commelina communis L. in relation to stomatal functioning. - Planta 134 : 77 - 81, 1977.

29313 - DITTRICH, P., RASCHKE, K. : Uptake and metabolism of carbohydrates by epidermal tissue. - Planta 134 : 83 - 90, 1977. [Ps.]

29314 - DMITROVSKIĬ, L.G., GOL'DFEL'D, M.G. : Fotofosforilirovanie v khloroplastakh, sopryazhennoe s reaktsiyami perenosa èlektrona na razlichnykh uchastkakh èlektrontransportnoĭ tsepi. [Photophosphorylation in chloroplasts connected with electron transfer reactions in different parts of the electron-transport chain.] - Dokl. Akad. Nauk SSSR 235 : 224 - 227, 1977. [In R.]

29315 - DOBBERSTEIN, B., BLOBEL, G., CHUA, N.-H. : In vitro synthesis and processing of a putative precursor for the small subunit of ribulose-1,5-bisphosphate carboxylase of Chlamydomonas reinhardtii. - Proc. nat. Acad. Sci. USA 74 : 1082 - 1085, 1977.

29316 - DOCKERTY, A., LORD, J.M., MERRETT, M.J. : Development of ribulose-1,5-diphosphate carboxylase in castor bean cotyledons. - Plant Physiol. 59 : 1125 - 1127, 1977.

29317 - DOCKERTY, A., LORD, J.M., MERRETT, M.J. : Ribulose biphosphate carboxylase in castor bean cotyledons. - Plant Physiol. 59 (6, Suppl.) : 44, 1977.

29318 - DOCKERTY, A., MERRETT, M.J. : The site of synthesis of chloroplast lipids in Euglena. - In : COOMBS, J. (ed.) : 4th International Congress on Photosynthesis. Pp. 95 - 96. UKISES, London 1977.

29319 - DOLEY, D. : Parthenium weed (Parthenium hysterophorus L.) : Gas exchange characteristics as a basis for prediction of its geographical distribution. - Aust. J. agr. Res. 28 : 449 - 460, 1977.

29320 - DOLPH, G.E. : The effect of different calculational techniques on the estimation of leaf area and the construction of leaf size distributions. - Bull. Torrey bot. Club 104 : 264 - 269, 1977.

29321 - DOLZHIKOVA, N.M., SLOVTSOV, R.I., GRUZDEV, G.S. : Vliyanie gerbitsidov eptama, TKhA, geksilura i ikh smeseĭ na rost i urozhaĭ sakharnoĭ svekly. [Effect of herbicides Eptam, TCA, hexilur and their combinations on the growth and yield of sugar beets.] - Izv. timiryazev. sel'.-khoz. Akad. *1977* (4) : 165 - 172, 1977. [Ps, Chl; in R, ab : E.]

29322 - DOMAN, N.G., KOSACOVSKAJA, I.K., KOMARNIZKI, I.K., CHERNYAD'EV, I.I., SITNIK, K.M. : Ribulosodiphosphate carboxylase in *Nicotiana* genus : Subunit structure and enzymatic activity. - In : COOMBS, J. (ed.) : 4th International Congress on Photosynthesis. P. 97. UKISES, London 1977.

*29323 - DOMAN'SKA, G., LENGOVYAK, Z., LESKA, L. : Vliyanie gerbitsidov na morfologicheskie svoĭstva i soderzhanie khlorofilla v bobovykh rasteniyakh. [Effect of herbicides on morphological properties and chlorophyll content in legumes.] - Tr. vsesoyuz. nauch.-issled. Inst. Zashchity Rast. *43* : 82 - 88, 1975. [In R, ab : E.]

29324 - DONCHEV, Kh. : Natrupvane novosintezirani karotinoidi pri s"zryavane i s"khranenie na nyakoi sortove yab"lki. [Accumulation of newly synthesized carotenoids during ripening and storage of some apple cultivars.] - Fiziol. Rast. (Sofia) *3* (1) : 24 - 33, 1977. [TLC; in Bulg., ab : E, R.]

29325 - DONOV, V., MAKEDONSKA, Ts., ĬOROVA, K. : Natrupvane na khranitelni elementi i obrazuvane na nyakoi asimilati v iglolistata na belborovi fidanki pri torene. [Nutrient element accumulation and the production of certain assimilates in the needles (acerose leaves) of white pine seedlings following mineral fertilizer application.] - Fiziol. Rast. (Sofia) *3* (1) : 69 - 77, 1977. [In Bulg., ab : E.]

29326 - DOOHAN, M.E., NEWCOMB, E.H. : Ultrastructural characteristics of representative and anomalous C_4 plants. - Plant Physiol. *59* (6, Suppl.) : 19, 1977. [Ps.]

29327 - DOR, I., SOURNIA, A., POR, F.D. : Preliminary data on the productivity of the mangrove environments of Sinai. - Rapp. Comm. int. Mer médit. *24* (4) : 193 - - 194, 1977.

29328 - DORAVARI, S., CANVIN, D.T. : Effect of butyl 2-OH-3 butyonate on photosynthesis and photorespiration. - Plant Physiol. *59* (6, Suppl.) : 42, 1977.

29329 - DORSMAN, A., ELGERSMA, O., MEIJER, G., STEUTEN, L. : Chlorophyll metabolism in etiolated gherkin seedlings. I. Photoinhibition of chlorophyll accumulation. - Photochem. Photobiol. *26* : 533 - 539, 1977.

29330 - DOTSON, L., HAGAR, W.G. : Cellular location of chlorophyll-protein 668 in *Atriplex hortensis*. - Plant Physiol. *59* (6, Suppl.) : 10, 1977.

29331 - DOWNTON, W.J.S. : Photosynthesis in salt-stressed grapevines. - Aust. J. Plant Physiol. *4* : 183 - 192, 1977.

29332 - DRAKE, B.G. : Community of photosynthetic efficiency in a salt marsh on the Chesapeake Bay. - In : COOMBS, J. (ed.) : 4th International Congress on Photosynthesis. P. 96. UKISES, London 1977.

29333 - DRAKE, B.G. : Incident photosynthetically active radiation and net CO_2 exchange in a salt marsh on the Chesapeake Bay. - Plant Physiol. *59* (6, Suppl.) : 99, 1977.

29334 - DRENNAN, D.S.H., JENNINGS, E.A. : Weed competition in irrigated cotton (*Gossypium barbadense* L.) and groundnut (*Arachis hypogaea* L.) in the Sudan Gezira. - Weed Res. *17* : 3 - 9, 1977. [Dry matter accumulation.]

29335 - DREW, A.P., FERRELL, W.K. : Morphological acclimation to light intensity in Douglas-fir seedlings. - Can. J. Bot. *55* : 2033 - 2042, 1977. [Growth analysis.]

29336 - DREW, E.A. : Seasonal variation in the photosynthetic potential of British *Laminaria* species. - J. Phycol. *13* (Suppl.) : 18, 1977.

29337 - DREW, E.A. : The physiology of photosynthesis and respiration in some Antarctic marine algae. - Brit. antarct. Surv. Bull. *46* : 59 - 76, 1977.

29338 - DREW, M.C., SISWORO, E.J. : Early effects of flooding on nitrogen deficiency and leaf chlorosis in barley. - New Phytol. *79* : 567 - 571, 1977.

29339 - DREWS, G. : The light-harvesting complex of *Rhodopseudomonas capsulata*. - In: OLSON, J.M., HIND, G. (ed.) : Chlorophyll-Proteins, Reaction Centers, and Photosynthetic Membranes. P. 366. Brookhaven nat. Lab., Upton 1977.

29340 - DRING, M.J. : Significance of enhancement for calculations of photosynthesis of red algae from action spectra. - J. Phycol. *13* (Suppl.) : 18, 1977.

29341 - DUBERTRET, G. : Control of the formation of photosynthetic units by light during greening of etiolated *Euglena gracilis* Z cells. - In : COOMBS, J. (ed.): 4th International Congress on Photosynthesis. P. 98. UKISES, London 1977.

29342 - DUBYNA, D.V., MOLYAKA, O.N., SOLOMAKHA, V.A. : Vmist deyakykh biologichno aktyvnykh rechovyn ta mikroelementiv v likars'kykh roslynakh milkovod' Seredn'ogo Dnipra. [Content of some biologically active substances and microelements in medical plants in shallow waters of the middle Dniepr.] - Introduk. Aklimat. Roslyn Ukr. *10* : 82 - 85, 118, 1977. [Chl, Car; in Ukr., ab : R.]

29343 - DUCLOUX, F., CARLIER, G. : Echanges de 3-0-methyl-D-glucose par les cellules isolees de feuilles de *Pelargonium zonale* L. AITON. - In : Exchanges Ioniques Transmembranaires chez les Végétaux. Colloque du C.N.R.S. *258*. Pp. 407 - 413. CNRS, Paris 1977. [Ps.]

29344 - DUDA, M. : Sezónna dynamika intenzity fotosyntézy pri dubových a bukových semenáčikoch. [Seasonal dynamic of photosynthetic rate in the oak and beach seedlings.] - In : HUZULÁK, J., MASAROVIČOVÁ, E. (ed.) : Fotosyntéza a Vodný Režim Drevín. Pp. 193 - 201. Modra-Piesky 1977. [In Slovak, ab : E, R.]

29345 - DUEDALL, I.W., O'CONNORS, H.B., PARKER, J.H., WILSON, R.E., ROBBINS, A.S. : The abundances, distribution and flux of nutrients and chlorophyll *a* in the New York Bight apex. - Estuar. coast. mar. Sci. *5* : 81 - 105, 1977.

*29346 - DUFOUR, P., SLEPOUKHA, M. : L'oxygène dissous en lagune Ebrie : Influences de l'hydroclimat et des pollutions. - Cent. Rech. Oceanograf. Abidjan Doc. Sci. *6* (2) : 75 - 118, 1975. [Ps.]

29347 - DUGGAN, J.X., GASSMAN, M.L. : Further studies on the oxidation of 5-aminolevulinic acid *in vitro* by extracts of etiolated barley shoots. - Plant Physiol. *59* (6, Suppl.) : 103, 1977.

29348 - DUJARDIN, E. : Transitory pigment-protein complexes similar to photosynthetic active centres during protochlorophyll(ide) photoreduction. - In : COOMBS, J. (ed.) : 4th International Congress on Photosynthesis. P. 99. UKISES, London 1977.

29349 - DUJARDIN, E., SIRONVAL, C. : Transitory pigment-protein complexes similar to photosynthesis active centres during protochlorophyll(ide) photoreduction. - Plant Sci. Lett. *10* : 347 - 355, 1977.

*29350 - DULIN, A.F. : Osobennosti énergeticheskogo obmena srezannykh list'ev yachmenya, obogashchennykh fitogormonami. [Peculiarities of energetic exchange in barley cut off leaves enriched by phytohormones.] - Nauch. Dokl. vyssh. Shkoly, biol. Nauki *19* (10) : 94 - 98, 1976. [Ps; in R.]

29351 - DUMBROFF, E.B., BROWN, D.C.W., THOMPSON, J.E. : Effect of senescence on levels of free abscisic acid and water potentials in cotyledons cf bean. - Bot. Gaz. *138* : 261 - 265, 1977. [Photosynthates.]

29352 - DUNIEC, J.T., THORNE, S.W. : The relation of light-induced slow absorbancy and scattering changes about 520 nm and structure of chloroplast thylakoids - A theoretical investigation. - J. Bioenerg. Biomembranes *9* : 223 - 235, 1977.

29353 - DURAINI, O., HAMLIN, L., TILLBERG, J.-E. : Effects of different buffers, DCMU and NaN_3 on menadione mediated photophosphorylation in isolated chloroplasts. - In : COOMBS, J. (ed.) : 4th International Congress on Photosynthesis. Pp. 99 - 100. UKISES, London 1977.

*29354 - DURBIN, E.G. : Studies on the autecology of the marine diatom *Thalassiosira nordenskiöldii* CLEVE. I. The influence of daylength, light intensity, and temperature on growth. - J. Phycol. *10* : 220 - 225, 1974. [Chl.]

29355 - DURBIN, E.G. : Studies on the autecology of the marine diatom *Thalassiosira nordenskioeldii*. II. The influence of cell size on growth rate, and carbon, nitrogen, chlorophyll *a* and silica content. - J. Phycol. *13* : 150 - 155, 1977.

29356 - DURBIN, R.D., UCHYTIL, T.F. : Cytoplasmic inheritance of chloroplast coupling factor 1 subunits. - Biochem. Genet. *15* : 1143 - 1146, 1977.

29357 - DUSHA, I., DĚNES, G. : A simplified assay of enzymes catalyzing ATP-pyrophosphate exchange reactions. - Anal. Biochem. *81* : 247 - 250, 1977.

29358 - DUTTON, P.L., PRINCE, R.C., TIEDE, D.M., PETTY, K.M., KAUFMANN, K.J., NETZEL, T.L., RENTZEPIS, P.M. : Electron transfer in the photosynthetic reaction center. - In : OLSON, J.M., HIND, G. (ed.) : Chlorophyll-Proteins, Reaction Centers, and Photosynthetic Membranes. Pp. 213 - 237. Brookhaven nat. Lab.,.Upton 1977.

28359 - DUTTON, P.L., PRINCE, R.C., van den BERG, W.H., TAKAMIYA, K. : Membrane electron transfer components in light energy conversion and in proton translocation; dynamic equilibria in coupled reactions of photosynthesis. - In : COOMBS, J. (ed.) : 4th International Congress on Photosynthesis. Pp. 100 - - 101. UKISES, London 1977.

*29360 - DUYSEN, M.E., FREEMAN, T.P. : The photosystem and ultrastructural modifications in wheat chloroplasts that developed under slight water stress. - Proc. North Dakota Acad. Sci. *30* : 16, 1976.

29361 - DUYSEN, M.E., FREEMAN, T.P. : Chloramphenicol impairment of Chl accumulation in CPI of wheat leaf sections. - Plant Physiol. *59* (6, Suppl.) : 8, 1977.

*29362 - DYKYJOVÁ, D., KVĚT, J. : Primary productivity of freshwater wetlands. - In : SMART, M. (ed.) : Proceedings of the International Conference on Conservation of Wetlands and Waterfowl. Pp. 173 - 179. Heiligenhafen 1974.

29363 - DYMOCK, I.J., HILL, B., BOWN, A.W. : An investigation into the influence of IAA and malate on *in vivo* and *in vitro* rates of dark carbon dioxide fixation in coleoptile tissue. - Can. J. Bot. *55* : 1641 - 1645, 1977.

*29364 - DYNESIUS, R.A., WALNE, P.L. : Ultrastructural and cytochemical comparisons of the euglenoid flagellates, *Phacus pleuronectes* and *Hyalophacus ocellatus*. - .J. Phycol. *10* (Suppl.) : 16, 1974. [Chloroplast.]

29365 - DYRSSEN, D. : The chemistry of plankton production and decomposition in seawater. - In : ANDERSEN, N.R., ZAHURANEC, B.J. (ed.) : Oceanic Sound Scattering Prediction. Pp. 65 - 84. Plenum Press, New York 1977.

*29366 - DZAGNIDZE, D.K., TAVADZE, P.G. : Produktivnost' fotosinteza i moshchnost' razvitiya kornevoĭ sistemy loz v zavisimosti ot biologicheskoĭ osobennosti privivaemykh komponentov. [Productivity of photosynthesis and development of root system of grapevine as dependent on biological peculiarities of grafts.] - Tr. nauch.-issled. Inst. Sadovod., Vinograd. Vinodel. (Tbilisi) *21* : 34 - - 39, 1972. [In R.]

*29367 - DZAGNIDZE, D.K., TAVADZE, P.G. : Vodnyĭ rezhim i vozdushnoe pitanie (fotosintez) vinogradnykh loz v zavisimosti ot privivaemykh komponentov i ėkologicheskikh usloviĭ   ikh proizrastaniya. [Water relations and air nutrition (photosynthesis) of grapevine in dependence on grafting components and ecological conditions of their growing.] - In : Voprosy Vinogradarstva i Vinodeliya. Pp. 189 - 191. Simferopol' 1971. [In R.]

29368 - DZEVYATAŬ, A.S., ANUCHKIN, P.D., GRAKOVICH, D.V., RAZVYAKOŬ, V.A. : Asablivastsi listavoga polagu yabloni ŭ intěnsiŭnykh sadakh u suvyazi z ikh svetlavym rězhymam. [Peculiarites of foliage in apple-trees in intensive orchards and its light regime.] - Vestsi Akad. Navuk belarus. SSR, Ser. biyal. Navuk *1977* (5) : 30 - 34, 138, 1977. [Chl; in Belorus., ab : R.]

29369 - DZHAGAROV, B.M., SAGUN, E.I., BONDAREV, S.L., GURINOVICH, G.P. : Vliyanie molekulyarnoĭ struktury na protsessy bezyzluchatel'noĭ dezaktivatsii nizshikh vozbuzhdennykh sostoyaniĭ porfirinov. [Effect of molecular structure on radiationless deactivation of lower excited states of porphyrins.] - Biofizika *22* : 565 - 570, 1977. [Chl; in R, ab : E.]

29370 - **EAKS, I.L.** : Physiology of degreening - summary and discussion of related topics. - Proc. int. Soc. Citric. *1* : 223 - 226, 1977. [Chl, Car.]

29371 - **EBRINGER, L.** : Súčasné názory na prokaryotický pôvod niektorých organel. [Present views on the prokaryotic origin of some organelles.] - Biol. Listy *42* : 275 - 295, 1977. [Chloroplast; in Slovak, ab : E.]

29372 - **ECKARDT, F.E.** : Physiological behaviour in relation to the environment. A comparison between a crop and various types of natural vegetation. - In : LANDSBERG, J.J., CUTTING, C.V. (ed.) : Environmental Effects on Crop Physiology. Pp. 157 - 171. Academic Press, London - New York - San Francisco 1977. [Ps.]

29373 - **ECKARDT, F.E., BERGER, A., MÉTHY, M., HEIM, G., SAUVEZON, R.** : Interception de l'énergie rayonnante, échanges de CO_2, régime hydrique et production chez différents types de végétation sous climat méditerranéen. - In : MOYSE, A. (ed.) : Les Processus de la Production Végétale Primaire. Pp. 1 - 75. Gauthier-Villars, Paris 1977.

29374 - **EDELMAN, M., SAGHER, D., REISFELD, A.** : *In vitro* translation of large subunit ribulosediphosphate carboxylase from *Euglena*. - In : BOGORAD, L., WEIL, J.H. (ed.) : Acides Nucléiques et Synthèse des Protéines chez les Végétaux. Coll. Int. C.N.R.S. No. 261. Pp. 305 - 311. Édit. CNRS, Paris 1977.

29375 - **EDMONDSON, W.T.** : Trophic equilibrium of Lake Washington. - Corvallis environm. Res. Lab. tech. Rep. *EPA-600/3-77-087* : I - VII, 1 - 36, 1977. [Chl.]

29376 - **EDWARDS, G., HUBER, S.** : Usefulness of isolated cells and protoplasts for photosynthetic studies. - In : COOMBS, J. (ed.) : 4th International Congress on Photosynthesis. Pp. 101 - 102. UKISES, London 1977.

29377 - **EDYE, L.A., WILLIAMS, W.T., BURT, R.L., GROF, B., STILLMAN, S.L., WINTER, W. H.** : The assessment of seasonal yield using some *Stylosanthes guyanensis* accessions in humid tropical and sub-tropical environments. - Aust. J. exp. Agr. anim. Husb. *17* : 425 - 434, 1977. [Dry matter production.]

29378 - **EFIMTSEV, E.I., BOĬCHENKO, V.A., GUSEV, M.V., NIKITINA, K.A., LITVIN, F.F.** : Issledovanie izmeneniĭ fotosinteticheskikh kharakteristik sine-zelenykh vodorosleĭ v zavisimosti ot dlitel'nosti temnovoĭ inkubatsii. [Changes in photosynthetic characteristics of blue-green algae depending on the duration of their incubation in the darkness.] - Fiziol. Rast. *24* : 23 - 29, 1977. [In R, ab : E.]

29379 - **EGAMI, F.** : Anaerobic respiration and photoautotrophy in the evolution of prokaryotes. - Orig. Life *8* : 169 - 171, 1977.

29380 - **EGARA, K., JONES, R.J.** : Effect of shading on the seedling growth of the leguminous shrub *Leucaena leucocephala*. - Aust. J. exp. Agr. anim. Husb. *17* : 976 - 981, 1977. [Growth analysis.]

*29381 - **EGLI, D.B., GOSSETT, D.R., LEGGETT, J.E.** : Effect of leaf and pod removal on distribution of ^{14}C labeled assimilate in soybeans. - Crop Sci. *16* : 791 - - 794, 1976.

29382 - **EGNÉUS, H., BLANCK, H.** : The effect of a quaternary amine (Aliquat 336) on growth and photosynthesis of the green alga *Chlorella emersonii*, and the effect on photosynthesis in isolated spinach chloroplasts. - Physiol. Plant. *41* : 73 - 78, 1977.

29383 - **EHLERINGER, J., BJÖRKMAN, O.** : Quantum yields for CO_2 uptake in C_3 and C_4 plants. Dependence on temperature, CO_2, and O_2 concentration. - Plant Physiol. *59* : 86 - 90, 1977.

29384 - **EICHENBERGER, W., SCHAFFNER, J.-C., BOSCHETTI, A.** : Characterization of proteins and lipids of photosystem I and II particles from *Chlamydomonas reinhardi*. - FEBS Lett. *84* : 144 - 148, 1977.

29385 - **EICHHORN, M., AUGSTEN, H.** : Der Einfluβ löβlicher Kohlenhydrate auf die Aktivität der Glukose-6-phosphat-Dehydrogenase verschiedenaltriger *Wolffia*-Populationen unter Berücksichtigung von energy charge, O_2-Austausch und Pyruvat-Gehalt. - Z. Pflanzenphysiol. *84* : 37 - 48, 1977. [Ps.]

29386 - EICHHORN, M., AUGSTEN, H. : Die Wirkung von Blau- und Rotlicht auf die Akti- vität der Glucose-6-phosphat-Dehydrogenase und das Adenylatsystem bei *Wolffia arrhiza* unter steady state Bedingungen. - Z. Pflanzenphysiol. *85* : 147 - 152, 1977. [Chl.]

29387 - EISENBACH, M., CAPLAN, S.R. : Light-induced ion transport in *Halobacterium halobium*. - Trends biochem. Sci. *2* : 245 - 247, 1977.

29388 - EISENBACH, M., COOPER, S., GARTY, H., JOHNSTONE, R.M., ROTTENBERG, H., CAPLAN, S.R. : Light-driven sodium transport in sub-bacterial particles of *Halobacte- rium halobium*. - Biochim. biophys. Acta *465* : 599 - 613, 1977.

29389 - EISENBACH, M., GARTY, H., KLEMPERER, G., WEISSMANN, C., TANNY, G., CAPLAN, S. R. : Light-induced pH changes in purple-membrane fragments of *Halobacterium halobium*. - In : PACKER, L., PAPAGEORGIOU, G.C., TREBST, A. (ed.) : Bioener- getics of Membranes. Pp. 119 - 128. Elsevier/North-Holland Biomedical Press, Amsterdam - Oxford - New York 1977.

29390 - ELFIMOV, E.I., VOZNYAK, V.M., KANTSEVA, T.A., EVSTIGNEEV, V.B. : Spektral'no- -kineticheskie kharakteristiki kation-radikala feofitina *a*. [Spectral kinetic characteristics of pheophytin *a* cation radical.] - Biofizika *22* : 1004 - 1009, 1977. [In R, ab : E.]

29391 - EL-GELANI, M., EL-GAMMUDI, A. : The degradation effect of high humidity on the chlorophylls in *Phaseolus vulgaris* plants. - Plant Physiol. *59* (Suppl.): 54, 1977.

29392 - ELGERSMA, O., MEIJER, G., STEUTEN, L. : Photoinhibition of chlorophyll accu- mulation in continuous light. - In : COOMBS, J. (ed.) : 4th International Congress on Photosynthesis. Pp. 102 - 103. UKISES, London 1977.

29393 - ELIAS, B.A., GIVAN, C.V. : *Alpha*-ketoglutarate supply for amino acid synthe- sis in higher plant chloroplasts. Intrachloroplastic localization of NADP- -specific isocitrate dehydrogenase. - Plant Physiol. *59* : 738 - 740, 1977.

29394 - ELIÁŠ, P. : Aktivita prieduchov duba zimného v prirodzenom prostredí. [Stomata activity of *Quercus petraea* in natural environment.] - In : HUZULÁK, J., MA- ŠAROVIČOVÁ, E. (ed.) : Fotosyntéza a Vodný Režim Drevín. Pp. 84 - 91. Modra- -Piesky 1977. [Stomatal resistance; in Slovak, ab : E, R.]

29395 - ELIÁŠ, P. : Vodivost prieduchov javora pol'ného v lesných podmienkach. [Sto- matal conductance of *Acer campestre* L. under forest conditions.] - Acta Mu- sei Silesiae, Ser. dendrol. *26* : 9 - 37, 1977. [In Slovak, ab : E.]

*29396 - ELIZAROVA, V.A., SIGAREVA, L.E. : Soderzhanie pigmentov fitoplanktona v mel- kovodnoĭ zone Rybinskogo vodokhranilishcha. [Pigment content of phytoplankton in the shallow water zone of the Rybinsk reservoir.] - Tr. Inst. Biol. vnutr. Vod Akad. Nauk SSSR *33* : 133 - 147, 1976. [In R.]

29397 - ELLEFSON, W., KROGMANN, D.W. : Isozymes of spinach ferredoxin-NADP oxidore- ductase. - Plant Physiol. *59* (6, Suppl.) : 22, 1977.

29398 - ELLER, B.M. : Beeinflussung der Energiebilanz von Blättern durch Straßenstaub. - Angew. Bot. *51* : 9 - 15, 1977.

29399 - ELLER, B.M. : Leaf pubescence : the significance of lower surface hairs for the spectral properties of the upper surface. - J. exp. Bot. *28* : 1054 - 1059, 1977.

29400 - ELLER, B.M., KOCH, W. : Globalstrahlung innerhalb und ausserhalb von Gass- wechselkammern. - Photosynthetica *11* : 268 - 275, 1977.

29401 - ELLER, B.M., WILLI, P. : Die Bedeutung der Wachsausblühungen auf Blättern von *Kalanchoë pumila* BAKER für die Absorbtion der Globalstrahlung. - Flora *166* : 461 - 474, 1977.

29402 - ELLER, B.M., WILLI, P. : The significance of leaf pubescence for the absorp- tion of global radiation by *Tussilago farfara* L. - Oecologia *29* : 179 - 187, 1977.

29403 - ELLIS, R.J. : Heterotrophic nutrition and its effects on chlorophyll synthe- sis in *Golenkinia (Chlorophyceae)*. - J. Phycol. *13* : 304 - 306, 1977.

29404 - **ELLIS, R.J.** : Protein synthesis by isolated chloroplasts. - Biochim. biophys. Acta *463* : 185 - 215, 1977.

29405 - **ELLIS, R.J.** : The synthesis of chloroplast proteins. - In : **BOGORAD, L.**, **WEIL, J.H.** (ed.) : Nucleic Acids and Protein Synthesis in Plants. Pp. 195 - - 212. Plenum Press, New York 1977.

29406 - **ELLIS, R.J.** : The synthesis of chloroplast proteins by subcellular systems. - In : **COOMBS, J.** (ed.) : 4th International Congress on Photosynthesis. P. 103. UKISES, London 1977.

29407 - **ELLIS, R.P.** : Distribution of the Kranz syndrome in the Southern African *Eragrostoideae* and *Panicoideae* according to bundle sheath anatomy and cytology. - Agroplantae *9* (3) : 73 - 109, 1977.

29408 - **ELSAHOOKIE, M.M.** : A new formula to estimate leaf area in corn (*Zea mays* L.). - Z. Acker- Pflanzenbau *145* : 79 - 83, 1977.

29409 - **EL-SHARKAWI, H.M., MICHEL, B.E.** : Effects of soil water matric potential and air humidity on CO_2 and water vapor exchange in two grasses. - Photosynthetica *11* : 176 - 182, 1977.

29410 - **EL-SHARKAWI, H.M., SALAMA, F.M.** : Effects of drought and salinity on some growth-contributing parameters in wheat and barley. - Plant Soil *46* : 423 - 433, 1977. [Chl.]

29411 - **EL-SHARKAWY, M.A., SOROUR, F.A., SHAALAN, M.I., SGAIER, K.** : The response of growth and yield of the semi-dwarf wheat cultivar "Sidi Misri 1" to water regime and cycocel. - Libyan J. Agr. *6* (1) : 35 - 45, 1977.

29412 - **ELSTNER, E.F.** : Mechanisms of formation and possible functions of oxygen radicals in chloroplasts. - In : **COOMBS, J.** (ed.) : 4th International Congress on Photosynthesis. P. 104. UKISES, London 1977.

*29413 - **ELSTON, J., MONTEITH, J.L.** : Micrometeorology and ecology. - In : **MONTEITH, J.L.** (ed.) : Vegetation and the Atmosphere. Vol. 1. Principles. Pp. 1 - 12. Academic Press, London - New York - San Francisco 1975. [Ps.]

29414 - **EL-ZEFTAWI, B.M.** : Factors affecting pigment levels during re-greening of Valencia orange. - J. hort. Sci. *52* : 127 - 134, 1977.

29415 - **EMMINGHAM, W.H., WARING, R.H.** : An index of photosynthesis for comparing forest sites in western Oregon. - Can. J. Forest Res. *7* : 165 - 174, 1977.

29416 - **ENAMI, I., FUKUDA, I.** : Mechanisms of the acido- and thermophily of *Cyanidium caldarium* GEITLER III. Loss of these characteristics due to detergent treatment. - Plant Cell Physiol. *18* : 671 - 680, 1977. [Ps.]

29417 - **ENAMI, I., FUKUDA, I.** : Mechanisms of the acido- and thermophily of *Cyanidium caldarium* GEITLER IV. Loss of these characteristics due to enzyme treatment. - Plant Cell Physiol. *18* : 707 - 710, 1977. [Ps.]

29418 - **ENDO, H., SANSAWA, H., NAKAJIMA, K.** : Studies on *Chlorella regularis*, heterotrophic fast-growing strain II. Mixotrophic growth in relation to light intensity and acetate concentration. - Plant Cell Physiol. *18* : 199 - 205, 1977. [Ps.]

29419 - **ENDRÉDI, L., HORVÁTH, I.** : Correlation between above-ground phytomass production and the chlorophyll content in the vegetation of a "löszpusztarét", in field experimentation and in conditioned situations. - Acta bot. Acad. Sci. hung. *23* : 63 - 75, 1977.

29420 - **ENGLANDER, J.J., ENGLANDER, S.W.** : Comparison of bacterial and animal rhodopsins by hydrogen exchange studies. - Nature *265* : 658 - 659, 1977.

29421 - **ENGLUND, B.** : The physiology of the lichen *Peltigera aphthosa*, with special reference to the bluegreen phycobiont (*Nostoc* sp.). - Physiol. Plant. *41* : 298 - 304, 1977. [Ps.]

29422 - **ENIKEEV, S.G., MESHKOVA, L.Z., FROLOV, N.S.** : Pigmentnyĭ sostav tritikale. [Pigment composition of *Triticale*.] - Fiziol. Biokhim. kul't. Rast. *9* : 32 - - 34, 1977. [In R, ab : E.]

29423 - ENMANJI, K. : [Studies of interaction of poly(odenylic acid) with Cu-chloro-phyllin by means of paramagnetic effects on ^{1}H relaxation rates.] - Nippon Kagaku Kaishi [J. chem. Soc. Jap., Chem. ind. Chem.] *3* : 438 - 440, 1977. [In Jap., ab : E.]

29424 - ENOCH, H.Z. : The effect of temporal variations of light on net photosynthe-sis. - In : COOMBS, J. (ed.) : 4th International Congress on Photosynthesis. P. 105. UKISES, London 1977.

29425 - ENOCH, H.Z., HURD, R.G. : Effect of light intensity, carbon dioxide concentra-tion, and leaf temperature on gas exchange of spray carnation plants. - J. exp. Bot. *28* : 84 - 95, 1977.

29426 - ENRIGHT, J.T. : Diurnal vertical migration : Adaptive significance and timing. Part 1. Selective advantage : A metabolic model. - Limnol. Oceanogr. *22* : 856 - 872, 1977. [Ps.]

29427 - ENYI, B.A.C. : Analysis of growth and tuber yield in sweet potato (*Ipomoea batatas*) cultivars. - J. agr. Sci. *88* : 421 - 430, 1977.

29428 - EPPLEY, R.W., HARRISON, W.G., CHISHOLM, S.W., STEWART, E. : Particulate orga-nic matter in surface waters off Southern California and its relationship to phytoplankton. - J. mar. Res. *35* : 671 - 696, 1977. [Chl.]

29429 - EREZ, J. : Influence of symbiotic algae on the stable isotope composition of hermatypic corals: A radioactive tracer approach. - In : Proceedings of the Third International Coral Reef Symposium (Miami) *2* : 563 - 569, 1977. [Ps.]

*29430 - ÉRGASHEV, A., KHODZHAEVA, Sh. : O vzaimosvyazi intensivnosti fotosinteza s aktivnost'yu fosforilazy krakhmala. [Interrelationship of photosynthetic rate and starch phosphorylase activity.] - Dokl. Akad. Nauk tadzh. SSR *18* (11) : 59 - 62, 1975. [In R, ab : Tajik.]

29431 - EROKHIN, Yu.E., CHUGUNOV, V.A., MAKHNEVA, Z.K., AGRIKOVA, I.M., SHANTUROVA, T.V. : Sravnitel'noe izuchenie svetosobirayushchikh kompleksov purpurnykh fotosinteziruyushchikh bakterii *Chromatium minutissimum* i *Rhodopseudomonas palustris*. [Comparative study of light-harvesting complexes of purple photo-synthetic bacteria *Chromatium minutissimum* and *Rhodopseudomonas palustris*.] - Biokhimiya *42* : 1817 - 1824, 1977. [In R, ab : E.]

29432 - EROKHIN, Yu.E., CHUGUNOV, V.A., MAKHNEVA, Z.K., AGRIKOVA, I.M., SHANTUROVA, T.V. : Sravnitel'noe izuchenie molekulyarnoi organizatsii svetosobirayush-chikh kompleksov *Chromatium minutissimum* i *Rhodopseudomonas palustris*. [Com-parative study of molecular organization of light-harvesting complexes from *Chromatium minutissimum* and *Rhodopseudomonas palustris*.] - Dokl. Akad. Nauk SSSR *234* : 709 - 712, 1977. [In R.]

29433 - ESASHI, Y., KATOH, H., HATA, Y., GOTŌ, N. : Dormancy and impotency of cockle-bur seeds. VII. Inability of dormant cotyledons to form chlorophyll. - Plant Physiol. *59* : 122 - 125, 1977.

29434 - ESKINS, K., SCHOLFIELD, C.R., DUTTON, H.J. : High-performance liquid chroma-tography of plant pigments. - J. Chromatogr. *135* : 217 - 220, 1977.

*29435 - ESTÉVEZ, M.P., VICENTE, C. : Inhibición por cloratranorina de la reducción de NADP$^+$ por ferredoxina de *Evernia prunastri*. [Chloratranorine inhibition of NADP$^+$ reduction by ferredoxin in *Evernia prunastri*.] - Bol. real. Soc. esp. Hist. nat. (Secc. biol.) *74* : 35 - 38, 1976. [In Span., ab : E.]

29436 - ETIENNE, A.L., LAVERGNE, J., van GORKOM, H.J., LAVOREL, J. : Fluorescence in-duction during a high intensity square pulse. - In : COOMBS, J. (ed.) : 4th International Congress on Photosynthesis. Pp. 105 - 106. UKISES, London 1977.

*29437 - ETTL, H. : Über den Teilungsverlauf des Chloroplasten bei *Chlamydomonas*. - Protoplasma *88* : 75 - 84, 1976.

29438 - EVANS, E.H., RUSH, J.D., JOHNSON, C.E. : Mössbauer spectroscopy of membrane fragments from a blue-green alga. - In : COOMBS, J. (ed.) : 4th International Congress on Photosynthesis. P. 106. UKISES, London 1977.

29439 - EVANS, M.C.W. : Electron paramagnetic resonance studies in photosynthesis. - In : BARBER, J. (ed.) : Primary Processes of Photosynthesis. Pp. 433 - 464. Elsevier, Amsterdam - New York - Oxford 1977.

29440 - **EVANS, M.C.W., CAMMACK, R., SLABAS, A.R.** : Redox properties of the Photosystem I reaction center. - In : OLSON, J.M., HIND, G. (ed.) : Chlorophyll-Proteins, Reaction Centers, and Photosynthetic Membranes. P. 365. Brookhaven nat. Lab., Upton 1977.

29441 - **EVANS, M.C.W., HEATHCOTE, P., WILLIAMS-SMITH, D.L.** : EPR determination of the quantitative relationships between the electron transport components of the photosystem I reaction centre. - In : PACKER, L., PAPAGEORGIOU, G.C., TREBST, A. (ed.) : Bioenergetics of Membranes. Pp. 217 - 224. Elsevier/North-Holland Biomedical Press, Amsterdam - Oxford - New York 1977.

29442 - **EVANS, M.C.W., SIHRA, C.K., SLABAS, A.R.** : The oxidation-reduction potential of the reaction-centre chlorophyll (*P700*) in Photosystem I. Evidence for multiple components in electron-paramagnetic-resonance signal 1 at low temperature. - Biochem. J. *162* : 75 - 85, 1977.

29443 - **EVENARI, M., LANGE, O.L., SCHULZE, E.-D., KAPPEN, L., BUSCHBOM, U.** : Net photosynthesis, dry matter production, and phenological development of apricot trees (*Prunus armeniaca* L.) cultivated in the Negev highlands (Israel). - Flora *166* : 383 - 414, 1977.

29444 - **EVSTIGNEEV, V.B.** : Utilization of the photosynthetic apparatus of green plants and algae for the production of gaseous hydrogen. - In : BUVET, R., ALLEN, M. J., MASSUÉ, J.P. (ed.) : Living Systems as Energy Converters. Pp. 275 - 284. North-Holland Publ. Co., Amsterdam - New York - Oxford 1977.

29445 - **EVSTIGNEEV, V.B., STOLOVITSKY, Ju.M.** : Electron transfer under illumination and charge separation in chlorophyll systems. - In : COOMBS, J. (ed.) : 4th International Congress on Photosynthesis. P. 107. UKISES, London 1977.

29446 - **EZE, J.M.O., BERRIE, G.K.** : Further investigations into the physiological relationship between an epiphyllous liverwort and its host leaves. - Ann. Bot. *41* : 351 - 358, 1977. [Ps, Chl.]

*29447 - **EZHOVA, T.A., GOSTIMSKIĬ, S.A.** : Analiz koriotipov khlorofill'nykh mutantov i iskhodnykh sortov gorokha. [Analysis of caryotypes of chlorophyll mutants and parental cultivars of pea.] - Nauch. Dokl. vyssh. Shkoly, biol. Nauki *19* (9) : 101 - 106, 1976. [In R.]

29448 - **FABRI, R.** : Végétation, production primaire et caractéristiques physico-chimiques d'une rivière de haute Ardenne (Belgique) : la Warche supérieure. - Lejeunia N.S. *87* : 1 - 43, 1977.

29449 - **FADEEVA, L.M., MATORIN, D.N., KRENDELEVA, T.E.** : Vliyanie gerbitsidov na pervichnye protsessy fotosinteza v izolirovannykh khloroplastakh gorokha. [Effect of herbicides on primary processes of photosynthesis in isolated pea leaves.] - Fiziol. Rast. *24* : 560 - 565, 1977. [In R, ab : E.]

*29450 - **FADIA, V.P., MEHTA, A.R.** : Tissue culture studies on cucurbits : chlorophyll development in *Cucumis* callus cultures. - Phytomorphology *26* : 170 - 175, 1976.

29451 - **FAGERGERG, W.R., MOON, R., TRUBY, E.** : Studies of the correlation between cytological structure and photosynthesis/respiration rates in the blade and stipe organs of *Sargassum filipendula*. - J. Phycol. *13* (Suppl.) : 21, 1977.

29452 - **FAJER, J., DAVIS, M.S., BRUNE, D.C., SPAULDING, L.D., BORG, D.C., FORMAN, A.**: Chlorophyll radicals and primary events. - In : OLSON, J.M., HIND, G. (ed.) : Chlorophyll-Proteins, Reaction Centers, and Photosynthetic Membranes. Pp. 74- - 104. Brookhaven nat. Lab., Upton 1977.

29453 - **FAJER, J., DAVIS, M.S., HOLTEN, J.D., PARSON, W.W., THORNBER, J.P., WINDSOR, M.W.** : Kinetic and paramagnetic studies of primary processes in *Rhodopseudomonas viridis*. - In : COOMBS, J. (ed.) : 4th International Congress on Photosynthesis. P. 108. UKISES, London 1977.

29454 - **FAJER, J., FORMAN, A., DAVIS, M.S., SPAULDING, L.D., BRUNE, D.C., FELTON, R. H.** : Anion radicals of bacteriochlorophyll *a* and bacteriopheophytin *a*. Electron spin resonance and electron nuclear double resonance studies. - J. amer. chem. Soc. *99* : 4134 - 4140, 1977.

29455 - FAKOREDE, M.A.B., MULAMBA, N.N., MOCK, J.J. :A comparative study of methods used for estimating leaf area of maize (*Zea mays* L.) from nondestructive measurements. - Maydica *22* : 37 - 46, 1977.

29456 - FALKOWSKI, M., KUKUŁKA, I. : Zawartość karotenu jako cecha charakterystyczna roślin łąkowych. [Carotene content as a characteristic feature of meadow plants.] - Rocz. Nauk roln. F *79* (3) : 97 - 104, 1977. [In Pol., ab : E, R.]

29457 - FALKOWSKI, M., KUKUŁKA, I. : Zawartość chlorofilu jako wskaźnik biologicznych właściwości roślin łąkowych. [Chlorophyll content as an indicator of biological properties of meadow plants.] - Rocz. Nauk roln. F *79* (3) : 105 - 112, 1977. [In Pol., ab : E, R.]

29458 - FALUDI-DÁNIEL, Á., MUSTÁRDY, L.A., ROUX, E. : Energization of granal and a-granal chloroplast membranes. - In : OLSON, J.M., HIND, G. (ed.) : Chlorophyll-Proteins, Reaction Centers, and Photosynthetic Membranes. P. 363. Brookhaven nat. Lab., Upton 1977.

29459 - FAM TKHAN' KHO : Gibridologicheskiĭ analiz mozaichnykh pigmentnykh mutantov *Chlamydomonas reinhardi*. [Hybridological analysis of mosaic pigment mutants of *Chlamydomonas reinhardi*.] - In : NASYROV, Yu.S. (ed.) : Genetika Fotosinteza. Pp. 120 - 127. Donish, Dushanbe 1977. [In R.]

29460 - FANKBONER, P.V., BURGH, M.E. DE : Diurnal exudation of ^{14}C-labelled compounds by the large kelp *Macrocystis integrifolia* BORY. - J. exp. mar. Biol. Ecol. *28* : 151 - 162, 1977.

29461 - FANTINET, M., LARRIEU, C. : Influence de la température, de la lumière et de la kinétine sur la sénescence de feuilles isolées d'Avoine. - Compt. rend. Acad. Sci. Paris, Sér. D *284* : 2495 - 2498, 1977. [Chl.]

29462 - FARINEAU, J., LAVAL-MARTIN, D. : Light *versus* dark carbon metabolism in cherry tomato fruits. II. Relationship between malate metabolism and photosynthetic activity. - Plant Physiol. *60* : 877 - 880, 1977.

29463 - FARINEAU, J., LAVAL-MARTIN, D. : Photoassimilation of CO_2 by tissues of green tomato fruit (*Lycopersicum esculentum* var.*cerasiforme* DUN A. GRAY). - In : COOMBS, J. (ed.) : 4th International Congress on Photosynthesis. P. 109. UKISES, London 1977.

29464 - FARKAS, D.L., MALKIN, S. : Cold storage of isolated chloroplasts. - In : COOMBS, J. (ed.) : 4th International Congress on Photosynthesis. P. 110. UKISES, London 1977.

29465 - FASULO, M.P., VANNINI, G.L., DALL'OLIO, G. : Inhibition by myomycin of light-induced chloroplast development in *Euglena gracilis*. - Caryologia *30* : 488 - 489, 1977.

29466 - FAUST, M.A., CORRELL, D.L. : Autoradiographic study to detect metabolically active phytoplankton and bacteria in the Rhode River estuary. - Mar. Biol. *41* : 293 - 305, 1977.

29467 - FEDERER, C.A. : Leaf resistance and xylem potential differ among broadleaved species. - Forest Sci. *23* : 411 - 418, 1977.

B29468 - FEDOROV, N.I. : Fotosintez i Mineral'noe Pitanie Rasteniĭ. [Photosynthesis and Mineral Nutrition of Plants.] - Saratov. sel'.-khoz. Inst., Saratov 1977. [In R.]

*29469 - FEDOROVA, E.I. : Dinamika litoral'nykh fitotsenozov (opyt metodicheskogo issledovaniya). [Dynamics of litoral phytocenoses (methodical study).] - In : Antropogennoe Évtrofirovanie Ozer. Pp. 45 - 81, 119. Nauka, Moskva 1976. [In R.]

29470 - FEDOSEEVA, G.P., BAGAUTDINOVA, R.I. : Osobennosti strukturnoĭ organizatsii i funktsional'noĭ aktivnosti fotosinteticheskogo apparata u kartofelya raznoĭ stepeni okul'turennosti. [Characteristics of the structural organization and functional activity of the photosynthetic apparatus in potatoes as a function of cultivars.] - Sel'skokhoz. Biol. *12* : 545 - 552, 1977. [In R, ab : E.]

29471 - FEDTKE, C. : Formation of nitrite in plants treated with herbicides that inhibit photosynthesis. - Pestic. Sci. *8* : 152 - 156, 1977.

29472 - FEDTKE, C. : Levels of malate and nitrate under conditions of reduced photosynthate production. - In : COOMBS, J. (ed.) : 4th International Congress on
Photosynthesis. Pp. 110 - 111. UKISES, London 1977.

29473 - FEDTKE, C., DEICHGRÄBER, G., SCHNEPF, E. : Herbicide induced changes in wheat
chloroplast ultrastructure and chlorophyll a/b ratio. - Biochem. Physiol.
Pflanzen 171 : 307 - 312, 1977.

29474 - FEHER, G., OKAMURA, M.Y. : Reaction centers from Rhodopseudomonas sphaeroides.
- In : OLSON, J.M., HIND, G. (ed.) : Chlorophyll-Proteins, Reaction Centers,
and Photosynthetic Membranes. Pp. 183 - 194. Brookhaven nat. Lab., Upton
1977.

29475 - FEHR, W.R., CAVINESS, C.E., VORST, J.J. : Response of indeterminate and determinate soybean cultivars to defoliation and half-plant cut-off. - Crop Sci.
17 : 913 - 917, 1977. [Dry-matter accumulation.]

29476 - FEIERABEND, J. : Capacity for chlorophyll synthesis in heat-bleached 70S ribo-
some-deficient rye leaves. - Planta 135 : 83 - 88, 1977.

29477 - FEIERABEND, J., BRASSEL, D. : Subcellular localization of shikimate dehydrogenase in higher plants. - Z. Pflanzenphysiol. 82 : 334 - 346, 1977. [Chl.]

29478 - FEIERABEND, J., MIKUS, M. : Occurrence of a high temperature sensitivity of
chloroplast ribosome formation in several higher plants. - Plant Physiol.
59 : 863 - 867, 1977.

29479 - FEINLEIB, M.E. : Photomovement in microorganisms : Strategies of response. -
In : CASTELLANI, A. (ed.) : Research in Photobiology. Pp. 71 - 84. Plenum
Press, New York - London 1977. [Chloroplast.]

29480 - FEKETE, G., TUBA, Z. : Supraindividual versus individual homogeneity of photosynthetic pigments : a study on community structure. - Acta bot. Acad. Sci.
hung. 23 : 319 - 331, 1977.

*29481 - FEKETE, M., KOZMA, L., HUSZKA, T. : Spectrophotometric method for determin-
ing the pigment content of ground paprika. - Z. Lebensmittel-Untersuch.-
Forsch. 161 : 31 - 33, 1976.

29482 - FELLOWS, R.J., EGLI, D.B., LEGGETT, J.E. : Application of pod bleeding technique to source-sink translocation studies in soybean (Glycine max). - Plant
Physiol. 59 (6, Suppl.) : 125, 1977.

29483 - FENNA, R.E., MATTHEWS, B.W. : Structure of a bacteriochlorophyll a-protein
from Prosthecochloris aestuarii. - In : OLSON, J.M., HIND, G. (ed.) : Chlorophyll-Protein, Reaction Centers, and Photosynthetic Membranes. Pp. 170 -
- 182. Brookhaven nat. Lab., Upton 1977.

29484 - FENNA, R.E., TEN EYCK, L.F., MATTHEWS, B.W. : Atomic coordinates for the
chlorophyll core of a bacteriochlorophyll a-protein from green photosynthetic bacteria. - Biochem. biophys. Res. Commun. 75 : 751 - 756, 1977.

29485 - FENSOM, D.S., WILLIAMS, E.J., AIKMAN, D., DALE, J.E., SCOBIE, J., LEDINGHAM,
K.W.O., DRINKWATER, A., MOORBY, J. : Translocation of ^{11}C from leaves of
Helianthus : preliminary results. - Can. J. Bot. 55 : 1787 - 1793, 1977.

29486 - FENTON, R., DAVIES, W.J., MANSFIELD, T.A. : The role of farnesol as a regulator of stomatal opening in Sorghum. - J. exp. Bot. 28 : 1043 - 1053, 1977.

29487 - FER, A. : Etude de la photosynthèse et de la migration des assimilats chez
Cuscuta lupuliformis KROCK. - Physiol. vég. 15 : 313 - 324, 1977.

29488 - FERHI, A., LETOLLE, R. : Transpiration and evaporation as principal factors
in oxygen isotope variations of organic matter in land plants. - Physiol.
vég. 15 : 363 - 370, 1977.

29489 - FERHI, A., LETOLLE, R. : Variation de la composition isotopique de l'oxygène
organique de quelques plantes en fonction de leur milieu de vie. - Compt.
rend. Acad. Sci. Paris, Sér. D 284 : 1887 - 1889, 1977. [Ps.]

29490 - FERNÁNDEZ, J. : Actividad fotosintética de las hojas de vid en un parral.
[Photosynthetic activity of leaves in a bower of grapevines.] - Fyton 35 :
61 - 64, 1977. [In Span., ab : E.]

29491 - FERNANDEZ, J., BALKAR, J., MEYER, L.H. : Influencia de la iluminación sobre la actividad fotosintética de las hojas de vid cultivada en espaldera. [Effect of illuminance on photosynthetic activity of leaves of grapevine cultivated as a trellis.] - Turrialba *27* : 3 - 6, 1977. [In Span., ab : E.]

29492 - FERNANDEZ, J., BALKAR, J., MEYER, L.H. : Distribución de la materia orgánica en un cultivo de vid conducido en espaldera. [Distribution of organic matter in grapevine cultivated as a trellis.] - Turrialba *27* : 233 - 238, 1977. [Growth analysis; in Span., ab : E.]

29493 - FERNÁNDEZ GONZÁLEZ, J. : Equipo para determinar la actividad fotosintetica en hojas de plantas superiores por medio de $^{14}CO_2$. [Equipment for determination of photosynthetic activity of leaves of higher plants by means of $^{14}CO_2$.] - J.E.N. (Madrid) *361* : 1 - 19, 1977. [In Span., ab : E.]

29494 - FERRARI, I., ASCOLINI, A., BELLAVERE, C. : Considerazioni conclusive sui risultati di ricerche pluriennali al Lago Santo Parmense. [Conclusive considerations upon the results of a five-year research at Lake Santo Parmense.] - Ateneo parmense, Acta nat. *13* : 433 - 444, 1977. [Chl; in Ital., ab : E.]

29495 - FERRON, F., COUDRET, A., GAUDILLERE, J.-P. : Effet de la salinité du milieu de culture sur les voies de carboxylation d'une halophyte (*Plantago maritima* L. Var. *graminaea*) et d'une glycophyte (*Plantago lanceolata* L.). - Compt. rend. Acad. Sci. Paris, Sér. D *285* : 323 - 326, 1977.

29496 - FERWERDA, J.-D. : Oil palm. - In : ALVIM, P. de T., KOZLOWSKI, T.T. (ed.) : Ecophysiology of Tropical Crops. Pp. 351 - 382. Academic Press, New York - San Francisco - London 1977. [Growth analysis.]

28497 - FETTERMAN, L.M., GALLOWAY, L., WINOGRAD, N., FONG, F.K. : The role of water on the photoactivity of chlorophyll *a*. *In vitro* experimental characterization of the PSI light reaction in photosynthesis. - J. amer. chem. Soc. *99* : 653 - - 655, 1977.

29498 - FEUILLADE, J., FEUILLADE, M. : Relations entre la croissance d'*Oscillatoria rubescens* D.C. et la teneur en cuivre de son milieu de culture. - Ann. hydrobiol. *8* : 389 - 399, 1977. [Chl, Car, biliproteins.]

29499 - FILIPPETTI, A., MARZANO, C.F., MONTI, L.M., SCARASCIA MUGNOZZA, G.T. : Ricerche di miglioramento genetico di varietà commerciali di pisello da industria mediante mutagenesi sperimentale I. Frequenze e tipi di mutazioni indotte con raggi X e dietilsolfato (DS). [Research on the improvement of commercial varieties of *Pisum sativum* using experimental mutagenesis : I. Frequency and types of mutations induced by X-radiation and diethylsulphate(DS).] - Genet. agrar. *31* : 295 - 307, 1977. [Chl; in Ital., ab : E.]

29500 - FILIPPOVA, L.A., MAMUSHINA, N.S., ZUBKOVA, E.K. : O vliyanii nakopleniya assimilyatov na fotosintez u kletok khlorelly. [Influence of accumulation of assimilates on photosynthesis in *Chlorella* cells.] - Bot. Zh. *62* : 179 - 184, 1977. [In R, ab : E.]

29501 - FILIPPOVA, R.I., STREL'NIKOVA, T.R. : Osobennosti pigmentnoi sistemy rastenii sakharnoi kukuruzy v svyazi s yavleniem geterozisa. [Peculiarities of pigment system of sweet maize plants in relation to heterosis.] - In : NASYROV, Yu.S. (ed.) : Genetika Fotosinteza. Pp. 257 - 260. Donish, Dushanbe 1977. [In R.]

29502 - FILIPPOVICH, I.I., ALINA, B.A., BEZSMERTNAYA, I.N., TONGUR, A.M., OPARIN, A.I. : Svyaz' beloksinteziruyushchei sistemy so strukturoi khloroplastov. [Interrelationship of protein synthesizing system with chloroplast structure.] - In : NASYROV, Yu.S. (ed.) : Genetika Fotosinteza. Pp. 28 - 33. Donish, Dushanbe 1977. [In R.]

29503 - FINDENEGG, G.R. : Adaptation of *Scenedesmus* photosynthesis to high and low CO_2 levels. - In : COOMBS, J. (ed.) : 4th International Congress on Photosynthesis. Pp. 111 - 112. UKISES, London 1977.

29504 - FINDENEGG, G.R. : Interactions of glycolate-, HCO_3^--, Cl^--, and H^+-balance of *Scenedesmus obliquus*. - Planta *135* : 33 - 38, 1977. [Ps.]

29505 - FIRSOW, N.N., DREWS, G. : Differentiation of the intracytoplasmic membrane of *Rhodopseudomonas palustris* induced by variations of oxygen partial pressure or light intensity. - Arch. Microbiol. *115* : 299 - 306, 1977. [Chl.]

29506 - FISCHER, R.A., AGUILAR, I., LAING, D.R. : Post-anthesis sink size in a high-
-yielding dwarf wheat : yield response to grain number. - Aust. J. agr. Res.
28 : 165 - 175, 1977. [Photosynthates.]

29507 - FISHER, K.A., STOECKENIUS, W. : Freeze-fractured purple membrane particles :
protein content. - Science *197* : 72 - 74, 1977.

29508 - FISHER, S.G. : Organic matter processing by a stream-segment ecosystem :
Fort River, Massachusetts, U.S.A. - Int. Rev. ges. Hydrobiol. *62* : 701 - 727,
1977.

29509 - FLINN, A.M., ATKINS, C.A., PATE, J.S. : Significance of photosynthetic and
respiratory exchanges in the carbon economy of the developing pea fruit. -
Plant Physiol. *60* : 412 - 418, 1977.

*29510 - FLINT, E.A., BULLOCK, S. : Fine structure of *Crucigenia truncata* G.M. SMITH
from Lake Pearson, Canterbury, New Zealand. - New Zeal. J. Bot. *14* : 261 -
- 270, 1976. [Chloroplast.]

29511 - FLINT, R.W., RICHARDS, R.C., GOLDMAN, C.R. : Adaptation of styrofoam substrate
to benthic algal productivity studies in Lake Tahoe, California-Nevada. - J.
Phycol. *13* : 407 - 409, 1977. [Ps.]

29512 - FLOYD, G.L., SALISBURY, J.L. : Glycolate dehydrogenase in primitive green
algae. - Amer. J. Bot. *64* : 1294 - 1296, 1977.

29513 - FLÜGGE, U.-I., HELDT, H.W. : Specific labelling of a protein involved in
phosphate transport of chloroplasts by pyridoxal-5'-phosphate. - FEBS Lett.
82 : 29 - 33, 1977.

29514 - FOCK, H., KLUG, K., CANVIN, D.T. : Influence of CO_2 and temperature on photosynthetic CO_2 uptake and photorespiratory CO_2 evolution in sunflower leaves.
- In : COOMBS, J. (ed.) : 4th International Congress on Photosynthesis. Pp.
112 - 113. UKISES, London 1977.

29515 - FOCKE, R. : Schaffung von optimalen Strahlungsverhältnissen für hohe Korn-
erträge des Getreides. - In : UNGER, K. (ed.) : Biophysikalische Analyse
pflanzlicher Systeme. Pp. 227 - 232. VEB Gustav Fischer Verlag, Jena 1977.

29516 - FONG, F., SCHIFF, J.A. : Blue-light induced absorbance changes in carotenoids
from *Euglena gracilis* var. *bacillaris* W_3BUL. - Plant Physiol. *59* (6, Suppl.):
48, 1977.

29517 - FONG, F., SCHIFF, J.A. : Mitochondrial respiration and chloroplast development in *Euglena gracilis*, var. *bacillaris*. - Plant Physiol. *59* (6, Suppl.):
92, 1977.

29518 - FONG, F.K. : Current developments in the investigation of the primary light
reactions in photosynthesis. - J. theor. Biol. *66* : 199 - 202, 1977.

29519 - FONG, F.K. : The energy upconversion model of photosynthesis. - In : COOMBS,
J. (ed.) : 4th International Congress on Photosynthesis. P. 113. UKISES, London 1977.

29520 - FONG, F.K., KOESTER, V.J., GALLOWAY, L. : Endo and exo carbomethoxy carbonyl
bonding in hydrated chlorophyll *a* dimers. Experimental criteria for the determination of the *P*700 structure in photosynthesis. - J. amer. chem. Soc.
99 : 2372 - 2375, 1977.

29521 - FONG, F.K., POLLES, J.S., GALLOWAY, L., FRUGE, D.R. : Far red photogalvanic
splitting of water by chlorophyll *a* dihydrate. A new model of plant photosynthesis. - J. amer. chem. Soc. *99* : 5802 - 5804, 1977.

29522 - FONG, F.K., WASSAM, W.A. : Molecular origin of long-wavelength forms of hydrated chlorophyll *a*. - J. amer. chem. Soc. *99* : 2375 - 2376, 1977.

29523 - FORD, M., BLACK, M., CHAPMAN, J.M. : Inter-organ synergism and the control of
chlorophyll accumulation in sunflower (*Helianthus annuus*) cotyledons. - J.
exp. Bot. *28* : 926 - 934, 1977.

29524 - FORDHAM, R. : Tea. - In : ALVIM, P. de T., KOZLOWSKI, T.T. (ed.) : Ecophysiology of Tropical Crops. Pp. 333 - 349. Academic Press, New York - San Francisco
- London 1977. [Ps.]

29525 - FORK, D.C. : Photosynthesis. - In : SMITH, K.C. (ed.) : The Science of Photobiology. Pp. 329 - 369. Plenum Press, New York - London 1977.

29526 - FORK, D.C., MURATA, N. : Studies on the effect of transition of the physical
phase of membrane lipids on electron transport in the extreme thermophile
Synechococcus lividus. - Carnegie Inst. Year Book 76 : 222 - 226, 1977.

29527 - FORK, D.C., MURATA, N. : The effect of temperature on the physical phase of
thylakoid lipids and photosynthesis in algae. - J. Phycol. 13 (Suppl.) : 22,
1977.

29528 - FORK, D.C., MURATA, N. : The relationship between changes of the physical
phase of membrane lipids and photosynthesis in the thermophilic alga Cyanidium
caldarium. - Plant Cell Physiol. 1977 (spec. Issue 3 - Photosynthetic Organelles. Structure and Function) : 427 - 436, 1977.

29529 - FORK, D.C., MURATA, N. : Transition of the physical phase of membrane lipids
and electron transport in the extreme thermophile Synechococcus lividus. -
In : COOMBS, J. (ed.) : 4th International Congress on Photosynthesis. P. 114.
UKISES, London 1977.

*29530 - FORTI, G. : Solar energy utilization in photosynthesis. - In : Energy and
Physics. Pp. 375 - 378. Europe. phys. Soc., Petit-Lancy 2 1975.

29531 - FORTI, G. : Flavoproteins. - In : TREBST, A., AVRON, M. (ed.) : Photosynthesis I. (Encycl. Plant Physiol. N.S. Vol. 5.) Pp. 222 - 226. Springer-Verlag,
Berlin - Heidelberg - New York 1977.

29532 - FORTI, G. : The effect of xanthine oxidase on chloroplasts electron transport:
a re-evaluation. - Plant Sci. Lett. 10 : 197 - 198, 1977.

29533 - FORTI, G., GEROLA, P. : Inhibition of photosynthesis by azide and cyanide and
the role of oxygen in photosynthesis. - Plant Physiol. 59 : 859 - 862, 1977.

29534 - FOSTER, A., BLACK, C.C. : Panicum maximum photosynthesis. - Plant Cell Physiol.
1977 (spec. Issue 3 - Photosynthetic Organelles. Structure and Function) :
325 - 340, 1977.

29535 - FOSTER, A., BLACK, C.C. : Pathways of photosynthetic carbon flow in Panicum
maximum : Response to oxygen concentration. - Plant Physiol. 59 (6, Suppl.):
91, 1977.

29536 - FOSTER, J.G., RUSS, P.N., WOLF, D.D., HESS, J.L. : Glycolate oxidase and
superoxide dismutase in cotton leaf tissue. - Plant Physiol. 59 (6, Suppl.) :
91, 1977.

29537 - FOULDS, W., YOUNG, L. : Effect of frosting, moisture stress and potassium
cyanide on the metabolism of cyanogenic and acyanogenic phenotypes of Lotus
corniculatus L. and Trifolium repens L. - Heredity 38 : 19 - 24, 1977. [Ps.]

29538 - FOURY, C., CADILHAC, B., AUBERT, S. : Observations sur les teneurs en chlorophylle et en cynarine, et sur la structure des chloroplastes d'un mutant d'
artichaut (Cynara scolymus L.). - Ann. Amélior. Plant. 27 : 587 - 602, 1977.

29539 - FOWLER, C.F. : Proton evolution from photosystem II. Stoichiometry and mecha-
nistic considerations. - Biochim. biophys. Acta 462 : 414 - 421, 1977.

29540 - FOWLER, C.F. : Proton translocation in chloroplasts and its relationship to
electron transport between the photosystems. - Biochim. biophys. Acta 459 :
351 - 363, 1977.

29541 - FOWLER, C.F. : Proton transport reactions associated with photoreduction of
lipophilic photosystem II, electron acceptors. - In : COOMBS, J. (ed.) : 4th
International Congress on Photosynthesis. Pp. 114 - 115. UKISES, London 1977.

29542 - FOY, C.D., VOIGT, P.W., SCHWARTZ, J.W. : Differential susceptibilities of
weeping lovegrass strains to an iron-related chlorosis on calcareous soils.
- Agron. J. 69 : 491 - 496, 1977.

29543 - FRACKOWIAK, D., BAUMAN, D., MANIKOWSKI, H., MARTYNSKI, T. : Spectral properties of chlorophyll-*a* in liquis crystal. - Acta phys. chem. (Szeged), N.S.
23 (1) : 183-188, 1977.

29544 - FRACKOWIAK, D., BAUMAN, D., MANIKOWSKI, H., MARTYŃSKI, T. : Spectral properties of chlorophyll *a* in liquid crystal. - Biophys. Chem. *6* : 369-377, 1977.

29545 - FRACKOWIAK, D., FIKSINSKI, K., PIENKOWSKA, H. : Phycoerythrin in stretched
PVA film. - Stud. biophys. *63* : 183 - 187, 1977.

29546 - FRADKIN, L.I., KALYAGA, V.M. : Farmiravanne submembrannykh chastsinak khlaraplastaŭ u khodze zelyanennya etyyaliravanykh listsyaŭ yachmenyu. [Formation
of submembrane particles of chloroplasts in the course of greening of etiolated barley leaves.] - Vestsi Akad. Navuk belarus. SSR, Ser. biyal. Navuk
1977 (6) : 19 - 23, 137, 1977. [In Belorus., ab : R.]

29547 - FRADKIN, L.I., SENKEVICH, G.S. : Vliyanie guanidina i 1,10-fenantrolina na
spektry fluorestsentsii submembrannykh chastits khloroplastov. [Effect of
guanidine and 1,10-phenanthroline on fluorescence spectra of submembrane
particles of chloroplasts.] - Zh. prikl. Spektrosk. *27*(2) : 253 - 258, 1977.
[In R.]

29548 - FRAGATA, M. : A far-red absorbing form of chlorophyll *a* detected in phospha-
tidylcholine vesicles. - Photosynthetica *11* : 296 - 301, 1977.

29549 - FRAGATA, M. : On the location of the tetrapyrrole macrocycle of chlorophyll *a*
in phospholipid vesicles and in hexadecane. - Experientia *33* : 177 - 179,
1977.

29550 - FRANK, A.B., HARRIS, D.G., WILLIS, W.O. : Growth and yield of spring wheat
as influenced by shelter and soil water. - Agron. J. *69* : 903 - 906, 1977.
[Growth analysis.]

29551 - FRANK, A.B., HARRIS, D.G., WILLIS, W.O. : Plant water relationships of spring
wheat as influenced by shelter and soil water. - Agron. J. *69* : 906 - 910,
1977. [Stomatal resistance.]

29552 - FRANK, G., SIDLER, W., WIDMER, H., ZUBER, H. : On the amino acid sequences
of *C*-phycocyanin and allophycocyanin from the blue-green algae *Mastigocladus
laminosus*. - In : COOMBS, J. (ed.) : 4th International Congress on Photosynthesis. Pp. 115 - 116. UKISES, London 1977.

29553 - FRANK, M.H. : Die Bewegungsreaktion des *Mougeotia*-Chloroplasten bei konti-
nuierlicher Belichtung mit linear polarisiertem längsschwingendem Rotlicht.-
Z. Pflanzenphysiol. *82* : 210 - 234, 1977.

29554 - FREEBERG, L.R., WILSON, W.B. : Photo-oxidation and degradation of chlorophyll
a in marine phytoplankton samples. - J. Phycol. *13* (Suppl.) : 22, 1977.

29555 - FREEMAN, H.C., RAMSHAW, J.A.M., WRIGHT, P.E. : High resolution proton magnetic resonance studies of plastocyanin. - In : COOMBS, J. (ed.) : 4th International Congress on Photosynthesis. P. 116. UKISES, London 1977.

29556 - FRENCH, C.S. : Action spectra. - Carnegie Inst. Year Book *76* : 212 - 217,
1977. [Ps.]

29557 - FRENCH, C.S. : Action spectrum for DCIP oxidation by *Nostoc* particles. -
In : COOMBS, J. (ed.) : 4th International Congress on Photosynthesis. P.117.
UKISES, London 1977.

29558 - FRENCH, C.S. : Sharper action spectra. - Photochem. Photobiol. *25* : 159 - 160,
1977. [Chl.]

29559 - FRENCH, S.A.W., HUMPHRIES, E.C. : The effect of partial defoliation on
yield of sugar beet. - Ann. appl. Biol. *87* : 201 - 212, 1977. [Growth analysis.]

29560 - FREYSSINET, G. : Characterization of cytoplasmic and chloroplast ribosomal
proteins of *Euglena gracilis*. - Biochimie *59* : 597 - 610, 1977.

*29561 - FRIDLAND, E.V., EREMINA, G.V., ORLOV, D.S., BIL'DEBAEVA, R.M. : Spektrofoto-
metricheskoe opredelenie v pochvakh khlorofilla i ego proizvodnykh. [Spectro-
photometric determination of chlorophyll and its derivatives in soils.] -
Nauch. Dokl. vyssh. Shkoly, biol. Nauki *19*(9) : 133 - 138, 1976. [In R.]

29562 - FRIEDRICH, J.W., SCHRADER, L.E. : Nitrate reductase, glutamine synthetase, glutamate dehydrogenases, and chlorophyll in young maize leaf blades as affected by sulfur deficiency. - Plant Physiol. *59* (6, Suppl.) : 127, 1977.

29563 - FRIEND, D.J.C., YAMAGUCHI, T. : Effect of leaf age on photosynthesis of coffee. - In : COOMBS, J. (ed.) : 4th International Congress on Photosynthesis. P. 118. UKISES, London 1977.

29564 - FRIER, V. : The relationship between photosynthesis and tuber growth in *Solanum tuberosum* L. - J. exp. Bot. *28* : 999 - 1007, 1977.

29565 - FRÖLICH, W.G., POLLMER, W.G., KLEIN, D. : Performance of upright-leaved liguleless-2 and liguleless-3 maize hybrids adapted to Central European climatic conditions. - Z. Pflanzenzücht. *79* : 134 - 144, 1977. [Growth analysis.]

29566 - FRÖLICH, W.G., POLLMER, W.G., KLEIN, D. : Performance of isogenic liguleless--2 maize hybrids as influenced by irrigation, row width, plant density and nitrogen fertilizer. - Z. Acker- Pflanzenbau *145* : 207 - 223, 1977. [Growth analysis.]

29567 - FROSCH, S., DRUMM, H., MOHR, H. : Regulation of enzyme levels by phytochrome in mustard cotyledons : multiple mechanisms ? - Planta *136* : 181 - 186, 1977. [RuBP carboxylase.]

29568 - FROSCH, S., MOHR, H. : Modulation by phytochrome of ribulose-bisphosphate carboxylase synthesis in mustard seedling cotyledons (*Sinapis alba* L.). - In : COOMBS, J. (ed.) : 4th International Congress on Photosynthesis. P.118. UKISES, London 1977.

29569 - FRY, D.J., PHILLIPS, I.D.J. : Photosynthesis of conifers in relation to annual growth cycles and dry matter production. II. Seasonal photosynthetic capacity and mesophyll ultrastructure in *Abies grandis*, *Picea sitchensis*, *Tsuga heterophylla* and *Larix leptolepis* growing in S.W.England. - Physiol. Plant. *40* : 300 - 306, 1977.

29570 - FRY, I., PAPAGEORGIOU, G., TEL-OR, E., PACKER, L. : Reconstitution of a system for H_2 evolution with chloroplasts, ferredoxin and hydrogenase. - Z. Naturforsch. *32C* : 110 - 117, 1977.

29571 - FRY, S.C., BIDWELL, R.G.S. : An investigation of photosynthetic sucrose production in bean leaves. - Can. J. Bot. *55* : 1457 - 1464, 1977.

29572 - FUCHS, M., SCHULZE, E.-D., FUCHS, M.I. : Spacial distribution of photosynthetic capacity and performance in a mountain spruce forest of Northern Germany. - Oecologia *29* : 329 - 340, 1977.

29573 - FUJII, Y., KUROKAWA, T., INOUE, Y., YAMAGUCHI, I., MISATO, T. : Inhibition of carotenoid biosynthesis as a possible mode of herbicidal action of 3, 3'-dimethyl-4-methoxybenzophenone (NK-049). - J. Pesticide Sci. *2* : 431 - 437, 1977.

29574 - FUKAI, S., DAVISON, L. : Estimation of radiation environments in row-planted communities by a leaf geometry model. - In : Second Australasian Conference on Heat and Mass Transfer. Pp. 21 - 28. The University of Sydney, Sydney 1977. [Growth analysis.]

29575 - FUKAI, S., SILSBURY, J.H. : Responses of subterranean clover communities to temperature. III. Effects of temperature on canopy photosynthesis. - Aust. J. Plant Physiol. *4* : 273 - 282, 1977.

29576 - FUKAI, S., SILSBURY, J.H. : Effects of irradiance and solar radiation on dry matter growth and net CO_2 exchange of *Trifolium subterraneum* L. swards at a constant temperature. - Aust. J. Plant Physiol. *4* : 485 - 497, 1977. [Ps.]

29577 - FYKSE, H. : Untersuchungen über *Sonchus arvensis* L., *Cirsium arvense* (L.) SCOP. und *Tussilago farfara* L. Entwicklung sowie Translokation von radioaktiv markierten Kohlenhydraten und MCPA. - Meld. Norges Landbrukshøgsk. *56* (27) : 1 - 22, 1977.

29578 - GÁBOR, A., JÁNOSSY, S., MUSTÁRDY, L.A., FALUDI-DÁNIEL, Á. : X-ray microana-
lytical study of Mn and Fe compartmentation in maize chloroplasts. - Acta
histochem. *58* : 317 - 323, 1977. [Ps.]

29579 - GADAL, P., VIDAL, J., JACQUOT, J.P. : Study of the NADP-MDH activation process by dithiothreitol and light. - In : COOMBS, J. (ed.) : 4th International
Congress on Photosynthesis. P. 119. UKISES, London 1977.

29580 - GAGLIANO, A.G., GEACINTOV, N.E., BRETON, J, : Orientation and linear dichroism of chloroplasts and sub-chloroplast fragments oriented in an electric
field. - Biochim. biophys. Acta *461* : 460 - 474, 1977.

29581 - GALLAHER, R.N., BROWN, R.H. : Starch storage in C_4 vs. C_3 grass leaf cells
as related to nitrogen deficiency. - Crop Sci. *17* : 85 - 88, 1977.

29582 - GALLEGOS, C.L., HORNBERGER, G.M., KELLY, M.G. : A model of river benthic
algal photosynthesis in response to rapid changes in light. - Limnol. Oceanogr.
22 : 226 - 233, 1977.

29583 - GALMICHE, J.M. : Post-illumination ATP formation. - In : TREBST, A., AVRON, M.
(ed.) : Photosynthesis I. (Encycl.Plant Physiol. N.S. Vol.5). Pp. 374 - 392.
Springer-Verlag, Berlin - Heidelberg - New York 1977.

29584 - GALMICHE, J.M., GIRAULT, G. : Effect of the bound nucleotides on the proper-
ties of the isolated coupling factor 1 from spinach chloroplasts. - In :
COOMBS, J. (ed.) : 4th International Congress on Photosynthesis. P. 119.
UKISES, London 1977.

29585 - GALUTVA, O.A., LOBODA, N.I., NEKRASOV, L.I. : Izuchenie fotosensibilizirovan-
noĭ khlorofillom reaktsii vosstanovleniya metilovogo krasnogo askorbinovoĭ
kislotoĭ v ětanole. [Photosensibilized by chlorophyll reaction of methyl red
reduction. with ascorbic acid in ethanol.] - Zh. fiz. Khim. *51*: 2634 - 2636,
1977. [In R.]

29586 - GALZIN, A.M., MONTIES, B. : Aerobic activation of glycolate-oxydase by low
m.w. compounds such chlorogenic acid. - In : COOMBS, J. (ed.) : 4th International Congress on Photosynthesis. P. 120. UKISES, London 1977.

29587 - GAMALEĬ, Yu.V., KULIKOV, G.V. : Vozrastnye izmeneniya kletok mezofilla listopadnykh i vechnozelenykh rasteniĭ. [Ontogenetic changes in mesophyll cells
in deciduous and evergreen plants.] - Tsitologiya *19* : 15 - 20, 1977. [In
R, ab: E.]

29588 - GAMBLE, J.C., DAVIES, J.M., STEELE, J.H. : Loch Ewe bag experiment, 1974. -
Bull. mar. Sci. *27* : 146 - 175, 1977. [Ps productivity.]

29589 - GANTT, E. : Recent contributions in phycobiliproteins and phycobilisomes. -
Photochem. Photobiol. *26* : 685 - 689, 1977.

29590 - GANTT, E., GRABOWSKI, J., LIPSCHULTZ, C.A. : Influence of wavelength on the
pigment composition, the quantum fluorescence yield, and structure of phycobilisomes in marine and fresh water prokaryotic and eukaryotic algae. -
J. Phycol. *13* (Suppl.) : 23, 1977.

29591 - GANTT, E., LIPSCHULTZ, C.A. : Probing phycobilisome structure by immuno-elec-
tron microscopy. - J. Phycol. *13* : 185 - 192, 1977.

29592 - GANTT, E., LIPSCHULTZ, C.A., ZILINSKAS, B.A. : Phycobilisomes in relation
to the thylakoid membrane. - In : OLSON, J.M., HIND, G. (ed.) : Chlorophyll-
-Proteins, Reaction Centers, and Photosynthetic Membranes. Pp. 347 - 357.
Brookhaven nat. Lab., Upton 1977.

29593 - GAPONENKA, V.I., BALEVA, E.F., SHAŬCHUK, S.M. : Abnaŭlenne khlarafilu i asi-
milyatsyĭnyya liki listsyaŭ yachmenyu roznaga ŭzrostu. [Chlorophyll regeneration and assimilation numbers of barley leaves of various age.]-Vestsi
Akad. Navuk belarus. SSR, Ser. biyal. Navuk *1977* (4) : 47 - 52, 140, 1977.
[In Belorus., ab : R.]

29594 - GAPONENKO, V.I., BALEVA, E.F., SHEVCHUK, S.N.: Korrelyatsiya mezhdu obnovle-

niem khlorofilla i assimilyatsionnymi chislami raznykh po vozrastu zon list'-
ev kukuruzy. [Correlation between chlorophyll turnover and assimilation numbers of different age zones of maize leaves.] - Dokl. Akad. Nauk belorus. SSR
21 : 749 - 752, 1977. [In R.]

29595 - **GARAB, G.I., BRETON, J.** : Low temperature fluorescence polarization of oriented spinach chloroplasts. - Acta phys. chem. *23* : 135 - 140, 1977.

29596 - **GARBER, M.P.** : Effect of light and chilling temperatures on chilling-sensitive and chilling-resistant plants. Pretreatment of cucumber and spinach thylakoids *in vivo* and *in vitro*. - Plant Physiol. *59* : 981 - 985, 1977.

29597 - **GARBER, M.P.** : Recovery of cucumber chloroplasts activity following chilling
injury in the light. - Plant Physiol. *59* (6, Suppl.) : 5, 1977.

29598 - **GARDESTRÖM, P., ERICSON, I., LARSSON, C.** : Preparation of mitochondria from
green leaves of spinach by differential centrifugation and phase partition.
- In : COOMBS, J. (ed.) : 4th International Congress on Photosynthesis.
P. 121. UKISES, London 1977. [Chl.]

29599 - **GARLASCHI, F.M., BENEDETTI, E. de, JENNINGS, R.C., FORTI, G.** : Influence
of ΔpH and phosphorylation substrates on the slow fluorescence decline of
isolated chloroplasts. - Plant Cell Physiol. *1977* (Spec. Issue 3 - Photosynthetic Organelles. Structure and Function) : 67 - 73, 1977.

29600 - **GARLASCHI, F.M., BENEDETTI, E. de, ROSSI, M., FORTI, G.** : The reversible
and irreversible components of slow fluorescence quenching in broken chloroplasts: The role of magnesium ion, electron transport and ΔpH. - In :
COOMBS, J. (ed.) : 4th International Congress on Photosynthesis. Pp. 121 -
122. UKISES, London 1977.

29601 - **GARLICK, S., OREN, A., PADAN, E.** : Occurrence of facultative anoxygenic photosynthesis among filamentous and unicellular cyanobacteria. - J. Bacteriol.
129 : 623 - 629, 1977.

29602 - **GARNIER, J., MAROC, J.** : Photochemical activities, cytochromes contents and
FCCP-induced photooxidation of cytochrome *b*-559 in new strains of non-photo-
synthetic mutants of *Chlamydomonas reinhardtii*. - In : COOMBS, J. (ed.) :
4th International Congress on Photosynthesis. Pp. 122 - 123. UKISES, London
1977.

29603 - **GARRARD, L.A., VAN, T.K., WEST, S.H.** : Plant response to middle ultraviolet
(UV-B) radiation : carbohydrate levels and chloroplast reactions. - Soil
Crop Sci. Soc. Fla. Proc. *36* : 184 - 188, 1977.

29604 - **GARRETT, M.K.** : Control of photorespiration at the level of RUDP carboxylase
in *Lolium*. - In : COOMBS, J. (ed.) : 4th International Congress on Photosynthesis. P.123. UKISES, London 1977.

29605 - **GÄRTNER, M.** : Biologische Grundlagen zu einer Modellierung der Stoffproduktion und Energienutzung in einem Wasserpflanzenbestand von *Typha latifolia*
und *Typha angustifolia*. - In : UNGER, K. (ed.) : Biophysikalische Analyse
pflanzlicher Systeme. Pp. 269 - 282. VEB Gustav Fischer Verlag, Jena 1977.

29606 - **GARTY, H., CAPLAN, S.R.** : Light-dependent rubidium transport in intact *Halobacterium halobium* cells. - Biochim. biophys. Acta *459* : 532 - 545, 1977.

29607 - **GARTY, H., KLEMPERER, G., EISENBACH, M., CAPLAN, S.R.** : The direction of
light-induced pH changes in purple membrane suspensions. Influence of pH
and temperature. - FEBS Lett. *81* : 238 - 242, 1977.

29608 - **GARTY, H., PASTERNAK, C., EISENBACH, M., CAPLAN, S.R.** : Effects of Triton
X-100 on the purple membrane of *Halobacterium halobium*. - In :COOMBS, J. (ed.)
: 4th International Congress on Photosynthesis. P. 125. UKISES, London 1977.

29609 - **GASANOV, R.A.** : Structural and functional relationships of chlorophyll complexes in chloroplast membranes. - In : COOMBS, J. (ed.) : 4th International
Congress on Photosynthesis. Pp. 123 - 124. UKISES, London 1977.

29610 - **GASSMAN, M., CASTELFRANCO, P., DUGGAN, J., WEZELMAN, B.** : Oxidation of 5-ami-
nolevulinic acid to carbon dioxide by extracts of barley shoots. - In :
COOMBS, J. (ed.) : 4th International Congress on Photosynthesis. Pp. 124-125.
UKISES, London 1977.

29611 - GASSMAN, M. L., CASTELFRANCO, P.A. : Oxidation of 5-aminolevulinic acid to
carbon dioxide by extracts of barley seedlings. - Plant Physiol. *59* (6, Suppl.)
: 103, 1977.

29612 - GATENBY, A.A., COCKING, E.C. : Polypeptide composition of fraction 1 protein
subunits in the genus *Petunia*. - Plant Sci. Lett. *10* : 97 - 101, 1977.

29613 - GAUDILLÈRE, J.P. : Effect of periodic oscillations of artificial light emission on photosynthetic activity. - Physiol. Plant. *41* : 95 - 98, 1977.

29614 - GAUDILLERE, J.P. : Photosynthetic quantum yield among various species. - In :
COOMBS, J. (ed.) : 4th International Congress on Photosynthesis. P. 126.
UKISES, London 1977.

29615 - GAUHL, E. : Photosynthesis of intact leaves and isolated chloroplasts of ecotypes adapted to contrasting light climates. - In : COOMBS, J. (ed.) : 4th
International Congress on Photosynthesis. P. 127. UKISES, London 1977.

29616 - GAUSMAN, H.W. : Reflectance of leaf components. - Remote Sensing Environ.
6 : 1 - 9, 1977. [Chloroplast.]

29617 - GAUSMAN, H.W., ESCOBAR, D.E., KNIPLING, E.B. : Relation of *Peperomia obtusi-
folia*'s anomalous leaf reflectance to its leaf anatomy. - Photogrammetr. Eng.
remote Sensing *43* : 1183 - 1185, 1977.

29618 - GAUTHERON, D.C., GODINOT, C. : Structure and function of ATP synthase. - In :
BUVET, R., ALLEN, M.J., MASSUÉ, J.-P. (ed.) : Living Systems as Energy Converters. Pp. 89 - 102. North-Holland Publ. Co., Amsterdam - New York - Oxford
1977.

29619 - GEACINTOV, N.E., BRETON, J. : Exciton annihilation in the two photosystems
in chloroplasts at 100 °K. - Biophys. J. *17* : 1 - 15, 1977.

29620 - GEACINTOV, N.E., BRETON, J., SWENBERG, C., CAMPILLO, A.J., HYER, R.C., SHAPI-
RO, S.L. : Picosecond and microsecond pulse laser studies of exciton quen-
ching and exciton distribution in spinach chloroplasts at low temperatures.
- Biochim. biophys. Acta *461* : 306 - 312, 1977.

29621 - GEACINTOV, N.E., BRETON, J., SWENBERG, C.E., PAILLOTIN, G. : A single pulse
picosecond laser study of exciton dynamics in chloroplasts. - Photochem.
Photobiol. *26* : 629 - 638, 1977.

29622 - GEBHARDT, S.E., ELKINS, E.R., HUMPHREY, J. : Comparison of two methods for
determining the vitamin A value of Clingstone peaches. - J. agr. Food Chem.
25 : 629 - 632, 1977. [Car.]

29623 - GEHRING, H., KASEMIR, H., MOHR, H. : The capacity of chlorophyll-*a* biosynthesis in the mustard seedling cotyledons as modulated by phytochrome and circadian rhythmicity. - Planta *133* : 295 - 302, 1977.

29624 - GELVIN, S., HEIZMANN, P., HOWELL, S.H. : Identification and cloning of the
chloroplast gene coding for the large subunit of ribulose-1,5-bisphosphate
carboxylase from *Chlamydomonas reinhardi*. - Proc. nat. Acad. Sci. USA *74* :
3193 - 3197, 1977.

29625 - GELVIN, S., HOWELL, S. : Isolation of mRNA and gene for RUBP carboxylase from
Chlamydomonas reinhardi. - Plant Physiol. *59* (6, Suppl.) : 105, 1977.

29626 - GELVIN, S., HOWELL, S.H. : Identification and precipitation of polyribosomes
in *Chlamydomonas reinhardi* involved in the synthesis of the large subunit
of D-ribulose-1,5-bisphosphate carboxylase. - Plant Physiol. *59* : 471 - 477,
1977.

29627 - GEMEINHARDT, F., KNOBLOCH, K. : Phospholipid content of membrane fractions
from the purple bacteria *Rhodopseudomonas palustris* and *Rhodopseudomonas
spheroides*. - Hoppe-Seyler's Z. physiol. Chem. *358* : 1205, 1977.

29628 - GENOV, A.P. : Vplyv umov mistsezrostannya na produktyvnist' i strukturu fito-
masy baïrachnykh lisiv Starobil'shchyny. [Effect of habitat conditions on
productivity and structure of phytomass of small forests in steppe ravines
of Starobel'shchina.] - Ukr. bot. Zh. *34* : 67 - 70, 112, 1977. [In Ukr., ab :
E, R.]

29629 - GEORGIEV, D., AVRAMOVA, S., NIKOLOV, N.N. : Fotosintetichno izpolzuvane na bikarbonatniya ĭon pri intenzivno kultivirane na *Scenedesmus acutus*. [Photosynthetic use of bicarbonate ions during intensive cultivation of *Scenedesmus acutus*.]- Rasteniev"d. Nauki *14* (8) : 38 - 45, 1977. [In Bulg., ab : E, R.]

29630 - GEPSHTEIN, A., CARMELI, C. : Properties of ATPase activity in coupling factor from *Chromatium* strain D chromatophores. - Europe. J. Biochem. *74* : 463 - 469, 1977.

29631 - GEPSHTEIN, A., CARMELI, C., NELSON, N. : Purification and properties of coupling factor from *Chromatium* strain D chromatophores. - In : COOMBS, J.(ed.) : 4th International Congress on Photosynthesis. P. 127. UKISES, London 1977.

29632 - GERBER, G.E., GRAY, C.P., WILDENAUER, D., KHORANA, H.G. : Orientation of bacteriorhodopsin in *Halobacterium halobium* as studied by selective proteolysis. - Proc. nat. Acad. Sci. USA *74* : 5426 - 5430, 1977.

29633 - GERIČ, I., ZLOKOLICA, M., GERIČ, C. : Effect of planting density on the activity of ribulose-1,5-bisphosphate carboxylase in various wheat varieties. - In : COOMBS, J. (ed.) : 4th International Congress on Photosynthesis. P. 128. UKISES, London 1977.

29634 - GEROLA, P., BENEDETTI, E. de, GARLASCHI, F.M., JENNINGS, R.C., RIZZI, S., FORTI, G. : Tryptic digestion of chloroplast membranes: Sequence of effects. - In : COOMBS, J. (ed.) : 4th International Congress on Photosynthesis. Pp. 128 - 129. UKISES, London 1977.

29635 - GEROLA, P., BENEDETTI, E. de, RIZZI, S., FORTI, G., GARLASCHI, F.M.: Effects of trypsin on chloroplast membranes. - In : PACKER, L., PAPAGEORGIOU, G.S., TREBST, A. (ed.) : Bioenergetics of Membranes. Pp. 361 - 369. Elsevier/North-Holland Biomedical Press, Amsterdam - Oxford - New York 1977.

29636 - GERSHONI, J.M., OHAD, I. : Cytoplasmic protein synthesis specifically required for 70 s activity involved in chloroplast membrane biosynthesis. - In : BOGORAD, L., WEIL, J.H. (ed.): Acides Nucléique et Synthèse des Protéines chez les Végétaux. Coll. Int. C.N.R.S. No. 261. Pp. 447 - 452. Édit. C.N.R.S., Paris 1977.

29637 - GERSTER, R., TOURNIER, P. : Metabolic pathway of oxygen during photorespiration : Incorporation of ^{18}O into glycolate, glycine and serine. - In : COOMBS, J. (ed.) : 4th International Congress on Photosynthesis. Pp. 129 - 130. UKISES, London 1977.

29638 - GERWICK, B.C., WILLIAMS, G.J. III : Environmental regulation of gas exchange in *Opuntia polyacantha* and its ecological implications in the short-grass prairie ecosystem. - Plant Physiol. *59*(6, Suppl.) : 115, 1977.

29639 - GERWICK, B.C., WILLIAMS, G.J. III, URIBE, E.G. : Effects of temperature on the Hill reaction and photophosphorylation in isolated cactus chloroplasts. - Plant Physiol. *60* : 430 - 432, 1977.

29640 - GEUNS, J.M.C. : Metyrapone and the inhibition of growth, anthocyanin, carotenoid and chlorophyll biosynthesis in mung bean seedlings. - Biochem. Physiol. Pflanzen *171* : 435 - 447, 1977.

29641 - GEZELIUS, K. : Ribulose-1,5-bisphosphate carboxylase in seedlings of *Pinus silvestris* L. : Variation in enzyme content during a simulated annual growth cycle. - In : COOMBS, J. (ed.) : 4th International Congress on Photosynthesis. P. 130. UKISES, London 1977.

29642 - GIAQUINTA, R. : Possible role of pH gradient and membrane ATPase in the loading of sucrose into the sieve tubes. - Nature *267* : 369 - 370, 1977.

29643 - GIAQUINTA, R. : Phloem loading of sucrose. pH dependence and selectivity. - Plant Physiology *59* : 750 - 755, 1977.

29644 - GIAQUINTA, R. : Possible role of protons and membrane ATPase in phloem loading of sucrose. - Plant Physiol. *59* (6, Suppl.) : 126, 1977.

29645 - GIAQUINTA, R. : Sink metabolism in relation to translocation. - Plant Physiol. *59* (6, Suppl.) : 126, 1977.

29646 - GIAQUINTA, R., BEYER, E.Jr. : $^{14}C_2H_4$: Distribution of ^{14}C-labeled tissue metabolites in pea seedlings. - Plant Cell Physiol. *18* : 141 - 148, 1977.

29647 - GIAQUINTA, R., DILLEY, R.A. : Chemical modification of chloroplast membranes. - In : TREBST, A., AVRON, M. (ed.) : Photosynthesis I. (Encycl. Plant Physiol. N.S. Vol. 5.) Pp. 297 - 303. Springer-Verlag, Berlin - Heidelberg - New York 1977.

29648 - GIAQUINTA, R., GEIGER, D.R. : Mechanism of cyanide inhibition of phloem translocation. - Plant Physiol. *59* : 178 - 180, 1977. [Ps.]

29649 - GIBOR, A. : The role of the cell-apex in elongation of *Acetabularia*. - Protoplasma *93* : 101 - 107, 1977. [Ps.]

29650 - GIBSON, J.L., TABITA, F.R. : Different molecular forms of D-ribulose-1,5--bisphosphate carboxylase from *Rhodopseudomonas sphaeroides*. - J. biol. Chem. *252* : 943 - 949, 1977.

29651 - GIBSON, J.L., TABITA, F.R. : Isolation and preliminary characterization of two forms of ribulose 1,5-bisphosphate carboxylase from *Rhodopseudomonas capsulata*. - J. Bacteriol. *132* : 818 - 823, 1977.

29652 - GIBSON, P.T., SCHERTZ, K.F. : Growth analysis of a sorghum hybrid and its parents. - Crop Sci. *17* : 387 - 391, 1977.

29653 - GIDDINGS, J.M. : Chemical composition and productivity of *Scenedesmus abundans* in nitrogen limited chemostat cultures. - Limnol. Oceanogr. *22* : 911 - 918, 1977.

29654 - GIERSCH, C. : A kinetic model for translocators in the chloroplast envelope as an element of computersimulation of the dark reaction of photosynthesis. - Z. Naturforsch. *32C* : 263 - 270, 1977.

29655 - GIESKES, .W.W.C., KRAAY, G.W. : Primary production and consumption of organic matter in the Southern North Sea during the spring bloom of 1975. - Neth. J. Sea Res. *11* : 146 - 167, 1977.

29656 - GIFFORD, R.M. : Growth pattern, carbon dioxide exchange and dry weight distribution in wheat growing under differing photosynthetic environments. - Aust. J. Plant Physiol. *4* : 99 - 110, 1977.

29657 - GILDERMAN, J.I. : Das Liebigsche Prinzip in Modellen für die Pflanze. - In : UNGER, K. (ed.) : Biophysikalische Analyse Pflanzlicher Systeme. Pp. 51 - 55. VEB·Gustav Fischer Verlag, Jena 1977.

29658 - GILL'BRIKHT-IL'KOVSKA, A., RYBAK, Ya., KAYAK, Z., VENGLEN'SKA, T., DYUSOZH, K., EÏSMONT-KARABINOVA, A., KARABIN, A., SPODNEVSKA, I., GODLEVSKA-LIPOVA, A. : Reaktsiya dvukh distrofnykh ozer na izvestkovanie i udobrenie. [Reaction of two dystrophic lakes on liming and fertilization.] - Gidrobiol. Zh. *13* (6) : 39 - 45, 1977. [In R, ab: E.]

29659 - GILLBRO, T., KRIEBEL, A.N. : Emission from secondary intermediates in the photocycle of bacteriorhodopsin at 77°K. - FEBS Lett. *79* : 29 - 32, 1977.

29660 - GILLBRO, T., KRIEBEL, A.N., WILD, U.P. : On the origin of the red emission of light adapted purple membrane of *Halobacterium halobium*. - FEBS Lett. *78* : 57 - 60, 1977.

29661 - GILLER, Yu.E., STOLBOVA, A.V., MUKHAMADIEV, B.T., VAKHIDOVA, L.R., YUKHANANOVA, L.N., KVITKO, K.V. : Osobennosti fotosinteticheskogo apparata svetochuvstvitel'nykh mutantov *Chlamydomonas reinhardi* s normal'nym pigmentnym sostavom plastid. [Characteristics of the photosynthetic apparatus of light--sensitive mutants of *Chlamydomonas reinhardi* with normal pigment composition of plastids.] - In : NASYROV, Yu.S. (ed.) : Genetika Fotosinteza. Pp. 201 - 208. Donish, Dushanbe 1977. [In R.]

29662 - GILLER, Yu.E., VAKHIDOVA, L.R., ASOEVA, L.M., YUKHANANOVA, L.N., ABDULLAEVA, S.K., LIPKIND, B.I., KRASICHKOVA, G.V., YUSUPOVA, G.A. : Iskusstvennye pigment-belkovolipoidnye kompleksy - model' molekulyarnoi organizatsii i nekotorykh funktsional'nykh svoĭstv pigmentnoĭ sistemy fotosinteticheskogo apparata. [Artificial pigment-protein-lipoid complexes - a model of molecular organisation and some functional properties of pigment system of the photosynthetic apparatus.] - In : NASYROV, Yu.S. (ed.) : Genetika Fotosynteza. Pp. 168 - 181. Donish, Dushanbe 1977. [In R.]

29663 - GILMANOV, T.G. : Plant submodel in the holistic model of grassland ecosystem (with special attention to the belowground part). - Ecol. Model. *3* : 149 - 163, 1977. [Ps.]

29664 - GIMENEZ-GALLEGO, G., del VALLE-TASCON, S., RAMIREZ, J.M. : The photooxidase system of *Rhodospirillum rubrum* and its relation to the redox state of the cyclic electron-transfer system. - In : COOMBS, J. (ed.) : 4th International Congress on Photosynthesis. P. 131. UKISES, London 1977.

29665 - GIMMLER, H. : Photophosphorylation *in vivo*. - In : TREBST, A., AVRON, M. (ed.) : Photosynthesis I. (Encycl. Plant Physiol. N.S. Vol. 5.) Pp. 448 - 472. Springer-Verlag, Berlin - Heidelberg - New York 1977.

29666 - GINGRAS, G., BOUCHER, F. : Carotenoids in bacterial photoreaction center. - In : COOMBS, J. (ed.) : 4th International Congress on Photosynthesis. P. 132. UKISES, London 1977.

29667 - GIRAULT, G., GALMICHE, J.-M. : Further study of nucleotide-binding site on chloroplast coupling factor 1. - Europe.J. Biochem. *77* : 501 - 510, 1977.

29668 - GIRAULT, G., GALMICHE, J.M. : Nucleotides effect of membrane potential in spinach chloroplasts. - In : COOMBS, J. (ed.) : 4th International Congress on Photosynthesis. Pp. - 132 - 133. UKISES, London 1977.

29669 - GIROLAMI, J.-P., CAVALIÉ, G. : PEP carboxykinase et photosynthèse chez *Chloris gayana* KUNTH. - Physiol. vég. *15* : 453 - 467, 1977.

29670 - GLAZER, A.N. : Structure and molecular organization of the photosynthetic accessory pigments of cyanobacteria and red algae. - Mol. cell. Biochem. *18* : 125 - 140, 1977.

29671 - GLAZER, A.N., APELL, G.S. : A common evolutionary origin for the biliproteins of cyanobacteria, rhodophyta and cryptophyta. - FEMS Microbiol. Lett. *1* : 113 - 116, 1977.

29672 - GLAZER, A.N., HIXSON, C.S. : Subunit structure and chromophore composition of rhodophytan phycoerythrins. *Porphyridium cruentum* B-phycoerythrin and b-phycoerythrin. - J. biol. Chem. *252* : 32 - 42, 1977.

29673 - GLOOSCHENKO, V., LOTT, J.N.A. : The effects of chlordane on the green algae *Scenedesmus quadricauda* and *Chlamydomonas* sp. - Can. J. Bot. *55* : 2866 - 2872, 1977. [Ps.]

29674 - GLOOSCHENKO, W.A., BLANTON, J.O. : Short-term variability of chlorophyll *a* concentrations in Lake Ontario. - Hydrobiologia *53* : 203 - 212, 1977.

29675 - GLOSER, J. : Characteristics of CO_2 exchange in *Phragmites communis* TRIN. derived from measurements *in situ*. - Photosynthetica *11* : 139 - 147, 1977.

29676 - GLOSER, J. : Photosynthesis and respiration of some alluvial meadow grasses : responses to soil water stress, diurnal and seasonal courses. - Přírodov. Pr. Ústavu Českoslov. Akad. Věd v Brně - Acta Sci. nat. (Brno) N.S. *11* (4) : 1 - 34, 1977.

29677 - GLOVER, H. : Effects of iron deficiency on *Isochrysis galbana (Chrysophyceae)* and *Phaeodactylum tricornutum (Bacillariophyceae)*. - J. Phycol. *13* : 208 - 212, 1977. [Ps, Chl.]

29678 - GOCKE, K. : 15. Heterotrophic activity. - In : RHEINHEIMER, G. (ed.) : Microbial Ecology of a Brackish Water Environment. Pp. 198 - 222. Springer--Verlag, Berlin - Heidelberg - New York 1977. [Chl.]

29679 - GODIK, V.I., BORISOV, A.Yu. : Excitation trapping by different states of photosynthetic reaction centres. - FEBS Lett. *82* : 355 - 358, 1977.

29680 - GOEDHEER, J.C.,KLEINEN HAMMANS, J.W. : Growth and chemical composition of the blue-green alga *Anacystis nidulans* cultured at high light intensities. - Acta bot. neerl. *26* : 273 - 284, 1977.

29681 - GOEDHEER, J.C., VOS, M. : On the origin of the photoconvertible chlorophyll-protein complex $C_p668 \rightarrow C_p743$ in *Chenopodium* and *Amaranthus* species. - Acta bot. neer. *26* : 289 - 298, 1977.

29682 - GOEDHEER, J.C., VOS, M. : Experiments with *Chenopodium* chlorophyll $C_{p668} \rightarrow C_{p473}$. - In : COOMBS, J. (ed.) : 4th International Congress on Photosynthesis. Pp. 133 - 134. UKISES, London 1977.

29683 - GOGOTOV, I.N. : Efficiency of hydrogen formation and hydrogenase stability in phototrophic microorganisms. - In : COOMBS, J. (ed.) : 4th International Congress on Photosynthesis. Pp. 134 - 135. UKISES, London 1977.

*29684 - GOGOTOV, I.N., KOSYAK, A.V., KRUPENKO, A.N. : Obrazovanie vodoroda tsianobakteriyami *Anabaena variabilis* v prisutstvii sveta. [Hydrogen production by the cyanobacterium *Anabaena variabilis* in light.] - Mikrobiologiya *45* : 941 - 945, 1976. [In R, ab : E.]

29685 - GÖKÇEOĞLU, M., REHDER, H.: Nutrient turnover studies in alpine ecosystems. III. Communities of lower altitudes dominated by *Carex sempervirens* VILL. and *Carex ferruginea* SCOP. - Oecologia *28* : 317 - 331, 1977.[Biomass production.]

29686 - GOLBECK, J.H., KOK, B. : Biochemical manipulation of the membrane-bound iron-sulfur proteins in Photosystem I. - Plant Physiol. *59* (6, Suppl.) : 90, 1977.

29687 - GOLBECK, J.H., LIEN, S., SAN PIETRO, A. : Isolation and characterization of a subchloroplast particle enriched in iron-sulfur protein and P700. - Arch. Biochem. Biophys. *178* : 140 - 150, 1977.

29688 - GOLBECK, J.H., LIEN, S., SAN PIETRO, A. : Electron transport in chloroplasts. - In : TREBST, A., AVRON, M. (ed.) : Photosynthesis I. (Encycl. Plant Physiol. N.S. Vol. 5.) Pp. 94 - 116. Springer-Verlag, Berlin - Heidelberg - New York 1977.

29689 - GOL'DFEL'D, M.G., KHALILOV, R.I., KHANGULOV, S.V. : Fotoindutsirovannye paramagnitnye tsentry v fotosisteme II khloroplastov. [Photoinduced paramagnetic centres in the photosystem II of chloroplasts.] - Dokl. Akad. Nauk SSSR *237* : 1494 - 1497, 1977. [In R.]

*29690 - GOL'DFEL'D, M.G., MIKOYAN, V.D., TSAPIN, A.I. : Kremniĭmolibdat kak konformatsionno-chuvstvitel'nyĭ aktseptor ėlektronov fotosistemy 2 khloroplastov. [Silicomolybdate as a conformation-sensitive acceptor of electrons of photosystem 2 of chloroplasts.] - Dokl. Akad. Nauk SSSR *229* : 1251 - 1254, 1976. [In R.]

29691 - GOLDMAN, J.C. : Biomass production in mass cultures of marine phytoplankton at varying temperatures. - J. exp. mar. Biol. Ecol. *27* : 161 - 169, 1977.

29692 - GOLDSCHMIDT, C.R., KALISKY, O., ROSENFELD, T., OTTOLENGHI, M. : The quantum efficiency of the bacteriorhodopsin photocycle. - Biophys. J. *17* : 179 - 183, 1977.

29693 - GOLOMAZOVA, G.M., KAVERZINA, L.N. : Intensivnost' fotosinteza i fotodykhaniya sosny obyknovennoĭ pri nizkikh kontsentratsiyakh CO_2. [The rate of photosynthesis and photorespiration of *Pinus silvestris* at low concentrations of carbon dioxide.] - Fiziol. Rast. *24* : 466 - 472, 1977. [In R, ab : E.]

29694 - GONCHAROVA, N.V., EVSTIGNEEV, V.B. : Fosforilirovanie pri kislotno-osnovnom perekhode v model'noĭ sisteme, soderzhashcheĭ khlorofill. [Phosphorylation at the acid-base transition in a model system containing chlorophyll.] - Dokl. Akad. Nauk SSSR *235* : 220 - 223, 1977. [In R.]

29695 - GONCHAROVA, N.V., EVSTIGNEEV, V.B. : Fotofosforilirovanie, sensibilizirovannoe khlorofillami *a* i *b*, feofitinom i β-karotinom v model'noĭ sisteme. [Photophosphorylation, sensibilized by chlorophylls *a* and *b*, pheophytin and β-carotene in a model system.]- Biokhimiya *42* : 963 - 970, 1977. [In R, ab : E.]

29696 - GOOD, N.E. : Uncoupling of electron transport from phosphorylation in chloroplasts. - In : TREBST, A., AVRON, M. (ed.) : Photosynthesis I. (Encycl. Plant Physiol. N.S. Vol. 5.) Pp. 429 - 436. Springer-Verlag, Berlin - Heidelberg - New York 1977.

*29697 - GOOD, N.E., IZAWA, S. : Inhibition of photosynthesis. - In : HOCHSTER, R.M., KATES, M., QUASTEL, J.H. (ed.) : Metabolic Inhibitors. Vol. 4. Pp. 179 - 214. Academic Press, New York - London 1973. [Ps.]

29698 - GOODWIN, T.W. : The prenyllipids of the membranes of higher plants. - In : TEVINI, M., LICHTENTHALER, H.K. (ed.) : Lipids and Lipid Polymers in Higher Plants. Pp. 29 - 45. Springer-Verlag, Berlin - Heidelberg - New York 1977. [Chl, Car.]

29699 - GORDON, A.J., RYLE, G.J.A., POWELL, C.E. : The strategy of carbon utilization in uniculm barley. I. The chemical fate of photosynthetically assimilated ^{14}C. - J. exp. Bot. *28* : 1258 - 1269, 1977.

29700 - GORDON, K.H.J., PEOPLES, M., MURRAY, D.R. : Ageing-linked changes in photosynthetic capacity and in fraction I protein content of the first leaf of the pea plant, *Pisum sativum* L. - In : COOMBS, J. (ed.) : 4th International Congress on Photosynthesis. Pp. 135 - 136. UKISES, London 1977.

*29701 - GORELIK, A.I. : Differentsial'nye spektrofotometry v biologii. [Differential spectrofotometers in biology.] - Nauch. Dokl. vyssh. Shkoly, biol. Nauki *19* (2) : 114 - 127, 1976. [In R.]

29702 - GORSKI, P.M., CREASY, L.L. : Color development in "Golden Delicious" apples. - J.amer. Soc. hort. Sci. *102* : 73 - 75, 1977.

*29703 - GORYSHINA, T.K. : Research of biological productivity of the herbaceous cover in the oak-wood of the forest-steppe zone. - Pol. ecol. Stud. *2* : 135 - 145, 1976. [Ps.]

29704 - GORYSHINA, T.K., ZABOTINA, L.N. : Issledovanie assimilyatsionnogo apparata nekotorykh vidov rasteniĭ pod pologom elovogo lesa i na subal'piĭskom lugu v Karpatakh. [Assimilatory apparatus in some plant species under the spruce forest canopy and in a subalpine meadow in the Carpathians.] - Lesovedenie *1977* (2) : 20 - 28, 1977. [In R, ab : E.]

29705 - GOSSETT, D.R., EGLI, D.B., LEGGETT, J.E. : The influence of calcium deficiency on the translocation of photosynthetically fixed ^{14}C in soybeans. - Plant Soil *48* : 243 - 251, 1977.

29706 - GOSTIMSKIĬ, S.A. : Vliyanie mutatsiĭ na organizatsiyu i funktsionirovanie fotosinteticheskogo apparata vysshikh rasteniĭ. [Effect of mutation on the organisation and functioning of the photosynthetic apparatus of higher plants.] - In : NASYROV, Yu.S. (ed.) : Genetika Fotosinteza. Pp. 155 - 160. Donish, Dushanbe 1977. [In R.]

B29707 - GOUDRIAAN, J. : Crop Micrometeorology : A Simulation Study. - Pudoc, Wageningen 1977.

29708 - GOUGH, S.P., KANNANGARA, G.C. : Synthesis of Δ-aminolevulinate by a chloroplast stroma preparation from greening barley leaves. - Carlsberg Res. Commun. *42* : 459 - 464, 1977.

29709 - GOUSHCHINA, L.M., FEDINA, I.S., BRESKOVSKA, Ts.P., VAKLINOVA, S.G. : Connection between the intensity of photosynthesis and the activity of the carbonic anhydrases in plants of C-3 and C-4 types. - Dokl. bolg. Akad. Nauk *30* : 575 - 578, 1977.

29710 - GOUTERMAN, M., HOLTEN, D. : Electron transfer from photoexcited singlet and triplet bacteriopheophytin-II. Theoretical. - Photochem. Photobiol. *25* : 85 - 92, 1977.

29711 - GOVINDARAJALU, T. : Photosynthesis in *Cercospora*-infected muskmelon varieties. - Indian J. exp. Biol. *15* : 332 - 334, 1977.

29712 - GOVINDJEE, DESAI, T.S., TATAKE, V.G., SANE, P.V. : A new glow peak in *Rhodopseudomonas sphaeroides*. - Photochem. Photobiol. *25* : 119 - 122, 1977.

29713 - GOVINDJEE, JURSINIC, P., WRAIGHT, C. : Photosystem II reactions in isolated thylakoid membranes after a single 10 ns flash. - In : COOMBS, J. (ed.) : 4thInternational Congress on Photosynthesis. P. 136. UKISES, London 1977.

29714 - GOVINDJEE, WARDEN, J.T. : Green plant photosynthesis. Upconversion or not ? - J. amer. chem. Soc. *99* : 8088 - 8090, 1977.

29715 - GOVINDJEE, WYDRZYNSKI, T., MARKS, S.B. : The role of manganese in the oxygen evolving mechanism of photosynthesis. - In : PACKER, L., PAPAGEORGIOU, G.C., TREBST, A. (ed.): Bioenergetics of Membranes. Pp. 305 -316. Elsevier/North--Holland Biomedical Press, Amsterdam - Oxford - New York 1977.

29716 - GOWER, J.F.R., NEVILLE, R.A. : A method for the remote measurement of the vertical distribution of phytoplankton in seawater. - In : 4th Canadian Symposium on Remote Sensing. Pp. 532 - 539. Can. Aeronautics & Space Inst., Québec 1977. [Chl.]

29717 - GOWIN, T., GÓRAL, I. : Chlorophyll and pheophytin content in needles of different age of trees growing under conditions of chronic industrial pollution. - Acta Soc. Bot. Pol. 46 : 151 - 159, 1977.

29718 - GOZZER, C., ZANETTI, G., GALLIANO, M., SACCHI, G.A., MINCHIOTTI, L., CURTI, B. : Molecular heterogeneity of ferredoxin-NADP+ reductase from spinach leaves. - Biochim. biophys. Acta 485 : 278 - 290, 1977.

29719 - GRÄBER, P. : On the turnover time of the chloroplasts ATPase. - In : COOMBS, J. (ed.) : 4th International Congress on Photosynthesis. P. 137. UKISES, London 1977.

29720 - GRÄBER, P., SCHLODDER, E., WITT, H.T. : Conformational change of the chloroplast ATPase induced by a transmembrane electric field and its correlation to phosphorylation. - Biochim. biophys. Acta 461 : 426 - 440, 1977.

29721 - GRABHERR, G. : Der CO_2-Gaswechsel des immergrünen Zwergstrauches Loiseleuria procumbens (L.) DESV. in Abhängigkeit von Strahlung, Temperatur, Wasserstreß und phänologischem Zustand. - Photosynthetica 11 : 302 - 310, 1977.

29722 - GRABHERR, G., CERNUSCA, A. : Influence of radiation, wind, and temperature on the CO_2 gas exchange of the alpine dwarf shrub community Loiseleurietum cetrariosum. - Photosynthetica 11 : 22 - 28, 1977.

29723 - GRABOWSKI, J., GANTT, E. : Excitation energy migration in phycobilisomes. - Plant Physiol. 59 (6, Suppl.) : 9, 1977.

B29724 - GRACE, J. : Plant Response to Wind.- Academic Press, London - New York - San Francisco 1977. [Ps.]

29725 - GRACE, J., RUSSELL, G. : The effect of wind on grasses III. Influence of continuous drought or wind on anatomy and water relations in Festuca arundinacea SCHREB. - J. exp. Bot. 28 : 268 - 278, 1977. [Resistances.]

29726 - GRAMATIKOVA, Kh., SALCHEVA, G. : Vliyanie na niskite temperaturi v"rkhu intenzivnostta na fotosintezata na zimna pshenitsa pri razlichno mineralno khranene. [Effect of low temperature on the photosynthetic rate of winter wheat grown at various mineral nutrition .] - Fiziol. Rast. (Sofia) 3 (1) : 47 - 55, 1977. [In Bulg., ab : E.]

29727 - GRANIN, A.V., PRONINA, N.D., VESELOVSKIĬ, V.A. : Vliyanie obezvozhivaniya i peregreva na poslesvechenie fotosinteticheskogo apparata poĭkilogidrovykh i gomeogidrovykh rasteniĭ. [The effect of dehydration and superheating on delayed light emission by the photosynthetic apparatus of poikilohydrous and homeohydrous plants.] - Fiziol. Rast. 24 : 1261 - 1268, 1977. [In R, ab : E.]

29728 - GRANT, B.R., GAYLER, K.R., HOWARD, R.J., WRIGHT, S.J. : Isolated chloroplasts from siphonous algae : Are their properties due to a unique envelope structure? - In : COOMBS, J. (ed.) : 4th International Congress on Photosynthesis. Pp. 137 - 138. UKISES, London 1977.

29729 - GRANT, N.G., WALSBY, A.E. : The contribution of photosynthate to turgor pressure rise in the planktonic blue-green alga Anabaena flos-aquae. - J.exp. Bot. 28 : 409 - 415, 1977.

29730 - GREBANIER, A.E., JAGENDORF, A.T. : Irreversible uncoupling of spinach chloroplasts by sulfate and ADP. - Plant Cell Physiol. 1977 (Spec.Issue 3 - Photosynthetic Organelles. Structure and Function) : 103 - 114, 1977.

29731 - GREBANIER, A.E., JAGENDORF, A.T. : Lack of site-specificity of spinach chloroplast coupling factor 1. - Biochim. biophys. Acta 459 : 1 - 9, 1977.

29732 - **GREEN, K., WRIGHT, R.** : Field response of photosynthesis to CO_2 enhancement in ponderosa pine. - Ecology *58* : 687 - 692, 1977.

29733 - **GREENBAUM, E.** : Photosynthetic oxygen evolution under varying redox conditions : new experimental and theoretical results. - Photochem. Photobiol. *25* : 293 - 298, 1977.

29734 - **GREENBAUM, E.** : The molecular mechanisms of photosynthetic hydrogen and oxygen production. - In : MITSUI, A., MIYACHI, S., SAN PIETRO, A., TAMURA, S. (ed.) : Biological Solar Energy Conversion. Pp. 101 - 107. Academic Press, New York - San Francisco - London 1977.

29735 - **GREENBAUM, E.** : The photosynthetic unit of hydrogen evolution. - Science *196* : 879 - 880, 1977.

29736 - **GREENBAUM, E., MAUZERALL, D.C.** : The photosynthetic units of hydrogen and oxygen production. - In : COOMBS, J. (ed.) : 4th International Congress on Photosynthesis. Pp. 138 - 139. UKISES, London 1977.

29737 - **GREENE, D.W., BUKOVAC, M.J.** : Foliar penetration of naphthaleneacetic acid : enhancement by light and role of stomata. - Amer. J. Bot. *64* : 96 - 101, 1977. [Ps.]

29738 - **GREGORY, R.P.F.** : Linear and circular dichroism : the organization of the photosynthetic mambrane. - In : BARBER, J. (ed.) : Primary Processes of Photosynthesis. Pp. 465 - 492. Elsevier, Amsterdam - New York - Oxford 1977.

29739 - **GREVE, W., PARSONS, T.R.** : Photosynthesis and fish production : Hypothetical effects of climatic change and pollution. - Helgoländer wiss. Meeresunters. *30* : 666 - 672, 1977.

29740 - **GRIBOVA, Z.P.** : O faktorakh, vliyayushchikh na stabil'nost' signala ÉPR II i Mn-vodorasshcheplyayushchego kompleksa khloroplastov. [Some factors influencing the stability of ESR II signal and Mn-water-splitting complex of chloroplasts.] - Biofizika *22* : 64 - 69, 1977. [In R, ab : E.]

29741 - **GRIBOVA, Z.P., TIKHONOV, A.N.** : Issledovanie sostoyaniya margantsa v khloroplastakh metodom ÉPR. [ESR study of mananese state in chloroplasts.] - Biofizika *22* : 651 - 655, 1977. [In R, ab : E.]

29742 - **GRIFFITHS, D.E., HYAMS, R.L., PARTIS, M.D.** : Studies of energy linked reactions : a role for lipoic acid in the purple membrane of *Halobacterium halobium*. - FEBS Lett. *78* : 155 - 160, 1977.

29743 - **GRILL, R.** : Influence of chlorophyll content on phytochrome measurements in turnip cotyledons. - Planta *134* : 11 - 16, 1977.

29744 - **GRIMWADE, S.** : Carbon fixation pathways. - Nature *269* : 201, 1977.

29745 - **GRODEN, D., BECK, E.** : Characterisation of a membrane bound, ascorbate specific peroxidase from spinach chloroplasts. - In : COOMBS, J. (ed.) : 4th International Congress on Photosynthesis. P. 139. UKISES, London 1977.

29746 - **GRODZINSKI, B.** : Oxidation of glyoxylate and formate in leaf peroxisomes. - Plant Physiol. *59* (6, Suppl.) : 20, 1977.

29747 - **GRODZINSKI, B., BUTT, V.S.** : The effect of temperature on glycollate decarboxylation in leaf peroxisomes. - Planta *133* : 261 - 266, 1977.

29748 - **GROMET-ELHANAN, Z.** : Electrochemical gradients and energy coupling in photosynthetic bacteria. - Trends biochem. Sci. *2* : 274 - 277, 1977.

29749 - **GROMET-ELHANAN, Z.** : Electron transport and photophosphorylation in photosynthetic bacteria. - In : TREBST, A., AVRON, M. (ed.) : Photosynthesis I. (Encycl. Plant Physiol. N.S. Vol. 5.) Pp. 637 - 662. Springer-Verlag, Berlin - Heidelberg - New York 1977.

29750 - **GROMET-ELHANAN, Z., OREN, R.** : Coupling factor ATPase of *Rhodospirillum rubrum* chromatophores : isolation and purification of an oligomycin and DCCD sensitive ATPase. - In : PACKER, L., PAPAGEORGIOU, G.C., TREBST, A. (ed.) : Bioenergetics of Membranes. Pp. 495 - 500. Elsevier/North-Holland Biomedical Press, Amsterdam - Oxford - New York 1977.

29751 - GROMET-ELHANAN, Z., OREN, R. : Purification and characterization of an oligomycin and DCCD sensitive Ca^{2+}, Mg^{2+} ATPase from *Rhodospirillum rubrum* chromatophores. - In : COOMBS, J. (ed.) : 4th International Congress on Photosynthesis. P. 140. UKISES, London 1977.

29752 - GROMOV, B.V., MAMKAEVA, K.A., KHABIL', M. : Osobennostl ul'trastruktury fotosinteticheskogo apparata *Nostoc calcicola* BREB. pri razlichnykh intensiv-
nostyakh osveshcheniya. [Peculiarities of ultrastructure of photosynthetic
apparatus of *Nostoc calcicola* BREB. under various irradiances.] - Tr. petergof. biol. Inst. *25* (Éksperimental'naya Al'gologiya) : 27 - 33, 1977. [In R,
ab : E.]

✲29753 - GRONEBAUM-TURCK, K., MATHÉ, P. : Der Einfluß von Fluorverbindungen auf den
Chlorophyll $a + b$ - Gehalt von Pappelblättern im Freiland bei verschiedener
Belastung. - Eur. J. Forest Pathol. *6* : 57 - 59, 1976.

29754 - GROSS, E.L., DAVIS, D.J. : Divalent-cation-mediated interactions between
chloroplast pigment-proteins. - In : OLSON, J.M., HIND, G. (ed.) : Chlorophyll-Proteins, Reaction Centers, and Photosynthetic Membranes. P. 362.
Brookhaven nat. Lab., Upton 1977.

29755 - GROSS, E.L., GRENIER, J. : Cation regulation of excitation energy transfer
within Photosystem I. - Plant Physiol. *59* (6, Suppl.) : 22, 1977.

29756 - GROSS, E.L., TAKAHASHI, M. : The use of immobilized light-harvesting chlorophyll *a/b* protein to determine the stoichiometry of its self-association. -
In : COOMBS, J. (ed.) : 4th International Congress on Photosynthesis. Pp.
140 - 141. UKISES, London 1977.

29757 - GROSS, E.L., TAKAHASHI, M.-A. : Use of affinity chromatography to determine
the stoiciometry of self association of the light-harvesting chlorophyll
a/b protein. - Plant Physiol. *59* (6, Suppl.) : 24, 1977.

29758 - GROUZIS, J.P., DURAND, M., PARIS-PIREYRE, N. : Caractéristique de la fixation du calcium par les chloroplastes isolés non photosynthétisants de plantes calcicoles (Fèverole) et calcifuges (Lupin). - In : THELLIER, M.,
MONNIER, A., DEMARTY, M., DAINTY, J. (ed.) : Échanges Ioniques Transmembranaires chez les Végétaux. Pp. 569 - 575. Édit. CNRS, Paris 1977.

29759 - GRUBER, P.J., FREDERICK, S.E. : Cytochemical localization of glycolate oxidase in microbodies of *Klebsormidium*. - Planta *135* : 45 - 49, 1977.

29760 - GRUMBACH, K.H., LICHTENTHALER, H.K. : Formation of prenylquinones and photosynthetic pigments during greening of barley etioplasts in continuous and
intermittent white light. - In : COOMBS, J. (ed.) : 4th International Congress on Photosynthesis. P. 141. UKISES, London 1977.

29761 - GRUM-HELLER, S. : Die inverse Schwachlichtbewegung des *Mougeotia*-Chloroplas-
ten: Die Induzierbarkeit wird durch zeitabhängige Dunkelreaktionen des Phy-
tochroms gehemmt. - Z. Pflanzenphysiol. *81* : 212 - 225, 1977.

29762 - GRUNWALD, C., SIMS, J.L., SHEEN, S.J. : Effects of nitrogen fertilization
and stalk position on chlorophyll, carotenoids, and certain lipids of three
tobacco genotypes. - Can. J. Plant Sci. *57* : 525 - 535, 1977.

B29763 - GUDERIAN, R. : Air Pollution. Springer-Verlag, Berlin - Heidelberg - New
York 1977. [Ps, Chl.]

29764 - GUILIZZONI, P. : Photosynthesis of the submergent macrophyte *Ceratophyllum
demersum* in Lake Wingra. - Trans. Wisconsin Acad. Sci., Arts Lett. *65* :
152 - 162, 1977.

29765 - GUILLOT-SALOMON, T., TUQUET, C., HALLAIS, M.-F., SIGNOL, M. : Effets d'un
éclairement continu sur l'ultrastructure et la composition lipidique des
plastes de feuilles étiolées d'orge verdies en éclairement intermittent. -
Biol. cell. *28* : 169 - 178, 1977.

29766 - GULYAEV, B.I. : Relaksatsionnye avtokolebaniya gazoobmena list'ev. [Self-ex-
cited relaxation oscillations of gas exchange in leaves.] - Fiziol. Biokhim.
kul't. Rast. *5* : 520 - 526, 1977. [In R, ab : E.]

*B29767 - GUNNING, B.E.S., ROBARDS, A.W.(ed.) : Intercellular Communication in Plants : Studies on Plasmodesmata. Springer-Verlag, Berlin - Heidelberg - New York 1976. [Ps, photosynthates.]

29768 - GUPTA, A.D., HALES, B.J. : Action of detergents and electrophoresis on spinach chloroplasts under high alkaline conditions. - Photochem. Photobiol. *26* : 421 - 425, 1977.

29769 - GUPTA, R.K. : A note on photosynthesis in relation to water content in liverworts : *Porella platyphylla* and *Scapania undulata*. - Aust. J. Bot. *25* : 363 - 365, 1977.

29770 - GUPTA, R.K. : A study of photosynthesis and leakage of solutes in relation to the desiccation effects in bryophytes. - Can. J. Bot. *55* : 1186 - 1194, 1977.

29771 - GUPTA, R.K., SHARMA, S.K. : Phenology and growth of a desert annual grass *Digitaria ciliaris* (RETZ.) KOEL. var. *criniformis* HENR. - Indian J. Ecol. *4* : 132 - 144, 1977. [Photosynthates.]

29772 - GUPTA, V.K., ANDERSON, L.E. : Light modulation of the activity of carbon metabolism enzymes in the CAM plant *Kalanchoe*. - Plant Physiol. *59* (6, Suppl.) : 115, 1977.

29773 - GUREVITZ, M., KRATZ, H., OHAD, I. : Polypeptides of chloroplastic and cytoplastic origin required for development of Photosystem II activity, and chlorophyll-protein complexes, in *Euglena gracilis* Z chloroplast membranes. - Biochim. biophys. Acta *461* : 475 - 488, 1977.

29774 - GURINOVICH, G.P., LOSEV, A.P., SAGUN, E.I. : Énergetika assotsiirovannykh khlorofillov "*a*" i "*b*" i bakteriokhlorofilla. [Energetics of associated molecules of chlorophyll *a* and *b* and of bacteriochlorophyll.] - Zh. prikl. Spektroskop. *26* : 1028 - 1034, 1977. [In R.]

29775 - GUTERSTAM, B., ITURRIAGA, R. : *In situ* primary production of *Fucus vesiculosus*. - J. Phycol. *13* (Suppl.) : 26, 1977.

29776 - GUTKNECHT, J., BISSON, M.A., TOSTESON, F.C. : Diffusion of carbon dioxide through lipid bilayer membranes. Effects of carbonic anhydrase, bicarbonate, and unstirred layers. - J. gen. Physiol. *69* : 779 - 794, 1977.

29777 - GYURJÁN, I., NAGY, A.H., KERESZTES, Á. : Structure and macromolecular composition of defected chloroplasts in variegated leaves of *Tradescantia albiflora*. - Photosynthetica *11* : 167 - 175, 1977.

29778 - HAAG, R.W., GORHAM, P.R. : Effects of thermal effluent on standing crop and net production of *Elodea canadensis* and other submerged macrophytes in Lake Wabamun, Alberta. - J. appl. Ecol. *14* : 835 - 851, 1977.

29779 - HABERKORN, R., MICHEL-BEYERLE, M.E. : Mechanism of triplet formation in photosynthesis via hyperfine interaction. - FEBS Lett. *75* : 5 - 8, 1977.

29780 - HACHTEL, W. : Dependency of chloroplast membrane proteins on both nuclear and plastid DNA in some *Oenothera*-species. - Biochem. Physiol. Pflanz. *171* : 75 - 83, 1977.

29781 - HACHTEL, W. : Isolierte Chloroplasten synthetisieren Proteine im Reagensglas. - Umschau Wissen. Techn. *77* : 180 - 181, 1977.

29782 - HACKERT, M.L., ABAD-ZAPATERO, C., STEVENS, S.E.,Jr., FOX, J.L. : Crystallization of C-phycocyanin from the marine blue-green alga *Agmenellum quadruplicatum*. - J. mol. Biol. *111* : 365 - 369, 1977.

29783 - HAEDER, H.E., BERINGER, H., MENGEL, K. : Assimilateinlagerung in das Korn bei zwei Sommerweizensorten. - Z. Pflanzenernähr. Bodenk. *140* : 409 - 419, 1977.

29784 - HAEHNEL, W. : Electron transfer between different photosystems I in chloroplasts. - In : PACKER, L., PAPAGEORGIOU, G.C., TREBST, A. (ed.) : Bioenergetics of Membranes. Pp. 317 - 328. Elsevier/North-Holland Biomedical Press, Amsterdam - Oxford - New York 1977.

29785 - **HAEHNEL, W.** : Electron transport between plastoquinone and chlorophyll A_1 in chloroplasts. II. Reaction kinetics and the function of plastocyanin *in situ.* - Biochim. biophys. Acta *459* : 418 - 441, 1977.

29786 - **HAEHNEL, W.** : Reaction kinetics of plastocyanin in spinach chloroplasts. - In : COOMBS, J. (ed.) : 4th International Congress on Photosynthesis. P. 142. UKISES, London 1977.

29787 - **HAGAR, W.G., HILL, A., O'CONNOR, E.** : Differences in chlorophyll protein 668 concentration in the leaves of *Atriplex hortensis* during development. - Plant Physiol. *59* (6, Suppl.) : 10, 1977.

29788 - **HAGAR, W.G., HIYAMA, T.** : Light-induced absorption changes of chlorophyll--protein 668 and its converted form. - In : OLSON, J.M., HIND, G. (ed.) : Chlorophyll-Proteins, Reaction Centers, and Photosynthetic Membranes. Pp. 359 - 360. Brookhaven nat. Lab., Upton 1977.

29789 - **HAHN, S.K.** : A quantitative approach to source potentials and sink capacities among reciprocal grafts of sweet potato varieties. - Crop Sci. *17* : 559 - 562, 1977.

29790 - **HAHN, S.K.** : Sweet potato. - In : ALVIM, P.de T., KOZLOWSKI, T.T. (ed.) : Ecophysiology of Tropical Crops. Pp. 237 - 248. Academic Press, New York - San Francisco - London 1977. [Ps.]

29791 - **HAINES, B., FOSTER, R.B.** : Energy flow through litter in Panamanian forest. - J. Ecol. *65* : 147 - 155, 1977.

29792 - **HAISSIG, B.E., SCHIPPER, A.L.** : Glyceraldehyde-3-phosphate dehydrogenase (G-3-PD) activity and NADH/NAD levels during early development of *Pinus banksiana.* - Plant Physiol. *59* (6, Suppl.) : 86, 1977.

29793 - **HALL, A.E., THOMSON, W.W., ASBELL, C.W., PLATT-ALOIA, K., LEONARD, R.T.** : Stomatal response to humidity and lanthanum. - Physiol. Plant. *41* : 89 - 94, 1977. [Ps.]

29794 - **HALL, A.J.** : Assimilate source-sink relationships in *Capsicum annuum* L. I. The dynamics of growth in fruiting and deflorated plants. - Aust. J. Plant Physiol. *4* : 623 - 636, 1977.

29795 - **HALL, A.J., BRADY, C.J.** : Assimilate source-sink relationships in *Capsicum annuum* L. II. Effects of fruiting and defloration on the photosynthetic capacity and senescence of the leaves. - Aust. J. Plant Physiol. *4* : 771 - 783, 1977.

29796 - **HALL, D.O.** : Iron-sulphur proteins and energy conversion systems. - In : BUVET, R., ALLEN, M.J., MASSUÉ, J.-P. (ed.) : Living Systems as Energy Converters. Pp. 67 - 80. North-Holland Publ. Co., Amsterdam - New York - Oxford 1977.

29797 - **HALL, D.O.** : Photosynthesis - a practical energy source ? - In : CASTELLANI, A. (ed.) : Research in Photobiology. Pp. 347 - 359. Plenum Press, New York - London 1977.

29798 - **HALL, D.O., LUMSDEN, J., TEL-OR, E.** : Iron-sulfur proteins and superoxide dismutases in the evolution of photosynthetic bacteria and algae. - In : PONNAMPERUMA, C. (ed.) : Chemical Evolution of the Early Precambrian. Pp. 191 - 210. Academic Press, New York - San Francisco - London 1977.

29799 - **HALL, D.O., RAO, K.K.** : Ferredoxin. - In : TREBST, A., AVRON, M. (ed.) : Photosynthesis I. (Encycl. Plant Physiol. N.S. Vol. 5.) Pp. 206 - 216. Springer-Verlag, Berlin - Heidelberg - New York 1977.

29800 - **HALL, N.P., MERRETT, M.J.** : Ribulose 1,5-diphosphate carboxylase protein during flag leaf senescence. - In : COOMBS, J. (ed.) : 4th International Congress on Photosynthesis. P. 143. UKISES, London 1977.

29801 - **HALL, N.P., MERRETT, M.J.** : Synthesis of ribulose biphosphate carboxylase subunits during the cell-cycle of *Euglena.* - Plant Physiol. *59* (6, Suppl.) : 44, 1977.

29802 - **HALLEGRAEFF, G.M.** : A comparison of different methods used for the quantitative evaluation of biomass of freshwater phytoplankton. - Hydrobiologia *55* : 145 - 165, 1977.

29803 - **HALLEGRAEFF, G.M.** : Pigment diversity in freshwater phytoplankton 2. Summer-succession in three Dutch lakes with different trophic characteristics. - Int. Rev. gesamten Hydrobiol. *62* : 19 - 39, 1977.

29804 - **HALLENSTVET, M., BUCHECKER, R., BORCH, G., LIAAEN-JENSEN, S.** : Absolute configuration of β,γ-carotene and biosynthetic implications. - Phytochemistry *16* : 583 - 585, 1977.

29805 - **HÄLLGREN, J.-E., BRUNES, L., ÖQUIST, G.** : An assembly for measuring photosynthetic quantum yields and quanta absorption spectra of intact plants. - In : COOMBS, J. (ed.) : 4th International Congress on Photosynthesis. P. 144. UKISES, London 1977.

29806 - **HÄLLGREN, J.-E., NYMAN, B.** : Observations on trees of Scots pine (*Pinus sylvestris* L.) and lichens around a HF and SO_2 emission source. - Stud. forest. suec. *137* : 1 - 40, 1977. [Ps.]

29807 - **HALLIER, U.W., HEBER, U.W.** : Cytochrome *f* deficient plastome mutants of *Oenothera*. - Plant Cell Physiol. *1977* (Spec. Issue 3 - Photosynthetic Organelles. Structure and Function) : 257 - 273, 1977.

29808 - **HALLIWELL, B., FOYER, C., de RYCKER, J.** : Thiol compounds and regulation of the Calvin cycle. - In : COOMBS, J. (ed.) : 4th International Congress on Photosynthesis. Pp. 144 - 145. UKISES, London 1977.

29809 - **HAMEEDI, M.J.** : Changes in specific photosynthetic rate of oceanic phytoplankton from the northeast Pacific Ocean. - Helgolander wiss. Meeresunters. *30* : 62 - 75, 1977.

29810 - **HAMLIN, L., TILLBERG, J.-E.** : Effects of DCMU and tricine and HEPES buffers on FMN mediated photophosphorylation in broken chloroplasts. - In : COOMBS, J. (ed.) : 4th International Congress on Photosynthesis. Pp. 145 - 146. UKISES, London 1977.

29811 - **HAMMERTON, J.L.** : Predicting dry weights of pigeon-pea plants from non-destructive measurements. - J. agr. Sci. *88* : 449 - 454, 1977.

29812 - **HAMMOND, A.L.** : Photosynthetic solar energy : rediscovering biomass fuels. - Science *197* : 745 - 746, 1977.

29813 - **HAMPP, R., SCHMIDT, H.W.** : Regulation of membrane properties of mitochondria and plastids during chloroplast development. 1. The action of phytochrome *in situ*. - Z. Pflanzenphysiol. *82* : 68 - 77, 1977.

29814 - **HAMPP, R., ZIEGLER, I.** : Sulfate and sulfite translocation via the phosphate translocator of the inner envelope membrane of chloroplasts. - Planta *137* : 309 - 312, 1977.

29815 - **HAND, W.G.** : Photomovement. - In : SMITH, K.C. (ed.) : The Science of Photobiology. Pp. 313 - 328. Plenum Press, New York - London 1977. [Chloroplasts.]

29816 - **HANSCOM III, Z., TING, I.P.** : Physiological responses to irrigation in *Opuntia basilaris* ENGELM. and BIGEL. - Bot. Gaz. *138* : 159 - 167, 1977. [Ps.]

29817 - **HANSCOM III, Z., TING, I.P.** : Responses of succulents to plant water stress. - Plant Physiol. *59* (6, Suppl.) : 115, 1977. [Ps.]

29818 - **HANSELMANN, K.W.** : Zur Struktur von Reaktionszentren in phototrophen Bakterien : eine Standortbestimmung. - Ber. deut. bot. Ges. *90* : 459 - 476, 1977.

29819 - **HANSEN, G.K.** : Adaptation to photosynthesis and diurnal oscillation of root respiration rates for *Lolium multiflorum*. - Physiol. Plant. *39* : 275 - 279, 1977.

29820 - **HANSEN, G.K., JENSEN, C.R.** : Growth and maintenance respiration in whole plants, tops, and roots of *Lolium multiflorum*. - Physiol. Plant. *39* : 155 - 164, 1977. [Ps.]

29821 - **HANSEN, J.E.** : Productivity of *Iridaea cordata* (Rhodophyta: Gigartinaceae). - J. Phycol. *13*(Suppl.) : 27, 1977. [Ps.]

29822 - HANSEN, P. : Carbohydrate allocation. - In: LANDSBERG, J.J., CUTTING, C.V.
(ed.) : Environmental Effects on Crop Physiology. Pp. 247 - 258. Academic
Press, London - New York - San Francisco 1977.

29823 - HANSON, A.D., NELSEN, C.E., EVERSON, E.H. : Evaluation of free proline accumulation as an index of drought resistance using two contrasting barley cultivars. - Crop Sci. *17* : 720 - 726, 1977. [Growth analysis.]

29824 - HANSON, M.R., BOGORAD, L. : Effects of erythromycin on membrane-bound chloroplast ribosomes from wild-type *Chlamydomonas reinhardi* and erythromycin-re-
sistant mutants. - Biochim. biophys. Acta *479* : 279 - 289, 1977.

29825 - HANSON, R.B. : Pelagic *Sargassum* community metabolism: Carbon and nitrogen.
- J. exp. mar. Biol. Ecol. *29* : 107 - 118, 1977. [Ps.]

29826 - HANSON, W.D. : Verification of chloroplast-associated processes with diver-
gent selections from *Zea mays* L. - Crop Sci. *17* : 300 - 304, 1977.

29827 - HARAYAMA, S., IINO, T. : Ferric ion as photoreceptor of photophobotaxis in
non-pigmented *Rhodospirillum rubrum*. - Photochem. Photobiol.*25* : 571 - 578,
1977.

29828 - HARAYAMA, S., IINO, T. : Phototaxis and membrane potential in the photosynthetic bacterium *Rhodospirillum rubrum*. - J. Bacteriol. *131* : 34 - 41, 1977.
[Car.]

29829 - HARBORNE, J.B. : Variations in pigment patterns in *Pyrrhopappus* and related
taxa of the *Cichorieae*. - Phytochemistry *16* : 927 - 928, 1977. [Car.]

29830 - HARDT, H., KOK, B. : O_2 evolution activity and Mn content in mesophyll and
bundle-sheath chloroplasts from *Zea mays*. - In: COOMBS, J. (ed.) : 4th
International Congress on Photosynthesis. P. 147. UKISES, London 1977.

29831 - HARDT, H., KOK, B. : Plastocyanin as the possible site of photosynthetic electron transport inhibition by glataraldehyde. - Plant Physiol. *60* : 225 - 229,
1977.

29832 - HARDY, R.W.F., HAVELKA, U.D., QUEBEDEAUX, B. : Rate-limiting steps in crop
productivity: Photosynthesis, assimilate partitioning, senescence and nitrogen fixation. - In : COOMBS, J. (ed.) : 4th International Congress on Photosynthesis. P. 146. UKISES, London 1977.

29833 - HAREL, E., LEA, P.J., MIFLIN, B.J. : The localisation of enzymes of nitrogen assimilation in maize leaves and their activities during greening. -
Planta *134* : 195 - 200, 1977.

29834 - HAREL, E., MELLER, E. : Synthesis of ^{14}C 5-aminolevulinic acid by cell-free
preparations from greening maize leaves. - In : COOMBS, J.(ed.) : 4th International Congress on Photosynthesis. P. 148. UKISES, London 1977.

29835 - HARIRI, M., BRANGEON, J. : Light-induced adaptive responses under greenhouse
and controlled conditions in the fern *Pteris cretica* var. *ouvrardii*. I.
Structural and infrastructural features. - Physiol. Plant. *41* : 280 - 288,
1977. [Chl.]

29836 - HARNISCHFEGER, G., CODD, G.A. : Liquid nitrogen fluorescence studies of the
photosynthetic apparatus of blue-green algae. - Brit. phycol. J. *12* : 225 -
232, 1977.

29837 - HARNISCHFEGER, G., SCHOPF,R. : A fluorescence method for measuring the retention of coupling factor (CF_1) in reconstitution experiments of photophosphorylation. - Z. Naturforsch. *32C* : 392 - 397, 1977.

29838 - HARRIS, D.A., CROFTS, A.R. : Interaction between the coupling ATPase and its
inhibitor protein - studies using single turnover flashes. - In : COOMBS, J.
(ed.) : 4th International Congress on Photosynthesis. P. 143. UKISES, London
1977.

29839 - HARRIS, E.H., BOYNTON, J.E., GILLHAM, N.W., TINGLE, C.L., FOX, S.B. : Mapping
of chloroplast genes involved in chloroplast ribosome biogenesis in *Chlamydomonas reinhardtii*. - Mol. gen. Genet. *155* : 249 - 265, 1977.

29840 - HARRIS, G.C., NASITIR, M., SCHIFF, J.A. : Presence of cytochrome 552 (C-552) in *Euglena gracilis*, var. *bacillaris* and mutants. - Plant Physiol. *59* (6, Suppl.) : 9, 1977.

29841 - HARRIS, G.C., STERN, A.I. : Isolation and some properties of ribulose-1,5-bisphosphate carboxylase-oxygenase from red kidney bean primary leaves. - Plant Physiol. *60* : 697 - 702, 1977.

29842 - HARRIS, G.P., PICCININ, B.B. : Photosynthesis by natural phytoplankton populations. - Arch. Hydrobiol. *80* : 405 - 457, 1977.

29843 - HARTGE, K.H., WIEBE, H.-J. : Der Wasserzustand von Pflanze und Boden, sein Einfluss auf die Ertragsbildung und seine Bestimmung. - Gartenbauwissenschaft *42* : 71 - 76, 1977. [Ps.]

29844 - HARTIG, P.R., BERTRAND, N.J., SAUER, K. : 5-Iodoacetamidofluorescein-labeled chloroplast coupling factor 1: Conformational dynamics and labeling-site characterization. - Biochemistry *16* : 4275 - 4282, 1977.

29845 - HARTLEY, M.R. : The synthesis and origin of low molecular weight ribosomal RNAs in *Spinacea oleracea* chloroplasts. - In : COOMBS, J. (ed.) : 4th International Congress on Photosynthesis. Pp. 148 - 149. UKISES, London 1977.

29846 - HARTMANN, C.J.R., BOULAY, M.P., DROUET, A.G. : Some properties of malic enzyme extracted from the cherry. - Physiol. vég. *15* : 567 - 574, 1977.

29847 - HARTMANN, R., SICKINGER, H.-D., OESTERHELT, D. : Quantitative aspects of energy conversion in *Halobacteria*. - FEBS Lett. *82* : 1 - 6, 1977.

29848 - HARVEY, D.M. : Photosynthesis and translocation. - In : SUTCLIFFE, J.F., PATE, J.S. (ed.) : The Physiology of the Garden Pea. Pp. 315 - 348. Academic Press, London - New York - San Francisco 1977.

29849 - HASCHKE, H.-P., LÜTTGE, U. : Action of auxin on CO_2 dark fixation in *Avena* coleoptile segments as related to elongation growth. - Plant Sci. Lett. *8* : 53 - 58, 1977.

29850 - HASCHKE, H.-P., LÜTTGE, U. : Auxin action on K^+-H^+-exchange and growth, $^{14}CO_2$ -fixation and malate accumulation in *Avena* coleoptile segments. - In : MARRÈ, E., CIFERRI, O. (ed.) : Regulation of Cell Membrane Activities in Plants. Pp. 243 - 248. North-Holland Publ. Comp., Amsterdam - Oxford - New York 1977.

29851 - HASE, T., MATSUBARA, H., EVANS, M.C.W. : Amino acid sequence of *Chromatium vinosum* ferredoxin: revisions. - J. Biochem. (Tokyo) *81* : 1745 - 1749, 1977.

29852 - HASE, T., WADA, K., MATSUBARA, H. : Horsetail (*Equisetum telmateia*) ferredoxins I and II. Amino acid sequences. - J. Biochem. (Tokyo) *82* : 267 - 276, 1977.

29853 - HASE, T., WADA, K., MATSUBARA, H. : Horsetail (*Equisetum arvense*) ferredoxins I and II. Amino acid sequences and gene duplication. - J. Biochem. (Tokyo) *82* : 277 - 286, 1977.

29854 - HASE, T., WAKABAYASHI, S., MATSUBARA, H., KERSCHER, L., OESTERHELT, D., RAO, K.K., HALL, D.O. : *Halobacterium halobium* ferredoxin. A homologous protein to chloroplast-type ferredoxins. - FEBS Lett. *77* : 308 - 310, 1977.

29855 - HASEBA, T. : [Forced-convection water-vapor transfer across the boundary layer on the flat leaf model of small dimensions.] - J. agr. Meteorol. *33* : 75 - 79, 1977. [In Jap., ab : E.]

29856 - HASEBE, H., YAMAZAKI, S., ALMAZAN, A.M., TAMAURA, Y., INADA, Y. : Inhibition of chloroplast adenosine triphosphate activity by adenosine triphosphatase inhibitor from beef heart mitochondria. - Biochem. biophys. Res. Commun. *77* : 932 - 938, 1977.

29857 - HASLETT, B.G., BOULTER, D. : Comparative amino acid sequence data of plant plastocyanins. - In : COOMBS, J. (ed.) : 4th International Congress on Photosynthesis. P. 149. UKISES, London 1977.

29858 - HASLETT, B.G., GLEAVES, T., BOULTER, D. : *N*-terminal amino acid sequences of plastocyanins from various members of the *Compositae*. - Phytochemistry *16* : 363 - 365, 1977.

29859 - HASPEL-HORVATOVIČ, E., HOLŮBKOVÁ, B. : Incorporation of ^{14}C into the chloro-
         phyll molecules of mildewed barley (*Hordeum vulgare* L., *Erysiphe graminis* f.
         sp. *hordei* MARCHAL). Changes of chlorophyll a. - Phytopathol. Z. *88* : 193 -
         198, 1977.

29860 - HATCH, M.D. : C_4 pathway photosynthesis : mechanism and physiological function.
         - Trends biochem. Sci. *2* : 199 - 202, 1977.

29861 - HATCH, M.D. : Light-dark mediated activation and inactivation of NADP malate
         dehydrogenase in isolated chloroplasts from *Zea mays*. - Plant Cell Physiol.
         1977 (Spec. Issue 3 - Photosynthetic Organelles. Structure and Function) :
         311 - 314, 1977.

29862 - HATCH, M.D., MAU, S.L. : Association of NADP- and NAD-linked malic enzyme
         activities in *Zea mays* : Relation to C_4 pathway photosynthesis. - Arch. Bio-
         chem. Biophys. *179* : 361 - 369, 1977.

29863 - HATCH, M.D., MAU, S.L. : Properties of phosphoenolpyruvate carboxykinase ope-
         rative in C_4 pathway photosynthesis. - Aust. J. Plant Physiol. *4* : 207 - 216,
         1977.

29864 - HATCHER, B.G. : An apparatus for measuring photosynthesis and respiration
         of intact large marine algae and comparison of results with those from expe-
         riments with tissue segments. - Marine Biol. *43* : 381 - 385, 1977.

29865 - HATTERSLEY, P.W., WATSON, L., OSMOND, C.B. : *In situ* immunofluorescent label-
         ling of ribulose-1,5-bisphosphate carboxylase in leaves of C_3 and C_4 plants.
         - Aust. J. Plant Physiol. *4* : 523 - 539, 1977.

29866 - HAURY, J.F., BOGORAD, L. : Action spectra for phycobiliprotein synthesis in
         a chromatically adapting cyanophyte, *Fremyella diplosiphon*. - Plant Physiol.
         60 : 835 - 839, 1977.

29867 - HAUSKA, G. : Artificial acceptors and donors. - In : TREBST, A., AVRON, M.
         (ed.) : Photosynthesis I. (Encycl. Plant Physiol. N.S. Vol.5). Pp. 253 - 265.
         Springer-Verlag, Berlin - Heidelberg - New York 1977.

29868 - HAUSKA, G. : Plasto- and ubiquinone as translocators of electrons and protons
         through membranes. A facilitating role of the isoprenoid side chain. - FEBS
         Lett. *79* : 345 - 347, 1977.

29869 - HAUSKA, G. : The permeability of quinones through membranes. - In : PACKER,
         L., PAPAGEORGIOU, G.C., TREBST, A. (ed.) : Bioenergetics of Membranes. Pp.
         177 - 187. Elsevier/North-Holland Biomedical Press, Amsterdam - Oxford -
         New York 1977.

29870 - HAUSKA, G. : Vectorial redox reactions of native and artificial quinoid com-
         pounds and the topography of photosynthetic membranes. - In : COOMBS, J. (ed.)
         : 4th International Congress on Photosynthesis. Pp. 150 - 151. UKISES, Lon-
         don 1977.

29871 - HAVELANGE, A. : Ultrastructure des chloroplastes des feuilles au cours de la
         croissance végétative et de la mise à fleurs de *Sinapis alba* L. - Physiol.
         vég. *15* : 723 - 734, 1977.

29872 - HAWKRIDGE, F.M., KE, B. : An electrochemical thin-layer cell for spectrosco-
         pic studies of photosynthetic electron-transport components. - Anal. Biochem.
         78 : 76 - 85, 1977.

29873 - HAWXBY, K., TUBEA, B., OWNBY, J., BASLER, E. : Effects of various classes of
         herbicides on four species of algae. - Pestic. Biochem. Physiol. *7* : 203 -
         209, 1977. [Ps.]

29874 - HAYASHI, K.-I., KAWAKAMI, J.-I., NAKAGAHRA, M., MIYAZAKI, S. : Effects of hea-
         vy metal on leaf photosynthesis and other growth attributes in F_2 population
         of rice. - In : Annual Report 1977. Pp. 2-3. Div. Genet. nat. Inst. agr. Sci.,
         Hiratsuka, Japan 1977.

29875 - HAYDEN, D.B., HOPKINS, W.G. : A second distinct chlorophyll *a*-protein com-
         plex in maize mesophyll chloroplasts. - In : COOMBS, J. (ed.) : 4th Interna-
         tional Congress on Photosynthesis. P. 151. UKISES, London 1977.

29876 - HAYDEN, D.B., HOPKINS, W.G. : A second distinct chlorophyll a-protein complex in maize mesophyll chloroplasts. - Can. J. Bot. 55 : 2525 - 2529, 1977.

29877 - HEATH, R.L., LEECH, R.M. : The stimulation of O_2 evolution by ammonium ion in intact chloroplasts. - Plant Physiol. 59 (6, Suppl.) : 129, 1977.

29878 - HEATHCOTE, P., CLAYTON, R.K. : Reconstitued energy transfer from antenna pigment-protein to reaction centres isolated from *Rhodopseudomonas sphaeroides*. - Biochim. biophys. Acta 459 : 506 - 515, 1977.

29879 - HEATHCOTE, P., VERMEGLIO, A., CLAYTON, R.K. : The carotenoid band shift in reaction centers from *Rhodopseudomonas sphaeroides*. - Biochim. biophys. Acta 461 : 358 - 364, 1977.

29880 - HEATHCOTE, P., WILLIAMS-SMITH, D.L., EVANS, M.C.W. : Determination of the quantitative relationships between the electron transport components of Photosystem I using E.P.R. spectrometry. - In : COOMBS, J. (ed.) : 4th International Congress on Photosynthesis. P. 152. UKISES, London 1977.

29881 - HEATHERLY, L.G., RUSSEL, W.J., HINCKLEY, T.M. : Water relations and growth of soybeans in drying soil. - Crop Sci. 17: 381 - 386, 1977. [Growth analysis.]

*29882 - HEATHERSHAW, A.D. : Measurements of turbulence in the Irish Sea benthic boundary layer. - In : McCAVE, I.N. (ed.) : The Benthic Boundary Layer. Pp. 11 - 31. Plenum Publ. Corp., New York 1976. [Energy production and dissipation.]

29883 - HEBBLETHWAITE, P.D. : Irrigation and nitrogen studies in S.23 ryegrass grown for seed. 1. Growth, development, seed yield components and seed yield. - J. agr. Sci. 88 : 605 - 614, 1977.

29884 - HEBER, U.: Photosynthetic metabolite fluxes across the chloroplast envelope. - In : COOMBS, J. (ed.) : 4th International Congress on Photosynthesis. P. 153. UKISES, London 1977.

29885 - HEBER, U., EGNEUS, H. : Coupling of ATP production to consumption in photosynthesis. - In : COOMBS, J. (ed.) : 4th International Congress on Photosynthesis. P. 154. UKISES, London 1977.

29886 - HEILMAN, J.L., KANEMASU, E.T., BAGLEY, J.O., RASMUSSEN, V.P. : Evaluating soil moisture and yield of winter wheat in the Great Plains using Landsat data. - Remote Sensing Environm. 6 : 315 - 326, 1977. [Growth analysis.]

29887 - HEILMAN, J.L., KANEMASU, E.T., PAULSEN, G.M. : Estimating dry-matter accumulation in soybean. - Can. J. Bot. 55 : 2196 - 2201, 1977.

29888 - HEISE, K.-P., HARNISCHFEGER, G. : On the correlation between photosynthesis and plant lipid composition. - In : COOMBS, J. (ed.) : 4th International Congress on Photosynthesis. Pp. 154 - 155. UKISES, London 1977.

29889 - HEISE, K.-P., KRAPF, G. : Comparison of lipid biosynthesis of normal and dark kept spinach leaves in photosynthetically active light. - Z. Naturforsch. 32C : 611 - 616, 1977.

29890 - HELDER, R.J., ZANSTRA, P.E. : Changes of the pH at the upper and lower surface of bicarbonate assimilating leaves of *Potamogeton lucens* L. - Proc. koninklijke nederl. Akad. Wetensch., Ser.C 80 : 421 - 436, 1977.

29891 - HELDT, H.W., CHON, C.J., MARONDE, D., HEROLD, A., STANKOVIC, Z.S., WALKER, D.A., KRAMINER, A., KIRK, M.R., HEBER, U. : Role of orthophosphate and other factors in regulation of starch formation in leaves and isolated chloroplasts. - Plant Physiol. 59 : 1146 - 1155, 1977.

29892 - HELDT, H.W., LILLEY, R. McC., PORTIS, A., CHON, C.J. : Control of CO_2 fixation by light dependent changes of the H^+ and Mg^{++} concentrations in the stroma. - In : COOMBS, J. (ed.) : 4th International Congress on Photosynthesis. P. 151. UKISES, London 1977.

29893 - HELLENBRAND, K. : The effect of pulp mill effluent on the productivity of seaweeds. - J. Phycol. 13 (Suppl.) : 29, 1977.

29894 - HELLGREN, N.O. : Influence of light intensities on growth, photosynthesis

and slow decay luminescence. - In : COOMBS, J. (ed.) : 4th International
Congress on Photosynthesis. Pp. 155 - 156. UKISES, London 1977.

29895 - HELMS, K., WARDLAW, I.F. : Effect of temperature on symptoms of tobacco mosaic virus and movement of photosynthate in *Nicotiana glutinosa*. - Phytopathology *67* : 344 - 350, 1977.

29896 - HEMENGER, R.P., LINDENBERG, K., PEARLSTEIN, R.M. : Combined effects of intra-
molecular vibrations and electronic excited-state interactions on the absorption spectra of chlorophyll aggregates. - In : OLSON, J.M., HIND, G. (ed.) :
Chlorophyll-Proteins, Reaction Centers, and Photosynthetic Membranes. P. 370.
Brookhaven nat. Lab., Upton 1977.

29897 - HEMLEY, R., KOHLER, B.E. : Electronic structure of polyenes related to the
visual chromophore. A simple model for the observed band shapes. - Biophys.
J. *20* : 377 - 382, 1977. [Car.]

29898 - HEMPHILL, J.K., VENKETESWARAN, S. : Greening of dark grown callus cultures
derived from three chlorophyllous soybean phenotypes (*Glycine max* L. MERRILL).
- Plant Physiol. *59* (6, Suppl.) : 3, 1977.

29899 - HEMPHILL, J.K., VENKETESWARAN, S. : Growth studies of three chlorophyllous
callus phenotypes of *Glycine max*. - Amer. J. Bot. *64* : 658 - 663, 1977.

29900 - HENDERSON, R. : The purple membrane from *Halobacterium halobium*. - Annu. Rev.
Biophys. Bioeng. *6* : 87 - 109, 1977.

29901 - HENDRICH, W. : Interaction of chlorophyll *a* with some amino acids. - In :
COOMBS, J. (ed.) : 4th International Congress on Photosynthesis. Pp. 156 - 157.
UKISES, London 1977.

29902 - HENDRIX, J.E. : Translocation species in two North American members of *Cucurbitaceae*. - Plant Physiol. *59* (6, Suppl.) : 125, 1977.

29903 - HENDRY, G.A.F., STOBART, A.K. : Haem and chlorophyll formation in etiolated
and greening leaves of barley. - Phytochemistry *16* : 1545 - 1548, 1977.

29904 - HENDRY, G.A.F., STOBART, A.K. : Glycine metabolism and chlorophyll synthesis
in barley leaves. - Phytochemistry *16*: 1567 - 1570, 1977.

29905 - HENDRY, G.A.F., STOBART, A.K. : Protochlorophyllide (P650) turnover in dark-
-grown barley leaves. - Phytochemistry *16* : 1663 - 1664, 1977.

29906 - HENKIN; B.M., SAUER, K. : Magnesium ion effects on chloroplast Photosystem II
fluorescence and photochemistry. - Photochem. Photobiol. *26* : 277 - 286,
1977.

29907 - HENRIQUES, F., PARK, R. : Polypeptide composition of chlorophyll-protein complexes from romaine lettuce. - Plant Physiol. *60* : 64 - 68, 1977.

29908 - HENRY, L.E.A., HALL, D.O. : Superoxide dismutases in algae. - In : COOMBS, J.
(ed.) : 4th International Congress on Photosynthesis. Pp. 157 - 158. UKISES,
London 1977.

29909 - HENRY, L.E.A., HALL, D.O. : Superoxide dismutases in green algae: An evolutionary survey. - Plant Cell Physiol. *1977* (Spec.Issue 3 - Photosynthetic
Organelles. Structure and Function) : 377 - 382, 1977.

29910 - HENRY, Y., WIESSNER, W., LEFORT-TRAN, M. : The significance of light for the
assimilation of carbon from acetate by two members of the genus *Gonium*: Differences between *G. multicoccum* and *G. octonarium*. - In : COOMBS, J. (ed.) :
4th International Congress on Photosynthesis. P.160. UKISES, London 1977.

29911 - HERATH, H.M.W., ORMROD, D.P. : Carbon dioxide compensation values in citro-
nella and lemongrass. - Plant Physiol. *59* : 771 - 772, 1977.

29912 - HERBERT, M., MÜHLBACH, H., SCHNARRENBERGER, C. : Intracellular distribution
of isoenzymes involved in the glucose-6-P metabolism of C_3-, C_4- and CAM-
-plants and algae. - In : COOMBS, J.(ed.) : 4th International Congress on
Photosynthesis. P. 159. UKISES, London 1977.

29913 - HERBERT, R.A., TANNER, A.C. : The isolation and some characteristics of photosynthetic bacteria (*Chromatiaceae* and *Chlorobiaceae*) from antarctic marine
sediments. - J. appl. Bacteriol. *43* : 437 - 445, 1977.

*29914 - HERBLAND, A. : Utilisation par la flore hétérotrophe de la matière organique
naturelle dans l'eau de mer. - J. exp. mar. Biol. Ecol. *19* : 19 - 31, 1975.
[Ps.]

29915 - HERBLAND, A. : The prevention of radiocarbon loss in liquid scintillation
counting of solutions containing ^{14}C-$NaHCO_3$. - Int. J. appl. Rad. Isotopes
28 : 795 - 796, 1977.

*29916 - HERBLAND, A., DANDONNEAU, Y. : Excretion organique du phytoplankton et acti-
vité bactérienne hétérotrophe dans le dôme de Guinée (Océan atlantique tropical est). - Doc. Sci. Cent. Rech. Oceanogr. Abidjan *8*(2) : 1 - 18, 1977.
[Chl.]

29917 - HERBLAND, A., VOITURIEZ, B. : Production primaire, nitrate et nitrite dans
l'Atlantique tropical. I.-Distribution du nitrate et production primaire. -
Cah. ORSTOM, Sér. Océanogr. *15* (1) : 47 - 55, 1977.

29918 - HERDMAN, M., STANIER, R.Y. : The cyanelle : chloroplast or endosymbiotic
prokaryote? - FEMS Lett. *1* : 7 - 12, 1977.

29919 - HERMAN, A.W., DENMAN, K.L. : Rapid underway profiling of chlorophyll with an
in situ fluorometer mounted on a "Batfish" vehicle. - Deep-Sea Res. *24* : 385
- 397, 1977.

29920 - HERNÁNDEZ-GIL, R., BAUTISTA, D. : Crecimiento y cambios bioquimicos durante
el proceso de maduracion de la mora (*Rubus glaucus* BENTH.). [Growth and biochemical changes during the maturation process of the blackberry (*Rubus glaucus* BENTH.).] - Agron. trop. *27* : 225 - 233, 1977. [Chl; in Span., ab : E.]

29921 - HEROLD, A. : Regulation of photosynthetic carbon assimilation by orthophosphate. - In : COOMBS, J. (ed.) : 4th International Congress on Photosynthesis.
P. 158. UKISES, London 1977.

29922 - HERRERA, A., BRADBEER, J.W., KEMBLE, R.J. : The effects of 2-(4-methyl-2,6-
-dinitroanilino)-N-methyl propionamide on bean chloroplast development. - In :
BOGORAD, L., WEIL, J.H. (ed.) : Acides Nucléiques et Synthèse des Protéines
chez les Végétaux. Coll. int. C.N.R.S. No. 261. Pp. 457 - 461. Edit. C.N.R.S.,
Paris 1977.

29923 - HERRIDGE, D.F., PATE, J.S. : Utilization of net photosynthate for nitrogen
fixation and protein production in an annual legume. - Plant Physiol. *60*:
759 - 764, 1977.

29924 - HERTZBERG, S., MORTENSEN, T., BORCH, G., SIEGELMAN, H.W., LIAAEN-JENSEN, S. :
On the absolute configuration of 19'-hexanoyloxyfucoxanthin. - Phytochemistry *16* : 587 - 590, 1977.

29925 - HERVO, G. : Optical properties of thin layer: The photosynthetic membrane. -
In : COOMBS, J. (ed.) : 4th International Congress on Photosynthesis. P. 158.
UKISES, London 1977.

29926 - HESS, B., KUSCHMITZ, D. : The photochemical reaction of the 412 nm chromophore of bacteriorhodopsin. - FEBS Lett. *74* : 20 - 24, 1977.

29927 - HESS, M. : Chlorophyll *a*-Lecithin-Wechselwirkungen. - Naturwissenschaften
64 : 94 - 95, 1977.

29928 - HESSE, M., KULANDAIVELU, G., BÖGER, P. : Characterization of synchronized
cultures of *Bumilleriopsis filiformis*. Changes in cytochrome-*f* photooxidation and fluorescence induction kinetics. - Arch. Microbiol. *112* : 141 -
145, 1977.

29929 - HEYN, M.P., CHERRY, R.J., MÜLLER, U. : Transient and linear dichroism studies
on bacteriorhodopsin: Determination of the orientation of the 568 nm all-
-*trans* retinal chromophore. - J. mol. Biol. *117* : 607 - 620, 1977.

29930 - HEYWOOD, P. : Chloroplast structure in the chloromonadophycean alga *Vacuola-
ria virescens*. - J. Phycol. *13* : 68 - 72, 1977.

29931 - HICKLENTON, P.R., OECHEL, W.C. : The influence of light intensity and temperature on the field carbon dioxide exchange of *Dicranum fuscenscens* in the
subarctic. - Arctic Alp. Res. *9* : 407 - 419, 1977.

29932 - **HIEKE, B.** : Die Bedeutung ausgewählter Elemente des Mineralstoffhaushaltes
für Struktur und Funktion der Chloroplasten. - Biol. Rundschau *15*: 222 -
239, 1977.

29933 - **HIEKE, B., SCHOTTE, J.** : Charakterisierung der PS I-Aktivität isolierter
Chloroplasten aus Primärblättern von *Triticum aestivum* L. nach Temperatur-
behandlung mit Hilfe der Entfärbung von Diphenylcarbazon (DPCN). - Biochem.
Physiol. Pflanzen *171* : 353 - 357, 1977.

29934 - **HIGHFIELD, P.E.** : Synthesis of the small subunit of fraction I protein. -
In : COOMBS, J. (ed.) : 4th International Congress on Photosynthesis. P.160.
UKISES, London 1977.

29935 - **HILDEBRAND, E.** : What does *Halobacterium* tell us about photoreception? - Biophys. Struct. Mech. *3* : 69 - 77, 1977.

29936 - **HILL, R.** : Some tentative questions about radiant energy and its conversion
in chloroplasts. - Plant Cell Physiol. *1977* (Spec. Issue 3 - Photosynthetic
Organelles. Structure and Function) : 47 - 54,1977.

29937 - **HILLBRICHT-ILKOWSKA, A., RYBAK, J.I., KAJAK, Z., DUSOGE, K., EJSMONT-KARABIN,
J., SPODNIEWSKA, I., WĘGLEŃSKA, T., GODLEWSKA-LIPOWA, W.A.** : Effect of liming
on a humic lake. - Ekol. pol. *25* : 379 - 420, 1977. [Ps.]

29938 - **HILLER, R.G.** : Chloroplast lamellar proteins with special reference to chlorophyll-protein complexes. - Proc. aust. biochem. Soc. *10*: Q10, 1977.

29939 - **HILLER, R.G., PILGER, T.B.G., GENGE, S.** : Effect of lincomycin on the chlorophyll protein complex I content and Photosystem I activity of greening
leaves. - Biochim. biophys. Acta *460* : 431 - 444, 1977.

29940 - **HILLMER, P., GEST, H.** : H_2 metabolism in the photosynthetic bacterium *Rhodopseudomonas. capsulata* : H_2 production by growing cultures. - J. Bacteriol.
129 : 724 - 731, 1977.

29941 - **HILLMER, P., GEST, H.** : H_2 metabolism in the photosynthetic bacterium *Rhodopseudomonas capsulata* : Production and utilization of H_2 by resting cells. -
J. Bacteriol. *129* : 732 - 739, 1977.

29942 - **HIND, G., MILLS, J., SLOVACEK, R.** : Cyclic electron transport in photosynthesis. - In : COOMBS, J. (ed.) : 4th International Congress on Photosynthesis.
P. 161. UKISES, London 1977.

29943 - **HINDMAN, J.C., KUGEL, R., SVIRMICKAS, A., KATZ, J.J.** : Chlorophyll lasers :
Stimulated light emission by chlorophylls and Mg-free chlorophyll derivatives. - Proc. nat. Acad. Sci. USA *74* : 5-9, 1977.

*29944 - **HINO, A., AMANO, S., SAWAMURA, Y., SASAKI, S., KURAOKA, T.** : [Studies on the
photosynthetic activity in several kinds of fruit trees. II. Seasonal changes in the rate of photosynthesis.] - J. jap. Soc. hort. Sci. *43* : 209 - 214,
1974. [In Jap., ab : E.]

29945 - **HIPKIN, C.R., SYRETT, P.J.** : Some effects of nitrogen-starvation on nitrogen
and carbohydrate metabolism in *Ankistrodesmus braunii*. - Planta *133* : 209 -
214, 1977.

29946 - **HIPKINS, M.F.** : The effect of chlorophyll fluorescence yield on delayed fluorescence from pea chloroplasts. - In : COOMBS, J. (ed.) : 4th International
Congress on Photosynthesis. Pp. 161 - 162. UKISES, London 1977.

*29947 - **HIRAYAMA, O., HARA, N., TANAKA, A., OKA, S.** : [Pigments of *Rhodopseudomonas
spheroides* S and effects of inorganic salts on the pigment formations.] -
J. agr. chem. Soc. Japan *50* : 41 - 47, 1976. [In Jap., ab : E.]

29948 - **HIRAYAMA, O., KABATA, K.** : Lipid extractions and reconstitution of lyophili-
zed chloroplasts. - Agr. biol. Chem. (Tokyo) *41* : 2423 - 2426, 1977.

29949 - **HIYAMA, T., McSWAIN, B.D., ARNON, D.I.** : Correlation of redox levels of component electron carriers with total electron flux in an electron-transport
system. *P*-700 and the photoreduction of $NADP^+$ in chloroplast fragments. -
Biochim. biophys. Acta *460* : 65 - 75, 1977.

29950 - HIYAMA, T., McSWAIN, B.D., ARNON, D.I. : Evidence for two types of P-700 in membrane fragments from a blue-green alga. - Biochim. biophys. Acta *460* : 76 - 84, 1977.

29951 - HO, L.C. : Effects of CO_2 enrichment on the rates of photosynthesis and translocation of tomato leaves. - Ann. apl. Biol. *87* : 191 - 200, 1977.

29952 - HO, L.C., NICHOLS, R. : Translocation of ^{14}C-sucrose in relation to changes in carbohydrate content in rose corollas cut at different stages of development. - Ann. Bot. *41* : 227 - 242, 1977.

29953 - HO, L.C., REES, A.R. : The contribution of current photosynthesis to growth and development in the tulip during flowering. - New Phytol. *78* : 65 - 69, 1977.

29954 - HO, L.C., SHAW, A.F. : Carbon economy and translocation of ^{14}C in leaflets of the seventh leaf of tomato during leaf expansion. - Ann. Bot. *41* : 833 - 848, 1977. [Ps.]

29955 - HOARAU, J., LECLERC, J.-C. : Variations spectroscopiques induites par illumination et par action chimique observées à -196 °C dans la chlorophylle *a* de *Porphyridium*. - Physiol. Plant. *39* : 13 - 20, 1977.

29956 - HOARAU, J., REMY, R., LECLERC, J.-C. : Hétérogénéité des variations spectrales photoinduites vers 700 nm observées sur les membranes chlorophylliennes et les complexes chlorophylle-protéines isolés de divers organismes photosynthétiques. - Biochim. biophys. Acta *462* : 659 - 670, 1977.

29957 - HOBSON, G.E., DAVIES, J.N. : Mitochondrial activity and carbohydrate levels in tulip bulbs in relation to cold treatment. - J. exp. Bot. *28* : 559 - 568, 1977.

29958 - HOCH, G.E. : *P-700*. - In : TREBST, A., AVRON, M. (ed.) : Photosynthesis I. (Encycl. Plant Physiol. N.S. Vol.5.). Pp. 136 - 146. Springer-Verlag, Berlin - Heidelberg - New York 1977.

29959 - HOCHMAN, A., BEN-HAYYIM, G., CARMELI, C. : Light-induced electron transport pathways in membrane preparations from *Rhodopseudomonas capsulata*. - Arch. Biochem. Biophys. *184* : 416 - 422, 1977.

29960 - HOCHMAN, A., CARMELI, C. : Reconstitution of photosynthetic electron transport and photophosphorylation in cytochrome-c_2-deficient membrane preparation of *Rhodopseudomonas capsulata*. - Arch. Biochem. Biophys. *179* : 349 - 359, 1977.

29961 - HOCHMAN, A., FRIDBERG, I., CARMELI, C. : The possible function of bacterial cell wall as a barrier for nucleotide-phosphates permeability. - In : COOMBS, J. (ed.) : 4th International Congress on Photosynthesis. P. 163. UKISES, London 1977. [Photophosphorylation.]

29962 - HOCHMAN, Y., CARMELI, C., LANIR, A., WERBER, M.M. : The effect of cobalt (III) nucleotides complexes on the kinetic properties of adenosine triphosphatase (CF_1) from chloroplasts. - In : COOMBS, J. (ed.) : 4th International Congress on Photosynthesis. P.162. UKISES, London 1977.

29963 - HODDINOTT, J. : Rates of transpiration and photosynthesis in *Mimosa pudica* L. - New Phytol. *79* : 269 - 272, 1977.

29964 - HODDINOTT, J., GORHAM, P.R. : Translocation and photosynthesis in *Phaseolus* and *Beta* with rapidly changing soil water potentials. - Plant Physiol. *59* (6, Suppl.) : 125, 1977.

29965 - HODGES, C.F. : Influence of irrigation on survival of *Poa pratensis* infected by *Ustilago striiformis* and *Urocystis agropyri*. - Can. J. Bot. *55* : 216 - 218, 1977. [Production.]

29966 - HODGES, C.F., ROBINSON, P.W. : Sugar and amino acid content of *Poa pratensis* infected with *Ustilago striiformis* and *Urocystis agropyri*. - Physiol. Plant. *41* : 25 - 28, 1977. [Photosynthates.]

29967 - HODGES, T., KANEMASU, E.T. : Modeling daily dry matter production of winter wheat. - Agron. J. *69* : 974 - 978, 1977.

29968 - HOFÄCKER, W. : Untersuchungen zur Stoffproduktion der Rebe unter dem Einfluss
wechselnder Bodenwasserversorgung. - Vitis *16* : 162 - 173, 1977.

29969 - HOFF, A.J., GAST, P. : Emissive light induced electron polarisation in reaction centers of *Rhodopseudomonas sphaeroides* wild type under reducing conditions. - In : COOMBS, J. (ed.) : 4th International Congress on Photosynthesis. Pp. 163 - 164. UKISES, London 1977.

29970 - HOFF, A.J., GAST, P., ROMIJN, J.C. : Time-resolved ESR and chemically induced
dynamic electron polarisation of the primary reaction in a reaction center
particle of *Rhodopseudomonas sphaeroides* wild type at low temperature. - FEBS
Lett. *73* : 185 - 190, 1977.

29971 - HOFF, A.J., GOVINDJEE, ROMIJN, J.C. : Electron spin resonance in zero magnetic field of triplet states of chloroplasts and subchloroplast particles. -
FEBS Lett. *73* : 191 - 196, 1977.

29972 - HOFF, A.J., RADEMAKER, H. : Light-induced magnetic polarization in photosynthesis. - In : MUUS, L.T., ATKINS, P.W., McLAUCHLAN, K.A., PEDERSEN, J.B.
(ed.) : Chemically Induced Magnetic Polarization. Pp. 399 - 404. D.Reidel
Publ. Comp., Dordrecht 1977.

29973 - HOFF, A.J., RADEMAKER, H., van GRONDELLE, R., DUYSENS, L.N.M. : On the magnetic field dependence of the yield of the triplet state in reaction centers
of photosynthetic bacteria. - Biochim. biophys. Acta *460* : 547 - 554, 1977.

29974 - HOFFMANN, D., THAUER, R., TREBST, A. : Photosynthetic hydrogen evolution by
spinach chloroplasts coupled to a *Clostridium* hydrogenase. - Z. Naturforsch.
32 C : 257 - 262, 1977.

29975 - HOFFMANN, F. : Untersuchungen zur Ertragsbildung und zu einigen Bedingungen
für die Entstehung hoher Erträge bei Zuckerrüben. - Arch. Acker- Pflanzenbau
Bodenk. *21* : 157 - 168, 1977. [Ps, growth analysis.]

29976 - HOFFMANN, P. : Energetische Aspekte der pflanzlichen Stoffproduktion. - Ta-
gungsber.Akad.Landwirtschaftswiss.DDR Berlin *158* (Züchtung und Züchtungsforschung bei Getreide) : 147 - 162, 1977.

29977 - HOFFMANN, P., KRAUSE, C. : Regulative Aspekte der Substanzverteilung auf die
einzelnen Organe in Keimpflanzen von *Triticum aestivum* L. im Verlaufe der
Entwicklung unter besonderer Berücksichtigung von Energiebilanzen. - In :
UNGER, K. (ed.) : Biophysikalische Analyse Pflanzlicher Systeme. Pp. 165 -
174. Gustav Fischer Verlag, Jena 1977.

29978 - HOFFMANN, P., LEUPOLD, D., HIEKE, B., VOIGT, B. : Laserspectroskopic characterization of the absorption behavior of chlorophyll *in vitro* and *in vivo*. -
In : COOMBS, J. (ed.) : 4th International Congress on Photosynthesis. Pp.
164 - 165. UKISES, London 1977.

29979 - HOFFMANN, P., PLESCHER, A., MEINL, G. : Pigment- und N-stoffwechselphysiolo-
gische Grundlagen im Verlauf der Lagerung von Kopfkohl (*Brassica oleracea* L.
var. *capitata*) unter besonderer Berücksichtigung resistenzphysiologischer
Aspekte. - Arch. Phytopathol. Pflanzenschutz *13* : 61 - 78, 1977.

29980 - HOFMANN, M.E., BACHOFEN, R. : Binding and reaction studies with adenine nu-
cleotides on purified coupling factor from *Rhodospirillum rubrum*. - In :
COOMBS, J. (ed.) : 4th International Congress on Photosynthesis. P. 165. UKI-
SES, London 1977.

29981 - HOFMANN, M.E., BACHOFEN, R. : Binding and reaction studies with adenine nucleotides on purified coupling factor from *Rhodospirillum rubrum*. - J. Bioenerg. Biomembranes *9* : 349 - 361, 1977

29982 - HÖFNER, W. : Wirkung von Ancymidol auf Längenwachstum und Chlorophyllgehalt
von *Helianthus annuus* L. in Wasserkultur. - Z. Pflanzenernähr. Bodenkunde
140 : 223 - 228, 1977.

29983 - HÖFNER, W., ORLOVIUS, K. : Einfluß der N-Düngung auf den ^{14}C-Einbau in die
Komponenten der äthanollöslichen Fraktion von Sommerweizen verschiedener
Entwicklungsstadien. - Z. Pflanzenernähr. Bodenkunde *140* : 491 - 504, 1977.

29984 - HOFSTRA, G., HESKETH, J.D., MYHRE, D.L. : A plastochron model for soybean
leaf and stem growth. - Can. J. Plant Sci. *57* : 167 - 175, 1977.

29985 - **HOFSTRA, J.J., STIENSTRA, A.W.** : Growth and photosynthesis of closely related C_3 and C_4 grasses, as influenced by light intensity and water supply. - Acta bot. neerl. *26* : 63 - 72, 1977.

29986 - **HOGETSU, D., MIYACHI, S.** : Effects of CO_2 concentration during growth on subsequent photosynthetic CO_2 fixation in *Chlorella*. - Plant Cell Physiol. *18* : 347 - 352, 1977.

29987 - **HÖHLER, T., SCHAUB, H.** : Influence of oxygen on dry matter production and daily changes of CO_2 uptake. 3. Investigations of the carbon metabolism of the leaves of the C_4-plant *Amaranthus paniculatus* grown in 4 % oxygen as compared with normal air. - In : COOMBS, J. (ed.) : 4th International Congress on Photosynthesis. P. 166. UKISES, London 1977.

29988 - **HØJERSLEV, N., JERLOV, N.G., KULLENBERG, G.** : Colour of the ocean as an indicator in photosynthetic studies. - J. Conseil int. Explor. Mer *37* : 313 - 316, 1977.

29989 - **HOLADAY, S., HALLER, W.T., BOWES, G.** : Variation in the CO_2 compensation point of aquatic plants. - Plant Physiol. *59* (6, Suppl.) : 65, 1977.

29990 - **HOLDSWORTH, E.S., BRUCK, K.** : Enzymes concerned with β-carboxylation in marine phytoplankter. Purification and properties of phosphoenolpyruvate carboxykinase. - Arch. Biochem. Biophys. *182* : 87 - 94, 1977.

29991 - **HOLDSWORTH, E.S., JUZU, H.A.** : A manganese-copper-pigment-protein complex isolated from the Photosystem II of *Phaeodactylum tricornutum*. - In : COOMBS, J. (ed.) : 4th International Congress on Photosynthesis. P. 167. UKISES, London 1977.

29992 - **HOLE, C.C.** : Productivity of pea pods; efficiency of assimilate use. - In : COOMBS, J. (ed.) : 4th International Congress on Photosynthesis. P. 168. UKISES, London 1977.

29993 - **HOLE, C.C.** : The effect of a reduction in carbon dioxide concentration on the loss of carbon dioxide from pea fruits. - Ann. Bot. *41* : 1367 - 1370, 1977.

29994 - **HOLMES, M.G., SMITH, H.** : The function of phytochrome in the natural environment - II. The influence of vegetation canopies on the spectral energy distribution of natural daylight. - Photochem. Photobiol. *25* : 539 - 545, 1977. [Chl.]

29995 - **HOLMES, N.G., CROFTS, A.R.** : The carotenoid shift in *Rhodopseudomonas sphaeroides*. The flash induced change. - Biochim. biophys. Acta *459* : 492 - 505, 1977.

29996 - **HOLMES, N.G., CROFTS, A.R.** : The carotenoid shift in *Rhodopseudomonas sphaeroides*. Change induced under continuous illumination. - Biochim. biophys. Acta *461* : 141 - 150, 1977.

29997 - **HOLMES, N.G., VAN GRONDELLE, R., DUYSENS, L.N.M.** : Flash-induced fluorescence yield changes due to reaction centre triplet states in carotenoid containing photosynthetic bacteria. - In : COOMBS, J. (ed.) : 4th International Congress on Photosynthesis. P. 169. UKISES, London 1977.

29998 - **HOLMGREN, A., BUCHANAN, B.B., WOLOSIUK, R.A.** : Photosynthetic regulatory protein from rabbit liver is identical with thioredoxin. - FEBS Lett. *82* : 351 - 354, 1977.

29999 - **HOLOWKA, D.A., HAMMES, G.G.** : Chemical modification and fluorescence studies of chloroplast coupling factor. - Biochemistry *16* : 5538 - 5545, 1977.

30000 - **HOLZAPFEL, C.** : Simulationsmodell des Elektronentransportes der photosynthetischen Primärreaktionen in der Thylakoidmembran grüner Pflanzen. - Ber. Kernforschungsanlage Jülich *JÜL 1390* : 1 - 66, 1977.

30001 - **HOMANN, P.H.** : Ion control of the interaction of N-methylphenazinium cations with illuminated thylakoids of isolated chloroplasts. - In : COOMBS, J. (ed.) : 4th International Congress on Photosynthesis. P. 170. UKISES, London 1977.

30002 - HOMANN,P.H. : The light dependent uptake of N-methylphenazinium cations by the thylakoids of isolated chloroplasts. - Biochim. biophys. Acta *460* : 1 - 16, 1977.

30003 - HONSELL, E., AVANZINI, A., GHIRARDELLI, L.A. : Preliminary notes on ultrastructural aspects of chloroplast multiple division in *Nitophyllum punctatum (Delesseriaceae, Rhodophyta)*. - Caryologia *30* : 490 - 491, 1977.

30004 - HOPE, A.B., CHOW, W.S. : The rates of onset and decay of photophosphorylation. - In : COOMBS, J. (ed.) : 4th International Congress on Photosynthesis. P. 171. UKISES, London 1977.

30005 - HOPFIELD, J.J. : Photo-induced charge transfer. A critical test of the mechanism and range of biological electron transfer processes. - Biophys. J. *18*: 311 - 321, 1977.

30006 - HOPKINS, W.G., HAYDEN, D.B. : Photochemistry and membrane composition of chloroplasts from a light intensity dependent chlorophyll b-deficient mutant of maize. - In : COOMBS, J. (ed.) : 4th International Congress on Photosynthesis. P. 171. UKISES, London 1977.

30007 - HOPKINS, W.G., WALDEN, D.B. : Temperature sensitivity of virescent mutants of maize. - J. Heredity *68* : 283 - 286, 1977. [Chl.]

*30008 - HOPPE, J.H., HEITEFUSS, R. : Untersuchungen zur Regulation des Kohlenhydratstoffwechsels in Weizenpflanzen nach Infektion mit *Puccinia graminis tritici*. - Phytopathol. Z. *86* : 37 - 55, 1976. [ATP.]

30009 - HORÁNSZKY, A., NAGY, A.H. : Study of assimilation types in species of a sand steppe community. - Acta bot. Acad. Sci. hung. *23* : 91 - 95, 1977.

30010 - HORI, Y., SHISHIDO, Y. : Studies on translocation and distribution of photosynthetic assimilates in tomato plants. I. Effects of feeding time and night temperature on the translocation and distribution of ^{14}C-assimilates. - Tohoku J. agr. Res. *28* : 26 - 40, 1977.

30011 - HORIE, T. : Simulation of sunflower growth. I. Formulation and parametrization of dry matter production, leaf photosynthesis, respiration and partitioning of photosynthates. - Bull. nat. Inst. agr. Sci. Ser. A (Tokyo) *24* : 45 - 70, 1977.

30012 - HORTON, P., CROZE, E. : Cytochrome b-559, a probe of Photosystem II oxidizing ability. - In : COOMBS, J. (ed.) : 4th International Congress on Photosynthesis. P. 172. UKISES, London 1977.

30013 - HORTON, P., CROZE, E. : Quenching of chlorophyll fluorescence in chloroplast photosystem II particles by magnesium ions. - FEBS Lett. *81* : 259 - 263, 1977.

30014 - HORTON, P., CROZE, E. : The relationship between the activity of chloroplast Photosystem II and the midpoint oxidation-reduction potential of cytochrome b-559. - Biochim. biophys. Acta *462* : 86 - 101, 1977.

30015 - HORTON, R.F. : Leaf senescence in *Maianthemum canadense* : the effect of cytokinins and gibberellin. - Can. J. Bot. *55* : 2272 - 2274, 1977. [Chl.]

30016 - HORVÁTH, G., DROPPA, M., MUSTÁRDY, L.A., FALUDI-DÁNIEL, A. : Isolation and characterization of intact chloroplasts from mesophyll protoplasts and bundle sheath cell of maize leaves. - In : COOMBS, J. (ed.) : 4th International Congress on Photosynthesis. P. 173. UKISES, London 1977.

30017 - HORVÁTH, M.M., FAZEKAS, M. : Vadalma fajok pigmentvizsgálata. [Pigment examination in crab-apple species.] - Bot. Közlem. *64* : 255 - 257, 1977. [In Hung., ab : E.]

30018 - HOSKINS, L.C., ALEXANDER, V. : Determination of carotenoid concentrations in marine phytoplankton by resonance Raman spectroscopy. - Anal. Chem. *49* : 695 - 697, 1977.

30019 - HOU, L.-Y., HILL, A.C., SOLEIMANI, A. : Influence of CO_2 on the effects of SO_2 and NO_2 on alfalfa. - Environm. Pollut. *12* : 7 - 16, 1977. [Ps.]

30020 - HOURSIANGOU-NEUBRUN, D., DUBACQ, J.P., PUISEUX-DAO, S. : Heterogeneity of the plastid population and chloroplast differentiation in *Acetabularia mediterranea*. - In : WOODCOCK, C.L.F. (ed.) : Progress in *Acetabularia* Research. Pp. 175 - 194. Academic Press, New York - San Francisco - London 1977.

30021 - HOUSLEY, T.L., FISHER, D.B. : Estimation of osmotic gradients in soybean sieve tubes by quantitative autoradiography.Qualified support for Münch hypothesis. - Plant Physiol. *59* : 701 - 706, 1977. [Photosynthate translocation.]

30022 - HOUSLEY, T.L., PETERSON, D.M., SCHRADER, L.E. : Long distance translocation of sucrose, serine, leucine, lysine, and CO_2 assimilates. 1. Soybean. - Plant Physiol. *59* : 217 - 220, 1977.

30023 - HOWARD, R.J., GRANT, B.R., FOCK, H. : Storage and structural products formed during photosynthesis in the siphonous alga *Caulerpa simpliciuscula (Chlorophyceae)*. - J. Phycol. *13* : 340 - 345, 1977.

30024 - HOWELL, S., HEIZMANN, P., GELVIN, S. : Properties of the mRNA and localization of the gene coding for the large subunit of ribulose bisphosphate carboxylase in *Chlamydomonas reinhardi*. - In : BOGORAD, L., WEIL, J.H. (ed.) : Acides Nucléiques et Synthèse des Protéines chez les Végétaux. Coll. Int. C.N.R.S. No. 261. Pp. 313 - 318. Edit. C.N.R.S., Paris 1977.

30025 - HOWELL, S.H., HEIZMANN, P., GELVIN, S., WALKER, L.L. : Identification and properties of the messenger RNA activity in *Chlamydomonas reinhardi* coding for the large subunit of D-ribulose-1,5-bisphosphate carboxylase. - Plant Physiol. *59* : 464 - 470, 1977.

30026 - HOXMARK, R.C., NORDBY, O. : A warning against using chloramphenicol in the light. - Plant Sci. Lett. *8* : 113 - 118, 1977. [Ps.]

30027 - HØYER-HANSEN, G. : The formation of thylakoid membrane proteins in wild-type barley and selected mutants. - In : COOMBS, J. (ed.) : 4th International Congress on Photosynthesis. Pp. 173 - 174. UKISES, London 1977.

30028 - HØYER-HANSEN, G., SIMPSON, D.J. : Changes in the polypeptide composition of internal membranes of barley plastids during greening.- Carlsberg Res. Commun. *42* : 379 - 389, 1977.

30029 - HOZUMI, K. : Ecological and mathematical considerations of self-thinning in even-aged pure stands. I. Mean plant weight-density trajectory during the course of self-thinning. - Bot. Mag. (Tokyo) *90* : 165 - 179, 1977.

30030 - HOZYO, Y. : The influences of source and sink on plant production of *Ipomoea* grafts. -   Jap. agr. Res. quart. *11* (2) : 77 - 83, 1977. [Photosynthates.]

30031 - HRUŠKA, L. : Vliv stavby rostliny bramboru na tvorbu a distribuci biomasy. [Influence of potato plant structure on formation and distribution of biomass.] - Rostl. Výroba (Praha) *23* : 1259 - 1266, 1977. [In Czech, ab : E, G, R.]

30032 - HSIAO, S.I.C., FOY, M.G., KITTLE, D.W. : Standing stock, community structure, species composition, distribution, and primary production of natural populations of phytoplankton in the southern Beaufort Sea. - Can. J. Bot. *55* : 685 - 694, 1977.

30033 - HUBAC, C., GUERRIER, D., FERRAN, J. : Résultats préliminaires sur le métabolisme de la proline en relation avec la résistance à la sécheresse. - Compt. rend. Acad. Sci. Paris, Sér. D *284* : 1397 - 1400, 1977. [Photorespiration.]

30034 - HUBER, S.C., EDWARDS, G.E. : Inhibition of NADP reduction and CO_2 fixation by mesophyll chloroplasts of *Pisum sativum* and *Hordeum vulgare* by chloramphenicol. - Plant Sci. Lett. *9* : 37 - 43, 1977.

30035 - HUBER, S.C., EDWARDS, G.E. : The importance of reducing conditions for the inhibitory action of DBMIB[+], antimycin A and EDAC[+] on cyclic photophosphorylation. - FEBS Lett. *79* : 207 - 211, 1977.

30036 - HUBER, S.C., EDWARDS, G.E. : Transport in C_4 mesophyll chloroplasts. Characterization of the pyruvate carrier. - Biochim. biophys. Acta *462* : 583 - 602, 1977.

30037 - HUBER, S.C., EDWARDS, G.E. : Transport in C_4 mesophyll chloroplasts. Evidence for an exchange of inorganic phosphate and phosphoenolpyruvate. - Biochim. biophys. Acta *462* : 603 - 612, 1977.

30038 - HUBER, S.C., HALL, T.C., EDWARDS, G.E. : Light dependent incorporation of $^{14}CO_2$ into protein by mesophyll protoplasts and chloroplasts isolated from *Pisum sativum*. - Z. Pflanzenphysiol. *85* : 153 - 163, 1977.

30039 - HUCHZERMEYER, B., STROTMANN, H. : Acid/base-induced exchange of adenine nucle- otides on chloroplasts coupling factor (CF_1). - Z. Naturforsch. *32C* : 803 - 809, 1977.

30040 - HUDÁK, J. : Štúdium morfogenézy chloroplastov u ihličnanov. [Chloroplast morphogenesis in conifers.] - In : HUZULÁK, J., MASAROVIČOVÁ, E. (ed.) : Fotosyntéza a Vodný Režim Drevín. Pp. 219 - 223. Modra-Piesky 1977. [In Slovak, ab : E, R.]

30041 - HUDSON, J.P. : Water shortage and plant growth. - Span *20*(2) : 86 - 87, 1977. [Ps.]

30042 - HUISMAN, J.G., BERNARDS, A., LIEBREGTS, P., GEBBING, M.G.T., STEGWEE, D. : Qualitative and quantitative immunofluorescence studies of chloroplast ferre- doxin. Application to investigations of ferredoxin inheritance in *Nicotiana* hybrids. - Planta *137* : 279 - 286, 1977.

30043 - HUISMAN, J.G., GEBBINK, M.G.T., MODDERMAN, P., STEGWEE, D. : The coding site of chloroplasts ferredoxin. - Planta *137* : 97 - 105, 1977.

30044 - HULL, J.C., MULLER, C.H. : The potential for dominance by *Stipa pulchra* in a California grassland. - Amer. Midland Nat. *97* : 147 - 175, 1977.

30045 - HUMPHREY, G.F. : The concentration of phytoplankton pigments in Australian waters. - Annu. Rep. mar. Biochem. Unit CSIRO *1976-77* : 7 - 9, 1976/77.

30046 - HUMPHREY, G.F. : Chlorophyll concentrations in marine unicellular algae. - Annu. Rep. mar. Biochem. Unit CSIRO *1976-77* : 10 - 11, 1976/77.

30047 - Hunan Agricultural College, Department of Chemistry : [Preliminary analysis of physiological and biochemical characteristics of the hybrid rice "Nan U-2".] - Acta bot. sin. *19* : 226 - 236, 1977. [Ps; in Chin., ab : E.]

30048 - HUNT, H.W. : A simulation model for decomposition in grasslands. - Ecology *58* : 469 - 484, 1977.

30049 - HUNT, R., PARSONS, I.T. : Plant growth-analysis: further applications of a recent curve-fitting program. - J. appl. Ecol. *14* : 965 - 968, 1977.

30050 - HUNTER, C.N., JONES, O.T.G. : Kinetic studies on photosynthetic electron flow in bacteriochlorophyll-less membranes of *Rhodopseudomonas sphaeroides* reconstituted with reaction centres. - In : COOMBS, J. (ed.) : 4th Interna- tional Congress on Photosynthesis. Pp. 174 - 175. UKISES, London 1977.

30051 - HUNTER, F., THORNBER, J.P. : Further characterization of the P700-chlorophyll a-protein from blue-green algae. - In : OLSON, J.M., HIND, G. (ed.) : Chlo- rophyll-Proteins, Reaction Centers, and Photosynthetic Membranes. P. 361. Brookhaven nat. Lab., Upton 1977.

30052 - HUNTSMAN, S.A., BARBER, R.T. : Primary production off northwest Africa : the relationship to wind and nutrient conditions. - Deep-Sea Res. *24* : 25 - 33, 1977.

30053 - HURD, R.G. : Vegetative plant growth analysis in controlled environments. - Ann. Bot. *41* : 779 - 787, 1977.

30054 - HURKMAN, W.J., KENNEDY, G.S. : Development and cytochemistry of the thylako- idal body in tobacco chloroplasts. - Amer. J. Bot. *64* : 86 - 95, 1977.

30055 - HURLEY, J.B., EBREY, T.G., HONIG, B., OTTOLENGHI, M. : Temperature and wave- length effects on the photochemistry of rhodopsin, isorhodopsin, bacteriorho- dopsin and their photoproducts. - Nature *270* : 540 - 542, 1977.

30056 - HÜSEMANN, W., BARZ, W. : Photoautotrophic growth and photosynthesis in cell suspension cultures of *Chenopodium rubrum*. - Physiol. Plant. *40* : 77 - 81, 1977.

*30057 - HUSZÁR, J. : Zmeny v ontogenetickej variabilite intenzity fotosyntézy (IFS)
u *Nicotiana tabacum* L. [Changes in ontogenetic variability of photosynthetic
rate (IFS) in *Nicotiana tabacum* L.] - Bull. tabak. Priemyslu *12* : 1 - 15,
1969. [In Slovak, ab : G.]

*30058 - HUSZÁR, J. : Sledovanie intenzity fotosyntézy a narastanie listovej plochy
u odrody Sabolčský (*N. tabacum* L.). [Photosynthetic rate and leaf-area incre-
ment in the cultivar Sabolčský (*N. tabacum* L.).] - Bull. tabak. Priemyslu *13* :
37 - 43, 1970. [In Slovak, ab : G.]

30059 - HUTBER, G.N., HUTSON, K.G., ROGERS, L.J. : Effect of iron deficiency on levels
of two ferredoxins and flavodoxin in a cyanobacterium. - FEMS Microbiol. Lett.
1 : 193 - 196, 1977.

30060 - HUTCHISON, B.A., MATT, D.R. : The annual cycle of solar radiation in a deciduous forest. - Agr. Meteorol. *18* : 255 - 265, 1977.

30061 - HUTCHISON, B.A., MATT, D.R. : The distribution of solar radiation within a
deciduous forest. - Ecol. Monogr. *47* : 185 - 207, 1977.

30062 - HUZISIGE, H. : Dependence of quantum requirement for ferricyanide photoreduction upon phosphorylating and uncoupled conditions. - Plant Cell Physiol.
1977 (Spec. Issue 3 - Photosynthetic Organelles. Structure and Function) :
211 - 218, 1977.

30063 - HWANG, S.-B., KORENBROT, J.I., STOECKENIUS, W. : Structural and spectroscopic characteristics of bacteriorhodopsin in air-water interface films. -
J. Membrane Biol. *36* : 115 - 135, 1977.

30064 - HWANG, S.-B., KORENBROT, J.I., STOECKENIUS, W. : Proton transport by bacteriorhodopsin through an interface film. - J. Membrane Biol. *36* : 137 - 158,
1977.

30065 - HWANG, S.-B., KORENBROT, J.I., STOECKENIUS, W. : Transient photovoltages ge-
nerated by charge displacements in intermediates of the bactoriorhodopsin
photoreaction cycle. - In : PACKER, L., PAPAGEORGIOU, G.C., TREBST, A. (ed.) :
Bioenergetics of Membranes. Pp. 137 - 147. Elsevier/North-Holland Biomedical
Press, Amsterdam - Oxford - New York 1977.

30066 - HYAMS, R.L., CARVER, M.A., PARTIS, M.D., GRIFFITHS, D.E. : Studies of energy-
-linked reactions: oleoyl phosphate-dependent ATP synthesis (oleoyl phospho-
kinase) activity of membrane ATPase and soluble ATPases from mitochondria,
chloroplasts, chromatophores and *Escherichia coli* plasma membrane. - FEBS
Lett. *82* : 307 - 313, 1977.

*30067 - IBRAGIMOV, M., AZIMOV, R.A. : O dejstvii estestvennogo zasoleniya na soder-
zhanie pigmentov v ontogeneze khlopchatnika. [Effect of natural salinization
on pigment contents during cotton ontogeny.] - Uz. biol. Zh. *1975* (1) :
22 - 24, 85, 1975. [In R, ab : Uz.]

30068 - IBRAHIM, R.K., PHAN, C.T. : Chloroplast ultrastructure and flavonid synthesis in flax callus and suspension cultures. - Plant Physiol. *59* (6, Suppl.) :
105, 1977.

30069 - IDLE, D.B. : The effects of leaf position and water vapour density deficit on
the transpiration rate of detached leaves. - Ann. Bot. *41* : 959 - 968, 1977.

30070 - IDSO, S.B., JACKSON, R.D., REGINATO, R.J. : Remote-sensing of crop yields. -
Science *196* : 19 - 25, 1977.

30071 - IDSO, S.B., REGINATO, R.J., JACKSON, R.D. : Albedo measurement for remote
sensing of crop yields. - Nature *266* : 625 - 628, 1977.

30072 - IGNATIADES, L. : *In situ* short term enrichment experiments and evaluation of
the ^{14}C method for testing oligotrophy in the sea. - Hydrobiologia *56* : 247 -
252, 1977.

30073 - IKEDIOBI, C.O., SNYDER, H.E. : Cooxidation of β-carotene by an isoenzyme of
soybean lipoxygenase. - J. agr. Food Chem. *25* : 124 - 127, 1977.

30074 - ILIEV, V., VANGELOVA, M. : Vliyanie na samostoyatelnoto azotno torene v"rkhu
rastezha i fotosintetichnata produktivnost na sl"nchogleda. [Effect of nitro-

gen fertilization on the growth and photosynthetic productivity of sunflower.]
- Rasteniev"dni Nauki *14* (1) : 102 - 109, 1977. [Dry-matter accumulation;
in Bulg., ab : F, R.]

30075 - ILMAVIRTA, V., JONES, R.I., KAIRESALO, T. : The structure and photosynthetic
activity of pelagial and littoral plankton communities in Lake Pääjärvi,
southern Finland. - Ann. bot. fenn. *14* : 7 - 16, 1977.

30076 - IL'YASHUK, E.M. : Uchastie kaliya v svetoindutsirovannykh dvizheniyakh ust'-
its u sakharnoĭ svekly. [Participation of potassium in light-induced movements of stomata in sugar beet.] - Fiziol. Biokhim. kul't. Rast. *9* : 285 -
290, 1977. [In R, ab : E.]

30077 - IMAI, K., MURATA, Y. : [Effect of carbon dioxide concentration on growth and
dry matter production of crop plants II. Specific and varietal differences
in the response of dry matter production.] - Jap.J.Crop Sci.*46*: 291 - 297,
1977. [In Jap., ab : E.]

30078 - IMBAMBA, S.K., TIESZEN, L.L. : Influence of light and temperature on photosynthesis and transpiration in some C_3 and C_4 vegetable plants from Kenya. -
Physiol. Plant. *39* : 311 - 316, 1977.

30079 - IMHOFF, J.F., TRÜPER, H.G. : *Ectothiorhodospira halochloris* sp. nov., a new
extremly halophilic phototrophic bacterium containing bacteriochlorophyll *b*.
- Arch. Microbiol. *114* : 115 - 121, 1977.

30080 - INADA, K. : Effects of leaf color and the light quality applied to leaf-de-
veloping period on the photosynthetic response spectra in crop plants. -
Jap. J. Crop Sci. *46* : 37 - 44, 1977.

30081 - INCOLL, L.D. : Field studies of photosynthesis: Monitoring with $^{14}CO_2$. -
In : LANDSBERG, J.J., CUTTING, C.V. (ed.) : Environmental Effects on Crop
Physiology. Pp. 137 - 155. Academic Press, London - New York - San Francisco
1977.

30082 - INDIRA, G.M., GNANAM, A. : Induction of chloroplast development - transla-
tional studies with temporally distinguished messengers. - In : BOGORAD, L.,
WEIL, J.H. (ed.) : Acides Nucléiques et Synthèse des Protéines chez les Végétaux. Coll. Int. C.N.R.S. No. 261. Pp. 617 - 621. Edit. C.N.R.S., Paris
1977.

30083 - INOSAKA, M., ITO, K., NUMAGUCHI, H., MISUMI, M. : [Studies on the productivity of some tropical grasses. III. The relation of heading property to productivity of rhodesgrass (*Chloris gayana* KUNTH) sown in interrow space of
maize (*Zea mays* L.).] - Jap. J. trop. Agr. *20* : 231 - 235, 1977. [In Jap.,
ab : E.]

30084 - INOUE, K. : Numerical experiments about three-dimensional transfer of CO_2
over a finite model rice field in relation to canopy photosynthesis. - Bull.
nat. Inst. agr. Sci. Ser. A (Tokyo) *24* : 19 - 44, 1977.

30085 - INOUE, Y., SHIBATA, K. : Development of oxygen-evolving system as described
in terms of thermoluminescence. - In : MITSUI, A., MIYACHI, S., SAN PIETRO,
A., TAMURA, S. (ed.) : Biological Solar Energy Conversion. Pp. 109 - 128.
Academic Press, New York - San Francisco - London 1977.

30086 - INOUE, Y., YAMASHITA, T., KOBAYASHI, Y., SHIBATA, K. : Thermoluminescence
changes during inactivation and reactivation of the oxygen-evolving system
in isolated chloroplasts. - FEBS Lett. *82* : 303 - 306, 1977.

30087 - IONESCU, A., ELIADE, G., CORBU, S. : The effects of continuous pollution
with fluorine on agricultural cultures and spontaneous flora. - Rev. roum.
Biol., Sér. Biol. vég. *22* : 157 - 163, 1977. [Chl.]

30088 - ĬORDANOV, I.T. : Vliyanie povyshennoĭ temperatury na fotosintez i raspredelenie C^{14} u list'ev razlichnogo fiziologicheskogo sostoyaniya. [The effect of
high temperature on photosynthesis and ^{14}C distribution in leaves of different physiological state.] - Bot. Zh. *62* : 93 - 100, 1977. [In R.]

*30089 - ĬORDANOV, I.T., MERAKCHIĬSKA, M.G. : Vliyanie na olovoto v"rkhu intenzivnost-
ta na fotosintezata, razpredelenieto na C^{14}, s"stava na strukturnite belt"tsi
na khloroplastite i spektralnite svoĭstva na PS I i PS II. [Effect of lead

on photosynthetic rate, ^{14}C distribution, chloroplast structural protein composition, and PS I and PS II spectral properties.] - Fiziol. Rast. (Sofia) 2 (3) : 3 - 9, 1976. [In Bulg., ab : E.]

*30090 - IOVVA, E.P., DOROKHOV, B.L. : Izmenenie fotosinteticheskoǐ deyatel'nosti tomatov pri razlichnom opylenii. [Changes in photosynthetic activity of tomato at various pollination.] - In : Genetika i Selektsiya v Moldavii. Pp. 138 - 139. Kishinev 1971. [In R.]

30091 - IRIYAMA, K., SHIRAKI, M., YOSHIURA, M. : Partial purification of chlorophyll extracted from spinach leaves before chromatographic separation and isolation. - Chem. Lett. 1977 : 787 - 788, 1977.

30092 - IRIYAMA, K., YOSHIURA, M. : Absorption spectroscopy of chlorophyll a and b in methanol dioxane and/or water. - Colloid Polym. Sci. 255 : 133 - 139, 1977.

30093 - ISAAKIDOU, J., FORTI, G. : Uncoupling of chloroplast electron transport by xanthine oxidase. - Plant Sci. Lett. 9 : 65 - 69, 1977.

30094 - ISAAKIDOU, J., PAPAGEORGIOU, G. : Molecular and functional properties of isolated chloroplasts after crosslinking with dimethylsuberimidate. - In : COOMBS, J. (ed.) : 4th International Congress on Photosynthesis. P.175. UKISES, London 1977.

*30095 - ISAEV, B.M., RUSTAMOV, K. : Vliyanie mikroêlementa medi na nekotorye fiziologo-biokhimicheskie protsessy i produktivnost' khlopchatnika. [Effect of the trace element copper on some physiological and biochemical processes and productivity of cotton.] - Uz. biol. Zh. 19 (2) : 27 - 30, 90, 1975. [Ps; in R, ab : Uz.]

30096 - ISÉPY, I. : Gyertyános-tölgyesek primér produkciója és az időjárásviszonyok hatása a lombavar bomlására. [Leaf production in Querco-Carpinetum forests and the effect of environmental factors on litter decomposition.] - MTA Biol. Oszt. Közl. 20 : 199 - 206, 1977. [In Hung.]

30097 - ISHII, R., OHSUGI, R., MURATA, Y. : The effect of temperature on the rates of photosynthesis, respiration and the activity of RuDP carboxylase in barley, rice and maize leaves. - Jap. J. Crop Sci. 46 : 516 - 523, 1977.

30098 - ISHII, R., SAMEJIMA, M., MURATA, Y. : Photosynthetic $^{14}CO_2$ fixation in the leaves of rice and some other species. - Jap. J. Crop Sci. 46 : 97 - 102, 1977.

30099 - ISHII, R., TAKEHARA, T., MURATA, Y., MIYACHI, S. : Effects of light intensity on the rates of photosynthesis and photorespiration in C_3 and C_4 plants.-In : MITSUI, A., MIYACHI, S., SAN PIETRO, A., TAMURA, S. (ed.) : Biological Solar Energy Conversion. Pp. 265 - 271. Academic Press, New York - San Francisco - London 1977.

30100 - ISHII, R., YAMAGISHI, T., MURATA, Y. : On a method for measuring photosynthesis and respiration of leaf slices with an oxygen electrode. - Jap. J. Crop Sci. 46 : 53 - 57, 1977.

30101 - ISHITANI, T., KIMURA, S. : [Photodegradation of chlorophylls and their derivatives.] - Nippon Shokuhin Kogyo Gakkaishi [J. Jap. Soc. Food Sci. Technol.]24 : 448 - 452, 1977. [In Jap., ab : E.]

30102 - ISLER, O. : Progress in the field of fat-soluble vitamins and carotenoids. - Experientia 33 : 555 - 573, 1977.

30103 - ISRAELSTAM, G.F. : Photosynthetic and chloroplastic activity of dwarf and tall cultivars of pea (Pisum sativum L.). - Plant Physiol. 59 (6, Suppl.) : 129, 1977.

*30104 - ITO, K., FUTATSUYA, F., HIBI, K., ISHIDA, S., YAMADA, O., MUNAKATA, K. : [Herbicidal activity of 3,3'-dimethyl-4-methoxybenzophenone (NK-049) in paddy field. 1. Herbicidal characteristics of NK-049 on weeds.] - Zasso Kenkyu 18 : 10 - 15, 1974. [Chl; in Jap., ab : E.]

*30105 - ITO, K., FUTATSUYA, F., HIBI, K., ISHIDA, S., YAMADA, O., MUNAKATA, K. : [Herbicidal activity of 3,3'-dimethyl-4-methoxybenzophenone (NK-049) in paddy

field. 2. Response of rice plant to NK-049.] - Zasso Kenkyu *18* : 16 - 20,
1974. [Chl; in Jap., ab: E.]

30106 - ITO, O., KUMAZAWA, K. : Amino acid metabolism in plant leaf. I. Amino acids
synthesis from $^{14}CO_2$ and $^{15}NH_4$ in detached sunflower leaves. - Soil Sci.
Plant Nutr. *23* : 365 - 372, 1977.

30107 - ITOH, S. : Temperature dependencies of the rate of electron flow and of the
formation of the high energy state in spinach chloroplasts and leaves. -
Plant Cell Physiol. *18* : 801 - 806, 1977.

30108 - ITOH, S., NISHIMURA, M. : pH dependent changes in the reactivity of the primary electron acceptor of System II in spinach chloroplasts to external oxi-
dant and reductant. - Biochim. biophys. Acta *460* : 381 - 392, 1977.

30109 - ITURRIAGA, R., HOPPE, H.-G. : Observations of heterotrophic activity of photoassimilated organic matter. - Mar. Biol. *40* : 101 - 108, 1977.

30110 - IVANCHANKA, V.M., MARSHAKOVA, M.I., KRUCHYNINA, S.S., URBANOVICH, T.A. :
Uplyŭ aprafenu na ab'ěm i fotakhimichnuyu aktyŭnasts' khlaraplastaŭ. [Effect
of aprophen on the volume and photochemical activity of chloroplasts.] -
Vestsi Akad. Navuk belarus. SSR, Ser. biyal. Navuk *1977* (4) : 43 - 46, 140,
1977. [In Belorus., ab : R.]

30111 - IVANCHENKO, V.M., KRUCHININA, S.S., MARSHAKOVA, M.I., URBANOVICH, T.A. :
K voprosu o mekhanizme fenomena Brilliant. [Mechanism of the Brilliant phenomenon.] - Fiziol. Rast. *24* : 416 - 418, 1977. [In R.]

30112 - IVANCHENKO, V.M., MARSHAKOVA, M.I. : O fotokhimicheskikh ěffektakh v sistemakh kofaktorov ělektronnogo transporta v khloroplastakh. [Photochemical
effects in systems of electron transport co-factors in chloroplasts.] -
Vestsi Akad. Navuk belarus. SSR, Ser. biyal. Navuk *1977* (6) : 98 - 100, 143,
1977. [In R.]

*30113 - IVANOV, A.F., BELOUSOV, A.M. : Korrelyativnye svyazi i zavisimosti produk-
tivnosti fotosinteza sortov ozimoǐ pshenitsy intensivnogo tipa v usloviyakh
orosheniya. [Correlative connections and dependences of productivity of photosynthesis of intensive types of winter wheat under irrigation.] - Nauch.
Dokl. vysch. Shkoly, biol. Nauki *19* (8) : 96 - 99, 1976. [In R.]

30114 - IVANOV, B.N. : Pogloshchenie protonov khloroplastami pri tsiklicheskom ělek-
tronnom transporte, kataliziruemom fotosistemoǐ I v anaěrobnykh usloviyakh.
[Proton absorption by chloroplasts in cyclic electron transport catalysed by
photosystem I in anaerobic conditions.] - Biokhimiya *42* : 2121 - 2130, 1977.
[In R, ab : E.]

30115 - IVANOV, B.N., RUZIEVA, R.Kh. : Vliyanie kvertsetina i kvertsetin-3-glyuko-
zil-P-kumarata na protonnyǐ obmen izolirovannykh khloroplastov gorokha.
[Effect of quercetin and quercetin-3-glucosyl-p-cumarate on proton exchange
of isolated pea chloroplasts.] - In : AKULOVA, E.A., MUZAFAROV, E.N. (ed.) :
Regulyatsiya Énergeticheskogo Obmena Khloroplastov i Mitokhondriǐ Éndogennymi Fenol'nymi Ingibitorami. Pp. 27 - 41, 127 - 132. Pushchino 1977.
[In R., ab : E.]

30116 - IVANOVSKY, R.N., ZHUKOV, V.G. : The role of sulfur compounds in the metabolism of some phototrophic bacteria. - In : COOMBS, J. (ed.) : 4th International
Congress on Photosynthesis. P. 176. UKISES, London 1977.

30117 - IVNITSKAYA, I.N., YAKOVENKO, G.M., MANUIL'SKAYA, S.V., DILUNG, I.I. : Priroda
temnovogo vzaimodeǐstviya khlorofilla s fosfolipidami. [Nature of dark interaction between chlorophyll and phospholipids.] - Dokl. Akad. Nauk ukr.
SSR, Ser. B *1977* : 640 - 642, 1977. [In R, ab : E.]

30118 - IZAWA, S. : Inhibitors of electron transport. - In : TREBST, A., AVRON, M.
(ed.) : Photosynthesis I. (Encycl. Plant Physiol. N.S. Vol. 5.) Pp. 266 -
282. Springer-Verlag, Berlin - Heidelberg - New York 1977.

30119 - IZAWA, S., PAN, R.-L. : Photosystem II energy coupling with H_2O_2 as electron
donor. - In : COOMBS, J. (ed.) : 4th International Congress on Photosynthesis. P. 177. UKISES, London 1977.

30120 - IZDEBSKI, K., KIMSA, T., KOZAK, K., MICHNA, E., POPIOŁEK, Z., STĄCZEK, A., ZINKIEWICZ, A. : The effect of habitats in two forest ecosystems on the productivity of pine stands in central Roztocze. III. Results. - Ekol. polska 25 : 89 - 105, 1977.

30121 - IZMEST'EVA, L.R., KOZHOVA, O.M., LOPATINA, N.I. : Produktsionnye kharakteristiki balaganskogo rasshireniya Bratskogo vodokhranilishcha. [Production characteristics of Balangansk extension of the Bratsk reservoir.] - In : Gidrobiologicheskie i Ikhtiologicheskie Issledovaniya v Vostochnoĭ Sibiri. Vol. I. Pp. 175 - 184. Irkutsk 1977. [Chi; in R.]

30122 - JACKSON, C., DENCH, J., MOORE, A.L. : Characterisation of superoxide dismutase in spinach leaves. - In : COOMBS, J. (ed.) : 4th International Congress on Photosynthesis. Pp. 177 - 178. UKISES, London 1977.

30123 - JACOBI, G. : Subchloroplast preparations. - In : TREBST, A., AVRON, M. (ed.): Photosynthesis. I. (Encycl. Plant Physiol. N.S. Vol. 5.) Pp. 543 - 562. Springer-Verlag, Berlin - Heidelberg - New York 1977.

30124 - JACQUES, T.G., PILSON, M.E.Q., CUMMINGS, C., MARSHALL, N. : Laboratory observations on respiration, photosynthesis and factors affecting calcification in the temperate coral Astrangia danae. - Proc. 3rd Int. Coral Reef Symp. 2 : 455 - 461, 1977.

30125 - JACQUOT, J.P., VIDAL, J., GADAL, P. : Evidence for chloroplastic localization of spinach leaf NADP malate dehydrogenase activating factors. - Planta 137 : 89 - 90, 1977.

30126 - JAGENDORF, A.T. : Photophosphorylation. - In : TREBST, A., AVRON, M. (ed.) : Photosynthesis I. (Encycl. Plant Physiol. N.S. Vol. 5.) Pp. 307 - 337. Springer-Verlag, Berlin - Heidelberg - New York 1977.

30127 - JAGENDORF, A.T., SCHMID, R. : Indications of a role for arginine residues in the function of chloroplast coupling factor. - In : COOMBS, J. (ed.) : 4th International Congress on Photosynthesis. P. 179. UKISES, London 1977.

30128 - JÄGER, H.-J. : Auswirkungen phytotoxischer Immissionen auf enzymatische Aktivitäten und Reaktionen. - Angew. Bot. 51 : 1 - 7, 1977. [Ps.]

30129 - JAGGER, J. : Phototechnology and biological experimentation. - In : SMITH, K.C. (ed.) : The Science of Photobiology. Pp. 1 - 26. Plenum Press, New York - London 1977. [Irradiance measurements.]

30130 - JAMRICH, V., CICÁK, A. : Fluór a fotosyntetická asimilácia. [Fluor and photosynthetic assimilation.] - In : HUZULÁK, J., MASAROVIČOVÁ, E. (ed.) : Fotosyntéza a Vodný Režim Drevín. Pp. 202 - 210. Modra-Piesky 1977. [In Slovak, ab : E, R.]

30131 - JANARDHAN, K.V., MURTY, K.S. : Association of some leaf characters with photosynthesis in rice. - Curr. Sci. 46 : 497 - 498, 1977.

30132 - JANJIČ, V., PLESNIČAR, M., BOGDANOVIČ, V. : Proučavanje dejstva Roneet-a na rastenje, količinu ukupnog azota i fotohemijsku aktivnost hloroplasta suncokreta, šećerne repe i ječma. [Effect of Roneet on the growth, total nitrogen content, and photochemical activity of chloroplasts of sunflower, sugar beet, and barley.] - Agrohemija 1977 (5-6) : 205 - 211, 1977. [In Croat., ab : E.]

30133 - JANSA, J. : Estudio preliminar del contenido en pigmentos fotosinteticos en el tubo digestivo de apendicularias y salpas. [Preliminary study of the photosynthetic pigment content of the digestive tract of salps and appendicularians.] - Bol.Inst. esp. Oceanogr. 1 : 5 - 29, 1977. [In Span., ab : E, F.]

30134 - JAUNEAU, E., REISS-HUSSON, F. : Bound cytochromes solubilization in wild type Rhodopseudomonas sphaeroides. - In : COOMBS, J. (ed.) : 4th International Congress on Photosynthesis. P. 178. UKISES, London 1977.

30135 - JAYNES, J.M., KLEIN, S.M., VERNON, L.P. : Formation and properties of small chlorophyll-containing vesicles from Anabaena flos-aquae membrane fragments and isolated chlorophyll-protein complex. - Plant Cell Physiol. 1977 (Spec. Issue 3 - Photosynthetic Organelles. Structure and Function) : 165 - 172, 1977.

30136 - JEANJEAN, R. : Influence of light conditions and inhibitors on ATP level and on phosphate uptake in *Chlorella pyrenoidosa*. - In : Echanges Ioniques Transmembranaires chez les Végétaux. Colloque Int. CNRS No. 258. Pp. 205 - 211. Édit. CNRS, Paris 1977.

30137 - JEFFREY, S.W., VESK, M. : Effect of blue-green light on photosynthetic pigments and chloroplast structure in the marine diatom *Stephanopyxis turris*. - J. Phycol. *13* : 271 - 279, 1977.

30138 - JENNINGS, J.V., EVANS, M.C.W. : The irreversible photoreduction of a low potential component at low temperatures in a preparation of the green photosynthetic bacterium *Chlorobium thiosulphatophilum*. - FEBS Lett. *75* : 33 - 36, 1977.

30139 - JENSEN, C.R. : Effects of salinity in the root medium. IV. Photosynthesis and leaf diffusive resistance in relation to CO_2-concentration. - Acta Agr. scand. *27* : 159 - 164, 1977.

30140 - JENSEN, K.I.N., STEPHENSON, G.R., HUNT, L.A. : Detoxification of atrazine in three *Gramineae* subfamilies. - Weed Sci. *25* : 212 - 220, 1977. [Ps.]

30141 - JENSEN, K.I.N., STEPHENSON, G.R., HUNT, L.A., BANDEEN, J.D. : The effect of atrazine, cyanazine and cyprazine on photosynthesis and growth of nine grasses . - Weed Res. *17* : 379 - 386, 1977. [Ps.]

30142 - JENSEN, N.H. : Pulse radiolysis of chlorophyll *a* in solution. - Res. Establ. Risø-M-*1914* : 1 - 21, 1977.

30143 - JENSEN, R.G., BAHR, J.T. : Ribulose 1,5-bisphosphate carboxylase-oxygenase. - Annu. Rev. Plant Physiol. *28* : 379 - 400, 1977.

30144 - JENSEN, R.G., BAHR, J.T., SICHER, R.C. : Regulation of chloroplast RuBP carboxylase activity by CO_2, light and RuBP. - In : COOMBS, J. (ed.) : 4th International Congress on Photosynthesis. Pp. 179 - 180. UKISES, London 1977.

30145 - JESAITIS, A.J., HENERS, P.R., HERTEL, R., BRIGGS, W.R. : Characterization of a membrane fraction containing a b-type cytochrome. - Plant Physiol. *59* : 941 - 947, 1977.

30146 - JESSOP, R.S. : Influence of time of sowing and plant density on the yield and oil content of dryland sunflowers. - Aust. J. exp. Agr. anim. Husb. *17* : 664 - 668, 1977.

30147 - JEWSON, D.H. : A comparison between *in situ* photosynthetic rates determined using ^{14}C uptake and oxygen evolution methods in Lough Neagh, Northern Ireland. - Proc. roy. Irish Acad. B *77* (3) : 87 - 99, 1977.

30148 - JEWSON, D.H. : Light penetration in relation to phytoplankton content of the euphotic zone of Longh Neagh, N. Ireland. - Oikos *28* : 74 - 83, 1977.

30149 - JOHNS, S.R., LESLIE, D.R., WILLING, R.I., BISHOP, D.G. : Studies on chloroplast membranes. I. ^{13}C chemical shifts and longitudinal relaxation times of carboxylic acids. - Aust. J. Chem. *30* : 813 - 822, 1977.

30150 - JOHNS, S.R., LESLIE, D.R., WILLING, R.I., BISHOP, D.G. : Studies on chloroplast membranes. II. ^{13}C chemical shifts and longitudinal relaxation times of 1,2-di[(9Z,12Z,15Z)-octadeca-9,12,15-trienoyl]-3-galactosyl-*sn*-glycerol and 1,2-di[(9Z,12Z,15Z)-octadeca-9,12,15-trienoyl]-3-digalactosyl-*sn*-glycerol. - Aust. J. Chem. *30* : 823 - 834, 1977.

30151 - JOHNSTON, C.S., JONES, R.G., HUNT, R.D. : A seasonal carbon budget for a laminarian population in a Scottish sea-loch. - Helgol. wiss. Meeresunters. *30* : 527 - 545, 1977. [Ps.]

30152 - JOHNSTON, T.D. : The measurement of mesophyll air space in kale and rape (*Brassica oleracea* L. and *B. napus* L.) leaves. - Photosynthetica *11* : 311 - 313, 1977.

30153 - JOLCHINE, G. : Isolation and purification of a coupling factor ATPase from *Rhodopseudomonas sphaeroides* chromatophores. - In : COOMBS, J. (ed.) : 4th International Congress on Photosynthesis. P. 181. UKISES, London 1977.

30154 - JOLIOT, A. : Flash induced fluorescence kinetics in chloroplasts in the 20 μs - 100 s time range in the presence of 3(3,4-dichlorophenyl)-1,1-dimethyl-urea. Effects of hydroxylamine. - Biochim. biophys. Acta *460* : 142 - 151, 1977.

30155 - JOLIOT, P. : Analysis of the structure of the photosynthetic apparatus based on functional experiments. - In : BUVET, R., ALLEN, M.J., MASSUÉ, J.-P. (ed.) : Living Systems as Energy Converters. Pp. 175 - 184. North-Holland Publ. Co., Amsterdam - New York - Oxford 1977.

30156 - JOLIOT, P. : Champ électrique intramembranaire et structure de la chaine de transporteurs d'electrons photosynthètiques. - In : ROUX, E. (ed.) : Electri-cal Phenomena at the Biological Membrane Level. Pp. 521 - 531. Elsevier, Amsterdam - London - New York 1977.

30157 - JOLIOT, P., DELOSME, R., JOLIOT, A. : 515 nm absorption changes in *Chlorella* at short times (4 - 100 μs) after a flash. - Biochim. biophys. Acta *459* : 47 - 57, 1977.

30158 - JOLIOT, P., JOLIOT, A. : Control of the electron transfer by the transmembra-ne electric field and structure of system II centers. - In : CASTELLANI, A. (ed.) : Research in Photobiology. Pp. 129 - 137. Plenum Press, New York - London 1977.

30159 - JOLIOT, P., JOLIOT, A. : Evidence for a double hit process in Photosystem II based on fluorescence studies. - In : COOMBS, J. (ed.) : 4th International Congress on Photosynthesis. P. 180. UKISES, London 1977.

30160 - JOLIOT, P., JOLIOT, A. : Evidence for a double hit process in photosystem II based on fluorescence studies. - Biochim. biophys. Acta *462* : 559 - 574, 1977.

30161 - JONES, E.P., SMITH, S.D. : A first measurement of sea-air CO_2 flux by eddy correlation. - J. geophys. Res. *82* : 5990 - 5991, 1977.

30162 - JONES, H.G. : Aspects of the water relations of spring wheat (*Triticum aesti-vum* L.) in response to induced drought. - J. agr. Sci. *88* : 267 - 282, 1977. [Resistances.]

30163 - JONES, H.G. : Transpiration in barley lines with differing stomatal frequen-cies. - J. exp. Bot. *28* : 162 - 168, 1977. [Resistances.]

30164 - JONES, I.D., WHITE, R.C., GIBBS, E., BUTLER, L.S. : Estimation of zinc pheo-phytins, chlorophylls, and pheophytins in mixtures in diethyl ether or 80 % acetone by spectrophotometry and fluorometry. - J. agr. Food Chem. *25* : 146 - 149, 1977.

30165 - JONES, I.D., WHITE, R.C., GIBBS, E., BUTLER, L.S., NELSON, L.A. : Experimen-tal formation of zinc and copper complexes of chlorophyll derivatives in ve-getable tissue by thermal processing. - J. agr. Food Chem. *25* : 149 - 153, 1977.

30166 - JONES, J.G., SIMON, B.M. : Increased sensitivity in the measurement of ATP in freshwater samples with a comment on the adverse effect of membrane fil-tration. - Freshwater Biol. *7* : 253 - 260, 1977.

30167 - JONES, M.B., HEARD, A.J., WOLEDGE, J., LEAFE, E.L., PLUMB, R.T. : The effect of ryegrass mosaic virus on carbon assimilation and growth of ryegrasses. - Ann. appl. Biol. *87* : 393 - 405, 1977.

30168 - JONES, M.B., LEAFE, E.L. : The relationship between photosynthesis and pro-ductivity of ryegrass under water stress. - In : COOMBS, J. (ed.) : 4th In-ternational Congress on Photosynthesis. Pp. 181 - 182. UKISES, London 1977.

30169 - JONES, R.I. : A comparison of acetone and methanol as solvents for estimating the chlorophyll *a* and phaeophytin *a* concentrations in phytoplankton. - Ann. bot. fenn. *14* : 55 - 69, 1977.

30170 - JONES, W.T., MANGAN, J.L. : Complexes of the condensed tannins of sainfoin (*Onobrychis viciifolia* SCOP.) with fraction 1 leaf protein and with submaxil-lary mucoprotein, and their reversal by polyethylene glycol and pH. - J. Sci. Food Agr. *28* : 126 - 136, 1977.

30171 - **JOPE, C.A., ATCHINSON, B.A., PRINGLE, R.C., WILDMAN, S.G.** : Spiral, string of grana, chloroplast model tested by computer simulation. - In : BOGORAD, L., WEIL, J.H. (ed.) : Acides Nucléiques et Synthèse des Protéines chez les Végétaux. Coll. C.N.R.S. No. 261. Pp. 153 - 158. Édit. CNRS, Paris 1977.

30172 - **JØRGENSEN, E.G,** : Photosynthesis. - In : WERNER, D. (ed.) : The Biology of Diatoms. Bot. Monogr. Vol. 13. Pp. 150 - 168. Blackwell Sci. Publ., Oxford - London - Edinburgh - Melbourne 1977.

30173 - **JOSE, A.M., VINCE-PRUE, D., HILTON, J.R.** : Chlorophyll interference with phytochrome measurement. Planta *135* : 119 - 123, 1977.

30174 - **JOSHI, G.V.** : CO_2 fixation in plants exposed to salt and water stresses. - In : COOMBS, J. (ed.) : 4th International Congress on Photosynthesis. Pp. 182 - 183. UKISES, London 1977.

30175 - **JOSHI, G.V., NAIK, G.R.** : Salinity effect on growth and photosynthetic productivity in sugarcane var. Co 740. - Indian Sugar *27* : 329 - 332, 1977.

30176 - **JOUY, M.** : Early use of the light energy absorbed by etiolated leaves following reduction of protochlorophyll(ide). - In : COOMBS, J. (ed.) : 4th International Congress on Photosynthesis. Pp. 183 - 184. UKISES, London 1977.

30177 - **JOYARD, J., DOUCE, R.** : Galactolipids synthesis in spinach chloroplasts. - In : COOMBS, J. (ed.) : 4th International Congress on Photosynthesis. P. 184. UKISES, London 1977.

30178 - **JUNGE, W.** : Membrane potentials in photosynthesis. - Annu. Rev. Plant Physiol. *28* : 503 - 536, 1977.

30179 - **JUNGE, W.** : Physical aspects of light harvesting, electron transport and electrochemical potential generation in photosynthesis of green plants. - In : TREBST, A., AVRON, M. (ed.) : Photosynthesis I. (Encycl. Plant Physiol. N.S. Vol.5.) Pp. 59 - 93. Springer-Verlag, Berlin - Heidelberg - New York 1977.

30180 - **JUNGE, W., RENGER, G., AUSLÄNDER, W.** : Proton release into the internal phase of thylakoids due to photosynthetic water oxidation. On the periodicity under flashing light. - FEBS Lett. *79* : 155 - 159, 1977.

30181 - **JUNGE, W., SCHAFFERNICHT, H., NELSON, N.** : On the mutual orientation of pigments in Photosystem I particles from green plants. - Biochim. biophys. Acta *462* : 73 - 85, 1977.

30182 - **JURSINIC, P., GOVINDJEE** : Delayed light emission decay in the 6 to 340 µs range after a single flash: Temperature effects. - Plant Physiol. *59* (6, Suppl.) : 24, 1977.

30183 - **JURSINIC, P., GOVINDJEE** : Temperature dependence of delayed light emission in the 6 to 340 microsecond range after a single flash in chloroplasts. - Photochem. Photobiol. *26* : 617 - 628, 1977.

30184 - **JURSINIC, P., GOVINDJEE** : The rise in chlorophyll *a* fluorescence yield and decay in delayed light emission in Tris-washed chloroplasts in the 6 - 100 µs time range after an excitation flash. - Biochim. biophys. Acta *461* : 253 - 267, 1977.

30185 - **JÜTTNER, F.** : Thirty liter tower-type pilot plant for the mass cultivation of light- and motion-sensitive planktonic algae. - Biotechnol. Bioeng. *19* : 1679 - 1687, 1977. [Ps.]

30186 - **KABUZENKO, S.M., PONOMAR'OVA, S.O.** : Vmist plastydnykh pigmentiv u lysti deyakykh roslyn zalezhno vid formy lystkovoï plastynky. [Content of plastid pigments in leaves of some plants as dependent on the form of leaf blade.] - Ukr. bot. Zh. *34* : 248 - 251, 335, 1977. [In Ukr., ab : E, R.]

30187 - **KADOYA, K.** : Studies on the hydrophysiological rhythms of citrus trees. II. Seasonal effects of carbon dioxide on the cyclic fluctuations of leaf thickness. - J. jap. Soc. hort. Sci. *46* : 153 - 157, 1977.

30188 - **KAFALIEVA-BOEVA, D.N., BUCHEVA, M.C.** : Photo-induced electron spin resonance spectra of *Vicia faba* chloroplasts and of digitonin chloroplast fragments. - Studia biophys. *66* : 65 - 73, 1977.

30189 - **KAGAN, N.E.** : A synthetic model compound for the specialized chlorophyll of the photosynthetic reaction center. - In : **OLSON, J.M., HIND, G.** (ed.) : Chlorophyll-Proteins, Reaction Centers, and Photosynthetic Membranes. P. 371. Brookhaven nat. Lab., Upton 1977.

30190 - **KAGAN-ZUR, V., FRIEDLANDER, M., LIPS, S.H.** : Changes in chloroplasts envelopes induced by light or dark pretreatment of pea plants. - In : **COOMBS, J.** (ed.) : 4th International Congress on Photosynthesis. P. 185. UKISES, London 1977.

30191 - **KAGAWA, T., HATCH, M.D.** : Regulation of C_4 photosynthesis: Characterization of a protein factor mediating the activation and inactivation of NADP-malate dehydrogenase. - Arch. Biochem. Biophys. *184* : 290 - 297, 1977.

30192 - **KAGAWA, Y., OHNO, K., YOSHIDA, M., TAKEUCHI, Y., SONE, N.** : Proton translocation by ATPase and bacteriorhodopsin. - Fed. Proc. *36* : 1815 - 1818, 1977.

30193 - **KAGEYAMA, A., YOKOHAMA, Y.** : [Pigments and photosynthesis of deep-water green algae.] - Bull. Jap. Soc. Phycol. [Sorni] *25* : 168 - 175, 1977. [In Jap., ab : E.]

30194 - **KAGEYAMA, A., YOKOHAMA, Y., SHIMURA, S., IKAWA, T.** : An efficient excitation energy transfer from a carotenoid, siphonaxanthin to chlorophyll *a* observed in a deep-water species of chlorophycean seaweed. - Plant Cell Physiol. *18* : 477 - 480, 1977.

30195 - **KAIGAMA, B.K., TEARE, I.D., STONE, L.R., POWERS, W.L.** : Root and top growth of irrigated and nonirrigated grain sorghum. - Crop Sci. *17* : 555 - 559, 1977. [Growth analysis.]

30196 - **KAIRESALO, T.** : On the production ecology of epipelic algae and littoral plankton communities in Lake Pääjärvi, southern Finland. - Ann. bot. fenn. *14* : 82 - 88, 1977.

30197 - **KAISER, W., URBACH, W.** : The effect of dihydroxyacetone phosphate and 3-phosphoglycerate on O_2 evolution and on the levels of ATP, ADP and P_i in isolated intact chloroplasts. - Biochim. biophys. Acta *459* : 337 - 346, 1977.

*B30198 - **KAKHNOVICH, L.V.** (ed.) : Optimizatsiya Fotosinteticheskogo Apparata Vozdeĭstviem Razlichnykh Faktorov. [Optimization of the Photosynthetic Apparatus by the Action of Different Factors.]-Izd. BGU, Minsk 1976. [In R.]

*30199 - **KAKHNOVICH, L.V.** : Formirovanie fotosinteticheskogo apparata nekotorykh rastenii pri nizkikh i nasyshchayushchikh intensivnost'yakh sveta. [Formation of the photosynthetic apparatus of some plants at low and saturating irradiances.] - In : **KAKHNOVICH, L.V.** (ed.) : Optimizatsiya Fotosinteticheskogo Apparata Vozdeĭstviem Razlichnykh Faktorov. Pp. 12 - 20. Izd. BGU, Minsk 1976. [In R.]

*30200 - **KAKHNOVICH, L.V.** : Sostoyanie pigmentnoĭ sistemy list'ev gorokha v zavisimosti ot intensivnosti sveta. [State of the pigment system of pea leaves in relation to illuminance.] - In : **KAKHNOVICH, L.V.** (ed.) : Optimizatsiya Fotosinteticheskogo Apparata Vozdeĭstviem Razlichnykh Faktorov. Pp. 20 - 24. Izd. BGU, Minsk 1976. [In R.]

30201 - **KALER, V.L.** : Metabolicheskoe i épigeneticheskoe upravlenie biosintezom khlorofilla. [Metabolic and epigenetic control of chlorophyll biosynthesis.]- In : **NASYROV, Yu.S.** (ed.) : Genetika Fotosinteza. Pp. 61 - 65. Donish, Dushanbe 1977. [In R.]

30202 - **KALER, V.L., KLINGER, Yu.E., LOKTEV, A.V., VECHER, A.S.** : Sopryazhenie metabolizma glikolata s biosintezom khlorofilla v rasteniyakh. [Coupling between glycolate metabolism and chlorophyll biosynthesis in plants.] - Fiziol. Rast. *24* : 30 - 34, 1977. [In R, ab : E.]

30203 - **KALINOWSKA-ZDUN, M.** : Sugar beet yielding in the light of the results of the correlation analyses applied for appretiation of biomass increase in the middle part of the vegetation period. - In : Produkce Biomasy a Tvorba Výnosů Polních Plodin. Vol. 1. Pp. 136 - 150. Česká vědeckotechnická Společnost zemědělská, Praha 1977. [Growth analysis.]

30204 - KALLIO, S., WILKINSON, R.E. : The effects of some herbicides on nitrogenase
activity and carbon fixation in two subarctic lichens. - Bot. Gaz. *138* :
468 - 473, 1977.

*30205 - KALMA, J.D., FUCHS, M. : Citrus orchards. - In : MONTEITH, J.L. (ed.) :
Vegetation and the Atmosphere. Vol. 2. Case Studies. Pp. 309 - 328. Academic
Press, London - New York - San Francisco 1976. [Canopy.]

30206 - KAMANINA, M.S., ANISIMOV, A.A. : Vykhod assimilyatov iz mezofilla v svobod-
noe prostranstvo list'ev pri raznykh usloviyakh mineral'nogo pitaniya. [Photo-
synthate transport from mesophyll into free space of leaves under different
conditions of mineral nutrition.] - Fiziol. Rast. *24* : 767 - 772, 1977.
[In R, ab : E.]

30207 - KAMEKE, E.von, WEGMANN, K. : Isolation of manganese-containing subchloroplast
fractions from *Dunaliella*. - Plant Sci. Lett. *8* : 1 - 5, 1977.

30208 - KAMEN, M.D. : Progress toward a comparative biochemistry of cytochrome *C*. -
Plant Cell Physiol. *1977* (Spec. Issue 3 - Photosynthetic Organelles. Structure
and Function) : 283 - 291, 1977.

30209 - KAMIMURA, Y., YAMASAKI, T., MATSUZAKI, E. : Cytochrome components of green
alga, *Bryopsis maxima*. - Plant Cell Physiol. *18* : 317 - 324, 1977.

30210 - KAMINSKI, K. : Photosynthetic control in chloroplast suspensions frozen in
liquid nitrogen in the presence of glycerol. - In : COOMBS, J. (ed.) : 4th
International Congress on Photosynthesis. P. 186. UKISES, London 1977.

30211 - KAMINSKI, K. : Photosynthetic control in chloroplasts suspensions frozen in
liquid nitrogen in the presence of glycerol. - Z. Naturforsch. *32 C* : 254 -
256, 1977.

30212 - KANA, T.M., MILLER, J.H. : Effect of photoperiod on stomatal opening in *Vicia
faba*. - Plant Physiol. *60*: 803 - 804, 1977. [Stomatal resistance.]

30213 - KANDLER, O., LUGINGER, C., SCHILLING, N. : Triose phosphate shuttle and re-
arrangement of the carbon skeleton during photoassimilation of glucose by
Chlorella and higher plants. - In : COOMBS, J. (ed.) : 4th International
Congress on Photosynthesis. Pp. 186 - 187. UKISES, London 1977.

30214 - KANIVETS, N.P., VOLKOVA, N.V., REĬNGARD, T.A., ZAĬTSEVA, N.A., VASILENOK, L.I.,
MUSHKETIK, L.S., OSTROVSKAYA, L.K., YASNIKOV, A.A. : Ob izmenenii kharaktera
svetozavisimogo transporta protona u khloroplastov gorokha i kukuruzy pri
uvelichenii ikh osveshchennosti. [Changes in the character of proton light-
-dependent transport in pea and maize chloroplasts at increased illumination.]
- Dokl. Akad. Nauk ukr. SSR, Ser. B *1977* (3) : 246 - 250, 1977. [In R, ab : E.]

30215 - KANNANGARA, C.G., GOUGH, S.P. : Synthesis of Δ-aminolevulinic acid and chlorophyll by isolated chloroplasts. - Carlsberg Res. Commun. *42* : 441 - 457,
1977.

30216 - KANNANGARA, C.G., GOUGH, S.P., HANSEN, B., RASMUSSEN, J.N., SIMPSON, D.J. :
A homogenizer with replaceable razor blades for bulk isolation of active barley plastids. - Carlsberg Res. Commun. *42* : 431 - 439, 1977.

30217 - KANNANGARA, T., DURLEY, R.C., STOUT, D.G. : Hormones in relation to water deficit stress in *Sorghum bicolor* L. MOENCH. - Plant Physiol. *59* (6, Suppl.) :
92, 1977. [Chl.]

30218 - KAPLAN, A., GALE, J., POLJAKOFF-MAYBER, A. : Effect of oxygen and carbon dioxide concentrations on gross dark CO_2 fixation and dark respiration in *Bryophyllum daigremontianum*. - Aust. J. Plant Physiol. *4* : 745 - 752, 1977.

30219 - KAPLAN, A., SCHREIBER, U. : A proton gradient in intact cells of *Dunaliella
salina*. - Carnegie Inst. Year Book *76* : 320 - 323, 1977.

30220 - KAPLAN, A., SCHREIBER, U., AVRON, M. : Induction of CO_2-independent photosynthetic O_2 evolution in *Dunaliella salina*. - Carnegie Inst. Year Book *76* :
316 - 319, 1977.

30221 - KAPLAN, S.L., KOLLER, H.R. : Leaf area and CO_2-exchange rate as determinants
of the rate of vegetative growth in soybean plants. - Crop Sci. *17* : 35 - 38,
1977.

30222 - **KAPLANOVÁ, M., SOCHA, J.**: The effect of herbicides on the photooxidation of chlorophyll a in solution. ~ Photosynthetica *11* : 276 - 281, 1977.

30223 - **KARABAEV, M.K., GLAGOLEVA, T.A., ZALENSKIĬ, O.V.** : Ob uchastii tsiklicheskogo fotofosforilirovaniya v tsikle Kal'vina. [The role of cyclic photophosphorylation in Calvin cycle.] - Fiziol. Rast. *24* :·677 - 683, 1977. [In R,ab :E.]

30224 - **KARABAEV, M.K., GLAGOLEVA, T.A., ZALENSKIĬ, O.V.** : Vliyanie kisloroda na fotosintez khlorelly. [Effect of oxygen on photosynthesis of *Chlorella*.] - Bot. Zh. *62* : 802 - 810, 1977. [In R, ab : E.]

30225 - **KARABASHEV, G.S.** : Osobennosti raspredeleniya fluorestsentsii i rasseyaniya sveta v okeane pri intensivnom vertikal'nom peremeshivanii i pod"eme vod. [Distribution peculiarities of fluorescence and light scattering in the ocean at a strong vertical mixing and upwelling.]-Okeanologiya *17* : 312 - 318, 1977. [In R, ab : E.]

30226 - **KARABASHEV, G.S.** : O vliyanii fitoplanktona na oslablenie korotkovolnovoĭ solnechnoĭ radiatsii v Baltiĭskom more. [The influence of phytoplankton on the attenuation of short-wave solar radiation in the Baltic Sea.]-Okeanologiya *17* : 434 - 439, 1977. [Chl; in R, ab : E.]

30227 - **KARANTH, N.G.K., NAIR, S., BHARATHI, P.A.L.** : Studies on photosynthetic bacteria isolated from an estuarine beach of Goa. - Indian J. mar. Sci. *6* : 94 - 96, 1977. [Ps, Chl.]

30228 - **KARAPETYAN, N.V., KLIMOV, V.V., LANG, F., KRASNOVSKIĬ, A.A.** : Induktsiya fluorestsentsii normal'nykh i mutantnykh rasteniĭ kukuruzy. [Induction of fluorescence of normal and mutant plants of maize.] - In : NASYROV, Yu.S. (ed.) : Genetika Fotosinteza. Pp. 189 - 194. Donish, Dushanbe 1977. [In R.]

30229 - **KARAPETYAN, N.V., RAKHIMBERDIEVA, M.G., KRASNOVSKY, A.A.** : Biosynthesis of Photosystem 2 of higher plants. - In : COOMBS, J. (ed.) : 4th International Congress on Photosynthesis. Pp. 187 - 188. UKISES, London 1977.

30230 - **KARCZMARCZYK, S., ZBIEĆ, I.** : Badania możliwości stosowania norflurazonu do zwalczania *Agropyron repens* L. Część I. Wpływ norflurazonu na zawartość chlorofilu i karotenu w perzu rozgłogowym. [Possibility of using norflurazon in control of *Agropyron repens* L. Part I. Influence of norflurazon on chlorophyll and carotene contents in *Agropyron repens*.] - Acta agrobot. *30* : 135 - 150, 1977. [In Pol., ab : E.]

30231 - **KARLSTRÖM, U., BACKLUND, S.** : Relationship between algal cell number, chlorophyll a and fine particulate organic matter in a river in Northern Sweden. - Arch. Hydrobiol. *80* : 192 - 199, 1977.

30232 - **KARPILOV, Yu.S., AVDEEVA, T.A., PERSANOV, V.M.** : Lokalizatsiya reaktsiĭ uglerodnogo metabolizma v dvukh assimilyatsionnykh tkanyakh lista kukuruzy. [Localization of reactions of carbon metabolism in two assimilatory tissues of a maize leaf.] - In : NASYROV, Yu.S. (ed.) : Genetika Fotosinteza. Pp. 78 - 84. Donish, Dushanbe 1977. [In R.]

30233 - **KARPILOV, Yu.S., NOVITSKAYA, I.L., KUZ'MIN, A.N., MASLOV, A.I., POPOVA, E.I.:** Issledovanie obratimosti glikoliza v list'yakh C_4-rasteniĭ. [Reversibility of glycolysis in leaves of C_4-plants.] - Biokhimiya *42* : 1810 - 1816, 1977. [In R, ab : E.]

30234 - **KARPILOV, Yu.S., OPARINA, L.A., KUZNETSOVA, L.G., BIL', K.Ya., KARPOVA, R.N.:** Izmenenie fotosinteticheskogo apparata pri perekhode kul'tury tkani ruty ot fotogeterotrofnogo pitaniya k avtotrofnomu. [Changes in photosynthetic apparatus during transition of the rue tissue culture from photoheterotrophic to autotrophic nutrition.] - Fiziol. Biokhim. kul't. Rast. *9* : 93 - 99, 1977. [In R, ab : E.]

30235 - **KARPILOV, Yu.S., PERSANOV, V.M.** : Aktivnost' i funktsii NADF-malatdegidrogenaz v tkanyakh C_3- i C_4-rasteniĭ. [Activity and functions of NADP-malate dehydrogenases in tissues of C_3- and C_4-plants.] - Fiziol. Biokhim.kul't. Rast. *9* : 511 - 516, 1977. [In R, ab : E.]

30236 - **KARPILOVA, I.F., CHERMNYKH, L.N., CHUGUNOVA, N.G.** : Nekotorye osobennosti produktov fotosinteza i ottoka assimilyatov u ogurtsov pri razlichnoĭ temperature v zone korneĭ. [Characteristics of photosynthates and the outflow of photosynthates at different temperatures in the root zone in cucumber.] - Sel'skokhoz. Biol. *12* : 434 - 438, 1977. [In R.]

30237 - KARUNEN, P., IHANTOLA, A. : Studies on moss spores V. Carotenoids of *Polytrichum commune*. - Bryologist *80* : 88 - 92, 1977.

30238 - KARUNEN, P., IHANTOLA, A. : Studies on moss spores VI. Production of carotenoids in germinating *Polytrichum commune* spores. - Bryologist *80* : 313 - 316, 1977.

30239 - KARVALY, B., DANCSHÁZY, Z. : Bacteriorhodopsin: a molecular photoelectric regulator. Quenching of photovoltaic effect of bimolecular lipid membranes containing bacteriorhodopsin by blue light. - FEBS Lett. *76* : 36 - 40, 1977.

30240 - KASEMIR, H. : Feedback control of protochlorophyll(ide) holochrome on δ-aminolevulinic acid synthesis. - In : COOMBS, J. (ed.) : 4th International Congress on Photosynthesis. P. 189. UKISES, London 1977.

30241 - KASPERBAUER, M.J., HAMILTON, J.L. : Free amino acids in chlorophyllous and nonchlorophyllous tobacco callus tissue cultured in light or in darkness. - Phyton *35* : 103 - 108, 1977.

30242 - KASS, L.B., PAOLILLO, D.J.Jr. : Autoradiographic evidence for the effects of light on RNA and DNA synthesis during chloroplast replication in spores of *Polytrichum*. - J. Cell Sci. *28* : 61 - 70, 1977.

30243 - KASTORI, R. : Uticaj svetlosti i nekih inhibitora disanja i fotosinteze na usvajanje fosfora i kalcijuma od strane odsečaka listova mladih biljaka kukuruza. [The effect of both light and some inhibitors of respiration and photosynthesis upon phosphorus and calcium uptake by leaf slices of young maize plants.] - Arhiv poljopr. Nauke *30* : 43 - 55, 1977. [In Croat., ab : E.]

30244 - KATOH, S. : Plastocyanin. - In : TREBST, A., AVRON, M. (ed.) : Photosynthesis I. (Encycl. Plant Physiol. N.S. Vol. 5.) Pp. 247 - 252. Springer-Verlag, Berlin - Heidelberg - New York 1977.

30245 - KATOH, S. : The oxidation of C-550 by exogenously added oxidants in the presence of dichlorophenyldimethylurea: The localization of the primary electron acceptor of photosystem II in the thylakoid membrane. - Plant Cell Physiol. *18* : 893 - 906, 1977.

30246 - KATOH, S., KOIKE, H., SATOH, K. : Effects of electron transport inhibitors of flash-induced cytochrome *f* photooxidation in spinach chloroplasts. - In : COOMBS, J. (ed.) : 4th International Congress on Photosynthesis. P. 190. UKISES, London 1977.

*30247 - KATTAWAR, G.W., HUMPHREYS, T.J. : Remote sensing of chlorophyll in an atmosphere-ocean environment: a theoretical study. - Appl. Opt. *15* : 273 - 282, 1976.

30248 - KATZ, J.J., NORRIS, J.R., SHIPMAN, L.L. : Models for reaction-center and antenna chlorophyll. - In : OLSON, J.M., HIND, G. (ed.) : Chlorophyll-Proteins, Reaction Centers, and Photosynthetic Membranes. Pp. 16 - 55. Brookhaven nat. Lab., Upton 1977.

30249 - KAUFMANN, M.R. : Soil temperature and drought effects on growth of Monterey pine. - Forest Sci. *23* : 317 - 325, 1977. [Ps.]

30250 - KAUFMANN, M.R., ECKARD, A.N. : A portable instrument for rapidly measuring conductance and transpiration of conifers and other species. - Forest Sci. *23* : 227 - 237, 1977. [Stomatal conductance.]

30251 - KAUROV, B.S. : Photochemical activity of pea chloroplasts under powerful illumination. - In : COOMBS, J. (ed.) : 4th International Congress on Photosynthesis. P. 188. UKISES, London 1977.

30252 - KAZAKOVA, A.S., KAĬRIS, A.V., NIKOLAEVA, L.F., TIMOFEEV, K.N., MATORIN, D.N., VENEDIKTOV, P.S., RUBIN, A.B. : O prirode promezhutochnogo produkta v reaktsii fotovosstanovleniya protokhlorofillida. [Nature of intermediate product in protochlorophyllide photoreduction.] - Vestn. mosk. Univ., Ser. biol. *1977* (4) : 73 - 75, 1977. [In R, ab : E.]

30253 - **KAZARYAN, V.O., DAVTYAN, V.A., GEVORKYAN, I.A., CHILINGARYAN, A.A.** :
Izmenenie fiziologicheskogo sostoyaniya list'ev pri uvelichenii ikh korne-
obespechennosti. [Changes in physiological state of leaves at increasing
their root supply.] - In : Trudy bot. Inst. Akad. Nauk arm. SSR *20* (Voprosy
Individual'nogo Razvitiya Vysshikh Rasteniĭ) : 33 - 41, 1977. [In R.]

30254 - **KE, B.** : Recent developments in Photosystem I (PSI). - In : **COOMBS, J.** (ed.)
: 4th International Congress on Photosynthesis. P. 191. UKISES, London 1977.

30255 - **KE, B., DOLAN, E., SUGAHARA, K., HAWRIDGE, F.M., DEMETER, S., SHAW, E.R.** :
Electrochemical and kinetic evidence for a transient electron acceptor in
the photochemical charge separation in photosystem I. - Plant Cell Physiol.
1977 (Spec. Issue 3 - Photosynthetic Organelles. Structure and Function) :
187 - 199, 1977.

30256 - **KE, B., HAWRIDGE, F.M., SAHU, S.** : Redox titration of Photosystem II fluo-
rescence yield. - In : **OLSON, J.M., HIND, G.** (ed.) : Chlorophyll-Proteins, |
Reaction Centers, and Photosynthetic Membranes. P. 364. Brookhaven nat. Lab.,
Upton 1977.

30257 - **KEATINGE, J.D.H., GARRETT, M.K.** : Climatic control of leaf extension in
Lolium perenne varieties differing in photosynthetic efficiency. - In :
COOMBS, J. (ed.) : 4th International Congress on Photosynthesis. Pp. 191 -
192. UKISES, London 1977.

30258 - **KEEFER, L.M., BRADSHAW, R.A.** : Structural studies on *Halobacterium halobium*
bacteriorhodopsin. - Fed. Proc. *36* : 1799 - 1804, 1977.

30259 - **KEIFER, D., SPANSWICK, R.** : Membrane potential and resistance of *Chara au-
stralis*. - Plant Physiol. *59* (6, Suppl.) : 85, 1977.

30260 - **KELLER, R.M., WÜTHRICH, K.** : ^{1}H NMR studies at 360 MHz of the aromatic amino
acid residues in ferrocytochrome *c*-552 from *Euglena gracilis*. - Biochim.
biophys. Acta *491* : 416 - 422, 1977.

30261 - **KELLER, R.M., WÜTHRICH, K., SCHEJTER, A.** : ^{1}H NMR studies of the heme iron
coordination in cytochrome *c*-552 from *Euglena gracilis*. - Biochim. biophys.
Acta *491* : 409 - 415, 1977.

30262 - **KELLER, T.** : Begriff und Bedeutung der "latenten Immissionsschädigung". -
Allg. Forst - Jagdzeit. *148* : 115 - 120, 1977. [Ps.]

30263 - **KELLER, T.** : Der Einfluß von Fluorimmissionen auf die Nettoassimilation von
Waldbaumarten. - Mitt. eidgenöss. Anstalt forstliche Versuchswesen *53* :
161 - 198, 1977.

30264 - **KELLER, T.** : The effect of long during, low SO_2 concentrations upon photosynthesis of conifers.- In: Proceedings IV.International Clean Air Congress
1977.Pp. 81 - 83. Tokyo 1977.

30265 - **KELLEY, B.C., MEYER, C.M., GANDY, C., VIGNAIS, P.M.** : Hydrogen recycling by
Rhodopseudomonas capsulata. - FEBS Lett. *81* : 281 - 285, 1977.

30266 - **KELLY, G.J.** : Principles of enzyme regulation with examples from the enzymes
of photosynthetic carbon metabolism. - In : **COOMBS, J.** (ed.) : 4th International Congress on Photosynthesis. P. 193. UKISES, London 1977.

30267 - **KELLY, G.J., LATZKO, E.** : Chloroplast phosphofructokinase. I. Proof of phosphofructokinase activity in chloroplasts. - Plant Physiol. *60* : 290 - 294,
1977.

30268 - **KELLY, G.J., LATZKO, E.** : Chloroplast phosphofructokinase. II. Partial purification, kinetic and regulatory properties. - Plant Physiol. *60* : 295 - 299,
1977.

*30269 - **KELLY, M., NORGÅRD, S., LIAAEN-JENSEN, S.** : Bacterial carotenoids. XXXI. C_{50}-
-carotenoids 5. Carotenoids of *Halobacterium salinarium*, especially bacterio-
ruberin. - Acta chem. scand. *24* : 2169 - 2182, 1970.

30270 - **KEMP, P.R., WILLIAMS, G.J.III.** : A comparison of the effects of water stress
on the gas exchange responses of *Agropyron smithii* and *Bouteloua gracilis*. -
Plant Physiol. *59* (6, Suppl.) : 97, 1977.

30271 - KEMP, P.R., WILLIAMS, G.J.III., MAY, D.S. : Temperature relations of gas exchange in altitudinal populations of *Taraxacum officinale*. - Can. J. Bot.
55 : 2496 - 2502, 1977.

30272 - KENNEDY, R.A. : The effects of NaCl-, polyethyleneglycol-, and naturally-in-
duced water stress on photosynthetic products, photosynthetic rates, and CO_2
compensation points in C_4 plants. - Z. Pflanzenphysiol. *83* : 11 - 24, 1977.

30273 - KENNEDY, R.A. : Variation in C_3 and C_4 characteristics within the genus *Mollugo*. - In : COOMBS, J. (ed.) : 4th International Congress on Photosynthesis.
P. 192. UKISES, London 1977.

30274 - KENNEDY, R.A., BARNES, J.E., LAETSCH, W.M. : Photosynthesis in C_4 plant tissue cultures. Significance of Kranz anatomy to C_4 acid metabolism in C_4
plants. - Plant Physiol. *59* : 600 - 603, 1977.

30275 - KENNEDY, R.A., WILLIAMS, L.E. : Effect of different killing techniques on
early labeled photosynthetic products in C_4 plants. - Plant Physiol. *59* :
207 - 210, 1977.

30276 - KENT, S.S. : A plant survey for anomalously labeled citrate during photosynthesis in CO_2. - Plant Physiol. *59* (6, Suppl.) : 44, 1977.

30277 - KENT, S.S. : On the metabolic relationship between the Calvin cycle and the
tricarboxylic acid cycle. IV. A plant survey for anomalous acetyl coenzyme
A. - Plant Physiol. *60* : 274 - 276, 1977.

30278 - KENT, S.S., ANDERSEN, W.R., RINEHART, C.A. : Nodular-sink products of formate
utilization by *Vicia faba* during photosynthesis. - Plant Physiol. *59* (6, Suppl.) : 44, 1977.

30279 - KENT, S.S., RINEHART, C.A., ANDERSEN, W.R. : A method for obtaining the
^{14}C-isotope distribution in malate(C-2,3). - Anal. Biochem. *80* : 176 - 182,
1977.

30280 - KERBER, N.L., PUCHEU, N.L., GARCIA, A.F. : Possible initial events of photophosphorylation in membranes of *Rhodopseudomonas viridis* and *Rhodopseudomonas capsulata* Ala+ r. - FEBS Lett. *80* : 49 - 52, 1977.

B30281 - KERCHER, J.R. :GROW1 : A Crop Growth Model for Assessing Impacts of Gaseous
Pollutants from Geothermal Technologies. Lawrence Livermore Lab. Re. UCRL -
52247. Pp. 1 - 36. Univ. Calif., Livermore 1977.

30282 - KEREKES, J.J. : Factors relating to annual planktonic primary production in
five small oligotrophic lakes in Terra Nova National Park, Newfoundland. -
Inter. Rev. ges. Hydrobiol. *62* : 345 - 370, 1977.

30283 - KERNER, H., GROSS, E., KOCH, W. : Structure of the assimilation system of
a dominating spruce tree (*Picea abies* (L.) KARST.) of closed stand: Computa-
tion of needle surface area by means of a variable geometric needle model. -
Flora *166* : 449 - 459, 1977.

30284 - KERSCHER, L., OESTERHELT, D. : Ferredoxin is the coenzyme of α-ketoacid oxi-
doreductases in *Halobacterium halobium*. - FEBS Lett. *83* : 197 - 201, 1977.

30285 - KERSHAW, K.A. : Physiological-environmental interactions in lichens. II. The
pattern of net photosynthetic acclimation in *Peltigera cavina* (L.) WILLD.
var. *praetextata* (FLOERKE in SOMM.) HUE and *P. polydactyla* (NECK.) HOFFM. -
New Phytol. *79* : 377 - 390, 1977.

30286 - KERSHAW, K.A. : Physiological-environmental interactions in lichens. III.
The rate of net photosynthetic acclimation in *Peltigera canina* (L.) WILLD.
var. *praetextata* (FLOERKE in SOMM.) HUE. and *P. polydactyla* (NECK.) HOFFM. -
New Phytol. *79* : 391 - 402, 1977.

30287 - KESTLER, D.P., KATTERMAN, F.R.H., ENDRIZZI, J.E. : Buoyant density determinations on chloroplast DNA in a variegated cytoplasmic mutant of *Gossypium
hirsutum* L. - Biochem. biophys. Res. Commun. *76* : 720 - 727, 1977.

30288 - KEYS, A.J., BIRD, I.F., CORNELIUS, M.J., KUMARASINGHE, S.K., SAMPAIO, E.S.V.
B., WAIDYANATHA, U.P.de S., WHITTINGHAM, C.P. : Photosynthesis, photorespira-
tion and productivity: Biochemistry and effects of temperature. - In :
COOMBS, J. (ed.) : 4th International Congress on Photosynthesis. P. 194.
UKISES, London 1977.

30289 - **KEYS, A.J., SAMPAIO, E.V.S.B., CORNELIUS, M.J., BIRD, I.F.** : Effect of temperature on photosynthesis and photorespiration of wheat leaves. - J. exp. Bot. *28* : 525 - 533, 1977.

30290 - **KHALIFA, M.A., AKASHA, M.H., SAID, M.B.** : Growth and N-uptake by wheat as affected by sowing date and nitrogen in irrigated semi-arid conditions. - J. agr. Sci. *89* : 35 - 42, 1977. [Growth analysis.]

30291 - **KHANNA, R., GOVINDJEE, WYDRZYNSKI, T.** : Site of bicarbonate effect in Hill reaction. Evidence from the use of artificial electron acceptors and donors.- Biochim. biophys. Acta *462* : 208 - 214, 1977.

30292 - **KHANOVA, L.A., TARASEVICH, M.R.** : Élektrokhimicheskoe povedenie adsorbtsionnykh plenok khlorofilla. [Electrochemical behaviour of adsorption layers of chlorophyll.] - Dokl. Akad. Nauk SSSR *234* : 211 - 214, 1977. [In R.]

30293 - **KHARKATS, Yu.I., VOLKOV, A.G., BOGUSLAVSKY, L.I.** : Transfer of ions and electrons across the interface between two immiscible liquids in functioning enzyme membrane systems. - J. theor. Biol. *65* : 379 - 391, 1977. [Chl.]

30294 - **KHAVARI-NEJAD, R.A.** : Effects of α-hydroxy-2-pyridinemethanesulfonic acid on photosynthetic carbon dioxide uptake and stomatal movements in excised tomato leaves. - Plant Physiol. *60* : 44 - 46, 1977.

30295 - **KHIKE, B.** : O stabil'nosti funktsional'nykh struktur khloroplastov razlichnogo vozrasta pri vozdeĭstvii digitonina. [Stability of functional structures of chloroplasts of different age under the action of digitonin.] - Fiziol. Rast. *24* : 251 - 260, 1977. [In R, ab : E.]

30296 - **KHIMENOV, G.P.** : Effect of physiologically active mixture upon the water relations of maize. - In : KUDREV, T., IVANOVA, I., KARANOV, E. (ed.) : Plant Growth Regulators. Pp. 692 - 695. Publishing House of the Bulgarian Academy of Sciences, Sofia 1977.

30297 - **KHISAMUTDINOVA, V.I., KUZ'MINA, G.G., VASIL'EVA, I.M.** : Énergeticheskie protsessy khloroplastov ozimoĭ pshenitsy pri obrabotke semyan preparatom TUR v svyazi s zakalivaniem. [Energy processes of winter wheat chloroplasts during hardening after seed treatment with the preparation TUR.] - Fiziol. Rast. *24* : 1032 - 1037, 1977. [In R, ab : E.]

✶30298 - **KHODORENKO, L.A., BARAĬ, A.T.** : Vliyanie razlichnykh doz ul'trafioletovogo oblucheniya semyan ogurtsov na formirovanie fotosinteticheskogo apparata. [Effect of irradiation of cucumber seeds with different doses of ultraviolet radiation on the formation of the photosynthetic apparatus.] - In : KAKHNOVICH, L.V. (ed.) : Optimizatsiya Fotosinteticheskogo Apparata Vozdeĭstviem Razlichnykh Faktorov. Pp. 32 - 39. Izd. BGU, Minsk 1976. [In R.]

30299 - **KHODZHAEV, M.N., NIKITINA, K.A., GUSEV, M.V.** : O povedenii vodorosli *Anacystis nidulans* v temnote pri raznykh temperaturakh. [Behavior of the alga *Anacystis nidulans* in the darkness at different temperatures.] - Vestnik mosk. Univ. *1977* (1) : 79 - 83, 1977. [Ps, Chl; in R, ab : E.]

30300 - **KHOKHLOVA, V.A.** : Deĭstvie tsitokinina na formirovanie plastid na svetu i v temnote v izolirovannykh semyadolyakh tykvy. [The effect of cytokinin on plastid formation in excised pumpkin cotyledons in the light and in the dark.] - Fiziol. Rast. *24* : 1189 - 1193, 1977. [In R, ab : E.]

✶30301 - **KHRAMOVA, G.A., YAKUBOVA, M.M.** : K voprosu vydeleniya khloroplastov iz list'ev khlopchatnika. [Chloroplast isolation from cotton leaves.] - Dokl. Akad. Nauk tadzh. SSR *18* (6) : 62 - 64, 1975. [In R, ab : Tajik.]

30302 - **KHRAMOVA, G.A., YAKUBOVA, M.M., KRENDELEVA, T.E.** : Issledovanie pervichnykh reaktsiĭ fotosinteza v list'yakh khlopchatnika sorta 108-F i ego pestrolistnogo mutanta. [Primary reactions of photosynthesis in leaves of cotton cultivar 108-F and its variegated mutant.] - Nauch. Dokl. vyssh. Shkoly, biol. Nauki *20* (7) : 42 - 47, 1977. [In R.]

30303 - **KHRISTIN, M.S., AKULOVA, E.A., SUROVTSEV, V.I.** : Immobilizovannyĭ na sefaroze 4B tsitokhrom *c* i ego uchastie v fotokhimicheskikh reaktsiyakh khloroplastov. [Cytochrome *c* immobilized on sepharose 4B and its participation in photochemical reactions of chloroplasts.] - Biokhimiya *42* : 124 - 128, 1977. [In R, ab : E.]

30304 - KHRISTIN, M.S., AKULOVA, E.A., SUROVTSEV, V.I. : Immobilizatsiya ferredoksina. [Immobilization of ferredoxin.] - Biokhimiya 42 : 306 - 310, 1977.
[In R,ab : E.]

*30305 - KIEFER, D.A., OLSON, R.J., HOLM-HANSEN, O. : Another look at the nitrite
and chlorophyll maxima in the central North Pacific. - Deep-Sea Res. 23 :
1199 - 1208, 1976.

30306 - KIERSTAN, M., BUCKE, C. : The immobilization of microbial cells, subcellular
organelles, and enzymes in calcium alginate cells. - Biotechnol. Bioeng.
19 : 387 - 397, 1977. [Chloroplasts.]

30307 - KIMURA, T., MATSON, R.S. : Chloroplast ferredoxin biosynthesis in *Euglena
gracilis*. - In : COOMBS, J. (ed.) : 4th International Congress on Photosynthesis. P. 195. UKISES, London 1977.

30308 - KING, D., ERBES, D.L., BEN-AMOTZ, A., GIBBS, M. : Hydrogen metabolism in
photosynthetic organisms, the mechanism of hydrogen photoevolution. - In :
CASTELLANI, A. (ed.) : Research in Photobiology. Pp. 329 - 334. Plenum Press,
New York - London 1977.

30309 - KING, D., ERBES, D.L., GIBBS, M. : Inhibition of ferricyanide reduction in
chloroplast particles by anaerobicity. - Biochem. biophys. Res. Commun. 78 :
734 - 738, 1977.

30310 - KING, R.W., EVANS, L.T. : Inhibition of flowering in *Lolium temulentum* L. by
water stress: a role for abscisic acid. - Aust. J. Plant Physiol. 4 : 225 -
233, 1977. [Ps.]

30311 - KIPE-NOLT, J.A., STEVENS, S.E.Jr. : δ-aminolevulinic acid production by the
blue-green alga *Agmenellum quadruplicatum* strain PR-6. - Plant Physiol. 59
(6, Suppl.) : 91, 1977.

30312 - KIRICHENKO, A.B., KIRICHENKO, E.B., CHEBOTAR', A.A. : Ul'trastructura pyl'-
nika *Hordeum vulgare* L. na stadii dvukhkletochnoĭ pyl'tsy: osobennosti dif-
ferentsirovki plastid. [Ultrastructure of anther of *Hordeum vulgare* L. at
the stage of bicellular pollen: characteristics of plastid differentiation.]
- Fiziol. Rast. 24 : 751 - 755, 1977. [In R, ab : E.]

*30313 - KIRICHENKO, E.B., SMOLYGINA, L.D., SERDYUK, O.P., ANDREEV, L.V. : Sostav ka-
rotinoidov golokhromnykh kompleksov v razvivayushchikhsya plastidakh mezofilla i obkladki *Zea mays* L. [Composition of carotenoids of holochrome complexes in developing mesophyll and bundle sheath plastids of *Zea mays* L.] -
Fiziol. Rast. 23 : 626 - 629, 1976. [In R.]

30314 - KIRK, J.T.O. : Thermal dissociation of fucoxanthin-protein binding in pigment
complexes from chloroplasts of *Hormosira (Phaeophyta)*. - Plant Sci. Lett.
9 : 373 - 380, 1977.

30315 - KIRK, J.T.O. : Use of a quanta meter to measure attenuation and underwater
reflectance of photosyntheticaly active radiation in some inland and coastal
South-eastern Australian waters. - Aust. J. mar. Freshwater Res. 28 : 9 -
21, 1977. [Chl.]

30316 - KLEE, R., FLEMMING, B.-U. : Wechselwirkung zwischen Atmung der Bodenorganis-
men und Photosynthese der grünen Pflanzen - ein Schülerversuch. - Prax. Naturwiss., Biol. 26 : 270 - 272, 1977.

30317 - KLEIN, D.A. : Seasonal carbon flow and decomposer parameter relationships in
a semiarid grassland soil. - Ecology 58 : 184 - 190, 1977. [Root biomass.]

30318 - KLEIN, O., SENGER, H. : Pathways of δ-ALA formation during the development
of the photosynthetic apparatus in pigment mutant C-24 of *Scenedesmus*. -
In : COOMBS, J. (ed.) : 4th International Congress on Photosynthesis. Pp.
195 - 196. UKISES, London 1977.

30319 - KLEIN, S., KATZ, E., NEEMAN, E. : Induction of δ-aminolevulinic acid formation in etiolated maize leaves controlled by two light systems. - Plant Physiol. 60 : 335 - 338, 1977.

30320 - KLEIN, S., KONIS, Y., OHAD, I. : The effect of levulinic acid on development
of PS I and PS II activities and on thylakoid proteins in greening maize lea-

ves. - In : COOMBS, J. (ed.) : 4th International Congress on Photosynthesis. P. 196. UKISES, London 1977.

30321 - KLEIN, S.M., VERNON, L.P. : Composition of a photosystem I chlorophyll protein complex from *Anabaena flos-aquae*. - Biochim. biophys. Acta *459* : 364 - 375, 1977.

30322 - KLEINEN HAMMANS, J.W. : The photoreduction of oxygen by *Anacystis nidulans:* Influence of added and endogenous reductans. - In : COOMBS, J. (ed.) : 4th International Congress on Photosynthesis. P. 197. UKISES, London 1977.

30323 - KLEINEN HAMMANS, J.W., HENDRIKS, G.M., TEERLINK, T. : Light dependent oxygen uptake by the blue green alga *Anacystis nidulans*. - Biochem. biophys. Res. Commun. *74* : 1560 - 1565, 1977.

30324 - KLEINEN HAMMANS, J.W., TEERLINK, T., VAN DEN BERG, J.T. : Light dependent oxygen uptake by *Anacystis nidulans*, studies with endogenous and added reductants. - Acta bot. neerl. *26* : 285 - 288, 1977. [Ps.]

B30325 - KLEINKOPF, G.E., STEEN, A.J., HARTSOCK, T.L., WALLACE, A. : Determination of Photosynthesis and Transpiration Using a Flow Through Gas Exchange Chamber. - Laboratory of Nuclear Medicine and Radiation Biology, Univ. California, Los Angeles 1977.

30326 - KLEVANIK, A.V., KLIMOV, V.V., SHUVALOV, V.A., KRASNOVSKIĬ, A.A. : Vosstanovlenie feofitina v svetovoĬ reaktsii fotosistemy II vysshikh rasteniĬ. [Reduction of pheophytin in the light reaction of the photosystem II of higher plants.] - Dokl. Akad. Nauk SSSR *236* : 241 - 244, 1977. [In R.]

30327 - KLIMOV, V.V., KLEVANIK, A.V., SHUVALOV, V.A., KRASNOVSKY, A.A. : Reduction of pheophytin in the primary light reaction of photosystem II. - FEBS Lett. *82* : 183 - 186, 1977.

30328 - KLIMOV, V.V., SHUVALOV, V.A., KRAKHMALEVA, I.N., KLEVANIK, A.V., KRASNOVSKIĬ, A.A. : Fotovosstanovlenie bakteriofeofitina *b* v pervichnoĬ svetovoĬ reaktsii khromatoforov *Rhodopseudomonas viridis*. [Photoreduction of bacteriopheophytin *b* in the primary light reaction of *Rhodopseudomonas viridis* chromatophores.] - Biokhimiya *42* : 519 - 530, 1977. [In R, ab : E.]

*30329 - KLIMOVICH, A.S. : Formirovanie fotosinteticheskogo apparata i ego fotokhimicheskaya aktivnost' v zavisimosti ot intensivnosti i kachestva sveta. [Formation of the photosynthetic apparatus and its photochemical activity in relation to illuminance and light quality.] - In : KAKHNOVICH, L.V. (ed.) : Optimizatsiya Fotosinteticheskogo Apparata VozdeĬstviem Razlichnykh Faktorov. Pp. 25 - 32. Izd. BGU, Minsk 1976. [In R.]

30330 - KLOCKARE, B. : Influence of far-red irradiation on the protochlorophyllide components in wheat leaves. - In : COOMBS, J. (ed.) : 4th International Congress on Photosynthesis. Pp. 197 - 198. UKISES, London 1977.

30331 - KLOCKARE, B., SUNDQVIST, C. : Shifts in absorption and fluorescence maxima of chlorophyll(ide) in spectra of dark grown wheat leaves after irradiation. - Photosynthetica *11* : 189 - 199, 1977.

30332 - KLUGE, M. : Is *Sedum acre* L. a CAM plant ? - Oecologia *29* : 77 - 83, 1977.

30333 - KLUGE, M. : Regulation of carbon dioxide fixation in plants. - In : JENNINGS, D.H. (ed.) : Integration of Activity in the Higher Plant. Pp. 155 - 175. Cambridge Univ. Press, London - New York 1977.

30334 - KLYUCHAROVA, A.A., CHAĬKA, M.Ts., SHLYK, A.A. : Uplyŭ khloramfenikolu i tsyklageksimidu na razmerkavanne novaŭtvoranykh malekul byalkoŭ u membrannaĬ sistéme khlaraplastaŭ. [Effect of chloramphenicol and cycloheximide on the distribution of newly formed protein molecules in the membrane system of chloroplasts.] - Vestsi Akad. Navuk BSSR, Ser. biyal. Navuk *1977* (6) : 11 - 18, 137, 1977. [In Belorus., ab : R.]

30335 - KNAFF, D.B. : The primary reaction of plant photosystem II. - Photochem. Photobiol. *26* : 327 - 340, 1977.

30336 - KNAFF, D.B. : The role of cytochrome b_6 and cytochrome f in cyclic electron flow in a blue-green alga. - Arch. Biochem. Biophys. *182* : 540 - 545, 1977.

30337 - **KNAFF, D.B., MALKIN, R., MYRON, J.C., STOLLER, M.** : The role of plastoquinone and β-carotene in the primary reaction of Photosystem II. - In : **COOMBS, J.** (ed.) : 4th International Congress on Photosynthesis. P. 198. UKISES, London 1977.

30338 - **KNAFF, D.B., MALKIN, R., MYRON, J.C., STOLLER, M.** : The role of plastoquinone and β-carotene in the primary reaction of plant photosystem II. - Biochim. biophys. Acta *459* : 402 - 411, 1977.

30339 - **KNAUER, G.A., AYERS, A.V.** : Changes in carbon, nitrogen, adenosine triphosphate, and chlorophyll *a* in decomposing *Thalassia testudinum* leaves. - Limnol. Oceanogr. *22* : 408 - 414, 1977.

30340 - **KNOBLOCH, K.** : Energy transformation and electron transfer in cell-free preparations from non-sulfur purple bacteria. - In : **COOMBS, J.** (ed.) : 4th International Congress on Photosynthesis. P. 199. UKISES, London 1977.

30341 - **KNOBLOCH, K.** : Lipoic acid involved as a substituent for ATP in *Rhodopseudomonas palustris*. - Hoppe-Seyler's Z. physiol. Chem. *358* : 262 - 263, 1977.

30342 - **KNOBLOCH, K., ACKER, G.** : Ferritin as a marker for open vesicular membranes in chemically fixed preparations from *Rhodopseudomonas palustris*. - Hoppe-Seyler's Z. physiol. Chem. *358* : 1232 - 1233, 1977.

30343 - **KNOF, G.** : Eine transportable Meßanordnung zur Erfassung der CO_2-Aufnahme, der Beleuchtungsstärke, der Temperatur und des relativen Wassergehaltes an Pflanzenblättern in Feldbeständen. - Arch. Acker- Pflanzenbau Bodenk. *21* : 35 - 44, 1977.

30344 - **KNOF, G.** : Kombinierte Bestimmung des Gesamtkohlenstoffs und des ^{14}C-Anteils in Bodenproben und organischen Substanzen. - Arch. Acker- Pflanzenbau Bodenk. *21* : 535 - 543, 1977.

30345 - **KNOTH, R., HAGEMANN, R.** : Struktur und Funktion der genetischen Information in den Plastiden. XVI. Die Feinstruktur der Plastiden und der elektronenmikroskopische Nachweis echter Mischzellen in Blättern der Plastommutationen auslösenden Genmutante albostrians vor *Hordeum vulgare* L. - Biol. Zentralbl. *96* : 141 - 150, 1977.

30346 - **KNOWLTON, L.L., SINK, K.C.Jr.** : The G locus in *Petunia hybrida* VILM : Negative relationship to endogenous chlorophyll, starch and sugar. - Euphytica *26* : 433 - 440, 1977.

30347 - **KNOX, R.S.** : Photosynthetic efficiency and exciton transfer and trapping. - In : **BARBER, J.** (ed.) : Primary Processes of Photosynthesis. Pp. 55 - 97. Elsevier, Amsterdam - New York - Oxford 1977.

30348 - **KNUDSON, L.L., TIBBITTS, T.W., EDWARDS, G.E.** : Measurement of ozone injury by determination of leaf chlorophyll concentration. - Plant Physiol. *60* : 606 - 608, 1977.

30349 - **KNÜPPEL, D.** : Methoden der prognostischen Schätzung von Ernteerträgen in der Pflanzenproduktion. - Arch. Acker- Pflanzenbau Bodenk. *21* : 649 - 657, 1977.

39350 - **KNYPL, J.S., KABZIŃSKA, E.** : Growth, phosphatase and ribonuclease activity in phosphate-deficient *Spirodela oligorrhiza* cultures. - Biochem. Physiol. Pflanzen *171* : 279 - 287, 1977. [Chl.]

*30351 - **KOBAYASHI, H., HITAKA, N.** : Studies on the lodging of rice plant. (6) On the carbon dioxide assimilation and translocation of assimilated products in the lodged plant. - J. agr. Meteorol. *24* : 15 - 23, 1968.

*30352 - **KOBAYASHI, H., HITAKA, N.** : Studies on the lodging of rice plant. (7) On the photosynthetic ability in lodged rice plant. - J. agr. Meteorol. *24* : 67 - 74, 1968.

*30353 - **KOBAYASHI, S.** : Growth analysis of plant as an assemblage of internodal segments - a case of sunflower plants in pure stands -. - Jap. J. Ecol. *25* : 61 - 70, 1975.

30354 - **KOBAYASHI, T., NISHIMURA, S., TANAKA, S.** : [Comparative growth responses of seven tropical and subtropical grasses to various controlled temperatures.] - Sci. Bull. Fac. Agr., Kyushu Univ. *2* (3) : 93 - 99, 1977. [Chl; in Jap., ab : E.]

30355 - KOBAYASHI, Y., INOUE, Y., SHIBATA, K. : Life time of the light-induced reactivity of chloroplasts to p-nitrothiophenol. - Plant Cell Physiol. _18_ :
453 - 457, 1977.

30356 - KOBLENTZ-MISHKE, O.J., PELEVIN, V.N., SEMENOVA, M.A. : Phytoplankton pigments and efficiency of photosynthesis. - Pol. Arch. Hydrobiol. _24_ (Suppl.)
: 185 - 199, 1977.

30357 - KOBLENTZ-MISHKE, O.J., SEMENOVA, M.A. : Phytoplankton pigments in the meso-
trophic and eutrophic regions of the tropical Pacific Ocean. - Pol. Arch.
Hydrobiol. _24_ (Suppl.) : 173 - 183, 1977.

30358 - KOCH, W., ELLER, B.M. : Tagesgang von Globalstrahlungsverlusten in Gaswechsel-
kammern bei verschiedenen Expositionen. - Flora _166_ : 279 - 288, 1977.

30359 - KOCHUBEĬ, S.M., SAMOKHVAL, E.G., SERIKOV, A.A., KHOMENKO, Yu.M. : Model'
donor-aktseptornogo ènergoperenosa v fotosisteme I vysshikh rasteniĭ. [Mo-
del of donor-acceptor energy transfer in Photosystem I of higher plants.] -
Dokl. Akad. Nauk ukr. SSR, Ser. A _1977_ : 363 - 366, 1977. [In R, ab : E.]

*30360 - KOCHUBEĬ, S.M., SAMOKHVAL, E.G., SHADCHINA, T.M. : Osobennosti temperaturnoĭ
zavisimosti spektrov fluorestsentsii fragmentov khloroplastov. [Temperature
dependence characteristics of the fluorescence spectra of chloroplast fragments.] - Molek. Genet. Biofiz. (Kiev) _1_ : 25 - 32, 1976. [In R, ab : E.]

30361 - KOENIG, F., MENKE, W., RADUNZ, A., SCHMID, G.H. : Localization and functional characterization of three thylakoid membrane polypeptides of the molecular weight 66 000. - Z. Naturforsch. _32 C_ : 817 - 827, 1977.

30362 - KOENIG, F., SCHMID, G.H., RADUNZ, A., MENKE, W. : Localization of a polypeptide fraction with the apparent molecular weight 66 000 in the thylakoid
membrane. - In : COOMBS, J. (ed.) : 4th International Congress on Photosynthesis. P. 200. UKISES, London 1977.

30363 - KOHNO, H., YOSHIDA, F. : Culture of chlorophyllous tobacco-cells not requi-
ring any organic additives except sucrose in the medium. I. Effects of light
and temperature on the growth of the cells. - Plant Cell Physiol. _18_ : 907 -
913, 1977. [Chl.]

30364 - KOIWAI, A., FUKUDA, M., KISAKI, T. : Effect of piperonyl butoxide and diphe-
nylamine on lipid peroxidation in ozonated chloroplasts. - Plant Cell Physiol.
_18_ : 127 - 139, 1977.

30365 - KOK, B., VELTHUYS, B. : Present status of the O_2 evolution model. - In :
CASTELLANI, A. (ed.) : Research in Photobiology. Pp. 111 - 119. Plenum
Press, New York - London 1977.

30366 - KOKA, P., SONG, P.-S. : The chromophore topography and binding environment
of peridinin · chlorophyll a · protein complexes from marine dinoflagellate
algae. - Biochim. biophys. Acta _495_ : 220 - 231, 1977.

30367 - KOLESNIKOV, M.P., EGOROV, I.A. : Metalloporfiriny v otlozheniyakh dokembri-
ya. [Metalloporphyrins in precambrian formations.] - Dokl. Akad. Nauk SSSR
_233_ (3) : 483 - 486, 1977. [Ps; in R.]

30368 - KOLESNIKOV, M.P., EGOROV, I.A. : Proizvodnye khlorofilla v sovremennykh
pochvakh v svyazi s problemoĭ khimicheskoĭ èvolyutsii i proiskhozhdeniya
zhizni na zemle. [Chlorophyll derivatives in recent soils in connection with
the problem of chemical evolution and the origin of life on earth.] -
Dokl. Akad. Nauk SSSR _235_ : 228 - 231, 1977. [In R.]

30369 - KOLESNIKOV, P.A., ZORÉ, S.V., PETROCHENKO, E.I. : Obrazovanie glioksalevoĭ
kisloty iz ribozo-5-fosfata v khloroplastakh. [Formation of glyoxalic acid
from ribose-5-phosphate in chloroplasts.] - Dokl. Akad. Nauk SSSR _233_ :
974 - 977, 1977. [In R.]

30370 - KOLESNIKOV, P.A., ZORÉ, S.V., PETROCHENKO, E.I. : Regulyatsiya obrazovaniya
3-fosfoglitserinovoĭ kisloty okislitel'nym putem iz ribozo-5-fosfata v èks-
traktakh iz khloroplastov. [Regulation of synthesis of 3-phosphoglyceric
acid _via_ oxidative pathway from ribose-5-phosphate in chloroplast extracts.]
- Fiziol. Rast. _24_ : 267 - 272, 1977. [In R, ab : E.]

30371 - **KOLEVA, S., MEKHANDZHIEVA, A.** : Vliyanie na meteorologichnite faktori i os-
novnite vodno-fizichni kharakteristiki na pochvata v"rkhu transpiratsiyata
i evapotranspiratsiyata. [Influence of the meteorological factors and basic
hydrophysical characteristics of soil on the transpiration and evapotranspiration.] - Rasteniev"dni Nauki *14* (1) : 134 - 143, 1977. [Dry-matter accumulation; in Bulg., ab : R, E.]

30372 - **KOLLER, K.P., MÖRSCHEL, E., WEHRMEYER, W.** : Properties of a stable energy
transfering phycoerythrin-phycocyanin aggregate from disc-shaped phycobilisomes of *Rhodella violacea*. - In : COOMBS, J. (ed.) : 4th International Congress on Photosynthesis. Pp. 200 - 201. UKISES, London 1977.

30373 - **KOLLER, K.P., WEHRMEYER, W., SCHNEIDER, H.** : Isolation and characterization
of disc-shaped phycobilisomes from the red alga *Rhodella violacea*. - Arch.
Microbiol. *112* : 61 - 67, 1977.

30374 - **KÖLLER, W., KINDL, H.** : Glyoxylate cycle enzymes of the glyoxysomal membrane
from cucumber cotyledons. - Arch. Biochem. Biophys. *181* : 236 - 248, 1977.

*30375 - **KOLOMEĬCHENKO, V.V.** : Puti povysheniya koêffitsienta ispol'zovaniya solnechnoĭ ênergii na formirovanie urozhaya. [Ways of increasing efficiency of solar energy utilization in yield formation.] - In : Radiatsionnye Protsessy
v Atmosfere i na Zemnoĭ Poverkhnosti. Pp. 428 - 430. Gidrometeoizdat, Leningrad 1974. [In R.]

*30376 - **KOLOMEĬCHENKO, V.V.** : Vliyanie srokov seva i norm vyseva na nekotorye fiziologicheskie pokazateli i urozhaĭ zernobobovykh kul'tur pri oroshenii.
[Effect of time of sowing and sowing norms on some physiological characteristics and yield of irrigated leguminous crops.] - Byul. nauchno-tekh. Inf.
vses. nauchno-issled. Inst. zernobob. krup. Kul't. (Orel) *9* : 27 - 32,
1974. [Growth analysis; in R.]

*30377 - **KOLOMEĬCHENKO, V.V.** : Vliyanie vysokikh doz mineral'nykh udobreniĭ na foto-
sinteticheskuyu deyatel'nost' i urozhaĭ tomatov pri oroshenii. [Effect of
increased supply of mineral fertilizers on photosynthetic activity and yield
of irrigated tomato.] - Agrokhimiya *1974* (1) : 98 - 101, 1974. [In R.]

30378 - **KOLOMEĬCHENKO, V.V.** : K teorii fotosinteticheskoĭ deyatel'nosti oroshaemykh
senokosov i pastbishch. [Theory of photosynthetic activity of irrigated
meadows and pastures.] - In : Upravlenie Kompleksom Faktorov Zhizni Rasteniĭ
na Melioriruemykh Zemlyakh. P. 80. Min. Melior. vod. Khoz., Frunze 1977.
[In R.]

30379 - **KOLOMEĬCHENKO, V.V.** : Novyĭ sposob opredeleniya ploshchadi list'ev v posevakh zlakovykh kul'tur. [A new method for determining leaf area in cereal
stands.] - In : Upravlenie Kompleksom Faktorov Zhizni Rasteniĭ na Melioriruemykh Zemlyakh. Pp. 109 - 111. Min. Melior. vod. Khoz., Frunze 1977. [In R.]

30380 - **KOLOMEĬCHENKO, V.V.** : Vliyanie mineral'nykh udobreniĭ na koêfitsient ispol'-
zovaniya solnechnoĭ ênergii senokosami i pastbishchami. [Effect of mineral
fertilizers on solar energy utilization by meadows and pastures.] - In :
Tezisy Dokladov regional'nogo Soveshchaniya "Itogi Raboty Geograficheskoĭ
Seti Opytov s Udobreniyami i Puti Povysheniya Êffektivnosti Primeneniya
Udobreniĭ v Nechernozemnoĭ Zone. Pp. 130 - 131. Min. sel'. Khoz., Moskva
1977. [In R.]

30381 - **KOMÁRKOVÁ, J., JAVORNICKÝ, P.** : Circadian changes in the photosynthetic capacity and chlorophyll content of phytoplankton in eutrophic waters. - Arch.
Hydrobiol., Suppl. *51* (Algol. Stud. 18) : 77 - 100, 1977.

30382 - **KONDRATIEVA, E.N.** : Pathways of CO_2 assimilation in purple bacteria. - In :
COOMBS, J. (ed.) : 4th International Congress on Photosynthesis. P. 201.
UKISES, London 1977.

30383 - **KONISHI, T., PACKER, L.** : Chemical modification of bacteriorhodopsin with
N-bromosuccinimide. - FEBS Lett. *79* : 369 - 373, 1977.

30384 - **KONISHI, T., PACKER, L.** : Hydrogen exchange of dark-adapted and illuminated
bacteriorhodopsin. - FEBS Lett. *80* : 455 - 458, 1977.

30385 - **KONONENKO, A.** : Charge separation in photosynthetic reaction centres (RC) as conformation-controlled act. - In : COOMBS, J. (ed.) : 4th International Congress on Photosynthesis. P. 202. UKISES, London 1977.

30386 - **KONONENKO, A.A., NIKOLAEV, G.M., TIMOFEEV, K.N., LUKASHEV, E.P., RUBIN, A.B., CHUDINA, V.I., OSNITSKAYA, L.K.** : Fotoindutsirovannyĭ perenos êlektronov i sostoyanie vody v kletkakh *Chromatium vinosum*. [Light-induced electron transport and water state in *Chromatium vinosum* cells.] - Izv. Akad. Nauk SSSR, Ser. biol. *1977* : 869 - 878, 1977. [In R, ab : E.]

30387 - **KONSTANTINOV, A.R.** : Ein komplexes biophysikalisches Modell für die Ertragsleistung landwirtschaftlich genutzter Kulturpflanzen. - In : UNGER, K. (ed.) : Biophysikalische Analyse Pflanzlicher Systeme. Pp. 70 - 79. VEB Gustav Fischer Verlag, Jena 1977.

30388 - **KOOYMAN, R.P.H., SCHAAFSMA, T.J., KLEIBEUKER, J.F.** : Fluorescence spectra and zero-field magnetic resonance of chlorophyll *a*-water complexes. - Photochem. Photobiol. *26* : 235 - 240, 1977.

30389 - **KOPELEVICH, O.V., BURENKOV, V.I.** : O svyazi mezhdu spektral'nymi znacheniyami pokazateleĭ pogloshcheniya sveta morskoĭ vodoĭ, pigmentami fitoplanktona, zheltym veshchestvom. [Correlation between spectral values of the absorption coefficient of sea water, phytoplankton pigments, yellow substance.] - Okeanologiya *17* : 427 - 433, 1977. [In R, ab : E.]

30390 - **KORENSTEIN, R., HESS, B.** : Hydration effects on *cis-trans* isomerization of bacteriorhodopsin. - FEBS Lett. *82* : 7 - 11, 1977.

30391 - **KORENSTEIN, R., HESS, B.** : Hydration effects on the photocycle of bacteriorhodopsin in thin layers of purple membrane. - Nature *270* : 184 - 186, 1977.

30392 - **KÖRNER, C.** : Blattdiffusionswiderstände verschiedener Pflanzen im alpinen Grasheidegürtel der Hohen Tauern. - In : CERNUSCA, A. (ed.) : Alpine Grasheide Hohe Tauern. Veröf. Österreich. MaB-Hochgebirgsprogrammes Hohe Tauern. Vol. 1. Pp. 69 - 81. Universitätsverlag Wagner, Innsbruck 1977.

30393 - **KÖRNER, C.** : Evapotranspiration und Transpiration verschiedener Pflanzenbestände im alpinen Grasheidegürtel der Hohen Tauern. - In : CERNUSCA, A. (ed.) : Alpine Grasheide Hohe Tauern. Veröf. Österreich. MaB-Hochgebirgsprogrammes Hohe Tauern. Vol. 1. Pp. 47 - 68. Universitätsverlag Wagner, Innsbruck 1977.

30394 - **KÖRNER, C.** : Der CO_2-Gaswechsel verschiedener Pflanzen im alpinen Grasheidegürtel. I. Der Einsatz einer neuen teilklimatisierten Messkammer für *in-situ*-Messungen an kleinwüchsigen Gebirgspflanzen. - In : CERNUSCA, A. (ed.) : Alpine Grasheide Hohe Tauern. Veröf. Österreich. MaB-Hochgebirgsprogrammes Hohe Tauern. Vol. 1. Pp. 133 - 139. Universitätsverlag Wagner, Innsbruck 1977.

30395 - **KORNYUSHENKO, G.A., SAPOZHNIKOV, D.I., EVDOKIMOVA, I.V.** : Issledovanie reaktsii violaksantinovogo tsikla v izolirovannykh khloroplastakh. [Investigation of reactions of the violaxanthin cycle in isolated chloroplasts.] - Fiziol. Rast. *24* : 710 - 717, 1977. [In R, ab : E.]

*30396 - **KOROL'KOVA, N.A.** : Soderzhanie khlorofilla i karotinoidov v list'yakh razlichnykh sortov ogurtsa v svyazi s rostom i produktivnost'yu. [Content of chlorophyll and carotenoids in the leaves of different cucumber varieties in relation to growth and productivity.]-In : DOSPEKHOV, B.A. (ed.) : Biologicheskie Osnovy Povysheniya Urozhaĭnosti Sel'sko-khozyaĭstvennykh Kul'tur. Vol. 2. Pp. 154 - 158. Moskva 1974. [In R.]

30397 - **KORSAK, M.N.** : Pervichnaya produktsiya v razlichnykh raĭonakh Belogo morya. [Primary production in different regions of the White Sea.] - Gidrobiol. Zh. *13* (4) : 13 - 16, 1977. [In R, ab : E.]

30398 - **KORSZUN, Z.R., SALEMME, F.R.** : Structure of cytochrome c_{555} of *Chlorobium thiosulfatophilum* : Primitive low-potential cytochrome *c*. - Proc. nat. Acad. Sci. USA *74* : 5244 - 5247, 1977.

30399 - **KORYAKIN, B.V., DZHABIEV, T.S., SHILOV, A.E.** : Fotosensibilizirovannoe vosstanovlenie vody v rastvorakh krasiteleĭ. Model' bakterial'nogo fotosinteza.

[Photosensitized reduction of water in dye solutions. Model of bacterial photosynthesis.] - Dokl. Akad. Nauk SSSR *233* : 620 - 622, 1977. [In R.]

30400 - KOSHALEVA, L.L., TSYARĖNTS'EŬ, V.M., BAKHNOVA, K.V. : Uplyŭ umoŭ azotnaga zhyŭlennya na farmiravanne fotasintétychnaga aparatu i praduktsyĭnasts' ras- lin il'nu-daŭguntsu. [Effect of nitrogen nutrition on the formation of pho- tosynthetic apparatus and productivity in flax.] - Vestsi Akad. Navuk belo- rus. SSR, Ser. biyal. Navuk *1977* (5) : 40 - 45, 138, 1977. [In Belorus., ab : R.]

30401 - KOSTYAEV, V.Ya., YAGODKA, S.N. : Fotosintez vodorosleĭ v ul'trafioletovom svete. [Algae photosynthesis in ultraviolet light.] - Dokl. Akad. Nauk SSSR *237* : 743 - 745, 1977. [In R.]

30402 - KOUCHKOVSKY, Y.de, PASQUIER, P. : Modulation of the transmembrane pH gradi- ent of chloroplasts with various photosynthetic redox chains. - In : COOMBS, J. (ed.) : 4th International Congress on Photosynthesis. P. 203. UKISES, London 1977.

30403 - KOVÁČ, J., HENSELOVÁ, M. : Detection of triazine herbicides in soil by a Hill-reaction inhibition technique after thin-layer chromatography. - J. Chromatogr. *133* : 420 - 422, 1977.

30404 - KOVÁCS, V., VIRÁGH, E., KOCSIS, E., GYURJÁN, S. : Study on the biological effect of fast neutrons. 1. Effect of fast neutrons on germinability of bar- ley seeds and on the chlorophyll content of the seedlings according to the dose used. - Acta biochim. biophys. Acad. Sci. hung. *12* : 49 - 55, 1977.

*B30405 - KOVDA, V.A. : Biosphere, Soils and their Utilization. (10th Internat.Congress Soil Sci.) - Inst. Agrochem. Soil Sci., Acad. Sci. USSR, Moskva 1974. [Ps.]

30406 - KOW, Y.W., ROBINSON, J.M., GIBBS, M. : Influence of pH upon Warburg effect in isolated intact spinach chloroplasts. II. Interdependency of glycolate synthesis upon pH and Calvin cycle intermediate concentration in the absence of carbon dioxide photoassimilation. - Plant Physiol. *60* : 492 - 495, 1977.

30407 - KOZHOVA, O.M., KAPLIN, V.M., IZMEST'EVA, L.R. : Primenenie fotometra-prozrach- nomera dlya otsenki vertikal'noĭ stratifikatsii vodnykh mass. [Use of the photometer-nephelometer for determining the vertical stratification of water masses.] - In : Biologicheskie Issledovaniya Vodoemov Vostochnoĭ Sibiri. Pp. 13 - 23. Irkutsk 1977. [In R.]

30408 - KOZLOV, I.A., SKULACHEV, V.P. : H+-adenosine triphosphatase and membrane energy coupling. - Biochim. biophys. Acta *463* : 29 - 89, 1977. [Bacteriorho- dopsin.]

30409 - KRAAYENHOF, R. : Energy-dependent conformational changes. - In : TREBST, A., AVRON, M. (ed.) : Photosynthesis I. (Encycl. Plant Physiol. N.S. Vol. 5.) Pp. 422 - 428. Springer-Verlag, Berlin-Heidelberg-New York 1977.

30410 - KRAAYENHOF, R., ARENTS, J.C. : Fluorescent probes for the chloroplast ener- gized state - energy-linked change of membrane-surface charge. - In : ROUX, E. (ed.) : Electrical Phenomena at the Biological Membrane Level. Pp. 493 - 504. Elsevier, Amsterdam 1977.

30411 - KRAAYENHOF, R., SCHUURMANS, J.J. : The light-induced structural rearrange- ments in the thylakoid membrane and in the ATPase detected by covalent fluo- rescent labels. - In : COOMBS, J. (ed.) : 4th International Congress on Pho- tosynthesis. P. 204. UKISES, London 1977.

30412 - KRÁĽOVIČ, J. : Regulácia úrod a technologickej kvality cukrovej repy. [Regu- lation of yield and technological quality of sugar beet.] - In : Produkce . Biomasy a Tvorba Výnosů Polních Plodin. Vol.1. Pp. 151 - 154. Česká vědecko- technická Společnost zemědělská, Praha 1977. [Ps; in Slovak, ab: E, G, R.]

30413 - KRÁĽOVIČ, J. : Využitie chemických látok v regulácii fyziologických procesov a produkcie rastlín. [Use of chemical substances in the regulation of physio- logical processes and plant production.] - In : REPKA, J. (ed.) : Zborník Referátov zo Seminára Fyziologicko-Genetické a Chemické Faktory Produktivity Rastlín. Pp. 77 - 88. Vysoká škola poľnohospodárska, Nitra 1977. [Ps; in Slo- vak.]

30414 - **KRAPF, G.** : The distribution of ^{14}C and the levels of metabolites after photosynthetic $^{14}CO_2$ fixation in leaf discs of spinach plants kept in darkness for prolonged periods. - In : **COOMBS, J.** (ed.) : 4th International Congress on Photosynthesis. P. 207. UKISES, London 1977.

30415 - **KRASICHKOVA, G.V., CHANDYLOVA, L.V., GILLER, Yu.E.** : Issledovanie fotokhimicheskoĭ aktivnosti khloroplastov gibridnykh form khlopchatnika. [Photochemical activity of chloroplasts of cotton hybrid forms.] - Dokl. Akad. Nauk tadzh. SSR *20* (9) : 54 - 56, 1977. [In R, ab : Tajik.]

30416 - **KRASNA, A.I.** : Catalytic and structural properties of the enzyme hydrogenase and its role in biophotolysis of water. - In : **MITSUI, A., MIYACHI, S., SAN PIETRO, A., TAMURA, S.** (ed.) : Biological Solar Energy Conversion. Pp. 53 - 60. Academic Press, New York - San Francisco - London 1977. [Ps.]

30417 - **KRASNOVSKIĬ, A.A.** : Problema fotosinteticheskogo vodoroda. [The problem of photosynthetic hydrogen.] - Izv. Akad. Nauk SSSR, Ser. biol. *1977* (5) : 650 - 662, 1977. [In R, ab : E.]

30418 - **KRASNOVSKIĬ, A.A., BYSTROVA, M.I., UMRIKHINA, A.V.** : Spektral'nye èffekty agregatsii bakteriofeofitina. [Spectral effects of bacteriopheophytin aggregation.] - Dokl. Akad. Nauk SSSR *235* : 232 - 235, 1977. [In R.]

*30419 - **KRASNOVSKIĬ, A.A., LUGANSKAYA, A.N.** : Fotosensibilizirovannoe vosstanovlenie metilviologena i ferredoksina v vodnykh sredakh v prisutstvii kisloroda. [Photosensitized reduction of methylviologen and ferredoxin in aqueous media in the presence of oxygen.] - Dokl. Akad. Nauk SSSR *223* : 229 - 232, 1975. [In R.]

30420 - **KRASNOVSKIĬ, A.A. ml.** : Fotolyuminestsentsiya singletnogo kisloroda v rastvorakh khlorofillov i feofitinov. [Photoluminescence of singlet oxygen in solutions of chlorophylls and pheophytins.] - Biofizika *22* : 927 - 928, 1977. [In R, ab : E.]

*30421 - **KRASNOVSKY, A.A.** : Chemical evolution of photosynthesis. - In : **NOVÁK, V.J.A., PACLTOVÁ, B.** (ed.) : Evolutionary Biology. Pp. 45 - 60. Czechoslovak Biological Society, Praha 1976.

30422 - **KRASNOVSKY, A.A.** : Photoproduction of hydrogen in photosynthetic system. - Plant Cell Physiol. *1977* (Spec.Issue 3 - Photosynthetic Organelles. Structure and Function) : 219 - 227, 1977.

30423 - **KRASNOVSKY, A.A.** : Photoproduction of hydrogen in photosynthetic systems. - In : **CASTELLANI, A.** (ed.) : Research in Photobiology. Pp. 361 - 370. Plenum Press, New York - London 1977.

30424 - **KRASNOVSKY, A.A. Jr.** : Investigation of phosphorescence of chlorophylls in solutions, etiolated and greening leaves, chloroplasts and chloroplast fragments. - In : **COOMBS, J.** (ed.) : 4th International Congress on Photosynthesis. P. 204a. UKISES, London 1977.

30425 - **KRASNOVSKY, A.A. Jr., LEBEDEV, N.N., LITVIN, F.F.** : Phosphorescence and delayed fluorescence of chlorophyll and its precursors in solutions, leaves and chloroplasts at 77 °K. - Stud. biophys. *65* : 81 - 89, 1977.

30426 - **KRATKY, C., DUNITZ, J.D.** : Ordered aggregation states of chlorophyll a and some derivatives. - J. mol. Biol. *113* : 431 - 442, 1977.

30427 - **KRATOCHVÍLOVÁ, H., FRYDRYCH, J.** : Vliv přípravků Retacel a B 995 na růstové, fotosyntetické a výnosové charakteristiky rajčat. [The effect of Retacel and B 995 on growth, photosynthesis and yield of tomato.] - Rostl. Výroba (Praha) *23* : 1115 - 1121, 1977. [In Czech, ab : E, R.]

30428 - **KRAUSE, G.H.** : Light-induced movement of magnesium ions in intact chloroplasts. Spectroscopic determination with Eriochrome Blue SE. - Biochim. biophys. Acta *460* : 500 - 510, 1977.

30429 - **KRAUSE, G.H., LORIMER, G.H.** : Photorespiratory energy dissipation in leaves and chloroplasts. - In : **COOMBS, J.** (ed.) : 4th International Congress on Photosynthesis. P. 205. UKISES, London 1977.

30430 - **KRAUSE, G.H., THORNE, S.W., LORIMER, G.H.** : Glycolate synthesis by intact chloroplasts. Studies with inhibitors of photophosphorylation. - Arch. Biochem. Biophys. *183* : 471 - 479, 1977.

30431 - **KRAVYAZH, K., KARAVAĬKO, N.N., KOF, É.M., KULAEVA, O.N.** : Vzaimodeĭstvie abstsizovoĭ kisloty i tsitokinina v regulyatsii rosta i pozeleneniya semyadoleĭ tykvy. [Interaction between abscisic acid and cytokinin in regulation of growth and greening of gourd cotyledons.] - Fiziol. Rast. *24* : 160 - 167, 1977. [In R, ab : E.]

B30432 - **KREEB, K.** : Methoden der Pflanzenökologie. - VEB Gustav Fischer Verlag, Jena 1977. [Ps.]

30433 - **KREMER, B.P.** : Algenplastiden leben in Meeresschnecken. - Umschau Wiss. Tech. *77* : 215 - 216, 1977.

30434 - **KREMER, B.P.** : Rotalgen-Chloroplasten als funktionelle Endosymbionten in einem marinen Opisthobranchier. - Naturwissenschaften *64* : 147 - 148, 1977.

30435 - **KREMER, B.P., KÜPPERS, U.** : Carboxylating enzymes and pathway of photosynthetic carbon assimilation in different marine algae - evidence for the C_4-pathway? - Planta *133* : 191 - 196, 1977.

30436 - **KRENDELEVA, T.E., KUKARSKIKH, G.P., NIZOVSKAYA, N.V., PASHCHENKO, V.Z., TIMOFEEV, K.N., TULBU, G.V., KHITROV, Yu.A.** : Pervichnye reaktsii fotosinteza v izolirovannykh khloroplastakh, fiksirovannykh glutarovym al'degidom. [Primary photosynthetic reactions in isolated chloroplasts fixed by glutaraldehyde.] - Biokhimiya *42* : 1965 - 1972, 1977. [In R, ab : E.]

30437 - **KRENDELEVA, T.E., KUKARSKIKH, G.P., NIZOVSKAYA, N.V., RUBIN, A.B.** : PS II induced absorption changes at 520 nm in pea chloroplasts. - Photosynthetica *11* : 183 - 188, 1977.

30438 - **KREUZ, E.** : Neue Ergebnisse über die Wechselbeziehungen zwischen Lichtstrahlung, Photosynthese und Maisbestand - Übersichtsbeitrag. - Arch. Acker-Pflanzenbau Bodenk. *21* : 827 - 835, 1977.

30439 - **KRIEG, K.** : Zur Erfassung zytochemischer Parameter mit der Durchflußfluorometrie. - Acta histochem. *19* (Suppl.) : 173 - 179, 1977. [Chl.]

30440 - **KRIEG, K., RENNER, H., RATHSACK, R., DRESSEL, H.** : Impulsfluorometrische Bestimmung von DNS, Protein und Chlorophyll in kultivierten Zellen - ein Beitrag zur Wirkstofftestung. - Biol. Zentralbl. *96* : 51 - 59, 1977.

30441 - **KRKOŠKOVÁ, B.** : Vplyv rýchlosti zmrazovania na akosť mrazenej fazuľky. [Effect of the rate of freezing on the quality of frozen bean.] - Průmysl Potravin *28* : 197 - 200, 1977. [Chl; in Slovak.]

30442 - **KROGMANN, D.W.** : Blue-green algae. - In : **TREBST, A., AVRON, M.** (ed.) : Photosynthesis I. (Encycl. Plant Physiol. N.S. Vol.5.) Pp. 625 - 636. Springer-Verlag, Berlin - Heidelberg - New York 1977. [Ps.]

30443 - **KROGMANN, D.W., ULRICH, E., GOMEZ-LOJERO, C.** : Cytochromes of blue-green algae. - Plant Physiol. *59* (6, Suppl.) : 130, 1977.

30444 - **KROGMANN, D.W., ULRICH, E., MARKLEY, J.** : The structure of plastocyanin from NMR spectroscopy. - In : **COOMBS, J.** (ed.) : 4th International Congress on Photosynthesis. P. 206. UKISES, London 1977.

30445 - **KR"STEV, K.K., SAVOV, S.G.** : Fiziologo-biokhimichni izmeneniya v zabolyaloto ot vertitsiliĭno uvyakhvane pamukovo rastenie. [Physiological and biochemical alterations in cotton plants attacked by *Verticillium* wilt.] - Rasteniev"d. Nauki *14* (3) : 134 - 139, 1977. [Growth analysis; in Bulg., ab : E, R.]

30446 - **KRSTICH, B., SARICH, M.** : Deĭstvie raznykh spektral'nykh uchastkov sveta na soderzhanie karotinoidov v rasteniyakh kukuruzy. [Effect of different spectral areas of radiation on the content of carotenoids in maize plants.] - Fiziol. Biokhim. kul't. Rast. *9* : 488 - 491, 1977. [In R, ab : E.]

30447 - **KRUG, H., WIEBE, H.J., ROSE, H.B.** : Gaswechselmessanlage mit CO_2-Kompensationsverfahren. - Gartenbauwissenschaft *42* : 105 - 108, 1977.

30448 - KRUPA, J. : The interdependence between transpiration intensity and the anatomical structure of moss leaves. - Acta Soc. Bot. Pol. *46* : 57 - 68, 1977.
[Ps.]

30449 - KRUPA, Z., BASZYŃSKI, T. : Participation of sulphoquinovosyl diacylglycerol
in the reconstitution of Photosystem I activity of heptane-extracted chloroplasts. - Bull. Acad. pol. Sci., Sér. Sci. biol. *25* : 409 - 413, 1977.

30450 - KRUPATKINA, D.K., KUZ'MENKO, L.V., MARKOVA, G.A. : Biologiya kholodnogo ringa
Gol'fstrima. [Gulf stream cold ring biology.] - Morskie gidrofiz. Issledovaniya *4* (79) : 82 - 90, 1977. [Chl; in R, ab : E.]

30451 - KRŮŽELA, J. : Dynamika tvorby kořenů řepky ozimé při rozdílné vlhkosti půdy.
[Dynamics of root formation in winter rape under different soil moisture
content.] - Rostlinná Výroba (Praha) *23* : 439 - 448, 1977. [In Czech, ab :
E, G, R.]

30452 - KRYWALSKA, M., SKRZYPCZYK, J. : Influence of nitrate concentration on pigment
content in *Chlorella pyrenoidosa*. - Acta Soc. Bot. Pol. *46* : 489 - 499, 1977.

30453 - KU, S.-B., EDWARDS, G.E. : Kinetic characteristics of photosynthesis and O_2
inhibition of photosynthesis in *Triticum aestivum* L. as affected by temperature. - Plant Physiol. *59* (6, Suppl.) : 43, 1977.

30454 - KU, S.-B., EDWARDS, G.E. : Temperature dependence of O_2 inhibition of photosynthesis and quantum yield in C_3 plants in relation to solubility ratio
of O_2/CO_2. - Plant Physiol. *59* (6, Suppl.) : 43, 1977.

30455 - KU, S.B., EDWARDS, G.E. : Oxygen inhibition of photosynthesis. 1. Temperature
dependence and relation to O_2/CO_2 solubility ratio. - Plant Physiol. *59* :
986 - 990, 1977.

30456 - KU, S.B., EDWARDS, G.E. : Oxygen inhibition of photosynthesis. 2. Kinetic
characteristics as affected by temperature. - Plant Physiol. *59* : 991 - 999,
1977.

30457 - KU, S.-B., EDWARDS, G.E., TANNER, C.B. : Effects of light, carbon dioxide,
and temperature on photosynthesis, oxygen inhibition of photosynthesis, and
transpiration in *Solanum tuberosum*. - Plant Physiol. *59* : 868 - 872, 1977.

30458 - KU, S. B., HUNT, L.A. : Effects of temperature on the photosynthesis-irradiance response curves of newly matured leaves of alfalfa. - Can. J. Bot. *55*:
872 - 879, 1977.

30459 - KUBÍČEK, F. : Energy values of selected species of the herbaceous layer and
organic litter in the forest ecosystem. - Biológia (Bratislava) *32* : 505 -
515, 1977.

30460 - KUBÍČEK, F. : Retrakčný a redukčný index u listov duba, hraba a javora. [Re-
traction and reduction index for oak, hornbeam and maple leaves.] - In :
HUZULÁK, J., MASAROVIČOVÁ, E. (ed.) : Fotosyntéza a Vodný Režim Drevín. Pp.
70 - 76. Modra-Piesky 1977. [In Slovak, ab : E, R.]

30461 - KUBOTA, F., KANEKO, K. : Effects of light and air temperature on dry matter
production of ten corn varieties (*Zea mays* L.) at early growth stage. - Jap.
J. Crop Sci. *46* : 75 - 81, 1977. [In Jap., ab E.]

30462 - KUBOWICZ, D., MALESZEWSKI, S., POSKUTA, J. : The effect of oxygen concentration on the light-dependent conversion of photosynthetic products in the
leaves of maize. - Z. Pflanzenphysiol. *81* : 141 - 146, 1977.

30463 - KUJIRA, Y., KANDA, M. : Competition among individual plants in crop population (3). Competition from the viewpoint in root behavior. - Rep. Inst. agr.
Res. Tohoku Univ. *28* : 29 - 39, 1977. [Growth analysis.]

30464 - KUKUŁKA, I., KOZŁOWSKI, S. : Nowe kryteria oceny odmian traw uprawnych na
przykładzie *Lolium multiflorum*. [New estimation criteria of cultivated grass
cultivars exemplified by *Lolium multiflorum*.] - Zesz. problem. Postępów Nauk
roln. *194* : 29 - 43, 1977. [Chl; in Polish, ab : E, R.]

30465 - KULIKOV, G.V., IVANTSOVA, Z.V. : Dinamika pigmentov v list'yakh vechnozelenykh i listopadnykh drevesnykh rastenii v Krymu. [Dynamics of pigments in
the leaves of evergreen and deciduous woody plants in the Crimea.] - Bot. Zh.
62 : 1053 - 1062, 1977. [In R.]

30466 – KUMAKOV, V.A. : Selektsiya na povyshenie fotosinteticheskoĭ produktivnosti
        rasteniĭ. [Selection for increasing plant photosynthetic productivity.] –
        Itogi Nauki Tekh., Ser. Fiziol. Rast. *3* : 108 – 125, 1977. [In R.]

30467 – KUMAR, A., BENDER, L., JESKE, C., NEUMANN, K.-H., SENGER, H., STRAßBERGER, G. :
        The development of the photosynthetic system of carrot tissue cultures. – In :
        COOMBS, J. (ed.) : 4th International Congress on Photosynthesis. P. 207.
        UKISES, London 1977.

30468 – KUMARASINGHE, K.S., KEYS, A.J., WHITTINGHAM, C.P. : Effects of certain inhibi-
        tors on photorespiration by wheat leaf segments. – J. exp. Bot. *28* : 1163 –
        1168, 1977.

30469 – KUMARASINGHE, K.S., KEYS, A.J., WHITTINGHAM, C.P. : The flux of carbon through
        the glycolate pathway during photosynthesis by wheat leaves. – J. exp. Bot.
        28 : 1247 – 1257, 1977.

30470 – KUMAZAWA, S., BARCIELA, S., SKJOLDAL, H., MITSUI, A. : Hydrogen production
        by marine blue-green algae : effects of inhibitors. – Plant Physiol. *59*
        (6, Suppl.) : 20, 1977.

*30471 – KUNG, M.C., DEVAULT, D., HESS, B., OESTERHELT, D. : Photolysis of bacterial
        rhodopsin. – Biophys. J. *15* : 907 – 911, 1975.

30472 – KUNG, S. : Expression of chloroplast genomes in higher plants. – Annu. Rev.
        Plant Physiol. *28* : 401 – 437, 1977.

30473 – KUNG, S., LEE, C., WOOD, D.D., MOSCARELLO, M.A. : Evolutionary conservation
        of chloroplast genes coding for large subunits of fraction 1 protein. – Plant
        Physiol. *60* : 89 – 94, 1977.

30474 – KUNG, S.D., WOOD, D.D., LEE, C.I., apRHYS, C. : Amino acid analysis and pep-
        tide mapping of Fraction 1 protein from several species of *Nicotiana*. – Plant
        Physiol. *59* (6, Suppl.) : 9, 1977.

30475 – KUNIFUJI, Y., NAKAYAMA, K., OKADA, M. : Distribution of the light-induced
        561-nm absorbance change in *Chlorophyceae* : Carotenoids responsible for the
        absorbance change. – Plant Cell Physiol. *1977* (Spec. Issue 3 – Photosynthetic
        Organelles. Structure and Function) : 173 – 178, 1977.

30476 – KÜNSTLE, E., MITSCHERLICH, G. : Photosynthese, Transpiration und Atmung in
        einem Mischbestand im Schwarzwald. IV. Teil: Bilanz. – Allg. Forst- Jagdzeit.
        148 : 227 – 239, 1977.

B30477 – KUPERMAN, I.A., KHITROVO, E.V. : Dykhatel'nyĭ Gazoobmen kak Élement Produk-
        tsionnogo Protsessa. [Respiration Gas Metabolism as an Element of Production
        Process in Plants.] – Nauka, Novosibirsk 1977. [Ps, biomass; in R.]

30478 – KUPKA, J., NOVÁK, V., LIPAVSKÝ, J. : Dynamika obsahu fosforu a tvorba sušiny
        v ontogenezi jarního ječmene. [Dynamics of phosphorus content and dry matter
        production in the ontogeny of spring barley.] – Rostl. Výroba (Praha) *23* :
        741 – 744, 1977. [In Czech, ab : E, R.]

30479 – KUPRIN, S.P., KUKUSHKIN, A.K., TIKHONOV, A.N. : Teoreticheskoe issledovanie
        protsessov migratsii ėnergii i ėlektronnogo transporta v fotosinteze vysshikh
        rasteniĭ. I. Osnovnye upravleniya i ikh reshenie. [Theoretical study of energy
        migration and electron transport in photosynthesis of higher plants. I. Basic
        equations and their solution.] – Biofizika *22* : 161 – 163, 1977. [In R, ab :
        E.]

30480 – KURTEV, P., TSANKOV, B., BRAĬKOV, D., PANDELIEV, S. : Prouchvane v"rkhu di-
        namikata na rastezha na listata i tyakhnata fiziologichna aktivnost v zavisi-
        most ot metamernoto im razpolozhenie i nachina na otglezhdane na lozite. I.
        Dinamika na rastezha na listata pri sorta Bolgar v zavisimost ot metamernoto
        im razpolozhenie i visochinata na st"bloto. [Leaf growth rate and physiologi-
        cal activity depending on metameric position and mode of grapevine cultivat-
        ion. I. Leaf growth rate in cv. Bolgar as related to metameric position and
        stem height.] – Fiziol. Rast. (Sofia) *3* (2) : 61 – 70, 1977. [In Bulg., ab :
        E, R.]

30481 - KUSHNIRENKO, M.D., PECHERSKAYA, S.N., KRYUKOVA, E.V., KANASH, E.V. : Vliyanie
vlazhnosti sredy na pigment-belkovyĭ kompleks list'ev i khloroplastov grushi.
[Effect of medium moisture on the pigment-protein complex in pear-tree leaves
and chloroplasts.] - Fiziol. Biokhim. kul't. Rast. *9* : 625 - 630, 1977. [In R,
ab : E.]

30482 - KUSUKI, Y.: [Growth of the Japanese oyster in relation to chlorophyll *a*.] -
Bull. Hiroshima Fisheries exp. Sta. *9* : 28 - 36, 1977. [In Jap.]

30483 - KUSUTANI, A., NAKASEKO, K., GOTOH, K. : [Canopy structure and dry matter production in grasses.] - Jap. J. Crop Sci. *46* : 205 - 211, 1977. [In Jap., ab :
E.]

30484 - KUZ'MENKO, L.V. : Nakoplenie i vydelenie radioaktivnogo ugleroda chernomor-
skimi planktonnymi vodoroslyami. [Accumulation and excretion of radioactive
carbon by Black Sea planktonic algae.] - In : Raspredelenie i Povedenie Mor-
skogo Planktona v Svyazi s Mikrostrukturoĭ Vod. Pp. 97 - 102. Naukova Dumka,
Kiev 1977. [In R.]

30485 - KUZ'MENKO, L.V., KRUPATKINA, D.K. : Pervichnaya produktsiya i khlorofill
Sargassova morya. [Primary production and chlorophyll in the Sargasso Sea.]
- Morskie gidrofiz. Issled. *4*(79) : 91 - 98, 1977. [In R, ab: E.]

30486 - KUZNETSOV, S.I., ROMANENKO, V.I., KUZNETSOVA, N.S., KARPOVA, N.S. : Kharakteristika mikrobiologicheskikh i gidrologicheskikh protsessov v Rybinskom vodo-
khranilishche v 1974 g. [Characteristics of microbiological and hydrobiologi-
cal processes in Rybinsk water reservoir in 1974.] - Tr. Inst. Biol. vnutr.
Vod Akad. Nauk SSSR *36* (Gidrologicheskie i gidrokhimicheskie Aspekty Izucheniya Vodokhranilishch) : 149 - 162, 1977.[Primary production; in R.]

30487 - KVITKO, K.V., BOYADZHIEV, P.Kh., CHUNAEV, A.S., MUKHAMADIEV, B.T., BARANOV,
A.A., SAAKOV, V.S. : Genotipicheskaya i fenotipicheskaya izmenchivost' pig-
ment-lipoproteidnogo kompleksa mutantov zelenykh vodorosleĭ. 2. Issledovanie
spektrov pogloshcheniya mutantov s izmenennoĭ reaktsieĭ na svet u *Chlamydomonas reinhardii* 137C. [Genotype and phenotype variability of the pigment-lipo-
proteid complex in mutants of green algae. 2. Study of absorption spectra of
mutants of *Chlamydomonas reinhardii* 137C with changed reaction to light.] -
Tr. petergof. biol. Inst. *25* (Éksperimental'naya Al'gologiya) : 106 - 132,
1977. [In R, ab : E.]

30488 - KVITKO, K.V., CHUNAEV, A.S., TUGARINOV, V.V. :Deĭstvie organellospetsifiches-
kikh antibiotikov na pigment-belkovolipidnyĭ kompleks membran khloroplasta u
khlamidomonady. [Effect of organelle-specific antibiotics on a pigment-protein-
-lipid complex of chloroplast membranes in *Chlamydomonas*.] - In : NEIFAKH, S.A.,
TROSHIN, ΙA.S. (ed.) : Molekulyarnaya Genetika Mitokhondriĭ. Pp. 44 - 55.
Nauka, Leningrad 1977. [In R.]

30489 - KVITKO, K.V., LAZEEVA, G.S., KHOMYAKOV, R.V., CHUNAEV, A.S., YAKUBOV, K.F.:
Primenenie spektral'no-izotopnogo metoda dlya izucheniya izmenchivosti pokazateleĭ fotosinteza i obmena organogennykh élementov u zelenykh vodorosleĭ. II.
Kharakteristika rosta i fotosinteza mutantov khlamidomonady po vklyucheniyu
^{13}C i ^{15}N v biomassu. [The use of the spectroisotopic method for studying variability of indices of photosynthesis and organogenic element metabolism in
green algae. II. Characteristics of the growth and photosynthesis of chlamy-
domonad mutants during assimilation of ^{13}C and ^{15}N into the biomass.] - Vestn.
leningr. Univ., Biol. *1977* (4) : 124 - 133, 1977. [In R, ab : E.]

30490 - LABUS, B., SCHUSTER, H., NOBEL, W., KOHLER, A. : Wirkung von toxischen Abwas-
serkomponenten auf submerse Makrophyten. - Angew. Bot. *51* : 17 - 36, 1977.
[Ps.]

30491 - LACH, H.J., BÖGER, P. : Some properties of plastidic cytochrome *b*-563. -
Z. Naturforsch. *32 C* : 877 - 879, 1977.

30492 - LACH, H.-J., BÖGER, P. : Isolation and some molecular properties of plastidic
algal cytochrome *b*-559. - Z. Naturforsch. *32 C* : 75 - 77, 1977.

30493 - LACH, H.-J., BÖHME, H., BÖGER, P. : Some photoreactions of isolated cytochrome *b*-559. - Biochim. biophys. Acta *462* : 12 - 19, 1977.

30494 - LADIGES, P.Y. : Differential susceptibility of two populations of *Eucalyptus viminalis* LABILL. to iron chlorosis. - Plant Soil *48* : 581 - 597, 1977.

30495 - LADYGIN, V.G. : Metod vydeleniya nefotosinteziruyushchikh mutantov vodoroslei po priznaku karlikovosti kolonii. [Isolation of non-photosynthesizing algae mutants for dwarf colony character.] - Genetika *13* : 905 - 910, 1977. [Ps, Chl; in R, ab : E.]

30496 - LADYGIN, V.G., SADOVNIKOVA, L.G. : Nasledovanie svetlo-zelenoi okraski kolonii u odnokletochnoi zelenoi vodorosli *Chlamydomonas reinhardi*. [Heritability of light-green colour of colonies of a unicellular green alga *Chlamydomonas reinhardi*.] - In : NASYROV, Yu.S. (ed.) : Genetika Fotosinteza. Pp. 128 - 132. Donish, Dushanbe 1977. [In R.]

30497 - LADYGIN, V.G., SEMENOVA, G.A., TAGEEVA, S.V. : Pigmentatsiya i struktura khloroplastov zigot *Chlamydomonas reinhardi* pri retsiprokhnykh skreshchivaniyakh khlorofil'nykh mutantov i kletok dikogo tipa. [Pigmentation and structure of zygote chloroplasts of *Chlamydomonas reinhardi* in reciprocal crossings of the chlorophyll mutants and the wild type cells.] - Tsitologiya *19* : 585 - 591, 1977. [In R, ab : E.]

30498 - LAETSCH, W.M., SPREY, B. : Structure of the peripheral reticulum and Mg^{2+}-dependent ATPase activity of C_4 plant chloroplast membrane fractions. - In : COOMBS, J. (ed.) : 4th International Congress on Photosynthesis. P. 208. UKISES, London 1977.

30499 - LAISK, A. : Modelling of the closed Calvin cycle. - In : UNGER, K. (ed.) : Biophysikalische Analyse pflanzlicher Systeme. Pp. 175 - 182. VEB Gustav Fischer Verlag, Jena 1977.

30500 - LAKATOS, Gy., TOKÁR, M. : Effect of some biocids on attached algae (biotecton). - Acta biol. debrecina *14* : 105 - 114, 1977. [Ps, Chl.]

30501 - LAKSHMINARAYANA, J.S.S., DEVI, J.S., DURVASULA, S.V. : Primary production in relation to tracemetal variations in the shelf area of Nova Scotia and Georges Bank. - J. Phycol. *13* (Suppl.) : 38, 1977.

30502 - LAMBERT, G.R., SMITH, G.D. : Hydrogen formation by marine blue-green algae. - FEBS Lett. *83* : 159 - 167, 1977.

30503 - LAMOLA, A.A., TURRO, N.J. : Spectroscopy. - In : SMITH, K.C. (ed.) : The Science of Photobiology. Pp. 27 - 61. Plenum Press, New York - London 1977. [Basic methods.]

*30504 - LAMONTAGNE, R.A., SMITH, W.D., SWINNERTON, J.W. : C_1-C_3 hydrocarbons and chlorophyll *a* concentrations in the Equatorial Pacific Ocean. - In : GIBB, T.R.P. (ed.) : Analytical Methods in Oceanography. Advances in Chemistry Series, No. 147. Pp. 163 - 171. American Chemical Society, Washington 1975.

30505 - LAMOTTE, M. : Observations preliminaires sur les flux d' énérgie dans un écosystème herbace tropical, la savane de Lamto (Côte d' Ivoire). - Géo-Eco-Trop. *1* : 45 - 63, 1977. [Primary production.]

30506 - LAMOTTE, M. : Première approche du bilan énergétique d'un écosystème herbacé tropical (Lamto, Côte-d'Ivoire): production primaire et consommation animale. - Compt. rend. Acad. Sci. Paris, Sér. D *284* : 1449 - 1452, 1977.

30507 - LANCASTER, J.E., MANN, J.D. : Extrastomatal control of transpiration in leaves of yellow lupin (*Lupinus luteus* var. Weiko III) after drought or abscisic acid treatment. - J. exp. Bot. *28* : 28 : 1373 - 1379, 1977. [Resistances.]

30508 - LANCASTER, J.E., MANN, J.D., PORTER, N.G. : Ineffectiveness of abscisic acid in stomatal closure of yellow lupin, *Lupinus luteus*, var. Weiko III. - J. exp. Bot. *28* : 184 - 191, 1977. [Stomatal resistance.]

30509 - LANDRUM, H.L., SALMON, R.T., HAWKRIDGE, F.M. : A surface-modified gold minigrid electrode which heterogeneously reduces spinach ferredoxin. - J. amer. chem. Soc. *99* : 3154 - 3158, 1977.

30510 - LANDSBERG, J.J. : Some useful equations for biological studies. - Exp. Agr. *13* : 273 - 286, 1977. [Ps.]

B30511 - LANDSBERG, J.J., CUTTING, C.V. (ed.) : Environmental Effects on Crop Physiology. (Proc. Symp. Long Ashton Res. Sta. Univ. Bristol, 1975). - Academic
Press, London - New York - San Francisco 1977. [Ps.]

30512 - LANDSBERG, P.T. : A note on the thermodynamics of energy conversion in
plants. - Photochem. Photobiol. 26 : 313 - 314, 1977.

30513 - LANE, H.C., HESKETH, J.D. : Cotyledon photosynthesis during seedling growth
of cotton, Gossypium hirsutum L. - Amer. J. Bot. 64 : 786 - 790, 1977.

30514 - LÃNG, F., BÖDDI, B., SOÕS, J. : Spectral properties of protochlorophyll
forms in solid films. - In : COOMBS, J. (ed.) : 4th International Congress
on Photosynthesis. Pp. 208 - 209. UKISES, London 1977.

30515 - LANGE, O.L., GEIGER, I.L., SCHULZE, E.-D, : Ecophysiological investigations
on lichens of the Negev desert. V. A model to simulate net photosynthesis
and respiration of Ramalina maciformis. - Oecologia 28 : 247 - 259, 1977.

30516 - LANGE, O.L., ZUBER, M. : Frerea indica, a stem succulent CAM plant with deciduous C_3 leaves.-Oecologia 31 : 67 - 72, 1977.

30517 - LÄNNERGREN, C. : A compact multisample filtration apparatus. - Sarsia 63 :
27 - 28, 1977. [Ps.]

30518 - LANNOYE, R., WOUTERS, J., DELHAYE, J.P. : Effects of sulfur on some photosynthetic characteristics and protein composition of isolated chloroplasts
from spinach. - In : COOMBS, J. (ed.) : 4th International Congress on Photosynthesis. P. 209. UKISES, London 1977.

30519 - LANYI, J.K. : Light-induced transport in Halobacterium halobium cell envelo-
pe vesicles. - In : PACKER, L., PAPAGEORGIOU, G.C., TREBST, A. (ed.) :
Bioenergetics of Membranes. Pp. 129 - 136. Elsevier/North-Holland Biomedical
Press, Amsterdam - Oxford - New York 1977.

30520 - LANYI, J.K. : Transport in Halobacterium halobium : Light-induced cation-gra-
dients, amino acid transport kinetics, and properties of transport carriers. -
J. supramol. Structure 6 : 169 - 177, 1977.

30521 - LANYI, J.K., MacDONALD, R.E. : Light-dependent cation gradients and electrical potential in Halobacterium halobium cell envelope vesicles. - Fed. Proc.
36 : 1824 - 1827, 1977.

*30522 - LAPTEVA, N.A., MONAKOVA, S.V. : Mikrobiologicheskaya kharakteristika ozer
Yaroslavskoĭ oblasti. [Microbiological characteristics of lakes in the Yaro-
slavl region.] - Mikrobiologiya 45 : 717 - 723, 1976. [Ps; in R, ab : E.]

30523 - LARCHER, W. : Ergebnisse des IBP-Projekts "Zwergstrauchheide Patscherkofel". -
Sitzungsber. österr. Akad. Wiss.,math. naturwiss. Klasse, Abt. I, 186 : 301 -
371, 1977. [Ps, Chl, Car.]

30524 - LARCHER, W. : Produktivität und Überlebensstrategien von Pflanzen und Pflanzenbeständen im Hochgebirge. Bericht über pflanzenökologische Forschungspro-
jekte des Instituts für Allgemeine Botanik der Universität Innsbruck (1966-
1976). - Sitzungsber. österr. Akad. Wiss. math. naturwiss. Klasse, Abt. I,
186 : 373 - 386, 1977.

30525 - LARKUM, A.W.D., WEYRAUCH, S.K. : Photosynthetic action spectra and light-harvesting in Griffithsia monilis (Rhodophyta). - Photochem. Photobiol. 25 :
65 - 72, 1977.

30526 - LARSSON, C., ALBERTSSON, P.-Å. : $^{14}CO_2$ fixation by a reconstituted chloroplast-
-cytoplasm system. - In : COOMBS, J. (ed.) : 4th International Congress on
Photosynthesis. P. 210. UKISES, London 1977.

30527 - LARSSON,C., ANDERSSON,B., ROOS, G. : Scanning electron microscopy of different populations of chloroplasts isolated by phase partition. - Plant Sci.
Lett. 8 : 291 - 298, 1977.

30528 - LARSSON, C.-M., TILLBERG, J.-E. : Light-induced phosphate uptake in relation
to photophosphorylation and levels of ATP, ADP and AMP in the green alga
Scenedesmus. - In : COOMBS, J. (ed.) : 4th International Congress on Photosynthesis. Pp. 210 - 211. UKISES, London 1977.

30529 - LASLEY, S.E., GARBER, M.P. : Factors influencing chilling injury in cucumber. - Plant Physiol. *59* (Suppl.) : 4, 1977. [Ps.]

30530 - LATIMORE, M.,Jr., GIDDENS, J., ASHLEY, D.A. : Effect of ammonium and nitrate nitrogen upon photosynthate supply and nitrogen fixation by soybeans. - Crop Sci. *17* : 399 - 404, 1977.

30531 - LAU, R.H., MacKENZIE, M.M., DOOLITTLE, W.F. : Phycocyanin synthesis and degradation in the blue-green bacterium *Anacystis nidulans*. - J. Bacteriol. *132* : 771 - 778, 1977.

30532 - LAUBIG, U., LUMMITSCH, M., HOBERG, W. : Ein Modell der Stoffproduktion bei Weizen in der Kornfüllungsperiode. - In : UNGER, K. (ed.) : Biophysikalische Analyse pflanzlicher Systeme. Pp. 193 - 200. VEB Gustav Fischer Verlag, Jena 1977.

30533 - LAUENROTH, W.K., WHITMAN, W.C. : Dynamics of dry matter production in a mixed--grass prairie in Western North Dakota. - Oecologia *27* : 339 - 351, 1977.

30534 - LAULHERE, J.P., DORNE, A.M. : Are cytoplasm ribosomes in chloroplast preparations functionally attached to chloroplast membrane? - Plant Sci. Lett. *8* : 251 - 256, 1977.

30535 - LAURINAVICHENE, T.V., GOGOTOV, I.N. : Sravnenie svoĭstv NADP-reduktazy u dvukh vidov purpurnykh bakteriĭ. [Comparison of NADP-reductase properties in two species of purple bacteria.] - Biokhimiya *42* : 1387 - 1391, 1977. [In R, ab : E.]

30536 - LAVAL-MARTIN, D., DUBERTRET, G., CALVAYRAC, R. : Photosynthetic properties of a DCMU resistant strain of *Euglena gracilis* Z. - Plant Sci. Lett. *10* : 185 - 195, 1977.

30537 - LAVAL-MARTIN, D., FARINEAU, J., DIAMOND, J. : Light *versus* dark carbon metabolism in cherry tomato fruits. I. Occurrence of photosynthesis. Study of the intermediates. - Plant Physiol. *60* : 872 - 876, 1977.

30538 - LAVERGNE, D. : Characters of carbon metabolism of cultivated and wild species of *Pennisetum*. - In : COOMBS, J. (ed.) : 4th International Congress on Photosynthesis. Pp. 211 - 212. UKISES, London 1977.

30539 - LAVOREL, J., ETIENNE, A.-L. : *In vivo* chlorophyll fluorescence. - In : BARBER, J. (ed.) : Primary Processes of Photosynthesis. Pp. 203 - 268. Elsevier, Amsterdam - New York - Oxford 1977.

30540 - LAWANSON, A.O. : Development of photochemical activity during greening of heat-stressed etiolated seedlings of *Zea mays*. - Experientia *33* : 200 - 201, 1977.

30541 - LAWANSON, A.O., OTUSANYA, O.O., AKOMOLEDE, D.A. : Mechanism of potassium deficiency-induced retardation of chlorophyll biosynthesis in *Zea mays*. - Experientia *33* : 1145 - 1146, 1977.

30542 - LAWLOR, D.W., FOCK, H. : Photosynthesis, photorespiration and productivity: effects of water stress. - In : COOMBS, J. (ed.) : 4th International Congress on Photosynthesis. Pp. 212 - 213. UKISES, London 1977.

30543 - LAWLOR, D.W., FOCK, H. : Photosynthetic assimilation of $^{14}CO_2$ by water-stressed sunflower leaves at two O_2 concentrations and the specific activity of products. - J. exp. Bot. *28* : 320 - 328, 1977.

30544 - LAWLOR, D.W., FOCK, H. : Water stress induced changes in the amounts of some photosynthetic assimilation products and respiratory metabolites of sunflower leaves. - J. exp. Bot. *28* : 329 - 337, 1977.

30545 - LAWLOR, D.W., MAHON, J.D., FOCK, H. : An assimilation chamber for rapid leaf sampling and a gas switching system for control of $^{12}CO_2$ and $^{14}CO_2$ supply. - Photosynthetica *11* : 322 - 326, 1977.

30546 - LAWYER, A.L., ZELITCH, I. : Inhibition of glutamate: glyoxylate aminotransferase in tobacco callus by glycidate, an inhibitor of photorespiration. - Plant Physiol. *59* (6, Suppl.) : 42, 1977.

30547 - **LEA, H.Z., DUNN, G.M., KOCH, D.W.** : Stomatal diffusion resistance in three ploidy levels of smooth bromegrass. - Crop Sci. *17* : 91 - 93, 1977.

30548 - **LEBEDEV, I.S., LITVINENKO, L.G., SHIYAN, L.T.** : Posledeĭstvie postoyannogo magnitnogo polya na fotokhimicheskuyu aktivnost' khloroplastov. [Aftereffect of stationary magnetic field on photochemical activity of chloroplasts.] - Fiziol. Rast. *24* : 491 - 495, 1977. [In R, ab : E.]

30549 - **LECHEVALLIER, D.** : Lipides, nucléotides pyridiniques et nucléotides adényliques de tissus et de plastes isolés de Spirodèles cultivées sur milieu enrichi en calcium. - Physiol. vég. *15* : 95 - 119, 1977.

30550 - **LECHEVALLIER, D.** : Effets du polyéthylène glycol sur les lipides et les lipochromes des colonies de Spirodèle. - Physiol. vég. *15* : 387 - 402, 1977. [Chl.]

30551 - **LECHEVALLIER, D., VERMEERSCH, J., MONÉGER, R.** : Micro-analyse du NADP et du NAD réduits et oxydés dans les tissus foliaires et dans les plastes isolés de Spirodèle et de Blé. 2. Méthode d'analyse des nucléotides pyridiniques de tissus végétaux. Application à l'étude des effets du saccharose, de la lumière rouge et de l'obscurité. - Physiol. vég. *15* : 63 - 93, 1977.

30552 - **LECLERC, J.C., TANDEAU de MARSAC, N.** : Phycobiliproteins and chlorophyll *a* forms variations in *Cyanobacteria*. - In : COOMBS, J. (ed.) : 4th International Congress on Photosynthesis. P. 213. UKISES, London 1977.

30553 - **LEDENT, J.F.** : Sur le calcul de coefficient d'extinction du rayonnement solaire incident direct dans un couvert végétale. - Oecol. Plant. *12* : 291 - 300, 1977.

30554 - **LEDENT, J.F.** : Effect of partial defoliation and vein cutting on grain growth and yield in winter wheat (*Triticum aestivum* L.). - Bull. Soc. roy. Bot. Belg. *110* : 239 - 250, 1977. [Foliage formation.]

30555 - **LEDENT, J.F., MOSS, D.N.** : Spatial orientation of wheat leaves. - Crop Sci. *17* : 873 - 879, 1977. [Canopy.]

30556 - **LEDIG, F.T., CLARK, J.G., DREW, A.P.** : The effects of temperature treatment on photosynthesis of pitch pine from northern and southern latitudes. - Bot. Gaz. *138* : 7 - 12, 1977.

30557 - **LEE, S.H., IKEDA, M., YAMADA, Y.** : Comparative studies on chloroplast development and photosynthetic activities in C_3- and C_4-plants. 1. Studies on ultrastructure of developing chloroplasts within vascular bundle sheaths and mesophyll cells of barley and maize leaves. - J. Fac. Agr. Kyushu Univ. *22* : 65 - 74, 1977.

30558 - **LEE, S.S.** : Effects of light intensity on photoreactions with chloroplasts types from corn leaves. - Virginia J. Sci. *28* (4) : 157 - 162, 1977.

30559 - **LEECH, R.M.** : Subcellular fractionation techniques in enzyme distribution studies. - In : SMITH, H. (ed.) : Regulation of Enzyme Synthesis and Activity in Higher Plants. Pp. 289 - 327. Academic Press, London - San Francisco - New York 1977. [Chloroplasts, peroxisomes.]

30560 - **LEECH, R.M.** : Etioplast structure and its relevance to chloroplast development. - Biochem. Soc. Trans. *5* : 81 - 84, 1977.

30561 - **LEECH, R.M., HEATH, R.L.** : The stimulation of O_2 evolution by ammonium ion in intact chloroplasts. - In : COOMBS, J. (ed.) : 4th International Congress on Photosynthesis. P. 214. UKISES, London 1977.

30562 - **LEESE, B.M., LEECH, R.M.** : Biosynthesis and structure of leaf lipids: Lipid changes during plastid and leaf development. - Biochem. Soc. Trans. *5* : 1266 - 1269, 1977.

30563 - **LEGGE, A.H., JAQUES, D.R., AMUNDSON, R.G., WALKER, R.B.** : Field studies on pine, spruce and aspen periodically subjected to sulfur gas emissions. - Water Air Soil Pollut. *8* : 105 - 129, 1977. [Ps.]

30564 - **LEHMANN, M., WÖBER, G.** : Preparation of [U-^{14}C]-labelled glycogen, maltosaccharides, maltose, and D-glucose by photoassimilation of $^{14}CO_2$ in *Anacystis nidulans* and selective enzymic degradation. - Carbohydr. Res. *56* : 357 - 362, 1977.

30565 - LEIHNER, D.E., COCK, J.H. : Causes for anomalous wet-dry season yield differences in lowland rice. - Crop Sci. *17* : 391 - 395, 1977. [Growth analysis.]

30566 - LEIMKUEHLER, W., JOSEPH, C., RANDALL, D.D., NELSON, C.J. : RuBP carboxylase
of *Festuca arundinacea* SCHREB. - Plant Physiol. *59* (6, Suppl.) : 90, 1977.

30567 - LEISER, M., GROMET-ELHANAN, Z. : Comparison of the electrochemical proton
gradient and phosphate potential maintained by *Rhodospirillum rubrum* chromatophores in the steady state. - Arch. Biochem. Biophys. *178* : 79 - 88, 1977.

30568 - LEMEZA, N.A., VECHER, A.S. : Glikolatoksidaznaya aktivnost' yader i khloroplastov v prorostkakh rzhi i yachmenya. [Glycolate oxidase activity of nuclei and chloroplasts in rye and barley seedlings.] - Fiziol. Rast. *24* :
453 - 457, 1977. [In R, ab : E.]

30569 - LEMOINE, Y., JOYARD, J., TREMOLIERES, A. : On the modifications of the photosynthetic apparatus in a photosensitive tobacco mutant with a deficiency
in the oxygen evolving side of Photosystem II. - In : COOMBS, J. (ed.) :
4th International Congress on Photosynthesis. Pp. 214 - 215. UKISES, London
1977.

30570 - LEMOINE, Y., JUPIN, H. : Modifications of the photosystems antennae in a photosensitive tobacco mutant. - In : COOMBS, J. (ed.) : 4th International Congress on Photosynthesis. Pp. 215 - 216. UKISES, London 1977.

30571 - LENZ, F. : Einfluß der Frucht auf Photosynthese und Atmung. - Z. Pflanzenernähr.
Bodenkunde   *140* : 51 - 61, 1977.

30572 - LENZ, J. : 5. Seston and its main components. - In : RHEINHEIMER, G. (ed.) :
Microbial Ecology of a Brackish Water Environment. Pp. 37 - 60. Springer-Verlag, Berlin - Heidelberg - New York 1977. [Chl.]

30573 - LESÁK, J., PAVLÍČEK, A. : Výsledky studia primární produkce nivních luk.
[Primary production in flood plain meadows.] - Rostl. Výroba *23* : 459 - 470,
1977. [In Czech, ab : E, G, R.]

30574 - LEVERENZ, J.W., JARVIS, P.G. : Light geometry and net photosynthetic rates
in individual Sitka spruce shoots and needles. - In : COOMBS, J. (ed.) :
4th International Congress on Photosynthesis. P. 217. UKISES, London 1977.

30575 - LEVI, I., BERNER, T., COHEN, Y. : Primary production in loess soil crusts
of the Negev. - Isr. J. Bot. *26* : 44, 1977. [Ps, Chl.]

30576 - LEVI, M., SUMMER, G., BELLINI, E. : Phytochrome control of chlorophyll *a* loss
in excised corn leaves requires calcium. - Experientia *33* : 1446 - 1447, 1977.

30577 - LEVITT, J. : Effect of environmental stress on transport of ions across mem-
branes. - In : MARRÈ, E., CIFERRI, O. (ed.) : Regulation of Cell Membrane
Activities in Plants. Pp. 103 - 119. North-Holland Publ.Corp., Amsterdam -
Oxford - New York 1977. [Ps.]

30578   LEVY, E.M., CUNNINGHAM, C.C., CONRAD, C.D.W., MOFFATT, J.D. : A titration
apparatus for the determination of dissolved oxygen in seawater. - J. Fish.
Res. Board Can. *34* : 2218 - 2220, 1977.

30579 - LEWANDOWSKA, M., HART, J.W., JARVIS, P.G. : Photosynthetic electron transport
in shoots of Sitka spruce from different levels in a forest canopy. - Physiol.
Plant. *41* : 124 - 128, 1977.

30580 - LEWANDOWSKA, M., JARVIS, P.G. : Photosynthetic adaptation to light in Sitka
spruce. - In : COOMBS, J. (ed.) : 4th International Congress on Photosynthesis. P. 217. UKISES, London 1977.

30581 - LEWANDOWSKA, M., JARVIS, P.G. : Changes in chlorophyll and carotenoid content, specific leaf area and dry weight fraction in Sitka spruce, in response
to shading and season. - New Phytol. *79* : 247 - 256, 1977.

30582 - LEWIN, J., NORRIS, R.E., JEFFREY, S.W., PEARSON, B.E.: An aberrant chrysophy-
cean alga *Pelagococcus subviridis* gen. nov. et sp. nov. from the North Pacific Ocean. - J. Phycol. *13* : 259 - 266, 1977. [Chl, Car.]

30583 - LEWIS, D.A., NOBEL, P.S. : Thermal energy exchange model and water loss of
a barrel cactus, *Ferocactus acanthodes*. - Plant Physiol. *60* : 609 - 616,
1977. [Resistances.]

30584 - LEWIS, O.A.M., PROBYN, T.A. : Effect of nitrate feeding level on incorporation of photosynthetically reduced nitrogen into leaf metabolism. - In :
COOMBS, J. (ed.) : 4th International Congress on Photosynthesis. Pp. 217 -
218. UKISES, London 1977.

30585 - LEWIS, W.M., Jr. : Net growth rate through time as an indicator of ecological similarity among phytoplankton species. - Ecology *58* : 149 - 157, 1977.

30586 - LEY, A.C., BUTLER, W.L. : Energy transfer from Photosystem II to Photosystem
I in *Porphyridium cruentum*. - Biochim. biophys. Acta *462* : 290 - 294, 1977.

30587 - LEY, A.C., BUTLER, W.L. : The distribution of excitation energy between Photosystem I and Photosystem II in *Porphyridium cruentum*. - Plant Cell Physiol.
1977 (Spec. Issue 3 - Photosynthetic Organelles. Structure and Function) :
33 - 46, 1977.

30588 - LEY, A.C., BUTLER, W.L., BRYANT, D.A., GLAZER, A.N. : Isolation and function
of allophycocyanin B of *Porphyridium cruentum*. - Plant Physiol. *59* : 974 -
980, 1977.

30589 - LI, J., BAYLEY, H. : Affinity and photoaffinity labeling of photosynthetic
membrane components. - In : OLSON, J.M., HIND, G. (ed.) : Chlorophyll-Proteins,
Reaction Centers, and Photosynthetic Membranes. Pp. 362 - 363. Brookhaven
nat. Lab., Upton 1977.

30590 - LI, Y.S. : Multiple effects of uncoupler on chloroplast fluorescence and luminescence. - In : COOMBS, J. (ed.) : 4th International Congress on Photosynthesis. Pp. 218 - 219. UKISES, London 1977.

30591 - LI, Y.-S. : Control of energy conservation of the Photosystem II reaction
center. - In : OLSON, J.M., HIND, G. (ed.) : Chlorophyll-Proteins, Reaction
Centers, and Photosynthetic Membranes. P. 364. Brookhaven nat. Lab., Upton
1977.

30592 - LIAAEN-JENSEN, S. : Algal carotenoids and chemosystematics. - In : FAULKNER,
D.J., FENICAL, W.H. (ed.) : Marine Natural Products Chemistry. Vol. 1.
Pp. 239 - 259. Plenum Publ. Corp., New York 1977.

30593 - LIANG, T., CHAN, W.M. : Estimating solar energy absorption potential for ma-
cadamia nut orchard design: A theoretical approach. - Trans. ASAE *20* : 1045 -
1049, 1977.

30594 - LICHTENTHALER, H.K. : Regulation of prenylquinone synthesis in higher plants. -
In : TEVINI, M., LICHTENTHALER, H.K. (ed.) : Lipids and Lipid Polymers in Higher Plants. Pp. 231 - 258. Springer-Verlag, Berlin - Heidelberg - New York
1977.

30595 - LICHTENTHALER, H.K., BUSCHMANN, C., MEIER, D. : Regulation of thylakoid com-
position, Hill-activity and chloroplast structure by blue and red light. -
In : COOMBS, J. (ed.) : 4th International Congress on Photosynthesis. P.220.
UKISES, London 1977.

30596 - LICHTENTHALER, H.K., KARUNEN, P., GRUMBACH, K.H. : Determination of prenylquinones in green photosynthetically active moss and liver moss tissues. - Physiol. Plant. *40* : 105 - 110, 1977.

30597 - LICHTENTHALER, H.K., KLEUDGEN, H.K. : Effect of the herbicide San 6706 on
biosynthesis of photosynthetic pigments and prenylquinones in *Raphanus* and
in *Hordeum* seedlings. - Z. Naturforsch. *32 C* : 236 - 240, 1977.

30598 - LICHTENTHALER, H.K., PRENZEL, U. : High-performance liquid chromatography of
natural prenylquinones. - J. Chromatogr. *135* : 493 - 498, 1977.

30599 - LIEN, S., McBRIDE, A.C., McBRIDE, J., TOGASAKI, R.K., SAN PIETRO, A. : The
differential action of PSII-specific action on a DCMU-resistant mutant
strain of *C. reinhardi*. - Plant Physiol. *59* (6, Suppl.) : 23, 1977.

30600 - LIEN, S.J., McBRIDE, J.C., McBRIDE, A.C., TOGASAKI, R.K., SAN PIETRO, A. :
A comparative study of photosystm II specific inhibitors : The differential
action on a DCMU resistant mutant strain of *C. reinhardi*. - Plant Cell Physiol. *1977* (Spec. Issue 3 - Photosynthetic Organelles. Structure and Function) : 243 - 256, 1977.

*30601 - **LIEN, T., KNUTSEN, G.** : Synchronized cultures of a cell wall-less mutant of
Chlamydomonas reinhardii. - Arch. Microbiol. *108* : 189 - 194, 1976.

30602 - **LILJENBERG, C.** : Chlorophyll formation : The phytylation step. - In :
TEVINI, M., LICHTENTHALER, H.K. (ed.) : Lipids and Lipid Polymers in Higher
Plants. Pp. 259 - 270. Springer-Verlag, Berlin - Heidelberg - New York 1977.

30603 - **LILJENBERG, C.** : The occurrence of phytylpyrophosphate and acyl esters of phytol in irradiated dark-grown barley seedlings and their possible role in biosynthesis of chlorophyll. - Physiol. Plant. *39* : 101 - 105, 1977.

30604 - **LILLEY, R.McC.,CHON,C.J., MOSBACH, A., HELDT, H.W.** : The distribution of metabolites between spinach chloroplasts and medium during photosynthesis
in vitro. - Biochim. biophys. Acta *460* : 259 - 272, 1977.

30605 - **LINDAHL, O.** : Studies on the production of phytoplankton and zooplankton in
the Baltic in 1975. - Medd. Havsfiskelab. Lysekil *217* : 1 - 43, 1977.

30606 - **LINDAHL, O.** : Studies on the production of phytoplankton and zooplankton in
the Baltic in 1976, and a summary of results from 1973 - 1976. - Medd. Havsfiskelab. Lysekil *220* : 1 - 52, 1977.

30607 - **LINDEMAN, W.** : Efficiency of cyclic photophosphorylation *in vivo*. - In :
COOMBS, J. (ed.) : 4th International Congress on Photosynthesis. P. 219.
UKISES, London 1977.

30608 - **LINDER, S., TROENG, E.** : The seasonal course of net photosynthesis and stem
respiration in a 20-year-old stand of Scots pine (*Pinus silvestris* L.). -
In : COOMBS, J. (ed.) : 4th International Congress on Photosynthesis. P. 221.
UKISES, London 1977.

30609 - **LIPS, S.H., KAGAN-ZUR, V.** : Quantitative aspects of photosynthesis enhance-
ment of isolated chloroplasts by cytoplasmic particulate factors. - In :
COOMBS, J. (ed.) : 4th International Congress on Photosynthesis. P. 222.
UKISES, London 1977.

30610 - **LIPSKAYA, G.A.** : Nakoplenie pigmentov i produktivnost' pervogo semennogo po-
koleniya yachmenya pri vnesenii kobal'ta pod materskie rasteniya. [Accumula-
tion of pigments and productivity of the first seed generation of barley as
a result of cobalt application to maternal plants.] - Sel'skokhoz. Biol. *12* :
510 - 514, 1977. [In R, ab : E.]

30611 - **LIPSKAYA, G.A., CHARNETSKI, S.S.** : Stan khlarafilu va ŭmovakh poŭnaga zatsya-
mnennya prarostkaŭ yachmenyu, vyrashchanykh z nasennya z roznaŭ kol'kastsyu
ŭ im kobal'tu. [State of chlorophyll in heavily shaded barley shoots grown
from seeds with different cobalt content.] - Vestsi Akad. Navuk belarus. SSR,
Ser. biyal.Navuk *1977* (2) : 27 - 33, 138, 1977. [In Belorus., ab : R.]

30612 - **LIPSKAYA, G.A., TITARENKO, N.O.** : Nakoplenie i geterogennost' khlorofillov
a i *b* v protsesse zeleneniya ètiolirovannykh prorostkov yachmenya pri neodi-
nakovom snabzhenii ikh kobal'tom. [Accumulation and heterogeneity of chlorophylls *a* and *b* during greening of etiolated barley seedlings supplied with
different amount of cobalt.] - Fiziol. Rast. *24* : 704 - 709, 1977. [In R,
ab : E.]

30613 - **LISOVSKIĬ, G.M., YAN, N.A.** :Produktivnost' i konkurentosposobnost' malokhlo-
rofil'nogo shtamma khlorelly. [Productivity and ability of competition in
a *Chlorella* strain with low amount of chlorophyll.] - In : NASYROV, Yu.S.
(ed.) : Genetika Fotosinteza. Pp. 261 - 265. Donish, Dushanbe 1977. [In R.]

30614 - **LITVIN, F.F., BALASHOV, S.P.** : Novye intermediaty v fotokhimicheskikh pre-
vrashcheniyakh bakteriorodopsina. [New intermediates in the photoconversions
of bacteriorhodopsin.] - Biofizika *22* : 1111 - 1114, 1977. [In R, ab : E.]

30615 - **LITVIN, F.F., BOĬCHENKO, V.A., BALASHOV, S.P., DUBROVSKIĬ, V.T.** : Fotoindutsirovannoe ingibirovanie i stimulirovanie dykhaniya v kletkakh *Halobacterium
halobium* : kinetika, spektry deĭstviya, svyaz' s fotoinduktsieĭ ΔpH.
[Photoinduced inhibition and stimulation of respiration in the cells of *Halobacterium halobium* : kinetics, action spectra, relation to photoinduction
of ΔpH.] - Biofizika *22* : 1062 - 1071, 1977. [In R, ab : E.]

30616 - LITVIN, F.F., STADNICHUK, I.N. : Spektry pogloshcheniya, fluorestsentsii i vozbuzhdeniya fluorestsentsii khlorofilla *c* v kletke pri -196 °C. [Absorption, fluorescence and fluorescence excitation spectra of chlorophyll *c in vivo* at -196 °C.] - Biofizika *22* : 541 - 542, 1977. [In R, ab : E.]

30617 - LIU, Han-Fan : [Preliminary introduction to chemical bionics.] - Hua Hsueh Tung Pao *5* : 291 - 299, 306, 1977. [Ps; in Chin.]

30618 - LIVESEY, D.L. : On the colorimetric method of assaying carbonic anhydrase. - Anal. Biochem. *77* : 552 - 561, 1977.

30619 - LIVINGSTON, A.L., KNOWLES, R.E., AMELLA, A., KOHLER, G.O. : Nutrient changes during alfalfa wilting and dehydration. - J. agr. Food Chem. *25* : 779 - 783, 1977. [Car.]

30620 - LIXANDRU, G., TĂRNĂUCEANU, E., CIUREA, G. : Zur Chlorose des Apfelbaumes. - Lucrări Ştiinţ. Inst. agron. "Ion Ionescu de la Brad" Iaşi, Ser. Horticult. *1977* : 13 - 16, 1977. [Chl.]

30621 - LJUBEŠIĆ, N. : The formation of chromoplasts in fruits of *Cucurbita maxima* DUCH. "turbaniformis". - Bot. Gaz. *138* : 286 - 290, 1977. [Chloroplast.]

30622 - LLOYD, N.D.H., CANVIN, D.T. : Photosynthesis and photorespiration in sunflower selections. - Can. J. Bot. *55* : 3006 - 3012, 1977.

30623 - LLOYD, N.D.H., CANVIN, D.T., BRISTOW, J.M. : Photosynthesis and photorespiration in submerged aquatic vascular plants. - Can. J. Bot. *55* : 3001 - 3005, 1977.

30624 - LLOYD, N.D.H., CANVIN, D.T., CULVER, D.A. : Photosynthesis and photorespiration in algae. - Plant Physiol. *59* : 936 - 940, 1977.

30625 - LOACH, P.A. : Primary photochemistry in photosynthesis. - Photochem. Photobiol. *26* : 87 - 94, 1977.

30626 - LOACH, P.A., MORRISON, L.E., BERING, C.L., RUNQUIST, J.A. : The role of ubiquinone in the primary events of bacterial photosynthesis. - In : COOMBS, J. (ed.) : 4th International Congress on Photosynthesis. P. 223. UKISES, London 1977.

30627 - LOBBAN, C.S. : Patterns of ^{14}C translocation in giant kelp. - Plant Physiol. *59* (6, Suppl.) : 126, 1977.

30628 - LO CASCIO, B., SAPORITO, C. : Ricerche preliminari sull'irrigazione del frumento in seconda coltura. [Preliminary research on the irrigation of wheat as second crop.] - Quaderni Ric. sci. *99* (3° Incontro sui Problemi Agronomici dell'Irrigazione) : 184 - 190, 1977. [Growth analysis; in Ital., ab : E.]

30629 - LOEBLICH, A.R. III. : Studies on synchronously dividing populations of *Cachonina niei*, a marine dinoflagellate. - Bull. jap. Soc. Phycol. *25* (Suppl.) : 119 - 128, 1977. [Chl, Car.]

30630 - LONERGAN, T.A., SARGENT, M.L. : Involvement of the light reactions in the photosynthetic circadian rhythm of *Euglena gracilis*. - Plant Physiol. *59* (6, Suppl.) : 130, 1977.

30631 - LONG, D.E., LONG, S.P., JONES, D.A., WOOLHOUSE, H.W. : The effect of temperature on steady-state carbon metabolism in species with the C_4 photosynthetic system. - In : COOMBS, J. (ed.) : 4th International Congress on Photosynthesis. P. 224. UKISES, London 1977.

30632 - LONG, M.M., URRY, D.W., STOECKENIUS, W. : Circular dichroism of biological membranes: purple membrane of *Halobacterium halobium*. - Biochem. biophys. Res. Commun. *75* : 725 - 731, 1977.

30633 - LONGSTRETH, D.J., STRAIN, B.R. : Effects of salinity and illumination on photosynthesis and water balance of *Spartina alterniflora* LOISEL. - Oecologia *31* : 191 - 199, 1977.

30634 - LONGWORTH, J.W. : Photon counting. - Photochem. Photobiol. *26* : 665 - 668, 1977.

30635 - **LORIMER, G.H.** : Determination of glycolic acid by the Eegriwe (Calkins) method. Interference by ethylenetetraacetic acid and formaldehyde. - Anal. Biochem. *83* : 785 - 787, 1977.

30636 - **LORIMER, G.H., BADGER, M.R., ANDREWS, T.J.** : D-ribulose-1,5-bisphosphate carboxylase-oxygenase. Improved methods for the activation and assay of catalytic activities. - Anal. Biochem. *78* : 66 - 75, 1977.

30637 - **LORIMER, G.H., BERRY, J.A., KRAUSE, G.H., OSMOND, C.B.** : Metabolism of [^{18}O] by the photorespiratory carbon oxidation cycle *in vivo*, and its significance. - Plant Physiol. *59* (6, Suppl.) : 43, 1977.

30638 - **LORIMER, G.H., KRAUSE, G.H., BERRY, J.A.** : Incorporation of [^{18}O] oxygen into glycolate by intact isolated chloroplasts. - Carnegie Inst. Year Book *76* : 314 - 316, 1977.

30639 - **LORIMER, G.H., KRAUSE, G.H., BERRY, J.A.** : The incorporation of [^{18}O] oxygen into glycolate by intact isolated chloroplasts. - FEBS Lett. *78* : 199 - 202, 1977.

30640 - **LORIMER, G.H., WOO, K.C., BERRY, J.A., OSMOND, C.B.** : The C_2 photorespiratory carbon oxidation cycle in leaves of higher plants: Pathway and function. - In : **COOMBS, J.** (ed.) : 4th International Congress on Photosynthesis. P. 225. UKISES, London 1977.

30641 - **LOSEV, A.P., ZEN'KEVICH, É.I., SAGUN, E.I.** : Kontsentratsionnoe tushenie fluorestsentsii i obrazovaniya tripletnykh sostoyaniĭ khlorofilla "a" i feofitina v rastvorakh. [Concentration quenching of fluorescence and formation of triplet states of chlorophyll *a* and pheophytin in solutions.] - Zh. prikl. Spektroskop. *27* : 244 - 247, 1977. [In R.]

30642 - **ŁOTOCKI, A.** : Effect of root aeration and form of nitrogen on photosynthetic productivity of Scots pine (*Pinus silvestris* L.). - Acta Soc. Bot. Pol. *46* : 303 - 316, 1977.

30643 - **LOUWERSE, W., ZWEERDE, W. v.d.** : Photosynthesis, transpiration and leaf morphology of *Phaseolus vulgaris* and *Zea mays* grown at different irradiances in artificial and sunlight. - Photosynthetica *11* : 11 - 21, 1977.

30644 - **LOVE, A., BARBER, J., MILLS, J.** : Influence of the electrical diffuse double layer associated with the thylakoid membrane on chlorophyll fluorescence yield. - In : **COOMBS, J.** (ed.) : 4th International Congress on Photosynthesis. P. 226. UKISES, London 1977.

30645 - **LOVE, R.J.R., ROBINSON, G.G.C.** : The primary productivity of submerged macrophytes in West Blue Lake, Manitoba. - Can. J. Bot. *55* : 118 - 127, 1977.

30646 - **LOVEYS, B.R.** : The intracellular location of abscisic acid in stressed and non-stressed leaf tissue. - Physiol. Plant. *40* : 6 - 10, 1977. [Chloroplast.]

30647 - **LOWE, R.H.** : Crystallization of Fraction I protein from tobacco by a simplified procedure. - FEBS Lett. *78* : 98 - 100, 1977.

30648 - **LOZANO, J., ABRUÑA, F.** : Effect of planting season on yields of eight short-grain varieties of rice under irrigation. - J. agr. Univ. Puerto Rico *61* : 6 -   , 1977.

30649 - **LOZIER, R.M., NIEDERBERGER, W.** : The photochemical cycle of bacteriorhodopsin. - Fed. Proc. *36* : 1805 - 1809, 1977.

30650 - **LU, Y.-C., WANG, S.-C., YANG, W.-J.** : [Studies on the spectral absorption properties of chlorophyll *d* in different solvents.] - Acta bot. sin. *19* : 283 - 290, 1977. [In Chin., ab : E.]

30651 - **LUCAS, W.J.** : Plasmalemma transport of HCO_3^- and OH^- in *Chara corallina* : Inhibitory effect of ammonium sulphate. - J.exp. Bot. *28* : 1307 - 1320, 1977.

30652 - **LUCAS, W.J.**: Analogue inhibition of the active HCO_3^- transport site in characean plasma membrane. - J. exp. Bot. *28* : 1321 - 1336, 1977.

30653 - **LUCAS, W.J., DAINTY, J.** : Bicarbonate assimilation and HCO_3^-/OH^- exchange in *Chara corallina*.- In : **MARRÈ, E., CIFERRI, O.** (ed.) : Regulation of Cell Membrane Activities in Plants. Pp. 43 - 53. North-Holland Publ. Corp., Amsterdam - Oxford - New York 1977.

30654 - LUCAS, W.J., DAINTY, J. : HCO_3^- influx across plasmalemma of *Chara corallina*. Divalent cation requirement. - Plant Physiol. *60* : 862 - 867, 1977.

30655 - LUCAS, W.J., DAINTY, J. : Spatial distribution of functional OH^- carriers along a characean internodal cell : determined by the effect of cytochalasin B on $H^{14}CO_3$ assimilation. - J. Membrane Biol. *32* : 75 - 92, 1977.

30656 - LUDWIG, L.J., WITHERS, A.C. : Leaf photosynthesis in tomato (*Lycopersicon esculentum*). - In : COOMBS, J. (ed.) : 4th International Congress on Photosynthesis. P. 227. UKISES, London 1977.

30657 - LUISETTI, J., MÖHWALD, H., GALLA, H.J. : Paramagnetic fluorescence quenching in chlorophyll *a* containing vesicles: evidence for the localization of chlorophyll. - Biochem. biophys. Res. Commun. *78* : 754 - 760, 1977.

30658 - LUNDIN, A., THORE, A., BALTSCHEFFSKY, M. : Sensitive measurement of flash induced photophosphorylation in bacterial chromatophores by firefly luciferase. - FEBS Lett. *79* : 73 - 76, 1977.

30659 - LURIE, S. : Photochemical properties of guard cell chloroplasts. - Plant Sci. Lett. *10* : 219 - 223, 1977.

30660 - LURIE, S. : Stomatal development in etiolated *Vicia faba* : relationship between structure and function. - Aust. J. Plant Physiol. *4* : 61 - 68, 1977. [Chl, chloroplast.]

30661 - LURIE, S. : Stomatal opening and photosynthesis in greening leaves of *Vicia faba* L. - Aust. J. Plant Physiol. *4* : 69 - 74, 1977.

30662 - LURIE, S., DUBLIN, B. : CO_2 fixation by guard cell chloroplasts and its relation to stomatal function. - In : COOMBS, J. (ed.) : 4th International Congress on Photosynthesis. P. 228. UKISES, London 1977.

30663 - LUTKOV, A.A. : Maksimal'nyĭ potentsial'nyĭ fotosintez i udel'naya plotnost' lista u intsukht-liniĭ sakharnoĭ svekly. [Maximum potential photosynthesis and specific density of the leaf in inbreeding lines of sugar beet.] - Fiziol. Rast. *24* : 473 - 477, 1977. [In R, ab : E.]

30664 - LÜTTGE, U., BALL, E. : Concentration and pH dependence of malate efflux and influx in leaf slices of CAM plants. - Z. Pflanzenphysiol. *83* : 43 - 54, 1977.

30665 - LÜTTGE, U., BALL, E. : Water relation parameters of the CAM plant *Kalanchoë daigremontiana* in relation to diurnal malate oscillations. - Oecologia *31* : 85 - 94, 1977.

30666 - LÜTTGE, U., BALL, E., GREENWAY, H. : Effects of water and turgor potential on malate efflux from leaf slices of *Kalanchoë daigremontiana*. - Plant Physiol. *60* : 521 - 523, 1977.

*30667 - LÜTTKE, A., RAHMSDORF, U., SCHMID, R. : Heterogeneity in chloroplasts of siphonacious algae as compared with higher plant chloroplasts. - Z. Naturforsch. *31C* : 108 - 110, 1976.

30668 - LÜTZ, C., KESSELMEIER, J., RUPPEL, H.G. : Biochemical and cytological observations on chloroplast development. IV. Reaggregations of solubilized prolamellar bodies from etioplasts of *Avena sativa* L. - Z. Pflanzenphysiol. *85* : 327 - 340, 1977.

30669 - LUTZ, M. : Antenna chlorophyll in photosynthetic membranes. A study by resonance Raman spectroscopy. - Biochim. biophys. Acta *460* : 408 - 430, 1977.

30670 - LUTZ, M., KLEO, J. : Bonding interactions of chlorophyll in photosynthetic membranes : Resonance Raman spectroscopy. - In : COOMBS, J. (ed.) : 4th International Congress on Photosynthesis. Pp. 228 - 229. UKISES, London 1977.

*30671 - LUTZ, M., KLEO, J., REISS-HUSSON, F. : Resonance Raman scattering of bacteriochlorophyll, bacteriopheophytin and spheroidene in reaction centers of *Rhodopseudomonas spheroides*. - Biochem. biophys. Res. Commun. *69* : 711 - 717, 1976.

30672 - LUTZ, M., KLEO, J., REISS-HUSSON, F. : Resonance Raman scattering of bacteriochlorophyll *a*, bacteriopheophytin *a*, and spheroidene in reaction centers of *Rhodopseudomonas sphaeroides*.- In : OLSON, J.M., HIND, G.(ed.) : Chlorophyll-Proteins, Reaction Centers, and Photosynthetic Membranes. P. 367. Brookhaven nat. Lab., Upton 1977.

30673 - LYMAN, H., SRINIVAS, U., LUEBBERT, J. : Regulation of chloroplast gene expression in *Euglena*. - In : COOMBS, J. (ed.) : 4th International Congress on Photosynthesis. Pp. 229 - 230. UKISES, London 1977.

30674 - LYNCH, D.R., ROWBERRY, R.G. : Population density studies with Russet Burbank II. The effect of fertilization and plant density on growth, development and yield. - Amer. Potato J. *54* : 57 - 71, 1977. [Growth analysis.]

30675 - LYSHEDE, O.B. : Anatomical features of some stem assimilating desert plants of Israel. - Bot. Tidsskr. *71* : 225 - 230, 1977.

30676 - MacCAULL, W.A., PLATT, T. : Diel variations in the photosynthetic parameters of coastal marine phytoplankton. - Limnol. Oceanogr. *22* : 723 - 731, 1977.

30677 - MacCOLL, D. : Growth and sugar accumulation of sugarcane. II. Dry weight increments and estimates of assimilation rate. - Exp. Agr. *13* : 61 - 69, 1977.

30678 - MacCOLL, R., BERNS, D.S. : Spectroscopic studies on biliproteins of the cryptomonad algae. - In : OLSON, J.M., HIND, G. (ed.) : Chlorophyll-Proteins, Reaction Centers, and Photosynthetic Membranes. P. 364. Brookhaven nat. Lab., Upton 1977.

30679 - MacDONALD, R.E., LANYI, J.K. : Light-activated amino acid transport in *Halobacterium halobium* envelope vesicles. - Fed. Proc. *36* : 1828 - 1832, 1977.

30680 - MACDOWALL, F.D.H. : Growth kinetics of Marquis wheat. VII. Dependence on photoperiod and light compensation point in vegetative phase. - Can. J. Bot. *55* : 639 - 643, 1977.

30681 - MACHADO, V.S., BANDEEN, J.D., STEPHENSON, G.R., JENSEN, K.I.N. : Differential atrazine interference with the Hill reaction of isolated chloroplasts from *Chenopodium album* L. biotypes. - Weed Res. *17* : 407 - 413, 1977.

30682 - MACHE, R. : The genetic system of plastids. - Egypt. J. Genet. Cytol. *6* : 178 - 187, 1977.

30683 - MACHIDA, H., OOISHI, A., HOSOI, T., KOMATSU, H., KAMOTA, F. : [Studies on photosynthesis in cuttings during propagation. I. Changes in the rate of apparent photosynthesis in the cuttings of several plants after planting.] - J. jap. Soc. hort. Sci. *46* : 274 - 282, 1977. [In Jap., ab : E.]

30684 - MÄCHLER, F., NÖSBERGER, J. : Temperature response of photosynthesis and photorespiration of altitudinal ecotypes of *Trifolium repens*. - In : COOMBS, J. (ed.) : 4th International Congress on Photosynthesis. Pp. 230 - 231. UKISES, London 1977.

30685 - MÄCHLER, F., NÖSBERGER, J. : Effect of light intensity and temperature on apparent photosynthesis of altitudinal ecotypes of *Trifolium repens* L. - Oecologia *31* : 73 - 78, 1977.

30686 - MÄCHLER, F., NÖSBERGER, J., ERISMANN, K.H. : Photosynthetic $^{14}CO_2$ fixation products in altitudinal ecotypes of *Trifolium repens* L. with different temperature requirements. - Oecologia *31* : 79 - 84, 1977.

30687 - MACHOLD, O., MEISTER, A., SAGROMSKY, H., HØYER-HANSEN, G., von WETTSTEIN, D. : Composition of photosynthetic membranes of wild-type barley and chlorophyll *b*-less mutants. - Photosynthetica *11* : 200 - 206, 1977.

30688 - MACHOLD, O., SIMPSON, D.J., HØYER-HANSEN, G. : Correlation between the freeze fracture appearance and polypeptide composition of thylakoid membranes in barley. - Carlsberg Res. Commun. *42* : 499 - 516, 1977.

30689 - MACK, A.R., SCHUBERT, J., GOODFELLOW, C., CHAGARLAMUDI, P., MOORE, H. : Global agricultural productivity estimation from LANDSAT data. - In : 4th Canadian Symposium on Remote Sensing. Pp. 8 - 18. Can. Aeronautics & Space Inst., Ottawa 1977.

30690 - MACKENDER, R.O. : The permeability of the envelopes of naturally differentiating chloroplasts from the leaves of *Avena sativa* L. - In : COOMBS, J. (ed.) : 4th International Congress on Photosynthesis. P. 231. UKISES, London 1977.

30691 - MACZEK, W. : Photosynthetic production of *Pinus sylvestris* (L.) in Niepoło-
mice forest within the range of industrial emission. - Bull. Acad. pol. Sci.,
Sér. Sci. biol. *25* : 685 - 693, 1977.

30692 - MADALAGERI, B.B., KOTESWARA RAO, S. : Leaf area estimation in sugarbeet. -
Mysore J. agr. Sci. *11* : 1 - 3, 1977.

30693 - MADER, P., CHLAD, F., CHLADOVÁ, J., NAUŠ, J., KOHLOVÁ, V., SOFROVÁ, D.,
MAKOVEC, P. : Development of photosynthetic apparatus of etiolated seedlings
of *Hordeum vulgare* grown under different nutrition conditions and water state. - In : COOMBS, J. (ed.) : 4th International Congress on Photosynthesis.
P. 232. UKISES, London 1977.

30694 - MADER, P., NOVÁK, V., PAVELEK, M., CHLAD, F., DEGENOVÁ, V., BURSOVÁ, M.,
JEŽKOVÁ, J. : Pigment-proteinové složení thylakoidů chloroplastů a fotosyntetická kapacita u jarního ječmene během ontogenese. [Pigment-protein composition of the chloroplast thylakoids and photosynthetic capacity of spring
barley during ontogenesis.] - In : Sborník z Vědecké Konference na Počest
30. Výročí Osvobození ČSSR Sovětskou Armádou. Sekce Fytotechnická. Pp. 223 -
234. Vysoká Škola zemědělská, Praha 1977. [In Czech.]

30695 - MADHUSUDANA RAO, I., SWAMY, P.M., DAS, V.S.R. : CAM-syndrome in some nonsuc-
culents and its inhibition by paraquat. - In : COOMBS, J. (ed.) : 4th International Congress on Photosynthesis. P. 233. UKISES, London 1977.

30696 - MADIGAN, M.T., BROCK, T.D. : Adaptation by hot spring phototrophs to reduced
light intensities. - Arch. Microbiol. *113* : 111 - 120, 1977. [Ps, Chl.]

30697 - MADIGAN, M.T., BROCK, T.D. : 'Chlorobium-type' vesicles of photosynthetically-grown *Chloroflexus aurantiacus* observed using negative staining techniques. - J. gen. Microbiol. *102* : 279 - 285, 1977.

30698 - MADIGAN, M.T., BROCK, T.D. : CO_2 fixation in photosynthetically-grown *Chloroflexus aurantiacus*. - FEMS Microbiol. Lett. *1* : 301 - 304, 1977.

30699 - MAESTRI, M., BARROS, R.S. : Coffee. - In : ALVIM, P. de T., KOZLOWSKI, T.T.
(ed.) : Ecophysiology of Tropical Crops. Pp. 249 - 279. Academic Press, New
York - San Francisco - London 1977. [Ps.]

30700 - MAGAMBO, M.J.S. : Canopy characteristics of seven clones of tea (*Camellia
sinensis* L.) estimated by the use of inclined point quadrats. - Trop. Agr.
54 : 205 - 211, 1977.

30701 - MAGNUSSON, A., VANDENDRIESSCHE, T. : Inhibitory effect of NaN_3 on dark and
light electron transport. - In : COOMBS, J. (ed.) : 4th International Congress
on Photosynthesis. P. 234. UKISES, London 1977.

30702 - MAGOMEDOV, I.M., AGAEV, M.G. : Évolyutsiya uglerodnogo metabolizma tsvetko-
vykh rasteniĭ s tsiklom C_4-dikarbonovykh kislot. [Evolution of carbon metabolism of flowering plants with the cycle of C_4-dicarbonic acids.] - In :
NASYROV, Yu.S. (ed.) : Genetika Fotosinteza. Pp. 85 - 89. Donish, Dushanbe
1977. [In R.]

*30703 - MAGRISO, Yu., SLAVCHEVA, T. : Vliyanie na pochvenata vlaga v"rkhu intenziteta
na fotosintezata i transpiratsiyata pri lozata. [Influence of soil moisture
on photosynthesis and transpiration of *Vitis vinifera*.] - Gradinarska lozar-
ska Nauka *13* (5) : 72 - 80, 1976. [In Bulg., ab : F, R.]

30704 - MAGYAROSY, A.C., SCHÜRMANN, P., MONTALBINI, P., BUCHANAN, B.B. : Effect of
infection by obligate parasites on photosynthesis. - In : KIRALY, Z. (ed.) :
Current Topics in Plant Pathology. Pp. 89 - 98. Akad. Kiadó, Budapest 1977.

30705 - MAHON, J.D., LOWE, S.B., HUNT, L.A. : Variation in the rate of photosynthetic
CO_2 uptake in cassava cultivars and related species of *Manihot*. - Photosynthetica *11* : 131 - 138, 1977.

30706 - MAHON, J.D., LOWE, S.B., HUNT, L.A., THIAGARAJAH, M. : Environmental effects
on photosynthesis and transpiration in attached leaves of cassava (*Manihot
esculenta* CRANTZ). - Photosynthetica *11* : 121 - 130, 1977.

30707 - MAIER, R. : Die Wirkung von Blei auf die NAD^+-abhängige Malat-Dehydrogenase
in *Medicago sativa* L. und *Zebrina pendula* SCHNIZL. - Z. Pflanzenphysiol. *85* :
319 - 326, 1977.

30708 - MAISON, B., LAVOREL, J. : Effect of temperature on photosystem II in *Chlorella* and chloroplasts. - Plant Cell Physiol. *1977* (Spec. Issue 3 - Photosynthetic Organelles. Structure and Function) : 55 - 65, 1977.

30709 - MAISON, B., MOYA, I. : Fluorescence lifetime of heat treated *Chlorella pyrenoidosa*. - In : COOMBS, J. (ed.) : 4th International Congress on Photosynthesis. Pp. 233 - 234. UKISES, London 1977.

30710 - MAISON-PETERI, B., ETIENNE,A.L. : Effects of sodium azide on Photosystem II of *Chlorella pyrenoidosa*. - Biochim. biophys. Acta *459* : 10 - 19, 1977.

30711 - MAJOR, D.J. : Analysis of growth of irrigated rape. - Can. J. Plant Sci. *57* : 193 - 197, 1977. [Dry-matter accumulation.]

30712 - MAKAROVA, N.N., VISHNYAKOVA, I.I. : Intensivnost' fotosinteza i napravlennost' ottoka assimilyatov u genetichecki raznokachestvennykh sortov ozimoĭ rzhi. [Photosynthetic rate and direction of photosynthate transport in winter wheat cultivars of different genetic quality.] - In : NASYROV, Yu.S. (ed.) : Genetika Fotosinteza. Pp. 238 - 240. Donish, Dushanbe 1977. [In R.]

30713 - MAKAROVA, N.N., VISHNYAKOVA, I.I. : K fiziologo-biokhimicheskoĭ diagnostike poleganiya ozimoĭ rzhi. [Physiological and biochemical diagnosis of winter rye lodging.] - Fiziol. Biokhim. kul't. Rast. *9* : 366 - 369, 1977. [Photosynthates; in R, ab : E.]

30714 - MAKOVCOVÁ, O., ŠINDELÁŘ, L. : Durch TMV-Infection hervorgerufene Veränderung der Intensität der Glykolyse und des Pentosephosphatcyclus bei Tabakpflanzen. - Biol. Plant. *19* : 253 - 258, 1977. [Ps.]

30715 - MALHOTRA, S.S. : Effects of aqueous sulphur dioxide on chlorophyll destruction in *Pinus contorta*. - New Phytol. *78* : 101 - 109, 1977.

30716 - MALKIN, R. : Primary electron acceptors. - In : TREBST, A., AVRON, M. (ed.) : Photosynthesis I. (Encycl. Plant Physiol. N.S. Vol. 5.) Pp. 179 - 186. Springer-Verlag, Berlin - Heidelberg - New York 1977.

30717 - MALKIN, R., BEARDEN, A.J. : Electron paramagnetic resonance studies of plastocyanin in chloroplast photosynthesis. - In : COOMBS, J. (ed.) : 4th International Congress on Photosynthesis. P. 235. UKISES, London 1977.

30718 - MALKIN, S. : Delayed luminescence. - In : BARBER, J. (ed.) : Primary Processes of Photosynthesis. Pp. 349 - 431. Elsevier, Amsterdam - New York - Oxford 1977.

30719 - MALKIN, S. : Delayed luminescence. - In : TREBST, A., AVRON, M. (ed.) : Photosynthesis I. (Encycl. Plant Physiol. N.S. Vol. 5.) PP. 473 - 491. Springer-Verlag, Berlin - Heidelberg - New York 1977.

30720 - MALKIN, S. : Modulating effects on the delayed luminescence from photosystem II of photosynthesis. - In : CASTELLANI, A. (ed.) : Research in Photobiology. Pp. 139 - 151. Plenum Press, New York - London 1977.

*30721 - MALKINA, I.S. : O stepeni uchastiya palisadnoĭ parenkhimy v fotosinteze list'-ev klena ostrolistnogo i berezy povisloĭ. [The degree of palisade parenchyma participation in photosynthesis of leaves of the Norway maple and the weeping birch.] - Lesovedenie *1976* (2) : 51 - 57, 1976. [In R, ab : E.]

30722 - MALKINA, I.S. : Vliyanie osveshchennosti mestoobitaniya na svetovye krivye fotosinteza klena i berezy. [The effect of site illumination on light response curves of maple and birch photosynthesis.] - Lesovedenie *1977* (3) : 21 - 25, 1977. [In R, ab : E.]

30723 - MALOFEEV, V.M. : Odnovremennaya i nepreryvnaya registratsiya O_2 i CO_2 pri issledovanii nestatsionarnykh sostoyaniĭ fotosinteza rasteniĭ. [Simultaneous and continuous registration of oxygen and carbon dioxide during studies of non-steady state photosynthesis of plants.] - Fiziol. Rast. *24* : 203 - 206, 1977. [In R, ab : E.]

30724 - MALONE, T.C. : Environmental regulation of phytoplankton productivity in the lower Hudson Estuary. - Estuar. coast. mar. Sci. *5* : 157 - 171, 1977.

30725 - MALONE, T.C. : Light-saturated photosynthesis by phytoplankton size fractions in the New-York Bight, USA. - Mar. Biol. *42* : 281 - 292, 1977.

30726 - **MAL'YAN, A.N., AKULOVA, E.A., MUZAFAROV, E.N.** : Kvertsetin - allostericheskiĭ regulyator aktivnosti sopryazhayushchego faktora fotofosforilirovaniya, CF_1. [Quercetin - an allosteric regulator of activity of the photophosphorylation coupling factor CF_1.] - Bioorg.Khim. *3*:639 - 645,   1977. [In R, ab : E.]

30727 - **MAL'YAN, A.N., MUZAFAROV, E.N., AKULOVA, E.A.** : Vzaimodeĭstvie flavonolov (aglikona, glikozidirovannykh i atsilproizvodnykh) s sopryagayushchim faktorom fotofosforilirovaniya CF_1. [Interaction of flavonols (aglycone, glycosidated and acyl derivatives) and photophosphorylation coupling factor CF_1.] - In : AKULOVA, E.A., MUZAFAROV, E.N. (ed.) : Regulyatsiya Énergeticheskogo Obmena Khloroplastov i Mitokhondriĭ Éndogennymi Fenol'nymi Ingibitorami. Pp. 41 - 55, 127 - 128, 132 - 133. Pushchino 1977. [In R, ab : E.]

30728 - **MAMLEEVA, N.A., NEKRASOV, L.I.** : Izuchenie struktury poverkhnostnykh sloev khlorofilla adsorbtsionnym metodom. [Structure of surface layers of chlorophyll studied by an adsorption method.] - Zh. fiz. Khim. *51* : 2340 - 2343, 1977. [In R.]

30729 - **MAMLEEVA, N.A., NEKRASOV, L.I.** : Spektry pogloshcheniya khlorofilla a, adsorbirovannogo na aĕrosile. [Absorption spectra of chlorophyll a adsorbed on aerosil.] - Biofizika *22* : 403 - 406, 1977. [In R, ab : E.]

30730 - **MANENTI, G., TEDESCO, G.** : Ultrastructure and pigments of ivy (*Hedera helix* L.) varieties with green and varigated leaves. - Caryologia *30* : 163 - 176, 1977.

30731 - **MANGAN, J.L., WEST, J.** : Ruminal digestion of chloroplasts and the protection of protein by glutaraldehyde treatment. - J. agr. Sci. *89* : 3 - 15, 1977.

30732 - **MANN, K.H.** : Some adaptations for high productivity in seaweeds. - In : COOMBS, J. (ed.) : 4th International Congress on Photosynthesis. Pp. 235 - 236. UKISES, London 1977.

30733 - **MANNING, C.E., MILLER, D.G., TEARE, I.D.** : Effect of moisture stress on leaf anatomy and water-use efficiency of peas. - J. amer. Soc. hort. Sci. *102* : 756 - 760, 1977. [Dry-matter accumulation.]

30734 - **MANSUROV, N.I., MECHISLAVSKIĬ, Yu.A., MEDVEDOVSKAYA, L.E.** : Izmeneniya fotosinteticheskoĭ deyatel'nosti khlopchatnika pri gibridizatsii. [Changes of photosynthetic activity of cotton at hybridization.] - In : NASYROV, Yu.S. (ed.) : Genetika Fotosinteza. Pp. 266 - 270. Donish, Dushanbe 1977. [In R.]

30735 - **MANZI, J.J., BURRELL, V.G., CARSON, W.Z.** : A comparison of growth and survival and subtidal *Crassostrea virginica* (GMELIN) in South Carolina salt marsh impoundments. - Aquaculture *12* : 293 - 310, 1977. [Ps.]

30736 - **MANZINI, M.L., LAUDI, G.** : Azione esercitata dal cianuro sulla crescita e sull' inverdimento di plantule di *Picea abies* (L.) KARSTEN e di *Larix decidua* MILLER. [Action of cyanide on growth and greening of seedlings of *Picea abies* (L.) KARSTEN and *Larix decidua* MILLER.] - G. bot. ital. *111* : 101 - 107, 1977. [In Ital., ab : E.]

30737 - **MAOTANI, T., MACHIDA, Y.** : [Changes in transpiration rate, leaf diffusion resistance and leaf water potential for satsuma mandarin (*Citrus unshiu* MARC.) trees during prolonged water stress and subsequent recovery.] - J. agr. Meteorol. *32* : 203 - 208, 1977. [In Jap., ab : E.]

30738 - **MAOTANI, T., MACHIDA, Y.** : [Studies on leaf diffusion resistance of fruit trees. I. Methods of measuring leaf diffusion resistance of satsuma mandarin trees and factors influencing it.] - J. jap. Soc. hort. Sci. *46* : 1 - 8, 1977. [In Jap., ab : E.]

30739 - **MAOTANI, T., MACHIDA, Y., YAMATSU, K.** : [Studies on leaf water stress in fruit trees VI. Effect of leaf water potential on growth of satsuma mandarin (*Citrus unshiu* MARC.) trees.] - J. jap. Soc. hort. Sci. *45* : 329 - 334, 1977. [Resistances; in Jap., ab : E.]

30740 - **MAPLESTON, R., GRIFFITHS, W.T.** : Some effects of the terminal stages of chlorophyll biosynthesis. - In : COOMBS, J. (ed.) : 4th International Congress on Photosynthesis. P. 237. UKISES, London 1977.

30741 - **MAPLESTON, R.E., GRIFFITHS, W.T.** : Effects of illumination of whole barley plants on the protochlorophyllide-activating system in the isolated plastids. - Biochem. Soc. Trans. *5* : 319 - 321, 1977.

30742 - **MAR, T., GINGRAS, G.** : Evidence for monomeric bacteriochlorophyll in P_{800} of the photoreaction center from *Rhodospirillum rubrum*. - Biochim. biophys. Acta *460* : 239 - 246, 1977.

30743 - **MARCELLOS, H.** : Wheat frost injury - freezing stress and photosynthesis. - Aust. J. agr. Res. *28* : 557 - 564, 1977.

30744 - **MARCHANT, R.H.** : Studies on the hydrolysis of ATP by relatively intact chloroplast membranes adopting a new experimental approach and with the use of chloroplasts isolated from spinach grown by the nutrient film technique of hydroponics. - In : COOMBS, J. (ed.) : 4th International Congress on Photosynthesis. P. 238. UKISES, London 1977.

30745 - **MARCO, G. DI, GRECO, S., TRICOLI, D., TURI, B.** : Carbon isotope ratios ($^{13}C/^{12}C$) in fractions of field-grown grape. - Physiol. Plant. *41* : 139 - 141, 1977.

30746 - **MARDANYAN, S.S., DEMIN, Yu.M., NALBANDYAN, R.M.** : Fluorestsentnye svoĭstva ferredoksinov tipa *b*. [Fluorescent properties of *b*-type ferredoxins.] - Biokhimiya *42* : 1024 - 1029, 1977. [In R, ab : E.]

30747 - **MAREK, J., HRAŠKA, Š., REPKA, J., PETROVIC, J.** : Structure of thylakoid system of leaf mesophyll chloroplasts from wheat, rye and *Triticale*. - Photosynthetica *11* : 211 - 213, 1977.

30748 - **MARENČÍK, A.** : Možnosti a problémy gravimetrického stanovenia fotosyntézy u jabloní. [Possibilities and problems concerning the gravimetric determination of apple tree photosynthesis.]-In : HUZULÁK, J., MASAROVIČOVÁ, E. (ed.) : Fotosyntéza a Vodný Režim Drevín. Pp. 185 - 192. Modra-Piesky 1977. [In Slovak, ab : E, R.]

30749 - **MAREŠ, J., LEBLOVÁ, S.** : Termostabilní fosfoenolpyruvátkarboxyláza listů a kořenů hrachu setého. [Thermostable phosphoenolpyruvate carboxylase of pea leaves and roots.] - Rostl. Výroba *23* : 1279 - 1285, 1977. [In Czech, ab : E, G, R.]

30750 - **MARGARIS, N.S.** : Photosynthesis and productivity in a phryganic (East mediterranean) ecosystem. - In : COOMBS, J. (ed.) : 4th International Congress on Photosynthesis. P. 239. UKISES, London 1977.

30751 - **MARGARIS, N.S.** : Physiological and biochemical observations in seasonal dimorphic leaves of *Sarcopoterium spinosum* and *Phlomis fruticosa*. - Oecol. Plant. *12* : 343 - 350, 1977.

30752 - **MARGULIES, M.M., MICHAELS, A.** : Biosynthesis of chloroplast membrane proteins in *Chlamydomonas reinhardtii*. In : BOGORAD, L., WEIL, J.H. (ed.) : Acides Nucléiques et Synthèse des Protéines chez les Végétaux. Coll. Int. C.N.R.S. No. 261. Pp. 395 - 401. Édit. CNRS, Paris 1977.

30753 - **MARIANI COLOMBO, P., ORSENIGO, M.** : Sea depth effects on the algal photosynthetic apparatus. II. An electron microscopic study of the photosynthetic apparatus of *Halimeda tuna* (*Chlorophyta, Siphonales*) at -0.5 m and -6.0 m sea depths. - Phycologia *16* : 9 - 17, 1977.

30754 - **MARIANI COLOMBO, P., RASCIO, N., ORSENIGO, M.** : An hypothesis on a possible regulation process of photosynthesis in maize plants exposed to continuous illumination. - Caryologia *30* : 492 - 493, 1977.

30755 - **MARKER, A.F.H.** : Some problems arising from the estimation of chlorophyll *a* and pheophytin *a* in methanol. - Limnol. Oceanogr. *22* : 578 - 579, 1977.

30756 - **MARKER, A.F.H., GUNN, R.J.M.** : The benthic algae of some streams in southern England. III. Seasonal variations in chlorophyll *a* in the seston. - J. Ecol. *65* : 223 - 234, 1977.

30757 - **MARKL, H.** : CO_2 transport and photosynthetic productivity of a continuous culture of algae. - Biotechnol. Bioeng. *19* : 1851 - 1862, 1977.

*30758 - MARKOWITZ, M.M., HOFFMAN, L.R. : Chloroplast inclusions in zoospores of *Oedocladium*. - J. Phycol. *10* (Suppl.) : 17, 1974.

30759 - MARKVART, T. : Exciton transport in the photosynthetic unit. - In : COOMBS, J. (ed.) : 4th International Congress on Photosynthesis. Pp. 239 - 240. UKISES, London 1977.

30760 - MARÓTI, I., PIPÁS, E. : Change in leaf pigments during autumn colouration. - Acta biol. Szeged *23* : 31 - 38, 1977.

30761 - MARRÈ, E. : Effects of fusicoccin and hormones on plant cell membrane activities : observations and hypotheses. - In : MARRÈ, E., CIFERRI, O. (ed.) : Regulation of Cell Membrane Activities in Plants. Pp. 185 - 202. Elsevier/North-Holland Biomedical Press, Amsterdam 1977. [Stomata, CO_2 transport.]

30762 - MARRS, B., WALL, J.D., GEST, H. : Emergence of the biochemical genetics and molecular biology of photosynthetic bacteria. - Trends biochem. Sci. *2* (5) : 105 - 108, 1977.

30763 - MARSHALL, B., BISCOE, P.V. : A mobile apparatus for measuring leaf photosynthesis in the field. - J. exp. Bot. *28* : 1008 - 1017, 1977.

30764 - MARSHALL, P.E., KOZLOWSKI, T.T. : Changes in structure and function of epigeous cotyledons of woody angiosperms during early seedling growth. - Can. J. Bot. *55* : 208 - 215, 1977. [Ps.]

30765 - MARSHO, T., RADMER, R., SOKOLOVE, P. : Oxygen exchange in type A spinach chloroplasts. - In : COOMBS, J. (ed.) : 4th International Congress on Photosynthesis. P. 236. UKISES, London 1977.

30766 - MARTIN, E.S., LARBALESTIER, G. : A membrane-bound plastid inclusion in the epidermis of leaves of *Taraxacum officinale*. - Can.J. Bot. *55* : 222 - 225, 1977.

30767 - MARTIN, G.E., SMITH, B.N. : Photosynthetic fractionation of oxygen isotopes by *Elodea canadiensis*. - Plant Physiol. *59* (6, Suppl.) : 66, 1977.

*30768 - MARTIN, T.C., WYATT, J.T. : Comparative physiology and morphology of six strains of Stigonematacean blue-green algae. - J. Phycol. *10* : 57 - 65, 1974. [Ps.]

30769 - MARTINOVIČ, B., PLESNIČAR, M. : The effect of SO_2 on CO_2 fixation and activity of ribulose-1,5-diphosphate carboxylase in isolated chloroplasts. - In : COOMBS, J. (ed.) : 4th International Congress on Photosynthesis. P. 240. UKISES, London 1977.

30770 - MARTON, A., PÉTERFI, S., CRĂCIUN, C. : The culture of some filamentous green algae under different conditions of light and nutritive media. II. Ultrastructural peculiarities of the algae *Stichococcus bacillaris* and *Gloeotila protogenita*. - Rev. roum. Biol., Sér. Biol. vég. *22* : 125 - 132, 1977. [Chloroplast.]

30771 - MARTY, D. : Localisation ultra-structurale des sites d'activité des photosystèmes I et II dans les chloroplastes *in situ*. - Compt. rend. Acad. Sci. Paris, Sér. D *285* : 27 - 30, 1977.

30772 - MASAMOTO, K., NISHIMURA, M. : Estimation of internal pH in cells of blue-green algae in the dark and under illumination. - J. Biochem. *82* : 483 - 487, 1977.

30773 - MASAMOTO, K., NISHIMURA, M. : H+ translocation in lysozyme-treated cells of the blue-green alga *Plectonema boryanum*. Difference between isotonically and hypotonically treated preparations and effects of divalent cations. - J. Biochem. *82* : 489 - 493, 1977.

30774 - MASKIEWICZ, R., BRUICE, T.C. : Dependence of the rates of dissolution of the Fe_4S_4 clusters of *Chromatium vinosum* high-potential iron protein and ferredoxin on cluster oxidation state. - Proc. nat. Acad. Sci. U S A *74* : 5231 - 5234, 1977.

30775 - MATEJKA, F. : Príspevok k problému merania teploty povrchu listov. [Leaf surface temperature measurement.]—In : HUZULÁK, J., MASAROVIČOVÁ, E. (ed.) : Fotosyntéza a Vodný Režim Drevín. Pp. 232 - 238. Modra-Plesky 1977. [In Slovak, ab : E, R.]

30776 - MATHEW, P.M. : Primary productivity of Govindgarh lake. - Indian J. anim.
        Sci. 47 : 349 - 356, 1977.

30777 - MATHIEU, Y., MOYSE, A. : Les effects de l'oxygène sur l'activité photosynthé-
        tique des chloroplastes. - In : MOYSE, A. (ed.) : Les Processus de la Produc-
        tion Végétale Primaire. Pp. 235 - 244. Gauthier-Villars, Paris 1977.

30778 - MATHIEU, Y., NATO, A. : Phosphoenolpyruvate carboxylase (PEPCase) and ribulo-
        se-biphosphate carboxylase (RUDPCase) activities during the photoheterotro-
        phic growth of tobacco cell suspensions. - In : COOMBS, J. (ed.) : 4th Inter-
        national Congress on Photosynthesis. P. 241. UKISES, London 1977.

30779 - MATHIS, P. : A study of Photosystem-2 reactions by flash absorption spectro-
        scopy. - In : COOMBS, J. (ed.) : 4th International Congress on Photosynthe-
        sis. P. 242. UKISES, London 1977.

30780 - MATHIS, P. : Fast absorption spectroscopy for studying the primary photo-
        reactions. - In : BARBER, J. (ed.) : Primary Processes of Photosynthesis.
        Pp. 269 - 302. Elsevier, Amsterdam - New York - Oxford 1977.

30781 - MATHIS, P., HAVEMAN, J. : Analysis of absorption changes in the ultraviolet
        related to charge-accumulating electron carriers in Photosystem II of chlo-
        roplasts. - Biochim. biophys. Acta 461 : 167 - 181, 1977.

30782 - MATHIS, P., HAVEMAN, J., YATES, M. : The reaction center of Photosystem II. -
        In : OLSON, J.M., HIND, G. (ed.) : Chlorophyll-Proteins, Reaction Centers,
        and Photosynthetic Membranes. Pp. 267 - 277. Brookhaven Nat. Lab., Upton
        1977.

30783 - MATSUMOTO, K., NISHIMURA, M., AKAZAWA, T. : Ribulose bisphosphate carboxylase
        in maize leaf cells. - Plant Physiol. 59 (6, Suppl.) : 64, 1977.

30784 - MATSUMOTO, K., NISHIMURA, M., AKAZAWA, T. : Ribulose-1,5-bisphosphate carbo-
        xylase in the bundle sheath cells of maize leaf. - Plant Cell Physiol. 18 :
        1281 - 1290, 1977.

30785 - MATSUNO-YAGI, A., MUKOHATA, Y. : Two possible roles of bacteriorhodopsin;
        a comparative study of strains of Halobacterium halobium differing in pigmen-
        tation. - Biochem. biophys. Res. Commun. 78 : 237 - 243, 1977.

30786 - MATSUSHIMA, J. : Sensitivities of plants to ethylene and nitrogen dioxide,
        and the characteristic changes in fine structure of the cell. - In :
        KASUGA, S., SUZUKI, N., YAMADA, T., KIMURA, G., INAGAKI, K., ONOE, K. (ed.) :
        Proceedings of the Fourth International Clean Air Congress. Pp. 112 - 115.
        The Jap. Union Air Pollution Prevention Ass., 1977. [Chloroplast.]

30787 - MATSUSHIMA, J., SAKO, S. : [Relationship between the absorption of sulphur di-
        oxide, degrees of visual injury, growth rate and photosynthesis in ornamental
        trees.] - Mie Daigaku Kankyo Kagaku Kenkyu Kiyo [Rep. Environm. Sci. Mie
        Univ.] 2 : 95 - 105, 1977. [In Jap., ab : E.]

30788 - MATSUURA, K., NISHIMURA, M. : Sidedness of membrane structures in Rhodopseudo-
        monas sphaeroides. Electrochemical titration of the spectrum changes of caro-
        tenoid in spheroplasts, spheroplast membrane vesicles and chromatophores. -
        Biochim. biophys. Acta 459 : 483 - 491, 1977.

30789 - MATSUURA, K., NISHIMURA, M. : Light- and diffusion-potential-induced shift
        of carotenoid spectrum in reconstituted vesicles of Rhodopseudomonas sphaeroi-
        des. - Biochim. biophys. Acta 462 : 700 - 705, 1977.

30790 - MATTHEIS, J.R., REBEIZ, C.A. : Chloroplast biogenesis. Metabolism of proto-
        chlorophyllide and protochlorophyllide ester in developing chloroplasts. -
        Arch. Biochem. Biophys. 184 : 189 - 196, 1977.

30791 - MATTHEIS, J.R., REBEIZ, C.A. : Chloroplast biogenesis. Net synthesis of pro-
        tochlorophyllide from magnesium-protoporphyrin monoester by developing chlo-
        roplasts. - J. biol. Chem. 252 : 4022 - 4024, 1977.

30792 - MATTHEIS, J.R., REBEIZ, C.A. : Chloroplast biogenesis. Net synthesis of proto-
        chlorophyllide from protoporphyrin IX by developing chloroplasts. - J. biol.
        Chem. 252 : 8347 - 8349, 1977.

30793 - MAU, A.W.H., PUZA, M. : Phosphorescence of chlorophylls. - Photochem. Photobiol. *25* : 601 - 603, 1977.

30794 - MAURO, S., BARBER, J., LANNOYE, R. : The relationship between the yield factors for prompt and 1 msec delayed fluorescence. - In : COOMBS, J. (ed.) :
4th International Congress on Photosynthesis. Pp. 242 - 243. UKISES, London
1977.

30795 - MAUZERALL, D. : Porphyrins, chlorophyll, and photosynthesis. - In : TREBST,
A., AVRON, M. (ed.) : Photosynthesis I. (Encycl. Plant Physiol. N.S. Vol. 5.)
Pp. 117 - 124. Springer-Verlag, Berlin - Heidelberg - New York 1977.

30796 - MAUZERALL, D. : Electron-transfer reactions and photoexcited phorphyrins. -
In : OLSON, J.M., HIND, G. : Chlorophyll-Proteins, Reaction Centers, and
Photosynthetic Membranes. Pp. 64 - 73. Brookhaven nat. Lab., Upton 1977.

30797 - MAWSON, B., COLMAN, B. : The effect of osmotic potential on photosynthetic
rates of enzymatically isolated leaf mesophyll cells. - Plant Physiol. *59*
(6, Suppl.) : 91, 1977.

30798 - MAXIMOVA, I.V. : Effects of O_2 and HCO_3^- concentration on glycolate excretion
in *Scenedesmus quadricauda*. - In : COOMBS, J. (ed.) : 4th International Congress on Photosynthesis. P. 244. UKISES, London 1977.

30799 - MAXWELL, P.C., BIGGINS, J. : The kinetic behavior of P-700 during the induction of photosynthesis in algae. - Biochim. biophys. Acta *459* : 442 - 450,
1977.

30800 - MAY, D.S., MEANS, K.S. : Chloroplast pigments as units of selection in *Picea
engelmannii*. - Amer. Midland Naturalist *98* : 283 - 295, 1977.

30801 - MAZLIAK, P. : Glyco- and phospholipids of biomembranes in higher plants. -
In : TEVINI, M., LICHTENTHALER, H.K. (ed.) : Lipids and Lipid Polymers in
Higher Plants. Pp. 48 - 74. Springer-Verlag, Berlin - Heidelberg - New York
1977. [Chloroplasts.]

30802 - MAZUR, T., ROGALSKI, L. : Wpływ nawożenia mineralnego na cechy morfologiczne
łodyg i zawartość barwników w liściach ziemniaków. [Effect of mineral fertilization on morphological features of potato shoots and content of pigments in
leaves.] - Acta agrobot. *30* : 71 - 83, 1977. [In Pol., ab : E.]

30803 - McBRIDE, J.C., McBRIDE, A.C., TOGASAKI, R.K. : Isolation of *Chlamydomonas
reinhardi* mutants resistant to the herbicide, DCMU. - Plant Cell Physiol.
1977 (Spec. Issue 3 - Photosynthetic Organelles. Structure and Function) :
239 - 241, 1977.

30804 - McCAIG, T.N., HILL, R.D. : Cyanide-insensitive respiration in wheat: cultivar differences and effects of temperature, carbon dioxide, and oxygen. -
Can. J. Bot. *55* : 549 - 555, 1977.

30805 - McCARTY, R.E. : Energy transfer inhibitors of photophosphorylation in chloro-
plasts. - In : TREBST, A., AVRON, M. (ed.) : Photosynthesis I. (Encycl. Plant
Physiol. N.S. Vol. 5.) Pp. 437 - 447. Springer-Verlag, Berlin - Heidelberg -
New York 1977.

30806 - McCARTY, R.E., WEISS, M.A. : Modification of coupling factor 1 in spinach
chloroplast thylakoids by a bifunctional maleimide. - In : COOMBS, J. (ed.) :
4th International Congress on Photosynthesis. Pp. 244 - 245. UKISES, London
1977.

30807 - McCONNELL, W.J., LEWIS, S., OLSON, J.E. : Gross photosynthesis as an estima-
tor of potential fish production. - Trans. amer. Fish. Soc. *106* : 417 - 423,
1977.

30808 - McCREE, K.J., VAN BAVEL, C.H.M. : Respiration and crop production : A case
study with two crops under water stress. - In : LANDSBERG, J.J., CUTTING,
C.V. (ed.) : Environmental Effects on Crop Physiology. Pp. 199 - 216. Academic Press, London - New York - San Francisco 1977. [Ps.]

30809 - McCURRY, S.D., PAECH, C., TOLBERT, N.E. : Inhibition of ribulose-1,5-bisphos-
phate carboxylase/oxygenase by xylulose-1,5-bisphosphate. - Plant Physiol.
59 (6, Suppl.) : 90, 1977.

30810 - McCURRY, S.D., TOLBERT, N.E. : Inhibition of ribulose-1,5-bisphosphate carbo-
xylase/oxygenase by xylulose 1,5-bisphosphate. - J. biol. Chem. *252* : 8344 -
8346, 1977.

30811 - McDAVID, C.R., OWORU, O.O., MacCOLL, D. : ^{14}C fixation and translocation in
two clones of sugar-cane with contrasting rates of sucrose uptake *in vitro*. -
Ann. Bot. *41* : 405 - 410, 1977.

30812 - McINTOSH, A.R., BLANCHARD, C., TAYLOR, C.P.S., WONG, S.K., BOLTON, J.R. :
Flash photolysis - electron spin resonance detection of transient   primary
photochemical species arising from Photosystem II in green plant and algal
photosynthesis. - In : COOMBS, J. (ed.) : 4th International Congress on Photosynthesis. P. 245. UKISES, London 1977.

30813 - McINTOSH, A.R., BOLTON, J.R. : Flash photolysis-electron spin resonance detection of a transient primary photochemical species arising from Photosystem II in green plant and algal photosynthesis. - In : OLSON, J.M., HIND, G.
(ed.) : Chlorophyll-Proteins, Reaction Centers, and Photosynthetic Membranes.
P. 365. Brookhaven nat. Lab., Upton 1977.

30814 - McKERSIE, B.D., THOMPSON, J.E. : Changes in membrane lipids during senescence.
- Plant Physiol. *59* (6, Suppl.) : 113, 1977. [Chloroplast.]

30815 - McKINLEY, K.R., WETZEL, R.G. : Tritium oxide uptake by algae: An independent
measure of phytoplankton photosynthesis. - Limnol. Oceanogr. *22*: 377 - 380,
1977.

30816 - McLAREN, J.S., BARBER, D.J. : Evidence for carrier-mediated transport of
L-leucine into isolated pea (*Pisum sativum* L.) chloroplasts. - Planta *136* :
147 - 151, 1977.

30817 - McLAREN, J.S., SMITH, H. : Effect of abscisic acid on photosynthetic products
of *Lemna minor*. - Phytochemistry *16* : 219 - 221, 1977.

30818 - McMICHAEL, B.L., HANNY, B.W. : Endogenous levels of abscisic acid in water-
-stressed cotton leaves. - Agron. J. *69* : 979 - 982, 1977.

30819 - McMULLEN, C.R., GARDNER, W.S., MYERS, G.A. : Ultrastructure of corn leaf tissue infected with the ND18 strain of barley stripe mosaic virus. - Proc. South
Dacota Acad. Sci. *56* : 100 - 104, 1977. [Chloroplast.]

30820 - McPHERSON, H.G., BOYER, J.S. : Regulation of grain yield by photosynthesis in
maize subjected to a water deficiency. - Agron. J. *69* : 714 - 718, 1977.

30821 - McPHERSON, H.G., BOYER, J.S. : Regulation of grain production by photosynthesis in maize subjected to a water deficiency. - Plant Physiol. *59* (6, Suppl.)
: 53, 1977.

30822 - MEDINA, E., DELGADO, M., TROUGHTON, J.H., MEDINA, J.D. : Physiological ecology of CO_2 fixation in *Bromeliaceae*. - Flora *166* : 137 - 152, 1977.

30823 - MEDINA, E., IZAGUIRRE, M.L. : Environmental control of CO_2 fixation in *Bromeliaceae* : CAM-induction. - In : COOMBS, J. (ed.) : 4th International Congress
on Photosynthesis. P. 246. UKISES, London 1977.

30824 - MEGO, V. : Vplyv vlhkosti substrátu na rast a metabolizmus základných živín
lucerny siatej (*Medicago sativa* L.) 2. Rast a vývin v kontrolovaných podmienkach prostredia. [Effect of substrate humidity on growth and metabolism of
basic nutrients of *Medicago sativa* L. 2. Growth and development under controlled conditions.]- Ved. Práce výsk. Úst. rast. Výroby Piešťanoch *14* : 137 -
145, 1977. [Dry-matter accumulation; in Slovak, ab : E, R.]

30825 - MEGURO, M., JOLY, C.A., BITTENCOURT, M.M. : Desiccation tolerant *Xerophyta
plicata* SPRENG. - *Velloziaceae*. - Plant Physiol. *59* (6, Suppl.) : 54, 1977.
[Ps, Chl.]

30826 - MEIER, D., LICHTENTHALER, H.K. : Differences in chloroplast ultrastructure
of barley seedlings grown in blue or red light. - In : COOMBS, J. (ed.) :
4th International Congress on Photosynthesis. Pp. 246 - 247. UKISES, London
1977.

30827 - MEISCH, H.-U., BENZSCHAWEL, H., BIELIG, H.-J. : The role of vanadium in
        green plants. II. Vanadium in green algae - two sites of action. - Arch. Mic-
        robiol. 114 : 67 - 70, 1977. [Chl.]

30828 - MEISTER, A., MACHOLD, O. : Stability of chlorophyll-protein complexes under
        different disintegration conditions. - In : COOMBS, J. (ed.) : 4th Interna-
        tional Congress on Photosynthesis. Pp. 247 - 248. UKISES, London 1977.

30829 - MELA, T., YOUNGNER, V.B. : Recovery of three temperate-climate grasses from
        drought stress. - Ann. agr. Fenn. 15 : 309 - 315, 1977. [Production.]

30830 - MELANDRI, B.A. : The high energy state. - In : TREBST, A., AVRON, M. (ed.) :
        Photosynthesis I. (Encycl. Plant Physiol. N.S. Vol 5.) Pp. 358 - 368. Sprin-
        ger-Verlag, Berlin - Heidelberg - New York 1977.

30831 - MELANDRI, B.A., BACCARINI MELANDRI, A., ZANNONI, D., CASADIO, R. : Energy le-
        vels and rate of photophosphorylation in bacterial chromatophores. - In :
        COOMBS, J. (ed.) : 4th International Congress on Photosynthesis. P. 245a.
        UKISES, London 1977.

30832 - MELANDRI, B.A., CASADIO, R., MELANDRI, A.B. : Bacterial photosynthetic phos-
        phorylation under conditions of limited electron flow. - Biochem. Soc. Trans.
        5 : 495 - 499, 1977.

30833 - MELCAREK, P.K., BROWN, G.N. : Effects of chill stress on prompt and delayed
        chlorophyll fluorescence from leaves. - Plant Physiol. 60 : 822 - 825, 1977.

30834 - MELCAREK, P.K., BROWN, G.N. : The effects of chilling stress on the chloro-
        phyll fluorescence of leaves. - Plant Cell Physiol. 18 : 1099 - 1107, 1977.

30835 - MELCAREK, P.K., CERNOHLAVEK, L.G., BROWN, G.N. : A solid-state device for the
        simultaneous measurement of prompt and delayed chlorophyll fluorescence in-
        duction transients in leaves. - Anal. Biochem. 82 : 473 - 484, 1977.

30836 - MELESHKO, G.I., LEBEDEVA, E.K., ANTONYAN, A.A., SIDORENKO, L.A. : Chlamido-
        monas reinhardii 449 v usloviyakh intensivnoĭ kul'tury. [Chlamydomonas rein-
        hardii 449 in an intense culture.] - Tr. petergof. biol. Inst. 25 (Eksperi-
        mental'naya Al'gologiya) : 47 - 66, 1977. [Chl; in R, ab : E.]

30837 - MELIS, A., AKOYUNOGLOU, G. : Development of the two heterogeneous photosystem
        II units in etiolated bean leaves. - Plant Physiol. 59 : 1156 - 1160, 1977.

30838 - MELIS, A., MALKIN, S. : Kinetic characterization of luminescence and fluores-
        cence induction transients in isolated chloroplasts. - In : COOMBS, J. (ed.) :
        4th International Congress on Photosynthesis. P. 243. UKISES, London 1977.

30839 - MENDE, D., WIESSNER, W. : Control of the efficiency of photosynthetic electron
        transport by cytochrome b-559. - In : COOMBS, J. (ed.) : 4th International Con-
        gress on Photosynthesis. Pp. 248 - 249. UKISES, London 1977.

30840 - MENGEL, K., HAEDER, H.-E. : Effect of potassium supply on the rate of phloem
        sap exudation and the composition of phloem sap of Ricinus communis. - Plant
        Physiol. 59 : 282 - 284, 1977.

30841 - MERBACH, W., SCHILLING, G. : Einfluß einiger herbizider Wirkstoffe auf die
        Photosynthese und ihre Teilschritte bei Sinapis alba L. - Biochem. Physiol.
        Pflanz. 171 : 171 - 186, 1977.

30842 - MERBACH, W., SCHILLING, G. : Ursachen der Unempfindlichkeit von Beta vulgaris
        L. gegenüber Pyrazon, Phenmedipham und Benzthiazuron. - Biochem. Physiol.
        Pflanz. 171 : 187 - 199, 1977. [Ps.]

30843 - MEREZHKO, A.I., RYABOV, A.K., TSYTSARIN, G.V. : Vliyanie makrofitov na neko-
        torye gidrokhimicheskie pokazateli melkovodiĭ Kremenchugskogo vodokhranilish-
        cha. [Effect of macrophytes on some hydrochemical indices of shallow areas
        of the Kremenchug reservoir.]- Gidrobiol. Zh. 13 (3) : 111 - 115, 1977.
        [Ps; in R, ab : E.]

30844 - METIVIER, J.R., DALE, J.E. : The effects of grain nitrogen and applied ni-
        trate on growth, photosynthesis and protein content of the first leaf of bar-
        ley cultivars. - Ann. Bot. 41 : 1287 - 1296, 1977.

30845 – METZLER, H. : Light triggered oscillations of reduced pyridine nucleotides, cytochrome oxidation/reduction, and energy change in *Anacystis nidulans*. – In : COOMBS, J. (ed.) : 4th International Congress on Photosynthesis. P. 249. UKISES, London 1977.

30846 – MEYER-REIL, L.-A. : 16. Bacterial growth rates and biomass production. – In : RHEINHEIMER, G. (ed.) : Microbial Ecology of a Brackish Water Environment. Pp. 223 – 236. Springer-Verlag, Berlin – Heidelberg – New Yoek 1977. [Primary production.]

30847 – MICHALK, D.L., HERBERT, P.K. : Assessment of four techniques for estimating yield on dryland pastures. – Agron. J. *69* : 864 – 868, 1977.

30848 – MICHEL, J.-M. : Formation of the photosynthetic apparatus in dark grown *Euglena gracilis* submitted to a flash regime. – In : COOMBS, J. (ed.) : 4th International Congress on Photosynthesis. P. 250. UKISES, London 1977.

30849 – MICHEL, J.M., SIRONVAL, C. : Shifts to $C_{675-670}$ and $C_{696-684}$ in etiolated leaves illuminated with series of brief flashes. – Plant Cell Physiol. *18* : 1223 – 1234, 1977.

30850 – MICHEL-WOLWERTZ, M.R. : Chlorophyll formation in cotyledons of *Pinus jeffreyi* during germination in the dark. Occasional accumulation of protochlorophyll (-ide) forms. – Plant Sci. Lett. *8* : 125 – 134, 1977.

30851 – MICU, M., ŞTIRBAN, M., BERCEA, V., ALBU, E. : Influenţa tratamentului cu ultrasunete asupra sintezei şi acumulării pigmenţilor asimilatori la *Thuja orientalis* L. [The influence of preemergent ultrasound treatment upon the quantum yield of assimilatory pigments in *Thuja orientalis* seedlings.]– Contr. bot. Cluj-Napoca *1977* : 243 – 248, 1977. [In Roum., ab : E.]

30852 – MIERNYK, J.A. : Phospholipid composition of sunflower chloroplasts. – Plant Physiol. *59* (6, Suppl.) : 32, 1977.

30853 – MIFLIN, B.J., ANDERSON, J.W., LEA, P.J., DONE, J., WALLSGROVE, R.M. : The role of chloroplasts in nitrogen metabolism. – In : COOMBS, J. (ed.) : 4th International Congress on Photosynthesis. Pp. 250 – 251. UKISES, London 1977.

30854 – MIGINIAC-MASLOW, M., HOARAU, A. : Étude des concentrations en ions Mg^{2+} et Ca^{2+} dans les chloroplastes intacts d'Épinard à l'aide de l'ionophore A23187 et de l'E.D.T.A. – Physiol. vég. *15* : 822 – 823, 1977.

30855 – MIGINIAC-MASLOW, M., HOARAU, A. : The adenine nucleotide levels and the energy charge values of different species of wheat. – In : COOMBS, J. (ed.) : 4th International Congress on Photosynthesis. Pp. 251 – 252. UKISES, London 1977.

30856 – MIGINIAC-MASLOW, M., HOARAU, A. : The effect of the ionophore A23187 on Mg^{2+} and Ca^{2+} movements and internal pH of isolated intact chloroplasts. – Plant Sci. Lett. *9* : 7 – 15, 1977.

30857 – MIKHAĬLOVA, A.V. : Nekotorye osobennosti raboty fotosinteticheskogo apparata list'ev yachmenya pod vliyaniem zatopleniya pochvy vodoĭ i vitamina PP. [Certain characteristics of activity of the photosynthetic apparatus of barley leaves under the effect of soil flooding and vitamin PP.] – Nauch. Dokl. vyssh. Shkoly, biol. Nauki *20* (9) : 104 – 108, 1977. [In R.]

30858 – MILBURN, J.A., ZIMMERMANN, M.H. : Preliminary studies on sapflow in *Cocos nucifera* L. II. Phloem transport. – New Phytol. *79* : 543 – 558, 1977. [Photosynthates.]

30859 – MILES, D., LETO, K. : Photosynthetic characteristics of isolated maize mesophyll chloroplasts bearing mutations of nuclear genes. – In : COOMBS, J. (ed.) : 4th International Congress on Photosynthesis. Pp. 252 – 253. UKISES, London 1977.

*30860 – MILES, G.E. : Gas exchange measurement of plant leaves. – CSIRO, Aust. Div. Land Res. Management tech. Pap. *1* : 1 – 7, 1976.

30861 – MILFORD, G.F.J., CORMACK, W.F., DURRANT, M.J. : Effects of sodium chloride on water status and growth of sugar beet. – J. exp. Bot. *28* : 1380 – 1388, 1977. [Ps.]

30862 - MILICĂ, C.I., RUSU, F., TOMA, L.-D., AIRINEI, A. : Capacitatea pigmenților asimilatori de a absorbi radiațiile din lumina monocromatică și relația cu productivitatea fotosintezei la fasole. [Assimilating pigments capacity of absorbing radiations from monochromatic light and relation with photosynthesis productivity in beans.] - Lucrări Științ. Inst. agron. "Ion Ionescu de la Brad" Iași, Ser. Horticult. *1977* : 69 - 72, 1977. [In Roum., ab : E.]

30863 - MILIUS, A., KÕVASK, V. : Seasonal variation of phytoplankton biomass, chlorophyll *a* content and alkaline phosphatase activity in lake Viitna Pikkjärv. - Izv. Akad. Nauk ėst. SSR, Biol. *26* (2) : 120 - 127, 1977.

30864 - MILIUS, A., PORK, M. : Seasonal variation of phytoplankton biomass, chlorophyll *a* content and alkaline phosphatase activity in lake Saadjärv. - Izv. Akad. Nauk ėst. SSR, Biol. *26* (1) : 36 - 48, 1977.

30865 - MILIUS, A., PORK, M. : Seasonal variation of phytoplankton biomass, chlorophyll *a* and alkaline phosphatase activity in lake Pangodi. - Izv. Akad. Nauk ėst. SSR, Biol. *26* (2) : 128 - 137, 1977.

30866 - MILIVOJEVIĆ, D., POPOVIĆ, R., VUČKOVIĆ, M., POPOVIĆ, Ž. : Uticaj temperature na fotosintetički aparat kukuruza. [Effect of temperature on photosynthetic apparatus in maize.] - Arhiv poljopr. Nauke *30* : 27 - 33, 1977. [In Croat, ab : E.]

30867 - MILKA, B., POPOVIĆ, R., KLJAJIĆ, R. : Effect of linuron (3-(3,4-dichlorophenyl)-1-metoxy-1-methylurea) on photosynthetic activity of *Pisum sativum*. - In : COOMBS, J. (ed.) : 4th International Congress on Photosynthesis. P. 253a. UKISES, London 1977.

30868 - MILLER, A.G., SPANSWICK, R.M. : ATP, NADH and NADPH levels in *Chara* and *Nitella*. - Plant Physiol. *59* (6, Suppl.) : 85, 1977.

30869 - MILLER, H.G., MILLER, J.D., BINNS, W.O. : Growth of Scots pine under nutritional and climatic stress. - Plant Soil *48* : 103 - 114, 1977. [Dry-matter accumulation.]

30870 - MILLER, K.R. : Location and structure of the system II reaction complex in the photosynthetic membrane. - In : OLSON, J.M., HIND, G. (ed.) : Chlorophyll-Proteins, Reaction Centers, and Photosynthetic Membranes. P. 363. Brookhaven nat. Lab., Upton 1977.

30871 - MILLER, K.R., MILLER, G.J., McINTYRE, K.R. : Organization of the photosynthetic membrane in maize mesophyll and bundle sheath chloroplasts. - Biochim. biophys. Acta *459* : 145 - 156, 1977.

30872 - MILLER, L.S., HOLT, S.C. : Effect of carbon dioxide on pigment and membrane content in *Synechococcus lividus*. - Arch. Microbiol. *115* : 185 - 198, 1977.

30873 - MILLER, S., ABELIOVICH, A., BELFORT, G. : Effects of high organic loading on mixed photosynthetic waste-water treatment. - J. Water Pollut. Contr. Fed. *49* : 436 - 440, 1977.

30874 - MILLINGTON, A.J., WHITING, M.I.K., WILLIAMS, W.T., BOUNDY, C.A.P. : The effect of sowing date on the growth and yield of three sorghum cultivars in the Ord River valley. I. Agronomic aspects.-Aust. J. agr. Res. *28* : 369 - 379, 1977.

30875 - MILLS, J.D., HIND, G. : Tb^{3+} as a probe of chloroplast thylakoid cation binding sites. - Plant Physiol. *59* (6, Suppl.) : 91, 1977.

30876 - MILNE, R., SMITH, S.K., FORD, E.D. : An automatic system for measuring shoot length in Sitka spruce and other plant species. - J. appl. Ecol. *14* : 523 - 529, 1977.

30877 - MIMURO, M., FUJITA, Y. : A native phycobiliprotein aggregate separated from blue-green algae *Anabaena cylindrica* and *Porphyridium cruentum*. - Plant Cell Physiol. *1977* (Spec. Issue 3 - Photosynthetic Organelles. Structure and Function) : 23 - 31, 1977.

30878 - MIMURO, M., FUJITA, Y. : Estimation of chlorophyll *a* distribution in the photosynthetic pigment systems I and II of the blue-green alga *Anabaena variabilis*. - Biochim. biophys. Acta *459* : 376 - 389, 1977.

30879 - MINKOV, I., KIMENOV, G., KALUCHEVA, I. : Structural characteristics of the chloroplasts of *Haberlea rhodopensis* FRIV. upon drying and restoration. - Dokl. bolg. Akad. Nauk *30* : 897 - 900, 1977.

30880 - MISHRA, S.D., GAUR, B.K. : Role of petiole in protein metabolism of senescing betel (*Piper betle* L.) leaves. - Plant Physiol. *59* : 961 - 964, 1977.[Chl.]

30881 - MISZALSKI, Z., WIĘCKOWSKI, S. : Photochemical activities of chloroplasts isolated from leaves of various zones of beech crown. - Bull. Acad. pol. Sci., Sér. Sci. biol. *25* : 141 - 145, 1977.

30882 - MITCHELL, B.W., DRURY, L.N. : Integrator-controller measures solar radiation at two tilt angles. - Agr. Eng. *58* : 40 - 41, 1977.

30883 - MITCHELL, C.A., DOSTAL, H.C., SEIPEL, T.M. : Dry weight reduction in mechanically-dwarfed tomato plants. - J. amer. Soc. hort. Sci. *102* : 605 - 608, 1977. [Resistances.]

30884 - MITCHELL, P. : A commentary on alternative hypotheses of protonic coupling in the membrane systems catalysing oxidative and photosynthetic phosphorylation. - FEBS Lett. *78* : 1 - 20, 1977.

30885 - MITTELHEUSER, C.J. : Rapid ultrastructural recovery of water stressed leaf tissue. - Z. Pflanzenphysiol. *82* : 458 - 461, 1977.

30886 - MIYACHI, S., KAMIYA, S., MUTO, S. : Isolation of peroxisomes from colorless *Chlorella* mutant cells and intracellular localization of isocitrate lyase, malate synthase and phosphoenolpyruvate carboxylase. - Plant Cell Physiol. *1977* (Spec. Issue 3 - Photosynthetic Organelles. Structure and Function) : 347 - 353, 1977.

30887 - MIYACHI, S., SAMEJIMA, M. : Light-enhanced dark CO_2 fixation in C_4 plant. - In : COOMBS, J. (ed.) : 4th International Congress on Photosynthesis. P. 253. UKISES, London 1977.

30888 - MLINKÓ, S., DOBIS, E., PAYER, K., OTTINGER, J., BÁNFI, D., PALÁGYI, T., TURI, A. : An automatic carbon-14 gas analyzer for organic compounds and biological samples. - Anal. Biochem. *83* : 7 - 19, 1977.

30889 - MŁODZIANOWSKI, F., BIAŁOBOK, S. : The effect of sulphur dioxide on ultrastructural organization of larch needles. - Acta Soc. Bot. Pol. *46* : 629 - 634, 1977.

30890 - MOE, R. : Effect of light, temperature and CO_2 on the growth of *Campanula isophylla* stock plants and on the subsequent growth and development of their cuttings. - Sci. Hort. *6* : 129 - 141, 1977. [Dry-matter accumulation.]

30891 - MOGENSEN, V.O. : Field measurements of dark respiration rates of roots and aerial parts in Italian ryegrass and barley. - J. appl. Ecol. *14* : 243 - 252, 1977.

30892 - MOHANTY, P. : The energetics of photosynthesis. - In : Symposium on Basic Sciences and Agriculture. Pp. 147 - 159. Indian nat. Sci. Acad., New Delhi 1977.

30893 - MOHR, H. : Phytochrome and chloroplast development. - Endeavour, new Ser. *1* : 107 - 114, 1977.

30894 - MOHR, H., KASEMIR, H. : Control of chloroplast development and chlorophyll synthesis by phytochrome. - In : CASTELLANI, A. (ed.) : Research in Photobiology. Pp. 501 - 509. Plenum Press, New York - London 1977.

30895 - MOHR, H., SCHOPFER, P. : The effect of light on RNA and protein synthesis in plants. - In : BOGORAD, L., WEIL, J.H. (ed.) : Nucleic Acids and Protein Synthesis in Plants. Pp. 239 - 260. Plenum Publ. Corp., New York 1977. [RuBP-carboxylase.]

30896 - MOKRONOSOV, A.T. : Raznye puti fenogeneza fotosinteticheskogo apparata v ramkakh odnogo genotipa. [Various ways of phenogenesis of photosynthetic apparatus in one genotype.] - In : NASYROV, Yu.S. (ed.) : Genetika Fotosinteza. Pp. 95 - 103. Donish, Dushanbe 1977. [In R.]

30897 - **MOKRONOSOV, A.T., NEKRASOVA, G.F.** : Ontogeneticheskiĭ aspekt fotosinteza
(na primere lista kartofelya). [Ontogenetic aspect of photosynthesis studied
with potato leaf.] - Fiziol. Rast. *24* : 458 - 465, 1977. [In R, ab : E.]

30898 - **MOLCHANOV, A.G.** : Zavisimost' fotosinteza sosny obyknovennoĭ ot usloviĭ okruzhayushcheĭ sredy. [Relation of *Pinus silvestris* (L.) photosynthesis to environmental conditions.] - Lesovedenie *1977* (1) : 48 - 54, 1977. [In R, ab : E.]

30899 - **MOLCHANOV, A.G.** : Dinamika CO_2 v kronakh sosnovogo nasazhdeniya v svyazi s
intensivnost'yu fotosinteza. [CO_2 dynamics in the crowns of a pine plantation
in relation to photosynthesis intensity.] - Lesovedenie *1977*(4):33-42, 1977.
[In R, ab : E.]

30900 - **MOLCHANOV, M.I., TRUSOVA, V.M.** : Vklyuchenie P^{32} v fosfolipidy membran pri
biogeneze khloroplastov. [Incorporation of ^{32}P into membrane phospholipids
during chloroplast biogenesis.] - Biokhimiya *42* : 2051 - 2057, 1977. [In R,
ab : E.]

30901 - **MOLCHANOV, M.I., TRUSOVA, V.M., KSENZENKO, S.M., KOTOVSKAYA, A.P.,
SULEĬMANOV, S.Yu.** : O prirode peptidov prochno assotsiirovannykh s belkovoĭ
chast'yu lipoproteidov membran pri biogeneze khloroplastov. [Nature of peptides
firmly associated with the protein part of membrane lipoproteins in chloroplast biogenesis.] - Biokhimiya *42* : 1516 - 1524, 1977. [In R, ab : E.]

30902 - **MOLDAU, H.** : Maximization of the plant reproductive yield under water stress.
- In : UNGER, K. (ed.) : Biophysikalische Analyse pflanzlicher Systeme.
Pp. 140 - 145. VEB Gustav Fischer Verlag, Jena 1977.

30903 - **MOLDAU, H., KAROLIN, A.** : CO_2-balance sheet of bean and its control by stomata. - In : COOMBS, J. (ed.) : 4th International Congress on Photosynthesis.
P. 254. UKISES, London 1977.

30904 - **MOLDAU, H., KAROLIN, A.** : Effect of the reserve pool on the relationship between respiration and photosynthesis. - Photosynthetica *11* : 38 - 47, 1977.

30905 - **MOLDAU, Kh.A.** : Ust'itsa - universal'nye regulyatory fotosinteza. [Stomata -
versatile controllers of photosynthesis.] - Fiziol. Rast. *24* : 969 - 975,
1977. [In R, ab : E.]

30906 - **MOLL, R.A.** : Phytoplankton in a temperate-zone salt marsh : net production
and exchanges with coastal waters. - Mar. Biol. *42* : 109 - 118, 1977.

30907 - **MÖLLER, G., STAMP, P., GEISLER, G.** : Fotometrische Messung der PEP-Carboxy-
lase-Aktivität in Maisblättern unter Berücksichtigung des Entwicklungszustan-
des der Pflanze. - Z. Pflanzenern. Bodenkunde *140* : 481 - 490, 1977.

30908 - **MOLNAR, M., KHEVESHI, Ya.** : Vliyanie temperatury na detergentnye sistemy,
soderzhashchie organicheskie krasiteli. [The effct of temperature on the organic dye-detergent systems.] - Acta phys. chem. *23* : 389 - 395, 1977. [Ps;
in R, ab : E.]

30909 - **MOLOTKOVSKIĬ,Yu.G., DZYUBENKO, V.S.** : Svetoindutsiruemoe razobshchenie fotofosforilirovaniya v khloroplastakh neorganicheskim fosfatom. Kharakteristika
éffekta. [Light-induced uncoupling of photophosphorylation in chloroplasts
by inorganic phosphate. Characteristics of the effect.] - Fiziol. Rast. *24* :
244 - 250, 1977. [In R, ab : E.]

30910 - **MOLOTKOVSKIĬ,Yu.G., DZYUBENKO, V.S., TIMONINA, V.N.** : Svetoindutsiruemoe razobshchenie fotofosforilirovaniya v khloroplastakh neorganicheskim fosfatom.
Issledovanie éffekta.[Light-induced uncoupling of photophosphorylation in
chloroplasts by inorganic phosphate. Study of the effect.] - Fiziol. Rast.
24 : 478 - 485, 1977. [In R, ab : E.]

30911 - **MOLYAKA, O.N., SOLOMAKHA, V.A.** : Vmist deyakykh biologichno aktyvnykh rechovyn v odnorichnomu bezlystomu pagoni introdukovanykh ta dykoroslykh vydiv
glodu (*Crataegus* L.). [Content of some biologically active substances in annual leafless shoots of introduced and wild *Crataegus* L. species.] - In :
Vykorystannya ta Zbagachennya Roslynnykh Resursiv Ukraïny. Pp. 91 - 96, 131.
Naukova Dumka, Kiev 1977. [Chl, Car; in Ukr., ab : R.]

30912 - MONDAL, M.H., BRUN, W.A., BRENNER, M.L. : Sink effects on photosynthesis and senescence in soybean leaves. - Plant Physiol. *59* (6, Suppl.) : 98, 1977.

30913 - MONDOVI, B., GRAZIANI, M.T., MORPURGO, L. : Properties of the copper site of plastocyanin. - In : COOMBS, J. (ed.) : 4th International Congress on Photosynthesis. P. 255. UKISES, London 1977.

30914 - MONÉGER, R., VERMEERSCH, J., LECHEVALLIER, D., RICHARD, C. : Micro-analyse du NADP et du NAD réduits et oxydés dans les tissus foliaires et dans les plastes isolés de Spirodèle et de Blé. 1. Problèmes posés par le dosage séparé du NADP et du NAD réduits et oxydés des extraits végétaux. - Physiol. vég. *15* : 29 - 62, 1977.

30915 - MONGER, T.G., PARSON, W.W. : Singlet-triplet fusion in *Rhodopseudomonas sphaeroides* chromatophores. A probe of the organization of the photosynthetic apparatus. - Biochim. biophys. Acta *460* : 393 - 407, 1977.

30916 - MONTEITH, J.L. : Climate. - In : ALVIM, P.de T., KOZLOWSKI, T.T. (ed.) : Ecophysiology of Tropical Crops. Pp. 1 - 27. Academic Press, New York - San Francisco - London 1977. [Ps.]

30917 - MONTEITH, J.L. : Climate and the efficiency of crop production in Britain. - Phil. Trans. roy. Soc. London *B 281* : 277 - 294, 1977. [Growth analysis.]

30918 - MONTEITH, J.L. : Resistance of a partially wet canopy : Whose equation fails? - Boundary-Layer Meteorol. *12* : 379 - 383, 1977.

30919 - MOONEY, H.A., BJÖRKMAN, O., COLLATZ, G.J. : Photosynthetic acclimation to temperature and water stress in the desert shrub *Larrea divaricata*. - Carnegie Inst. Year Book *76* : 328 - 335, 1977.

30920 - MOORBY, J. : Integration and regulation of translocation within the whole plant. - In : JENNINGS, D.H. (ed.) : Integration of Activity in the Higher Plant. Pp. 425 - 454. Cambridge University Press, London - New York - Melbourne 1977.

30921 - MOORE, A.L. : Photosynthesis·and photorespiration. - Nature *267* : 307 - 308, 1977.

30922 - MOORE, A.L., JACKSON, C., DENCH, J. : Coupling of glycine metabolism to electron transport in spinach leaf mitochondria. - In : COOMBS, J. (ed.) : 4th International Congress on Photosynthesis. P. 256. UKISES, London 1977.

30923 - MOORE, F.D., SHEPHARD, D.C. : Biosynthesis in isolated *Acetabularia* chloroplasts. II. Plastid pigments. - Protoplasma *92* : 167 - 175, 1977.

30924 - MOORE, F.D., TSCHISMADIA, I. : Biosynthesis in isolated *Acetabularia* chloroplasts. III. Complex lipids. - In : WOODCOCK, C.L.F. (ed.) : Progress in *Acetabularia* Research. Pp. 159 - 173. Academic Press, New York - San Francisco - London 1977.

30925 - MOORE, J.E., LOVE, R.J. : Effect of a pulp and paper mill effluent on the productivity of periphyton and phytoplankton. - J. Fish. Res. Board Can. *34* : 856 - 862, 1977. [Ps.]

30926 - MOORE, T.A. : Optically detected magnetic resonance in biomolecules. - Photochem. Photobiol. *26* : 75 - 78, 1977. [Chl.]

30927 - MORADSHAHI, A., VINES, H.M., BLACK, C.C.Jr. : CO_2 exchange and acidity levels in detached pineapple, *Ananas comosus* (L.) MERR., leaves during the day at various temperatures, O_2 and CO_2 concentrations. - Plant Physiol. *59* : 274 - 278, 1977.

30928 - MORADSHAHI, A., VINES, H.M., BLACK, C.C. : Pathway of CO_2 fixation in CAM plants in light. - Plant Physiol. *59* (6, Suppl.) : 115, 1977.

30929 - MORAES, V.H.F. : Rubber. - In : ALVIM, P.de T., KOZLOWSKI, T.T. (ed.) : Ecophysiology of Tropical Crops. Pp. 315 - 331. Academic Press, New York - San Francisco - London 1977. [Photosynthates.]

30930 - MORESHET, S., STANHILL, G., FUCHS, M. : Effect of increasing foliage reflectance on CO_2 uptake and transpiration resistance of a grain sorghum crop. - Agron. J. *69* : 246 - 250, 1977.

30931 - MORGAN, J.M. : Changes in diffusive conductance and water potential of wheat plants before and after anthesis. - Aust. J. Plant Physiol. *4* : 75 - 86, 1977.

30932 - MORGAN, W.C., PARBERY, D.G. : Effects of *Pseudopeziza* leaf spot disease on growth and yield in lucerne. - Aust. J. agr. Res. *28* : 1029 - 1040, 1977.

30933 - MORITA, K., KONO, M. : Developmental patterns of lamellae and stroma fractions of chloroplasts in rice plants. - Soil Sci. Plant Nutr. *23* : 381 - 389, 1977.

30934 - MORITA, S., HAYASHI, H., MIYAZAKI, T., TAKAICHI, S. : Hyperchromicity of bacteriochlorophyll in organella of photosynthetic bacteria. - Plant Cell Physiol. *1977* (Spec. Issue 3 - Photosynthetic Organelles. Structure and Function) : 97 - 99, 1977.

30935 - MORLÉ, F., FREYSSINET, G., NIGON, V. : Analysis of a new streptomycin-resistant mutant of *Euglena gracilis*. - In : COOMBS, J. (ed.) : 4th International Congress on Photosynthesis. P. 257. UKISES, London 1977.

30936 - MOROT-GAUDRY, J.F., FARINEAU, J., JOLIVET, E. : A study of photosynthetic pathways in maize (*Zea mays*) seedlings differing only by the presence or absence of opaque 2 gene (o2 or +). - In : COOMBS, J. (ed.) : 4th International Congress on Photosynthesis. P. 258. UKISES, London 1977.

30937 - MORRIS, C.B., COCKBURN, W. : Comparative studies on the distribution of the photosynthetic products of isolated chloroplasts. - In : COOMBS, J. (ed.) : 4th International Congress on Photosynthesis. P. 259. UKISES, London 1977.

30938 - MORRIS, P., NASH, G.V., HALL, D.O. : Chloroplast survival and photosynthesis *in vitro*. - In : COOMBS, J. (ed.) : 4th International Congress on Photosynthesis. Pp. 259 - 260. UKISES, London 1977.

30939 - MORRISON, L., RUNQUIST, J., LOACH, P. : Ubiquinone and photochemical activity in *Rhodospirillum rubrum*. - Photochem. Photobiol. *25* : 73 - 84, 1977.

30940 - MÖRSCHEL, E., KOLLER, K.P., WEHRMEYER, W. : The phycobilisome of *Rhodella violacea* : Studies on the fine structure of a light harvesting center. - In : COOMBS, J. (ed.) : 4th International Congress on Photosynthesis.P. 260. UKISES, London 1977.

30941 - MÖRSCHEL, E., KOLLER, K.P., WEHRMEYER, W., SCHNEIDER, H. : Biliprotein assembly in disc-shaped phycobilisomes of *Rhodella violacea*. I. Electron microscopy of phycobilisomes *in situ* and analysis of their architecture after isolation and negative staining. - Cytobiologie *16* : 118 - 129, 1977.

30942 - MÖRSCHEL, E., WEHRMEYER, W. : Multiple forms of phycoerythrin-545 from *Cryptomonas maculata*. - Arch. Microbiol. *113* : 83 - 89, 1977.

30943 - MOSER, W., BRZOSKA, W., ZACHHUBER, K., LARCHER, W. : Ergebnisse des IBP-Projekts "Hoher Nebelkogel 3184 m". - Sitzungbericht. österr. Akad. Wiss.,math.-naturw. Kl., Abt. I, *186* : 387 - 419, 1977.

30944 - MOSSER, J.L., MOSSER, A.G., BROCK, T.D. : Photosynthesis in the snow : the alga *Chlamydomonas nivalis* (*Chlorophyceae*). - J. Phycol. *13* : 22 - 27, 1977.

30945 - MOTT, G.O., POPENOE, H.L. : Grasslands. - In : ALVIM, P.de T., KOZLOWSKI, T.T. (ed.) : Ecophysiology of Tropical Crops. Pp. 158 - 186. Academic Press, New York - San Francisco - London 1977. [Ps.]

30946 - MOUSSEAU, M. : Adaptation de la photosynthèse d'une plante tolérante à l'ombrage, le *Teucrium scorodonia*. - In : MOYSE, A. (ed.) : Les Processus de la Production Végétale Primaire. Pp. 157 - 181. Gauthier-Villars, Paris 1977.

30947 - MOUSSEAU, M. : Night respiration in relation to growth, photosynthesis and development of *Chenopodium polyspermum* in long and short days. - Plant Sci. Lett. *9* : 339 - 346, 1977.

30948 - MOUSSEAU, M., MATHIEU, Y., NATO, A., DULIEU, H.L., MOUSSEAU, J. : Analysis of the effects of two homoeologous genes affecting the photosynthetic apparatus of *Nicotiana tabacum*. - In : COOMBS, J. (ed.) : 4th International Congress on Photosynthesis. P. 261. UKISES, London 1977.

30949 - **MOYA, I.** : La fluorescence de la chlorophylle: nano- ou picoseconde? - Recherche *8* : 180 - 182, 1977.

30950 - **MOYA, I., GOVINDJEE, VERNOTTE, C., BRIANTAIS, J.-M.** : Antagonistic effect of mono- and divalent-cations on lifetime (τ) and quantum yield of fluorescence (ϕ) in isolated chloroplasts. - FEBS Lett. *75* : 13 - 18, 1977.

*30951 - **MOYSE, A.** : Le métabolisme C_4. - In : JACQUES, R. (ed.) : Etudes de Biologie Végétale. Hommage au Professeur P. Chouard. Pp. 537 - 559. Paris 1976.

30952 - **MOYSE, A.** : Quelques aspects des activités ribulose-diphosphate et phosphoénolpyruvate carboxylases des plantes : effets de la carence en N, de divers substrats et cofacteurs, et discrimination isotopique à l'égard du ^{13}C. - In : MOYSE, A. (ed.) : Les Processus de la Production Végétale Primaire. Pp. 245 - 261. Gauthier-Villars, Paris 1977.

30953 - **MÓZSIK, L.** : Az agrotechnikai tényezők hatásának produkciós biológiai értékelése. I. Szárazanyag felhalmozás. [Production biological evaluation of some factors of cultural practices I. Dry matter accumulation.] - Növénytermelés *26* : 39 - 47, 1977. [In Hung., ab : E.]

30954 - **MSHIGENI, K.E.** : Seasonal changes in the standing crops of three *Hypnea* species (*Rhodophyta : Gigartinales*) in Hawaii. - Bot. mar. *20* : 303 - 306, 1977.

30955 - **MUALLEM, A., MALKIN, S.** : "Anomalous" oxygen uptake signal in measurements of photosynthetic oxygen exchange by the modulated oxygen electrode. - In : COOMBS, J. (ed.) : 4th International Congress on Photosynthesis. P. 262. UKISES, London 1977.

30956 - **MUCHOW, R.C., KERVEN, G.L.** : A low cost instrument for measurement of photosynthetically active radiation in field canopies. - Agr. Meteorol. *18* : 187 - 195, 1977.

30957 - **MUELLER, P., FELLER, U., ERISMANN, K.H.** : Einfluß verschiedener CO_2-Konzentrationen auf Wachstum und stoffliche Zusammensetzung von *Lemna minor* L. bei Nitrat- und Ammoniumernährung. - Z. Pflanzenphysiol. *85* : 233 - 241, 1977.

30958 - **MÜHLETHALER, K.** : Introduction to structure and function of the photosynthesis apparatus. - In : TREBST, A., AVRON, M. (ed.) : Photosynthesis I. (Encycl. Plant Physiol. N.S. Vol.5.) Pp. 503 - 521. Springer-Verlag, Berlin - Heidelberg - New York 1977.

30959 - **MUKERJI, S.K.** : Corn leaf phosphoenolpyruvate carboxylases. Purification and properties of two isoenzymes. - Arch. Biochem. Biophys. *182* : 343 - 351, 1977.

30960 - **MUKERJI, S.K.** : Corn leaf phosphoenolpyruvate carboxylases. The effect of divalent cations on activity. - Arch. Biochem. Biophys. *182* : 352 - 359, 1977.

30961 - **MUKERJI, S.K.** : Corn leaf phosphoenolpyruvate carboxylases. Inhibition of $^{14}CO_2$ fixation by SO_3^{2-} and activation by glucose 6-phosphate. - Arch. Biochem. Biophys. *182* : 360 - 365, 1977.

30962 - **MUKHAMADIEV, B.T., KVITKO, K.V.** : O svyazi pigmentnogo sostava kletok mutantov khlorelly s ikh ustoĭchivost'yu k ingibitoram reaktsii fotofosforilirovaniya. [Relation of pigment composition of cells of *Chlorella* mutants with their resistance to inhibitors of photophosphorylation.]-In : NASYROV, Yu.S. (ed.) : Genetika Fotosinteza. Pp. 182 - 188. Donish, Dushanbe 1977. [In R.]

30963 - **MUKHIN, E.N., CHERMNIKH, R.M.** : Regulation of ATP:NADPH ratio in chloroplasts by environmental conditions. - In : COOMBS, J. (ed.) : 4th International Congress on Photosynthesis. Pp. 262 - 263. UKISES, London 1977.

30964 - **MUKHIN, E.N., CHERMNYKH, R.M.** : Znachenie faktorov vneshneĭ sredy v opredelenii otnosheniya ATF:NADF·H v khloroplastakh. [The effect of environmental factors on the ATP:NADPH ratio in chloroplasts.] - Fiziol. Rast. *24* : 1148 - 1153, 1977. [In R, ab : E.]

30965 - **MUKOHATA, Y.** : Non-productive energy transduction in isolated chloroplasts in the presence of calcium ion. - In : COOMBS, J. (ed.) : 4th International Congress on Photosynthesis. Pp. 263 - 264. UKISES, London 1977.

30966 - **MUKOHATA, Y.** : [The mechanism of photophosphorylation; a view through regulation of electron transport.] - Tampakushitsu Kakusan Koso *22* : 1036 - 1046, 1977. [In Jap.]

30967 - **MULDOON, D.K., PEARSON, C.J.** : Hybrid pennisetum in a warm temperate climate : productivity span and effects of nitrogen fertilizer and irrigation on
summer production and survival. - Aust. J. exp. Agr. anim. Husb. *17* : 982 -
990, 1977.

30968 - **MUNDA, I.M., KREMER, B.P.** : Chemical composition and physiological properties of fucoids under conditions of reduced salinity. - Mar. Biol. *42* :
9 - 15, 1977. [Ps.]

30969 - **MURAKAMI, S., KUNIEDA, R.** : Location of coupling factor particles on the
thylakoid of spinach chloroplasts. - Plant Cell Physiol. *1977* (Spec. Issue
3 - Photosynthetic Organelles. Structure and Function) : 403 - 414, 1977.

30970 - **MURATA, N.** : Uphill energy transfer from chlorophyll *a* to phycobilins in the
blue-green algae *Anabaena variabilis* and *Anacystis nidulans*. - Plant Cell
Physiol. *1977* (Spec. Issue 3 - Photosynthetic Organelles. Structure and
Function) : 9 - 13, 1977.

30971 - **MURATA, N., FORK, D.C.** : Lipid phase changes in lettuce and spinach chloroplasts at sub-zero temperatures. - Carnegie Inst. Year Book *76* : 220 - 222,
1977.

30972 - **MURATA, N., FORK, D.C.** : The light-induced carotenoid shift in *Cyanidium*
and in higher plant leaves as an indicator of phase changes in chloroplast
membrane lipids. - Carnegie Inst. Year Book *76* : 226 - 231, 1977.

30973 - **MURATA, N., FORK, D.C.** : Temperature dependence of chlorophyll *a* fluorescence in lettuce and spinach chloroplasts at sub-zero temperatures. - Plant
Cell Physiol. *18* : 1265 - 1271, 1977.

30974 - **MURATA, N., FORK, D.C.** : Temperature dependence of the light-induced spectral shift of carotenoids in *Cyanidium caldarium* and higher plant leaves.
Evidence for an effect of the physical phase of chloroplast membrane lipids
on the permeability of the membrane to ions. - Biochim. biophys. Acta *461* :
365 - 378, 1977.

30975 - **MURATA, T.** : Water soluble chlorophyll-proteins of *Lepidium virginicum*. -
In : OLSON, J.M., HIND, G. (ed.) : Chlorophyll-Proteins, Reaction Centers,
and Photosynthetic Membranes. P. 359. Brookhaven nat. Lab., Upton 1977.

30976 - **MUREĬ, I.A., SHUL'GIN, I.A.** : Sostavlyayushchie balansa istinnogo fotosinteza tomatov v period vegetativnoĭ fazy ich rosta. [Components of true photosynthesis balance in tomato plants during the vegetative phase of their
growth.] - Fiziol. Rast. *24* : 1140 - 1147, 1977. [In R, ab : E.]

30977 - **MURPHY, D.J., LEECH, R.M.** : Lipid biosynthesis by isolated spinach chloroplasts during short term photosynthesis. - In : COOMBS, J. (ed.) : 4th International Congress on Photosynthesis. Pp. 264 - 265. UKISES, London 1977.

30978 - **MURPHY, D.J., LEECH, R.M.** : Lipid biosynthesis from [^{14}C]bicarbonate ,
[2-^{14}C]pyruvate and [1-^{14}C]acetate during photosynthesis by isolated spinach chloroplasts. - FEBS Lett. *77* : 164 - 168, 1977.

30979 - **MURRAY, D.B.** : Coconut palm. - In : ALVIM, P.de T., KOZLOWSKI, T.T. (ed.) :
Ecophysiology of Tropical Crops. Pp. 384 - 407. Academic Press, New York -
San Francisco - London 1977. [Production.]

30980 - **MURRAY, D.R., PEOPLES, M., GORDON, K.H.J.** : Unusual properties of fraction
I protein from *Pisum sativum* L. - In : COOMBS, J. (ed.) : 4th International
Congress on Photosynthesis. P. 266. UKISES, London 1977.

30981 - **MUSZBEK, L., SZABÓ, T., FÉSÜS, L.** : A highly sensitive method for the measurement of ATPase activity. - Anal. Biochem. *77* : 286 - 288, 1977.

30982 - **MUTUSKIN, A.A., MAKOVKINA, L.E., PSHENOVA, K.V.** : Vliyanie polienovykh antibiotikov na aktivnost' fotosistemy I khloroplastov. [Effect of polyene antibiotics on the activity of photosystem I of chloroplasts.] - Fiziol. Rast.
24 : 855 - 858, 1977. [In R.]

30983 - **MUTUSKIN, A.A., MAKOVKINA, L.E., PSHENOVA, K.V., VOSTROKNUTOVA, G.N.** :
Deĭstvie metalloproteidov na fotokhimicheskuyu aktivnost' khloroplastov,
obrabotannykh polienovymi antibiotikami. [Effect of metalloproteins on the

photochemical activity of chloroplasts, treated with polyene antibiotics.] - Biokhimiya *42* : 653 - 658, 1977. [In R, ab : E.]

30984 - MUZAFAROV, E.N., AKULOVA, E.A., MOSHKOV, D.A. : Svetoindutsirovannye izmene-niya ob"ema i struktury izolirovannykh khloroplastov pod vliyaniem razobsh-chayushchikh i ingibitornykh agentov. [Light induced changes in volume and structure of isolated chloroplasts under the influence of uncouplers and inhibitors.] - In : AKULOVA, E.A., MUZAFAROV, E.N. (ed.) : Regulyatsiya Energeticheskogo Obmena Khloroplastov i Mitokhondriĭ Ėndogennymi Fenol'nymi Ingibitorami. Pp. 55 - 73, 128, 133 - 134. Pushchino 1977. [In R, ab : E.]

30985 - MUZAFAROV, E.N., ZALETSKAYA, O.Yu. : Mekhanizm deĭstviya flavonolov na ėlektronnyĭ transport i fotofosforilirovanie v svyazi s uslozhneniem ikh struktury. [Mechanism of action of flavonols on electron transfer and photo-phosphorylation in connection with complication of their structture.]-In : AKULOVA, E.A., MUZAFAROV, E.N. (ed.) : Regulyatsiya Ėnergeticheskogo Obmena Khloroplastov i Mitokhondriĭ Ėndogennymi Fenol'nymi Ingibitorami. Pp. 7 - 27, 126, 131, Pushchino 1977. [In R, ab : E.]

30986 - MVE AKAMBA, L., SIEGENTHALER, P.A. : Influence of linolenic acid on photo-synthesis in intact spinach chloroplasts. - In : COOMBS, J. (ed.) : 4th International Congress on Photosynthesis. Pp. 265 - 266. UKISES, London 1977.

*30987 - MYKLESTAD, S. : Production of carbohydrates by marine planktonic diatoms. I. Comparison of nine different species in culture. - J. exp. mar. Biol. Ecol. *15* : 261 - 274, 1974. [Chl.]

30988 - MYKLESTAD, S. : Production of carbohydrates by marine planktonic diatoms. II. Influence of the N/P ratio in the growth medium on the assimilation ratio, growth rate, and production of cellular and extracellular carbohydrates by *Chaetoceros affinis* var. *willei* (GRAN) HUSTEDT and *Skeletonema costatum* (GREV.) CLEVE. - J. exp. mar. Biol. Ecol. *29* : 161 - 179, 1977.

30989 - NABEDRYK-VIALA, E., CALVET, P., THIÉRY, J.M., GALMICHE, J.M., GIRAULT, G. : Interaction of adenine nucleotides with the coupling factor of spinach chloroplasts. A hydrogen-deuterium exchange study. - FEBS Lett. *79* : 139 - 143, 1977.

30990 - NAD', A., BOKANI, A., ILLIK, M., BACH, B., DOMAN, N.G. : O nekotorykh gene-ticheskikh svoĭstvakh karboksiliruyushcheĭ sposobnosti fermentov v rasteni-yakh s C4-putem fotosinteza. [Some genetic properties of carboxylating ability of enzymes in C4-plants.] - In : NASYROV, Yu.S. (ed.) : Genetika Foto-sinteza. Pp. 90 - 94. Donish, Dushanbe 1977. [In R.]

30991 - NAD', A., DYUR'YAN, I., KOMAROVA, Yu.M., BELYAEVA, E.V., DOMAN, N.G. : Aktivnost' i raspredelenie karboangidrazy v normal'nykh i mutantnykh raste-niyakh kukuruzy pri razlichnom osveshchenii. [Activity and distribution of carboanhydrase in normal and mutant maize plants at different illuminance.] - Fiziol. Rast. *24* : 65 - 69, 1977. [In R, ab : E.]

30992 - NADAKAVUKAREN, M.J., McCRACKEN, D.A. : Effect of 2,4-dichlorophenoxyacetic acid on the structure and function of developing chloroplasts. - Planta *137* : 65 - 69, 1977. [Chl, Ps.]

30993 - NADAKAVUKAREN, M.J., McCRACKEN, D.A. : Effect of 2,4-D on structure and function of developing chloroplasts.- Plant Physiol. *59* (6, Suppl.) : 8, 1977.

30994 - NAIR, P.K.R. : Multispecies crop combinations with tree crops for increased productivity in the tropics. - Gartenbauwissenschaft *42* : 145 - 150, 1977.

30995 - NAIRIZI, S., RYDZEWSKI, J.R. : Effects of dated soil moisture stress on crop yields. - Exp. Agr. *13* : 51 - 59, 1977. [Dry-matter accumulation.]

30996 - NAKANISHI, M. : [Cold storing and freezing of foods. Cold storage of citrons.] - Kochi-ken Kogyo Shikenyo Hokoku *8* : 99 - 112, 1977. [Chl; In Jap.]

30997 - NAKATA, J., IMURA, T., KAWABE, K. : Energy levels and charge carrier genera-tion in crystalline chlorophyll-*a* studied by photoconduction and fluorescen-ce. - J. phys. Soc. Jap. *42* : 146 - 151, 1977.

30998 - NAKATANI, H.Y., BARBER, J. : An improved method for isolating chloroplasts retaining their outer membranes. - Biochim. biophys. Acta *461* : 510 - 512, 1977.

30999 - NAKATANI, H.Y., BARBER, J. : Ionic analyses of chloroplasts using neutron activation. - In : COOMBS, J. (ed.) : 4th International Congress on Photosynthesis. P. 267. UKISES, London 1977.

31000 - NAKAYAMA, K., OKADA, M., WAGO, K., YAKUSHIJI, E. : Studies on the cytochrome *f* of *Bryopsis maxima in vivo* and *in vitro*. - Bot. Mag. (Tokyo) *90* : 247 - 252, 1977.

31001 - NAKAZAWA, F., TAMAI, F., KANEKI, Y. : [Studies on the effect of ultraviolet radiation of different wavelength on the growth and photosynthesis of cucumber plant.] - Bull. Fac. Agr. Meiji Univ. *40* : 7 - 15, 1977. [In Jap., ab : E.]

31002 - NAKAZAWA, F., TAMAI, F., KANEKI, Y. : [Studies on the effect of radiation of GL-light on the growth and photosynthesis of kidney bean plant.] - Bull. Fac. Agr. Meiji Univ. *40* : 17 - 23, 1977. [In Jap., ab : E.]

31003 - NALBORCZYK, E. : Light/dark carboxylation ratio and the photosynthetic productivity of plants. - In : COOMBS, J. (ed.) : 4th International Congress on Photosynthesis. Pp. 267 - 268. UKISES, London 1977.

31004 - NAMBIAR, M.C. : Cashew.-In : ALVIM, P.de T., KOZLOWSKI, T.T. (ed.) : Ecophysiology of Tropical Crops. Pp. 461 - 478. Academic Press, New York - San Francisco - London 1977. [Production.]

31005 - NANDI, D.L., SHEMIN, D. : Quaternary structure of δ-aminolevulinic acid synthase from *Rhodopseudomonas spheroides*. - J. biol. Chem. *252* : 2278 - 2280, 1977.

*31006 - NASH, T.H.III. : Sensitivity of lichens to nitrogen dioxide fumigations. - Bryologist *79* : 103 - 106, 1976. [Chl.]

31007 - NASRULHAQ-BOYCE, A., JONES, O.T.G. : The light-induced development of nitrate-reductase in etiolated barley shoots : An inhibitory effect of laevulinic acid. - Planta *137* : 77 - 84, 1977. [Ps.]

31008 - NASYROV, Yu.S. : The regulatory sites of ribulose-1,5-bisphosphate carboxylase-oxygenase subunits synthesis. - In : COOMBS, J. (ed.) : 4th International Congress on Photosynthesis. Pp. 268 - 269. UKISES, London 1977.

*31009 - NASYROV, Yu.S., GILLER, Yu.E. : Molekulyarnaya anatomiya fotosinteticheskikh membran. [Molecular anatomy of photosynthetic membranes.] - Uspekhi sovrem. Biol. *81* (2) : 178 - 192, 1976. [In R.]

31010 - NASYROV, Yu.S., KVITKO, K.V., SHESTAK, Z. : Problemy i perspektivy issledovaniǐ genetiki fotosinteza. [Problems and perspectives of studying genetics of photosynthesis.] - In : NASYROV, Yu.S. (ed.) : Genetika Fotosinteza. Pp. 285 - 289. Donish, Dushanbe 1977. [In R.]

31011 - NATHANSON, B., WHITE, J.E. : Partial analysis of the bands seen in circular dichroism spectra of green and blue-green algal cells and thylakoids. - Plant Physiol. *59* : 196 - 202, 1977.

31012 - NATO, A., BAZETOUX, S., MATHIEU, Y. : Photosynthetic capacities and growth characteristics of *Nicotiana tabacum* (cv. Xanthi) cell suspension cultures. - Physiol. Plant. *41* : 116 - 123, 1977.

31013 - NÁTR, L., KOUSALOVÁ, I., APEL, P. : Popis akumulace sušiny, dusíku a fosforu ve vyvíjejícì se obilce jarního ječmene a ozimé pšenice. [Description of accumulation of dry matter, nitrogen and phosphorus in developing grain of spring barley and winter wheat.] - Rostl.Výroba(Praha) *23* : 705 - 714, 1977. [In Czech, ab : E, G, R.]

31014 - NEDYALKOV, N., FLOROV, R.J., STOYANOV, Zh.V. : Effect of retardant CCC on bean plants. - In : KUDREV, T., IVANOVA, I., KARANOV, E. (ed.) : Plant Growth Regulators. Pp. 686 - 691. Publishing House of the Bulgarian Academy of Sciences, Sofia 1977. [Car, Chl, Ps.]

31015 - **NEDYALKOV, N., STOYNOVA, E.** : On the yield from maize grown under various
conditions of soil moisture and at herbicide treatment. - In : KUDREV, T.,
IVANOVA, I., KARANOV, E. (ed.) : Plant Growth Regulators. Pp. 681 - 685.
Publishing House of the Bulgarian Academy of Sciences, Sofia 1977. [Chl, Ps.]

31016 - **NEGIEVICH, L.A., KASPERSKIĬ, V.A., KACHAN, A.A.** : Issledovanie vzaimodeĭstvi-
ya NAD s vodoĭ (ionami OH⁻) v geterogennykh usloviyakh. [Study od NAD interaction with water (OH⁻ ions) under heterogenous conditions.] - Dokl. Akad.
Nauk ukr. SSR, Ser. B *1977* : 343 - 346, 1977. [Role in Ps; in R, ab : E.]

31017 - **NEGIEVICH, L.A., KASPERSKIĬ, V.A., KACHAN, A.A.** : O komplekse s perenosom
zaryada mezhdu vodoĭ (ionami OH⁻) i *n*-benzilnikotinamidom na Aĕrosile.
[Complex with charge transfer between water (OH⁻ ions) and *N*-benzilnicotin-
amide on Aerosil.] - Ukr. khim. Zh. *43* : 1200 - 1203, 1977. [Ps model; in R.]

31018 - **NEGISI, K.** : Respiration in forest trees. - In : SHIDEI, T., KIRA, T. (ed.) :
Primary Productivity of Japanese Forests - Productivity of Terrestrial Communities. (JIBP Synthesis 1977.) Vol. 16. Pp. 86 - 99. Univ. Tokyo Press,Tokyo
1977. [Ps.]

31019 - **NEGISI, K.** : Seasonal changes in rate of photosynthesis and growth of *Pinus
densiflora, Cryptomeria japonica* and *Chamaecyparis obtusa* seedlings in their
second vegetation season. - In : Bicentenary Celebration of C.P. Thunberg's
Visit to Japan. Pp. 77 - 89. Roy. Swed. Emb., Bot. Soc. Jap., Tokyo 1977.

31020 - **NEILSON, R.E.** : A technique for measuring photosynthesis in conifers by
$^{14}CO_2$ uptake. - Photosynthetica *11* : 241 - 250, 1977.

31021 - **NELSON, C.J., ASAY, K.H., SLEPER, D.A.** : Mechanisms of canopy development
of tall fescue genotypes. - Crop Sci. *17* : 449 - 452, 1977.

31022 - **NELSON, C.J., DUNN, J.H., COUTTS, J.H.** : Growth responses of tall fescue and
bermudagrass to leaf application of ancymidol. - Agron. J. *69* : 61 - 64,
1977. [Ps.]

31023 - **NELSON, D.E., KORTELING, R.G., STOTT, W.R.** : Carbon-14 : Direct detection
at natural concentrations. - Science *198* : 507 - 508, 1977.

31024 - **NELSON, N.** : Chloroplast coupling factor. - In : TREBST, A., AVRON, M. (ed.) :
Photosynthesis I. (Encycl. Plant Physiol. N.S. Vol. 5.) Pp. 393 - 404. Springer-Verlag, Berlin - Heidelberg - New York 1977.

31025 - **NELSON, N.** : The function of individual polypeptides in chloroplast photosystem I reaction center. - Plant Physiol. *59* (6, Suppl.) : 22, 1977.

31026 - **NELSON, N.** : The function of individual polypeptides in photosynthetic energy transduction. - In : COOMBS, J. (ed.) : 4ᵗʰ International Congress on
Photosynthesis. P. 269. UKISES, London 1977.

31027 - **NELSON, N., BENGIS, C.** : Structure and function of chloroplast photosystem I
reaction center. - Ber. deut. bot. Ges. *90* : 477 - 484, 1977.

31028 - **NELSON, N., EYTAN, E.** : Isolation of an active proton channel-proteolipid
from chloroplasts. - Plant Physiol. *59* (6, Suppl.) : 23, 1977.

31029 - **NELSON, N., EYTAN, E., NOTSANI, B.-E., SIGRIST, H., SIGRIST-NELSON, K.,
GITLER, C.** : Isolation of a chloroplast N,N'-dicyclohexylcarbodiimide-binding
proteolipid, active in proton translocation. - Proc. nat. Acad. Sci. USA *74* :
2375 - 2378, 1977.

31030 - **NELSON, N., NOTSANI, B.-E.** : Function and organization of individual polypeptides in chloroplast photosystem I reaction center. - In : PACKER, L.,
PAPAGEORGIOU, G.C., TREBST, A. (ed.) : Bioenergetics of Membranes. Pp. 233 -
244. Elsevier/North-Holland Biomedical Press, Amsterdam - Oxford - New York
1977.

31031 - **NÉMETH, J., VIZKELETY, É.** : Ecological investigations on the algal communities in the catchment area of river Zala. I. - Acta bot. Acad. Sci. hung.
23 : 143 - 166, 1977. [Chl.]

*31032 - **NEMETH, J.C., DAVEY, C.B.** : Site factors and the net primary productivity of
young loblolly pine and slash pine plantations. - Soil Sci. Soc. Amer. Proc.
38 : 968 - 970, 1974.

31033 - NETZEL, T.L., RENTZEPIS, P.M., TIEDE, D.M., PRINCE, R.C., DUTTON, P.L. :
Effect of reduction of the reaction center intermediate upon the picosecond
oxidation reaction of the bacteriochlorophyll dimer in *Chromatium vinosum*
and *Rhodopseudomonas viridis*. - Biochim. biophys. Acta *460* : 467 - 479, 1977.

31034 - NEUBURGER, M., DOUCE, R. : Oxydation du malate, du NADH et de la glycine par
les mitochondries de plantes en C_3 et C_4. - Compt. rend. Acad. Sci. Paris,
Sér. D *285* : 881 - 884, 1977.

31035 - NEUBURGER, M., JOYARD, J., DOUCE, R. : Strong binding of cytochrome *c* on the
envelope of spinach chloroplasts. - Plant Physiol. *59* : 1178 - 1181, 1977.

31036 - NEUGEBAUER, D.-C., BLAUROCK, A.E., WORCESTER, D.L. : Magnetic orientation
of purple membranes demonstrated by optical measurements and neutron scattering. - FEBS Lett. *78* : 31 - 35, 1977.

*31037 - NEUMANN, D. : Zur Darstellung pflanzlicher Gewebe nach Gefriersubstitution
unter besonderer Berücksichtigung der Strukturerhaltung der Plastiden. -
Acta histochem. *47* : 278 - 288, 1973. [Chloroplast.]

31038 - NEUMANN, K.-H., KUMAR, A., BENDER, L. : Autotrophic growth and the photosynthetic system of carrot and pea nut tissue culture. - In : COOMBS, J. (ed.) :
4th International Congress on Photosynthesis. P. 270. UKISES, London 1977.

31039 - NEWTON, R.J., SCOTT, J.R., BENEDICT, C.R. : Leaf structure and $\delta^{13}C$ values
of the aquatic *Hydrilla verticillata*. - Plant Physiol. *59* (6, Suppl.) : 65,
1977.

31040 - NICHIPOROVICH, A.A. : Teoriya fotosinteticheskoĭ produktivnosti rasteniĭ.
[Theory of photosynthetic productivity of plants.] - Itogi Nauki Tekh., Ser.
Fiziol. Rast. *3* : 11 - 54, 1977. [In R.]

31041 - NICKELL, L.G. : Sugarcane. - In : ALVIM, P.de T., KOZLOWSKI, T.T. (ed.) :
Ecophysiology of Tropical Crops. Pp. 89 - 111. Academic Press, New York -
San Francisco - London 1977. [Ps.]

31042.- NICKLISCH, A., TSENOVA, E.N., HOFFMANN, P. : Enzymologische Untersuchungen
über regulative Beziehungen zwischen N- und C-Stoffwechsel während der Belichtung etiolierter Primärblätter von *Triticum aestivum* L. - Biochem. Physiol.
Pflanz. *171* : 375 - 384, 1977.

31043 - NICKSON, D., BARBER, J. : The effect of cations on the kinetics of the flash
induced 518 nm signal. - In : COOMBS, J. (ed.) : 4th International Congress
on Photosynthesis. Pp. 270 - 271. UKISES, London 1977.

31044 - NICOLAS, P., NIGON, V. : Somatic segregation of bleached mutants during the
multiplication of *Euglena gracilis* after ultraviolet irradiation. - In :
COOMBS, J. (ed.) : 4th International Congress on Photosynthesis. Pp. 271 -
272. UKISES, London 1977.

31045 - NICOLIC, D., POPOVIC, R., TYSZKIEWICZ, E., SARIC, M. : Photophosphorylation
and ultrastructure development in *Pinus nigra* chloroplasts grown under different spectral composition of light. - In : COOMBS, J. (ed.) : 4th International Congress on Photosynthesis. P. 272. UKISES, London 1977.

31046 - NIEDERMAN, R.A., HALL, R.L. : Photosynthetic membranes of *Rhodopseudomonas
sphaeroides* : assembly and orientation of light-harvesting bacteriochlorophyll *a*-protein complexes. - In : OLSON, J.M., HIND, G. (ed.) : Chlorophyll-
-Proteins, Reaction Centers, and Photosynthetic Membranes. Pp. 366 - 367.
Brookhaven nat. Lab., Upton 1977.

31047 - NIELL, F.X. : Rocky intertidal benthic systems in temperate seas : A synthesis of their functional performances. - Helgol. wiss. Meeresunters. *30* :
315 - 333, 1977. [Chl.]

31048 - NIELSEN, P.E., NISHIMURA, H., OTVOS, J.W., CALVIN, M. : Plant crops as a
source of fuel and hydrocarbon-like materials. - Science *198* : 942 - 944,
1977.

31049 - NIEMI, H., HORVÁTH, G., DROPPA, M., FALUDI-DÁNIEL, A. : Kinetic pattern of
the 515-nm absorbance change characteristic for intact chloroplasts. - In :
COOMBS, J. (ed.) : 4th International Congress on Photosynthesis. P. 273.
UKISES, London 1977.

31050 - NIKIFOROVA, T.A., BALNOKIN, Yu.V. : Lokalizatsiya peroksidazy v khloroplastakh shpinata. [Localization of peroxidase in spinach chloroplasts.] - Fiziol. Rast. *24* : 635 - 637, 1977. [In R.]

31051 - NIKITINA, K.A., GUSEV, M.V. : On nature of a donor for photoreduction of exogenous acceptors of electrons in blue-green algae cells. - In : COOMBS, J. (ed.) : 4th International Congress on Photosynthesis. Pp. 273 - 274. UKISES, London 1977.

31052 - NIKOLAEV, G.M., MATORIN, D.N., AKSENOV, S.I. : Sostoyanie vody i funktsionirovanie pervichnykh reaktsiĭ fotosinteza u lishaĭnika *Placoleconora melanophthalma* iz kholodnykh pustyn' vostochnogo Pamira. [State of water and primary photosynthetic reactions in lichen *Placoleconora melanophthalma* from cold deserts of eastern Pamir.] - Nauch. Dokl. vyssh. Shkoly, biol. Nauki *20* (1) : 101 - 105, 1977. [Chl; In R.]

31053 - NIKOLAEVA, L.F., FLOROVA, N.B., VERKHOTUROV, V.N., RUBIN, A.B. : O razvitii funktsional'no aktivnykh fotosistem I i II u prorostkov khvoĭnykh, vyrashchennykh v otsutstvie sveta. [Development of functionally active photosystems I and II in conifer seedlings grown without light.] - Nauch. Dokl. vyssh. Shkoly, biol. Nauki *20* (12) : 97 - 102, 1977. [In R.]

31054 - NIKOLAEVA, L.F., RASKIN, V.I., VERKHOTUROV, V.N., RUBIN, A.B. : Indutsirovannye svetom okislitel'no-vosstanovitel'nye reaktsii tsitokhromov v ėtiolirovannykh i postėtiolirovannykh list'yakh. [Light-induced redox reactions in cytochromes of etiolated and post-etiolated leaves.] - Nauch. Dokl. vyssh. Shkoly, biol. Nauki *20* (2) : 86 - 89, 1977. [In R.]

31055 - NIKOL'SKI, Yu.K., RASHĖTNIKAŬ, U.N. : Uzaemadzeyanne yadra i khlaraplastaŭ u paliploidnykh form raslin. [Interaction of the nucleus and chloroplasts of polyploid plants.] - Vestsi Akad. Navuk belorus. SSR, Ser. biyal. Navuk *1977* (1) : 44 - 49, 139, 1977. [In Belorus., ab : R.]

31056 - NISHI, N., SAKATA-SOGAWA, K., SOE, G., YAMASHITA, J. : Light-induced pH changes and changes in absorbance of pH indicators in *Rhodospirillum rubrum* chromatophores. - J. Biochem. (Tokyo) *82* : 1267 - 1279, 1977.

31057 - NISHIDA, K. : CO_2 fixation in leaves of a CAM plant without lower epidermis and the effect of CO_2 on their deacidification. - Plant Cell Physiol. *18* : 927 - 930, 1977.

31058 - NISHIDA, K., SANADA, Y. : Carbon dioxide fixation in chloroplasts isolated from CAM plants. - Plant Cell Physiol. *1977* (Spec. Issue 3 - Photosynthetic Organelles. Structure and Function) : 341 - 346, 1977.

*31059 - NISHIMURA, M. : Changes in the physical parameters of photosynthetic membranes associated with energy transduction - some temperature-sensitive processes. - J. Biochem. (Tokyo) *79* : 51p - 52p, 1976.

31060 - NISHIMURA, M. : [Physical and chemical properties of photosynthesis.] - Kagaku (Kyoto) *32* : 332 - 334, 1977. [In Jap.]

31061 - NISHIMURA, M., YAMAMOTO, Y., TAKAHAMA, U., SHIMIZU, M., MATSUURA, K. : Biological conversion of light energy into electrochemical potential. - In : MITSUI, A., MIYACHI, S., SAN PIETRO, A. (ed.) : Biological Solar Energy Conversion. Pp. 143 - 148. Academic Press, New York - San Francisco - London 1977.

31062 - NISIZAWA, K., YAMADA, T., IKAWA, T. : Circadian rhythm of several enzyme activities related to the photoassimilation of CO_2 in brown algae. - J. Phycol. *13* (Suppl.) : 49, 1977.

31063 - NISSANI, E., SCHERZ, A., LEVANON, H. : The photoexcited triplet state of tetraphenyl chlorin, magnesium tetraphenyl porphyrin and whole cells of *Chlamydomonas reinhardi*. A light modulation-EPR study. - Photochem. Photobiol. *25* : 93 - 101, 1977.

31064 - NITTA, T., SHIROYA, T. : Ultraviolet light irradiation effects on *Euglena gracilis* cultured under various bleaching conditions. - J. Fac. Sci., Univ. Tokyo, Sect. IV, *14* (1) : 61 - 69, 1977. [Chl.]

31065 - **NOAILLES, M.-C.** : Quelques aspects cytologiques et physiologiques de la re-
viviscence chez les *Bryophytes*. - In : Congrès International de Bryologie.
Bryophytorum Bibliotheca. Vol. 13. Bordeaux 1977. [Ps.]

31066 - **NOBEL, P.S.** : CO_2 uptake and mesophyll resistance of a CAM succulent, *Agave
deserti*. - Plant Physiol. *59* (6, Suppl.) : 115, 1977.

31067 - **NOBEL, P.S.** : Internal leaf area and cellular CO_2 resistance : Photosynthetic implications of variations with growth conditions and plant species. -
Physiol. Plant. *40* : 137 - 144, 1977.

31068 - **NOBEL, P.S.** : Water relations and photosynthesis of a barrel cactus, *Ferocactus acanthodes*, in the Colorado desert. - Oecologia *27* : 117 - 133, 1977.

31069 - **NOBEL, P.S.** : Water relations of flowering of *Agave deserti*. - Bot. Gaz.
138 : 1 - 6, 1977. [Chl, resistances.]

31070 - **NOBLE, I.R.** : Long-term biomass dynamics in an arid chenopod shrub community
at Koonamore, South Australia. - Aust. J. Bot. *25* : 639 - 653, 1977.

31071 - **NOBLE, R.D., CZARNOTA, C.D., CAPPY, J.J.** : Morphological and physiological
characteristics of an achlorophyllous mutant soybean variety sustained to
maturation via grafting. - Amer. J. Bot. *64* : 1042 - 1045, 1977. [Chl.]

31072 - **NOKS, P.P., KONONENKO, A.A., RUBIN, A.B.** : Nemonotonnyĭ kharakter sorbtsii
i desorbtsii vody preparatami khromatoforov fotosinteziruyushchikh purpurnykh bakteriĭ. [Non-monotonous character of water sorption and desorption
by chromatophore preparations from photosynthesizing bacteria.] - Biofizika
22 : 721 - 723, 1977. [In R, ab : E.]

31073 - **NOKS, P.P., KONONENKO, A.A., RUBIN, A.B.** : Spektral'naya pozitsiya osnovnoĭ
polosy pogloshcheniya pigmentnogo kompleksa *P*870 i kinetika fotoindutsiro-
vannykh oksidoreduktsiĭ v reaktsionnykh tsentrakh i khromatoforakh purpurnykh bakteriĭ pri razlichnoĭ temperature i gidratatsii preparatov. [Spectral
position of the main absorption band of pigment complex *P*870 and kinetic of
photoinduced oxidation-reductions in reaction centres and chromatophores of
purple bacteria at different temperatures and extents of hydration.] - Mol.
Biol. (Moskva) *11* : 933 - 940, 1977. [In R, ab : E.]

31074 - **NOKS, P.P., LUKASHEV, E.P., KONONENKO, A.A., VENEDIKTOV, P.S., RUBIN, A.B.** :
O vozmozhnoĭ roli makromolekulyarnykh komponentov v funktsionirovanii fotosinteticheskikh reaktsionnykh tsentrov purpurnykh bakteriĭ. [Possible role
of macromolecular components in the functioning of the photosynthetic reaction centres of purple bacteria.] - Mol. Biol. (Moskva) *11* : 1090 - 1099,
1977. [In R, ab : E.]

31075 - **NOLAN, W.G., SMILLIE, R.M.** : Temperature-induced changes in Hill activity
of chloroplasts isolated from chilling-sensitive and chilling-resistant
plants. - Plant Physiol. *59* : 1141 - 1145, 1977.

31076 - **NORDÉN, B.** : Absorption statistics in linear dichroism. - In : **NORDÉN, B.**
(ed.) : Linear Dichroism Spectroscopy. Pp. 229 - 234. Lund Univ. Press, Lund
1977. [Car.]

31077 - **NOSE, A., SHIROMA, M., MIYAZATO, K., MURAYAMA, S.** : Studies on matter production in pineapple plants. I. Effects of light intensity in light period
on the CO_2 exchange and CO_2 balance of pineapple plants. - Jap. J. Crop Sci.
46 : 580 - 587, 1977.

31078 - **NOTSANI, B.-E., NELSON, N.** : Studies on the subunit structure of chloroplast
Photosystem I reaction center. - In : **COOMBS, J.** (ed.) : 4th International
Congress on Photosynthesis. Pp. 274 - 275. UKISES, London 1977.

31079 - **NOVAK-HOFER, I., SIEGENTHALER, P.A.** : Chemical cross-linking of neighboring
thylakoid membrane polypeptides. - In : **COOMBS, J.** (ed.) : 4th International
Congress on Photosynthesis. Pp. 275 - 276. UKISES, London 1977.

31080 - **NOVAK-HOFER, I., SIEGENTHALER, P.-A.** : Two-dimensional separation of chloroplast membrane proteins by isoelectric focusing and electrophoresis in sodium
dodecyl sulphate. - Biochim. biophys. Acta *468* : 461 - 471, 1977.

31081 - **NOVIKAVA, A.A.** : Fotasintèz seyantsaŭ nekatorykh drèvavykh raslin pry roznaĭ pratsyaglastsi asvyatlennya. [Photosynthesis of some tree seedlings illuminated for different lengths of time.] - Vestsi Akad. Navuk belarus. SSR, Ser. biyal. Navuk *1977* (3) : 39 - 43, 140, 1977. [In Belorus., ab : R.]

31082 - **NOVITSKAYA, G.V., RUTSKAYA, L.A., MOLOTOVSKIĬ, Yu.G.** : Vozrastnye izmeneniya lipidnogo sostava i aktivnosti membran khloroplastov bobov. [Changes in lipid composition and the activity of broad bean chloroplast membranes with aging.] - Fiziol. Rast. *24* : 35 - 43, 1977. [In R, ab : E.]

31083 - **NOZDRACHEV, V.Ya.** : Vliyanie mineral'nykh udobreniĭ na soderzhanie khlorofilla v list'yakh duba. [Effect of mineral nutrients on chlorophyll content in oak leaves.] - Voprosy gornogo Lesoved. Lesovod. Gruzii *25* : 170 - 174, 1977. [In R, ab : E.]

31084 - **NULTSCH, W.** : Do correlations exist between photosynthesis and light induced chromatophore displacements in seaweeds ? - J. Phycol. *13* (Suppl.) : 50, 1977.

31085 - **NULTSCH, W.** : Effect of external factors on phototaxis of *Chlamydomonas reinhardtii*. II. Carbon dioxide, oxygen and pH. - Arch. Microbiol. *112* : 179 - 185, 1977.

31086 - **NUNES, M.A., DIAS, M.A., PINTO, E.** : Efeitos da água disponível e salinidade do solo no crescimento, trocas gasosas e açúcares solúveis em variedades de beterraba-sacarina. [Effects of available water and soil salinity on growth, gas exchanges, and soluble sugars in sugar beet varieties.] - Agronomia lusit. *38* : 229 - 255, 1977. [In Port., ab : E.]

31087 - **OBEN, G., MARCELLE, R.** : Photosynthesis and growth regulators: Where is the key for control? - In : KUDREV, T., IVANOVA, I., KARANOV, E. (ed.): Plant Growth Regulators. Pp. 487 - 490. Publ. House Bulg. Acad. Sci., Sofia 1977.

31088 - **OBEN, G., MARCELLE, R.** : The effects of gibberellic acid and CCC on the activities of some enzymes concerned with photorespiration. - In : COOMBS, J. (ed.) : 4th International Congress on Photosynthesis. Pp. 276 - 277. UKISES, London 1977.

31089 - **O'BRIEN, M.J., EASTERBY, J.S., POWLS, R.** : Glyceraldehyde-3-phosphate dehydrogenase of *Scenedesmus obliquus*. Effects of dithiothreitol and nucleotide on coenzyme specificity. - Biochim. biophys. Acta *481* : 348 - 358, 1977.

31090 - **OBYDENNYĬ, P.T.** : Svyaz' intensivnosti fotosinteza drevesnykh rasteniĭ s faktorami sredy v usloviyakh promyshlennogo zagryazneniya atmosfery. [Relation of photosynthetic rate of woody plants with environmental factors under industrial pollution of atmosphere.] - In : NASYROV, Yu.S. (ed.) : Genetika Fotosinteza. Pp. 271 - 277. Donish, Dushanbe 1977. [In R.]

31091 - **OCHIAI, H., SHIBATA, H., MATSUO, T., HASHINOKUCHI, K., YUKAWA, M.** : Immobilization of chloroplast photosystems. - Agr. biol. Chem. (Tokyo) *41* : 721 - 722, 1977.

31092 - **ODINTSOVA, M.S., YURINA, N.P.** : K voprosu o proiskhozhdenii plastid. [On the plastid origin.] - In : NASYROV, Yu.S. (ed.) : Genetika Fotosinteza. Pp. 17 - 20. Donish, Dushanbe 1977. [In R.]

*31093 - **OELZE, J., GOLECKI, J.R.** : Properties of reaction center depleted membranes of *Rhodospirillum rubrum*. - Arch. Microbiol. *102* : 59 - 64, 1975.

31094 - **OELZE, J., MECHLER, B.** : The effect of light intensity on the composition of the photosynthetic apparatus in *Chromatium* D. - In : COOMBS, J. (ed.) : 4th International Congress on Photosynthesis. P. 277. UKISES, London 1977.

*31095 - **OELZE, J., PAHLKE, W., BOHM, S.** : Ubiquinone 10 formation in *Rhodospirillum rubrum* under different culture conditions. - Arch.Microbiol. *102* : 65 - 69, 1975.

31096 - **OELZE-KAROW, H., MOHR, H.** : Control by phytochrome of the photophosphorylating capacity. - In : COOMBS, J. (ed.) : 4th International Congress on Photosynthesis. Pp. 277 - 278. UKISES, London 1977.

31097 - OETTMEIER, W., NORRIS, J.R., KATZ, J.J. : Evidence for the localization of chlorophyll in lipid vesicles: a spin label study. - In : OLSON, J.M., HIND, G. (ed.) : Chlorophyll-Proteins, Reaction Centers, and Photosynthetic Membranes. Pp. 371 - 372. Brookhaven nat. Lab., Upton 1977.

31098 - OETTMEIER, W., REIMER, S., TREBST, A. : Structure-activity relationship of halogenated quinones as inhibitors of photosynthetic electron transport. - In : COOMBS, J. (ed.) : 4th International Congress on Photosynthesis. P. 279. UKISES, London 1977.

31099 - OGANEZOVA, E.P., NALBANDYAN, R.M. : Reduction of plastocyanin by solvated electrons in non-aqueous media. - FEBS Lett. 82 : 147 - 150, 1977.

31100 - OGAWA, K., TSUKIHARA, T., TAHARA, H., KATSUBE, Y., MATSU-URA, Y., TANAKA, N., KAKUDO, M., WADA, K., MATSUBARA, H. : Location of the iron-sulfur cluster in *Spirulina platensis* ferredoxin by X-ray analysis. - J. Biochem. (Tokyo) 81 : 529 - 531, 1977.

31101 - OGAWA, M., KONISHI, M. : Effects of illumination on absorption peak shifts in spectra of intact etiolated cotyledons of *Pharbitis nil*. I. Existence of two kinds of shift patterns. - Plant Cell Physiol. 18 : 303 - 307, 1977.

31102 - OGREN, W.L. : Increasing carbon dioxide fixation by crop plants. - In : COOMBS, J. (ed.) : 4th International Congress on Photosynthesis. P. 278. UKISES, London 1977.

*31103 - OHAD, I. : Biogenesis of chloroplast membranes. - In : TZAGALOFF, A. (ed.) : Membrane Biogenesis. Pp. 279 - 350. Plenum, New York - London - Boston - Washington 1975.

31104 - OHAD, I. : Biogenesis of chloroplast membranes in algae. - In : PACKER, L., PAPAGEORGIOU, G.C., TREBST, A. (ed.) : Bioenergetics of Membranes. Pp. 61 - 70. Elsevier/North Holland Biomedical Press, Amsterdam - Oxford - New York 1977.

31105 - OHAD, I., BAR-NUN, S., CAHEN, D., GERSHONI, J., GUREVITZ, M., KRETZER, F., SCHANTZ, R., SCHONAT, S. : Biogenesis of chloroplast membranes in algae. - In : CASTELLANI, A. (ed.) : Research in Photobiology. Pp. 531 - 537. Plenum Press, New York - London 1977.

31106 - OHIWA, T. : Response of *Spirogyra* chloroplast to local illumination. - Planta 136: 7 - 11, 1977.

31107 - OHKI, K. : Interrelationship of growth, photosynthesis, respiration, and carbonic anhydrase activity to Zn status in soybean. - Plant Physiol. 59 (6, Suppl.) : 13, 1977.

31108 - OHKI, K., KATOH, T. : Excitation transfer from dark-synthesized phycocyanin directly to pigment system I chlorophyll in *Anabaena variabilis*. - Plant Cell Physiol. 1977 (Spec. Issue 3 - Photosynthetic Organelles. Structure and Function) : 15 - 21, 1977.

31109 - OHNISHI, J., YAMADA, M. : Lipid synthesis in *Bryopsis maxima*. - Plant Cell Physiol. 1977 (Spec. Issue 3 - Photosynthetic Organelles. Structure and Function) : 355 - 364, 1977.

31110 - OHNO, K., TAKEUCHI, Y., YOSHIDA, M. : Effect of light-adaptation on the photoreaction of bacteriorhodopsin from *Halobacterium halobium*. - Biochim. biophys. Acta 462 : 575 - 582, 1977.

31111 - OHNO, K., TAKEUCHI, Y., YOSHIDA, M. : Light-induced formation of the 410 nm intermediate from reconstituted bacteriorhodopsin. - J. Biochem. (Tokyo) 82 : 1177 - 1180, 1977.

31112 - OHTA, Y., KATOH, K., MIYAKE, K. : Establishment and growth characteristics of a cell suspension culture of *Marchantia polymorpha* L. with high chlorophyll content. - Planta 136 : 229 - 232, 1977.

31113 - OIKAWA, T. : Light regime in relation to plant population geometry II. Light penetration in a square-planted population. - Bot. Mag. (Tokyo) 90 : 11 - 22, 1977.

31114 - OIKAWA, T. : Light regime in relation to plant population geometry III. Eco-
        logical implications of a square-planted population from the viewpoint of
        utilization efficiency of solar energy. - Bot. Mag. (Tokyo) *90* : 301 - 311,
        1977.

31115 - OIKAWA, T., SAEKI, T. : Light regime in relation to population geometry I.
        A Monte Carlo simulation of light microclimates within a random distribution
        foliage. - Bot. Mag. (Tokyo) *90* : 1 - 10, 1977.

31116 - OJIMA, K., YAMADA, M., YAMAYA, T., OHIRA, K. : Studies on the greening of
        cultured soybean and *Ruta* cells. III. Effects of minorelement deficiency on
        the growth and photosynthetic activities of *Ruta* cells. - Soil Sci. Plant
        Nutr. *23* : 67 - 75, 1977.

31117 - OKABE, K. : Properties of ribulose diphosphate carboxylase/oxygenase in the
        tobacco urea mutant su/su var. *aurea*. - Z. Naturforsch. *32C* : 781 - 785,
        1977.

31118 - OKABE, K., SCHMID, G.H. : Genetic characterization, photosynthetic and pho-
        torespiratory activity of a new aurea mutant of tobacco. - In : COOMBS, J.
        (ed.) : 4th International Congress on Photosynthesis. P. 280. UKISES, London
        1977.

31119 - OKABE, K., SCHMID, G.H., STRAUB, J. : Genetic characterization and high effi-
        ciency photosynthesis of an aurea mutant of tobacco. - Plant Physiol. *60* :
        150 - 156, 1977.

31120 - OKAMOTO, T., KATOH, S. : Linolenic acid binding by chloroplasts. - Plant Cell
        Physiol. *18* : 539 - 550, 1977.

31121 - OKAMOTO, T., KATOH, S., MURAKAMI, S. : Effects of linolenic acid on spinach
        chloroplast structure. - Plant Cell Physiol. *18* : 551 - 560, 1977.

31122 - OKE, M.S., SHRIKHANDE, A.J. : Rapid method for detection of adulteration of
        tomato ketchup with red pumpkin. - J. Food Sci. Technol. *14* : 280 - 281,
        1977. [Car.]

31123 - O'KEEFE, D., DILLEY, R.A. : The role of membrane potential in the tight bin-
        ding of adenine nucleotides of chloroplast CF_1. - FEBS Lett. *81* : 105 - 110,
        1977.

31124 - O'KEEFE, D.P., DILLEY, R.A. : The effect of chloroplast coupling factor re-
        moval on thylakoid membrane ion permeability. - Biochim. biophys. Acta *461* :
        48 - 60, 1977.

31125 - O'KELLEY, J.C., HARDMAN, J.K. : A blue light reaction involving flavin nucle-
        otides and plastocyanin from *Protosiphon botryoides*. - Photochem. Photobiol.
        25 : 559 - 564, 1977.

31126 - OKITA, T.W., VOLCANI, B.E. : Isolation and characterization of cytoplasmic
        and chloroplastic ribosomes and their ribosomal RNAs from the diatom *Cylin-
        drotheca fusiformis*. - Arch. Microbiol. *111* : 247 - 253, 1977.

31127 - ØKLAND, K.A. : Testing a method for measuring dissolved oxygen concentrations
        in microhabitats in fresh water. - Hydrobiologia *56* : 253 - 257, 1977.

31128 - OKU, T., FURUTA, S., SHIMADA, K., KOBAYASHI, Y., OGAWA, T., INOUE, Y., SHIBATA,
        K. : Development of photosynthetic apparatus in spruce leaves. - Plant Cell
        Physiol. *1977* (Spec. Issue 3 - Photosynthetic Organelles. Structure and Func-
        tion) : 437 - 444, 1977.

31129 - OKU, T., TOMITA, G. : The photoconversion of *Chenopodium* chlorophyll protein.
        - Photochem. Photobiol. *25* : 199 - 202, 1977.

31130 - OKUBO, T., HIROSAKI, S., TAKAHASHI, S., AKIYAMA, T. : A model for the seasonal
        changes in productivity and solar energy utilization of grazing pasture. -
        Jap. agr. Res. quart.   *11* : 221 - 227, 1977.

31131 - ŌKUBO, T., TAKAHASHI, S., AKIYAMA, T. : Ecological efficiencies of energy con-
        version in pasture. II. Seasonal changes of matter production and light uti-
        lization in *Zoysia*-type grassland under one-year protection from grazing. -
        J. jap. Soc. Grassland Sci. *23* : 30 - 42, 1977. [In Jap., ab : E.]

31132 - OLIVEIRA, L., BISALPUTRA, T. : Studies in the brown alga *Ectocarpus* in culture: The chloroplast. - J. submicrosc. Cytol. *9* : 229 - 237, 1977.

31133 - OLIVEIRA, L., BISALPUTRA, T. : Ultrastructural studies in the brown alga *Ectocarpus* in culture: ageing. - New Phytol. *78* : 131 - 138, 1977.

31134 - OLIVER, D.J., ZELITCH, I. : Increasing photosynthesis by inhibiting photorespiration with glyoxylate. - Science *196* : 1450 - 1451, 1977.

31135 - OLIVER, D.J., ZELITCH, I. : Metabolic regulation of glycolate synthesis, photorespiration, and net photosynthesis in tobacco by L-glutamate. - Plant Physiol. *59* : 688 - 694, 1977.

31136 - OLOFFSON, J.A. Jr., WOODARD, F.E. : Effects of pH and inorganic carbon concentrations upon competition between *Anabaena flos-aquae* and *Selenastrum capricornutum*. - Completion Rep. Project A-034-ME. Pp. 1 - 55. Univ. Maine, Orono 1977.

31137 - OLSON, J.M., LEDBETTER, M.C., SHAW, E.K., SCANDELLA, C.J.: Reaction-center containing unit-membrane vesicles from green bacteria. - In : COOMBS, J. (ed.) : 4th International Congress on Photosynthesis. P. 281. UKISES, London 1977.

31138 - OLSON, J.M., PRINCE, R.C., BRUNE, D.C. : Reaction-center complexes from green bacteria. - In : OLSON, J.M., HIND, G. (ed.) : Chlorophyll-Proteins, Reaction Centers, and Photosynthetic Membranes. Pp. 238 - 246. Brookhaven nat. Lab., Upton 1977.

31139 - OMURA, T., SATOH, H., AIGA, I., NAGAO, N. : Studies on the character manifestation in chlorophyll mutants of rice. I. Virescent mutants sensitive to low temperature. - J. Fac. Agr. Kyushu Univ. *21* : 129 - 140, 1977.

B31140 - ONDOK, J.P. : Regime of Global and Photosynthetically Active Radiation in Helophyte Stands. - Academia, Praha 1977.

31141 - ONO, T.-A., MURATA, N. : Temperature dependence of the delayed fluorescence of chlorophyll *a* in blue-green algae. - Biochim. biophys. Acta *460* : 220 - 229, 1977.

31142 - OOHUSA, T., ARAKI, S., SAKURAI, T., SAITOH, M. : [Physiological studies on diurnal biological rhythms of *Porphyra*. I. The cell size, the physiological activity and the content of photosynthetic pigments in the thallus cultured in the laboratory.] - Bull. jap. Soc. sci. Fish. *43* : 245 - 249, 1977. [In Jap., ab : E.]

31143 - ÖQUIST, G., MARTIN, B., MÅRTENSSON, O. : Climatic effects on the photosynthetic apparatus of *Pinus silvestris*. - In : COOMBS, J.(ed.) : 4th International Congress on Photosynthesis. P. 282. UKISES, London 1977.

31144 - ÖREN, A. : Einfluß von Gibberellin A$_3$ auf die Chloroplastenfarbstoffe genetisch verschiedenartiger Inzuchtlinien des Roggens. - Angew. Bot. *51* : 57 - 68, 1977.

31145 - OREN, A., PADAN, E. : Photoautotrophic physiology of the cyanobacterium (blue-green alga), *Oscillatoria limnetica*, capable of both oxygenic and anoxygenic photosynthesis. - In : COOMBS, J. (ed.) : 4th International Congress on Photosynthesis. P. 283. UKISES, London 1977.

31146 - OREN, A., PADAN, E., AVRON, M. : Quantum yields for oxygenic and anoxygenic photosynthesis in cyanobacterium *Oscillatoria limnetica*. - Proc. nat. Acad. Sci. USA *74* : 2152 - 2156, 1977.

31147 - OREN, R., GROMET-ELHANAN, Z. : Coupling factor adenosine triphosphatase-complex of *Rhodospirillum rubrum*. Isolation of an oligomycin-sensitive Ca^{2+}, Mg^{2+}-ATPase. - FEBS Lett. *79* : 147 - 150, 1977.

31148 - ORIANS, G.H., SOLBRIG, O.T. : A cost-income model of leaves and roots with special reference to arid and semiarid areas. - Amer. Natur. *111* : 677 - 690, 1977.

31149 - ORT, D., DILLEY, R., GOOD, N. : The role of proton gradients in initiating photophosphorylation and in slowing electron transport. - In : COOMBS, J. (ed.) : 4th International Congress on Photosynthesis. P. 284. UKISES, London 1977.

31150 - ORT, D.R., PARSON, W.W. : A study of the mechanism and kinetics of proton
movement by bacteriorhodopsin-containing membrane pigments. - In : COOMBS, J.
(ed.) : 4th International Congress on Photosynthesis. Pp. 284 - 285. UKISES,
London 1977.

*31151 - ORTEGA, T., CASTILLO, F., CÁRDENAS, J. : Photolysis of water coupled to nit-
rate reduction by *Nostoc muscorum* subcellular particles. - Biochem. biophys.
Res. Commun. *71* : 885 - 891, 1976.

31152 - ORTEGA, T., CASTILLO, F., CÁRDENAS, J., LOSADA, M. : Inactivation by ammonia
of the photosynthetic reduction of nitrate in *Nostoc muscorum* particles. -
Biochem. biophys. Res. Commun. *75* : 823 - 831, 1977.

31153 - ORTEGA, T., RIVAS, J., CÁRDENAS, J., LOSADA, M. : Metabolic interconversion
of ferredoxin-nitrate reductase and NADP reductase of *Nostoc muscorum*. -
Biochem. biophys. Res. Commun. *78* : 185 - 193, 1977.

31154 - OSIPOV, V.I., SHELEMETOVA, L.I. : Vliyanie protsessa fotosinteza na obmen
gidroaromaticheskikh kislot v organakh i tkanyakh khvoĭnykh. [Effect of photosynthesis on the metabolism of hydroaromatic acids in conifer organs and
tissues.] - In : Obmer Veshchestv i Produktivnost' Khvoĭnykh. Pp. 21 - 29.
Nauka, Novosibirsk 1977. [In R.]

31155 - OSIPOVA, O.P., NIKOLAEVA, M.K., SEVERINA, I.A., ROMANKO, E.G. : Perestroĭka
fotosinteticheskogo apparata pri smene svetovogo rezhima. [Changes in the
photosynthetic apparatus caused by different light regime.] - Fiziol. Rast.
24 : 229 - 236, 1977. [In R, ab : E.]

31156 - OSMAN, A.M., GOODMAN, P.J., COOPER, J.P. : The effects of nitrogen, phosphorus and potassium on rates of growth and photosynthesis of wheat. - Photosynthetica *11* : 66 - 75, 1977.

31157 - OSNITSKAYA, L., CHUDINA, V. : Photosynthesis of purple sulphur bacteria in
green light. - In : COOMBS, J. (ed.) : 4th International Congress on Photosynthesis. P. 285. UKISES, London 1977.

31158 - OSNITSKAYA, L.K., CHUDINA, V.I. : Fotosinteticheskoe razvitie purpurnykh
sernykh bakteriĭ pri osveshchenii zelenym svetom. [Photosynthetic growth of
purple sulphur bacteria during illumination with green light.] - Mikrobiologiya *46* : 55 - 61, 1977. [In R, ab : E.]

31159 - OSNITSKAYA, L.K., CHUDINA, V.I. : Vozmozhnost' ispol'zovaniya sveta raznoĭ
dliny volny dlya razvitiya *Chromatium vinosum* v geterotrofnykh usloviyakh.
[Possible application of light of different wavelengths for growth of *Chromatium vinosum* under heterotrophic conditions.] - Mikrobiologiya *46* : 612 -
618, 1977. [In R, ab : E.]

31160 - OSTROUMOV, S.A. : Participation of chloroplasts and mitochondria in virus
reproduction and the evolution of the eukaryotic cell. - J. theor. Biol. *67* :
287 - 297, 1977.

31161 - OSTROVSKAYA, L.K., GAMAYUNOVA, M.S., SILAEVA, A.M., MANUILSKAYA, S.V., GRIGO-
RA, M.Y. : Structure differences between grana and intergrana thylakoids. -
In : COOMBS, J. (ed.) : 4th International Congress on Photosynthesis. P. 286.
UKISES, London 1977.

31162 - OSWALD, W.J. : Determinants of feasibility in bioconversion of solar energy.
- In : CASTELLANI, A. (ed.) : Research in Photobiology. Pp. 371 - 383. Plenum Press, New York - London 1977.

31163 - OSZLÁNYI, J. : Vertikálna distribúcia listov v korunách stromov s rôznym bio-
sociologickým postavením - príklad z dubovo-hrabového porastu. [Vertical
distribution of leaf biomass in the canopy of trees with various biosociolo-
gical position - an example from an oak-hornbeam forest.] - In : HUZULÁK, J.,
MASAROVIČOVÁ, E. (ed.) : Fotosyntéza a Vodný Režim Drevín. Pp. 211 - 218.
Modra-Piesky 1977. [In Slovak, ab : E, R.]

31164 - O'TOOLE, J.C., LUDFORD, P.M., OZBUN, J.L. : Gas exchange and enzyme activity
during leaf expansion in *Phaseolus vulgaris* L. - New Phytol. *78* : 565 - 571,
1977.

31165 - O'TOOLE, J.C., OZBUN, J.L., WALLACE, D.H. : Photosynthetic response to water stress in *Phaseolus vulgaris*. - Physiol. Plant. *40* : 111 - 114, 1977.

31166 - OULTON, K., WILLIAMS III, G.J. : Growth temperature effects on carboxylase enzymes from *Agropyron smithii* and *Bouteloua gracilis*. - Plant Physiol. *59* (6, Suppl.) : 65, 1977.

31167 - OVCHINNIKOV, Yu.A., ABDULAEV, N.G., FEIGINA, M.Yu., KISELEV, A.V., LOBANOV, N.A. : Recent findings in the structure-functional characteristics of bacterio-rhodopsin. - FEBS Lett. *84* : 1 - 4, 1977.

31168 - OVERDIECK, D. : Schwankungen des Trockensubstanz- und Gesamtzuckergehaltes von unterschiedlich intensiv bestrahlten Sonnenblumenblättern (*Helianthus annuus* L.). - Angew. Bot. *51* : 197 - 212, 1977.

31169 - OWEN, W.J., DELANEY, M.E., ROGERS, L.J. : Studies on the sites of inhibition of photosynthetic electron transport in barley by DDT (1,1,1-trichloro-2,2- -bis(*p*-chlorophenyl)ethane). - J. exp. Bot. *28* : 986 - 998, 1977.

31170 - PAAU, A.S., ORO, J., COWLES, J.R. : Flow-microfluorometric analysis of blue- -green and green algae. - Plant Physiol. *59* (6, Suppl.) : 20, 1977. [Chl.]

31171 - PACE, F., FERRARA, R., DEL CARRATORE, G. : Effects of sub-lethal doses of copper sulphate and lead nitrate on growth and pigment composition of *Dunaliella salina* TEOD. - Bull. environ. Contam. Toxicol. *17* : 679 - 685, 1977.

*31172 - PACKER, L. : Molecular architecture of energy transducing membranes related to function. - In : PACKER, L. (ed.) : Biomembranes: Architecture, Biogenesis, Bioenergetics, and Differentiation. Pp. 113 - 128. Academic Press, New York - San Francisco - London 1974.

31173 - PACKER, L., KONISHI, T., SHIEH, P. : Conformational changes in bacteriorho-dopsin accompanying ionophore activity. - Fed. Proc. *36* : 1819 - 1823, 1977.

31174 - PACKER, L., KONISHI, T., SHIEH, P. : Model systems reconstructed from bacterio-rhodopsin. - In : BUVET, R., ALLEN, M.J., MASSUÉ, J.-P. (ed.) : Living Systems as Energy Converters. Pp. 119 - 128. North-Holland Publ. Co., Amsterdam - New York - Oxford 1977.

31175 - PACKER, L., SHIEH, P.K., LANYI, J.K., CRIDDLE, R.S. : Use of lipid impregnated millipore filters for the direct measurement of photopotentials across envelope vesicles of *Halobacterium halobium* and for assay of ionophore activity of the oligomycin binding subunit 9 of the yeast mitochondrial ATPase. - In : PACKER, L., PAPAGEORGIOU, G.C., TREBST, A. (ed.) : Bioenergetics of Membranes. Pp. 149 - 159. Elsevier/North-Holland Biomedical Press, Amsterdam - Oxford - New York 1977.

31176 - PACKER, L., TEL-OR, E., LUIJK, L.W. : H_2 increases N_2 and CO_2 fixation in hy-drogenase induced cyanobacteria. - In : COOMBS, J. (ed.) : 4th International Congress on Photosynthesis. P. 287. UKISES, London 1977.

31177 - PACKER, L., TEL-OR, E., PAPAGEORGIOU, G. : The potential of H_2 production by photosynthetic preparations from chloroplasts and cyanobacteria. - In : BUVET, R., ALLEN, M.J., MASSUÉ, J.-P. (ed.) :Living Systems as Energy Conver-ters. Pp. 129 - 134. North-Holland Publ. Co., Amsterdam - New York - Oxford 1977.

31178 - PACKHAM, J.R., WILLIS, A.J. : The effects of shading on *Oxalis acetosella*. - J. Ecol. *65* : 619 - 624, 1977. [Chl, growth analysis.]

31179 - PACKHAM, N.K., JACKSON, J.B. : The requirement of the mebrane potential for electrically neutral proton transport catalysed by local anaesthetics across the chromatophore membrane. - In : COOMBS, J. (ed.) : 4th International Con-gress on Photosynthesis. Pp. 287 - 288. UKISES, London 1977.

31180 - PAČUTA, M., MARENČIK, A. : Vplyv rôznej organizácie porastu a minerálnej vý-živy na produkčné vlastnosti bôbu. [Effect of canopy organization and mineral nutrition on the production properties of broad-bean.] - In : REPKA, J. (ed.) : Zborník Referátov zo Seminára Fyziologicko-Genetické a Chemické Faktory Produktivity Rastlín. Pp. 160 - 170. Vysoká Škola poľnohospodárska, Nitra 1977. [In Slovak.]

*31181 - PADMANABHAN, D., VIDHYASEKARAN, P., SOUMINI RAJAGOPALAN, C.K. : Changes in photosynthesis and carbohydrates content in canker and halo regions in *Xanthomonas citri* infected *Citrus* leaves. - Indian Phytopathol. *27* : 215 - 217, 1974.

31182 - PAECH, C., McCURRY, S.D., TOLBERT, N.E. : Active site studies of ribulose-1,5-bisphosphate carboxylase/oxygenase from spinach. - In : COOMBS, J. (ed.) : 4th International Congress on Photosynthesis. P. 289. UKISES, London 1977.

31183 - PAERL, H.W. : Ultraphytoplankton biomass and production in some New Zealand lakes. - New Zeal. J. mar. Freshwater Res. *11* : 297 - 305, 1977. [Ps.]

31184 - PAERL, H.W., MACKENZIE, L.A. : A comparative study of the diurnal carbon fixation patterns of nannoplankton and net plankton. - Limnol. Oceanogr. *22* : 732 - 738, 1977.

31185 - PAILLOTIN, G. : Structure of light-haversting system and its relations with excitation energy transport. - In : COOMBS, J. (ed.) : 4th International Congress on Photosynthesis. P. 290. UKISES, London 1977.

31186 - PAILLOTIN, G., BRETON, J. : Orientation of chlorophylls within chloroplasts as shown by optical and electrochromic properties of the photosynthetic membrane. - Biophys. J. *18* : 63 - 79, 1977.

31187 - PAIS, I., FEHÉR, M., FARKAS, E., SZABÓ, Z., CORNIDES, I. : Titanium as a new trace element. - Commun. Soil Sci. Plant Anal. *8* : 407 - 410, 1977. [Chl.]

31188 - PAKHLAVUNI, I.K. : Metabolizm karotinoidov pri temnovoĭ i fotookislitel'noĭ degradatsii *Anabaena variabilis*. [Metabolism of carotenoids during dark and photooxidative degradation in *Anabaena variabilis*.] - Mikrobiologiya *46* : 981 - 987, 1977. [In R, ab : E.]

31189 - PAKHLAVUNI, I.K., VASIL'EVA, V.E. : Gipofaznye karotinoidy i struktura nekotorykh ėpifaznykh karotinoidov *Anabaena variabilis*. [Hypophasic carotenoids and structure of some epiphasic carotenoids of *Anabaena variabilis*.] - Mikrobiologiya *46* : 683 - 688, 1977. [In R, ab : E.]

31190 - PAKHLAVUNI, I.K., VASIL'EVA, V.E., GUSEV, M.V. : Ėpifaznye karotinoidy sine-zelenoĭ vodorosli *Anabaena variabilis*. [Epiphasic carotenoids of the blue-green alga *Anabaena variabilis*.] - Mikrobiologiya *46* : 482 - 489, 1977. [In R, ab : E.]

31191 - PAKSHINA, E.V., SHAPOSHNIKOVA, M.G., KRASNOVSKIĬ, A.A. : Svetozavisimoe pogloshchenie ionov vodoroda v khloroplastakh i khromatoforakh: deĭstvie nagrevaniya, rastvoriteleĭ, detergentov. [Light-dependent uptake of hydrogen ions in chloroplasts and chromatophores: effects of heating, solvents and detergents.] - Biokhimiya *42* : 1953 - 1959, 1977. [In R, ab : E.]

31192 - PALIKARPAVA, N.N. : Uplyŭ levulinavaŭ kislaty na kol'kasts' khlarafilu ŭ zyalënykh listsyakh yachmenyu. [Effect of levulinic acid on chlorophyll content in green barley leaves.] - Vestsi Akad. Navuk belarus. SSR, Ser. biyal. Navuk *1977* (6) : 39 - 43, 139, 1977. [In Belorus., ab : R.]

31193 - PALLAS, J.E. Jr. : An apparent anomaly in peanut leaf conductance. - Plant Physiol. *59* (6, Suppl.) : 97, 1977.

31194 - PALLETT, K.E., DODGE, A.D. : Photosystem activity of chloroplasts isolated from CMU treated leaves. - In : COOMBS, J. (ed.) : 4th International Congress on Photosynthesis. Pp. 290 - 291. UKISES, London 1977.

31195 - PALLETT, K.E., DODGE, A.D. : Silicomolybdate reduction by isolated pea chloroplasts. - Phytochemistry *16* : 427 - 429, 1977. [Ps.]

*31196 - PALMER, G. : Current insights into the active center of spinach ferredoxin and other iron-sulfur proteins. - In : LOVENBERG, W. (ed.) : Iron-Sulfur Proteins. Vol.2. Pp. 285 - 325. Academic Press, New York - London 1973.

31197 - PALMER, J.W. : Light transmittance by apple leaves and canopies. - J. appl. Ecol. *14* : 505 - 513, 1977.

31198 - PALMER, J.W. : Diurnal light interception and a computer model of light interception by hedgerow apple orchards. - J. appl. Ecol. *14* : 601 - 614, 1977.

31199 - PALMER, J.W., JACKSON, J.E. : Seasonal light interception and canopy development in hedgerow and bed system apple orchards. - J. appl. Ecol. *14* : 539 -
549, 1977.

31200 - PANDEY, R.M., FARMAHAN, H.L. : Changes in rate of photosynthesis and respiration in leaves and berries of *Vitis vinifera* grapevines at various stages
of berry development. - Vitis *16* : 106 - 111, 1977.

31201 - PANETTA, F.D. : The effects of shade upon seedling growth in groundsel bush
(*Baccharis halimifolia* L.). - Aust. J. agr. Res. *28* : 681 - 690, 1977.
[Growth analysis.]

31202 - PANT, A., BALASUBRAMANIAN, T., RAJAGOPAL, M.D., WAFAR, M.V.M., SUMITRA-VIJA-
YARAGHAVAN : Contribution of extracellular production by nannoplankton to
$^{14}CO_2$ fixation in a tropical estuary. - Indian J. mar. Sci. *6* : 147 - 150,
1977.

*31203 - PAOLETTI, C., FLORENZANO, G., BALLONI, W. : Ricerche sui pigmenti carotenoidi
di *Spirulina platensis* impiegata in coltura massiva. [Carotenoid pigments in
Spirulina platensis maintained in mass culture.] - Ann. Microbiol. Enzimol.
(Milano) *21* : 71 - 75, 1971. [In Ital., ab : E.]

31204 - PAOLILLO, D.J.Jr., KASS, L.B. : The relationship between cell size and chloroplast number in the spores of a moss, *Polytrichum*. - J. exp. Bot. *28* : 457 -
467, 1977.

31205 - PAPAGEORGIOU, G.C. : Destabilization of the photosynthetic activities of the
cyanobacterium *Anacystis nidulans* following treatment with protein crosslinking compounds. - In : COOMBS, J. (ed.) : 4th International Congress on Photosynthesis. Pp. 291 - 292. UKISES, London 1977.

31206 - PAPAGEORGIOU, G.C. : Photosynthetic activity of diimidoester-modified cells,
permeaplasts and cell-free membrane fragments of the blue-green alga *Anacystis nidulans*. - Biochim. biophys. Acta *461* : 379 - 391, 1977.

31207 - PAPAGEORGIOU, G.C., ISAAKIDOU, J. : A comparative study of glutaraldehyde
and dimethylsubermidate as protein crosslinking agents for chloroplasts membranes. - In : PACKER, L., PAPAGEORGIOU, G.C., TREBST, A. (ed.) : Bioenergetics of Membranes. Pp. 257 - 268. Elsevier/North-Holland Biomedical Press,
Amsterdam - Oxford - New York 1977.

31208 - PAPP, M. : Changes in the phytomass and production of the herbaceous layer
in the *Quercetum petraeae-cerris* forest after selecting by foresters. - Acta
bot. Acad. Sci. hung. *23* : 179 - 192, 1977.

31209 - PARAKHNEVICH, N.V., TSIMASHÈNKO, M.K., IVANOŬ, N.P., KONEVA, I.I. : Mikra-
struktura fotasintėtychnaga aparatu adzinki ploshchy lista, nakaplenne khlarafilu i praduktsyĭnasts' yachmenyu na novaasvoĭvaemaĭ dzyarnova-padzolistaĭ
zabalochanaĭ glebe pry roznym zabespyachènni tsynkam. [Microstructure of the
photosynthetic apparatus of a leaf area unit, accumulation of chlorophyll,
and productivity of barley on newly reclaimed sod-podzolic swampy soil supplied with different amounts of zinc.] - Vestsy Akad. Navuk belarus. SSR,
Ser. biyal. Navuk *1977*(3) : 59 - 62, 141, 1977. [In Belorus., ab : E.]

31210 - PARAMONOVA, N.V. : Osobennosti pinotsitozopodobnogo yavleniya v khloroplastakh.
[Peculiarities of pinocytosis-like phenomena in chloroplasts.] - In : VIKLIC-
KÝ, V., LUDVÍK, J. (ed.) : Proceedings of the XVth Czechoslovak Conference
on Electron Microscopy with International Participation. Vol.A. P.235. Inst.
Microbiol. Czechoslov. Acad. Sci., Praha 1977. [In R.]

31211 - PARDO, A.D., SCHIFF, J.A. : Plastid and seedling development in SAN 9789-trea-
ted etiolated bean seedlings. - Plant Physiol. *59* (6, Suppl.) : 122, 1977.

*31212 - PARDUE, J.W., SCALAN, R.S., VAN BAALEN, C., PARKER, P.L. : Maximum carbon
isotope fractionation in photosynthesis by blue-green algae and a green alga.
- Geochim. cosmochim. Acta *40* : 309 - 312, 1976.

31213 - PÂRJOL, L., PICU, I. : Modificarea unor însușiri morfofiziologice la grîul
irigat în funcție de sistemul de fertilizare. [Modification of certain mor-
phophysiological functions of irrigated wheat in dependence on the fertilization system.] - An. Inst. Cercet. Cereale Plante teh. Fundulea *42* : 415 - 425,
1977. [Chl; in Roum., ab : E, R.]

31214 - **PARKHURST, D.F.** : A three-dimensional model for CO_2 uptake by continuously distributed mesophyll in leaves. - J. theor. Biol. *67* : 471 - 488, 1977.

31215 - **PARPAROV, A.S.** : Dinamika soderzhaniya khlorofilla *a* v fitoplanktone ozera Sevan. [Dynamics of the chlorophyll *a* content in phytoplankton of Lake Sevan.] - Biol. Zh. Armenii *30* (7) : 95 - 96, 1977. [In R, ab : Arm.]

31216 - **PARPAROV, A.S.** : Novye dannye o pervichnoĭ produktsii fitoplanktona ozera Sevan. [New data on primary production of phytoplankton of the Lake Sevan.] - Biol. Zh. Armenii *30* (8) : 72 - 77, 1977. [In R, ab : Arm.]

31217 - **PARRONDO, R.T., GOSSELINK, J.G., HOPKINSON, C.S.** : Responses of 2 saltmarsh grasses from Louisiana to increased salinity in the root medium. - Plant Physiol. *59*(6, Suppl.) : 54, 1977. [Ps.]

31218 - **PARSHYKOV, V.M.** : Vplyv umov zhyvlennya na intensyvnist' protsesiv obminu ta aktyvnist' nitratreduktazy *Ankistrodesmus braunii* BRUNNTH. [Effect of nutrient conditions on the intensity of metabolic processes and activity of nitrate reductase of *Ankistrodesmus braunii* BRUNNTH.] - Ukr. bot. Zh. *34* : 244 - 247, 335, 1977. [Ps; in Ukr., ab : E, R.]

31219 - **PARSON, W.W., MONGER, T.G.** : Interrelationships among excited states in bacterial reaction centers. - In : OLSON, J.M., HIND, G. (ed.) : Chlorophyll-Proteins, Reaction Centers, and Photosynthetic Membranes. Pp. 195 - 212. Brookhaven nat. Lab., Upton 1977.

31220 - **PARSONS, A.J., ROBSON, M.J.** : Carbon fixation and utilization of a ryegrass community during reproductive development. - In : COOMBS, J. (ed.) : 4th International Congress on Photosynthesis. P. 293. UKISES, London 1977.

*31221 - **PARTHASARATHI, K., GUPTA, S.K., RANGASWAMY, C.R.** : Distribution of chlorophyllase activity and levels of chlorophylls *a* and *b* in sandal (*Santalum album* L.) affected by spike disease. - Experientia *32* : 1262 - 1264, 1976.

31222 - **PARTHIER, B.** : Light-induced chloroplast differentiation in *Euglena gracilis.* - In : NOVER, L., MOTHES, K. (ed.) : Cell Differentiation in Microorganisms, Plants and Animals. Pp. 602 - 624. North-Holland, Amsterdam - New York - Oxford 1977.

31223 - **PARTHIER, B., NEUMANN, D.** : Structural and functional analysis of some plastid mutants of *Euglena gracilis.* - Biochem. Physiol. Pflanzen *171* : 547 - 562, 1977.

31224 - **PARTYKOVÁ, E., ZRŮST, J.** : Výpočet růstově analytických charakteristik pomocí vyrovnaných růstových křivek u bramborů. [Calculation of the analytical characteristics of growth by means of smoothed growth curves for potatoes.] - Věd. Práce výzkum. šlecht. Ústavu brambor. Havl. Brodě *6* : 9 - 18, 1977. [Growth analysis; in Czech, ab : E, G, R.]

31225 - **PASCHENKO, V.Z., KONONENKO, A.A., PROTASOV, S.P., RUBIN, A.B., RUBIN, L.B., USPENSKAYA, N.Ya.** : Probing the fluorescence emission kinetics of the photosynthetic apparatus of *Rhodopseudomonas sphaeroides,* strain 1760-1, on a picosecond pulse fluorometer. - Biochim. biophys. Acta *461* : 403 - 412, 1977.

31226 - **PASSERA, C., ALBUZIO, A.** : Fluttuazioni ritmiche della ribulosio 1,5 difosfato carbossilasi-ossigenasi. [Rhythmic fluctuations of ribulose-1,5-bisphosphate carboxylase/oxygenase.] - Boll. Soc. ital. Biol. sper. *53* : 834 - 839, 1977. [In Ital.]

31227 - **PASSERA, C., ALBUZIO, A.** : Source of glycolate and cyclic changes in photosynthetic and photorespiratory activity during the development of barley leaves. - Biol. Plant. *19* : 448 - 452, 1977.

31228 - **PATE, J.S., SHARKEY, P.J., ATKINS, C.A.** : Nutrition of a developing legume fruit. Functional economy in terms of carbon, nitrogen, water. - Plant Physiol. *59* : 506 - 510, 1977. [Ps.]

31229 - **PATRICK, T.W., HALL, R., FLETCHER, R.A.** : Cytokinin levels in healthy and *Verticillium*-infected tomato plants. - Can. J. Bot. *55* : 377 - 382, 1977. [Chl.]

31230 - **PATTERSON, D.T., BUNCE, J.A., ALBERTE, R.S., VAN VOLKENBURGH, E.** : Photosynthe-

sis in relation to leaf characteristics of cotton from controlled and field
environments. - Plant Physiol. *59* : 384 - 387, 1977.

31231 - **PATTERSON, D.T., DUKE, S.O.** : Influence of irradiance during growth on photosynthetic characteristics of cotton (*Gossypium hirsutum* L.) and velvetleaf (*Abutilon theophrasti* MEDIC.). - Plant Physiol. *59* (6, Suppl.) : 98,
1977.

31232 - **PATTERSON, D.T., LONGSTRETH, D.J., PEET, M.M.** : Photosynthetic adaptation
to light intensity in Sakhalin knotweed (*Polygonum sachalinense*). - Weed Sci.
25 : 319 - 323, 1977.

31233 - **PATTERSON, D.T., MEYER, C.R.** : Analysis of the growth of velvetleaf (*Abutilon theophrasti* MEDIC.) and cotton (*Gossypium hirsutum* L.) at three irradiances. - Plant Physiol. *59* (6, Suppl.) : 98, 1977.

31234 - **PAUL, A.K., MUKHERJI, S.** : Germination behaviour and metabolism of rice seeds
under water-logged condition. - Z. Pflanzenphysiol. *82* : 117 - 124, 1977.
[Ps.]

31235 - **PAUL, J.S., BASSHAM, J.A.** : Maintenance of high photosynthetic rates in mesophyll cells isolated from *Papaver somniferum*. - Plant Physiol. *60* : 775 -
778, 1977.

31236 - **PAULE, L.** : Content of pigments in assimilatory organs of silver fir (*Abies
alba* MILL.). - Biológia (Bratislava) *32* : 729 - 737, 1977.

31237 - **PAVLOV, P., KRUMOVA, Z.** : Vliyanie na svetlinata v"rkhu rastezha, razvitieto
i dobiva na lena. [Effect of light on the growth, development and yield of
flax.] - Rasteniev. Nauki *14* (7) : 3 - 10, 1977. [Chl, Car; in Bulg., ab :
E, R.]

31238 - **PAVLOVA, I.E.** : Vliyanie usloviĭ osveshcheniya na fotokhimicheskuyu aktivnost'
khloroplastov iz list'ev drevesnykh rasteniĭ. [The effect of illumination
conditions on photochemical activity of chloroplasts from leaves of trees.] -
Fiziol. Rast. *24* : 237 - 243, 1977. [In R, ab : E.]

31239 - **PAZOUREK, J.** : The volumes of anatomical components in leaves of *Typha angustifolia* L. and *Typha latifolia* L. - Biol. Plant. *19* : 129 - 135, 1977. [Ps
tissue.]

31240 - **PEARCY, R.W.** : Acclimation of photosynthetic and respiratory carbon dioxide
exchange to growth temperature in *Atriplex lentiformis* (TORR.) WATS. - Plant
Physiol. *59* : 795 - 799, 1977.

31241 - **PEARCY, R.W., BERRY, J.A., FORK, D.C.** : Effects of growth temperature on thermal stability of photosynthetic apparatus of *Atriplex lentiformis* (TORR.)
WATS. - Plant Physiol. *59* : 873 - 878, 1977.

31242 - **PEARMAN, I., THOMAS, S.M., THORNE, G.N.** : Effects of nitrogen fertilizer on
growth and yield of spring wheat. - Ann. Bot. *41* : 93 - 108, 1977.

31243 - **PEARSON, C.J., BISHOP, D.G., VESK, M.** : Thermal adaptation of *Pennisetum:*
leaf structure and composition. - Aust. J. Plant Physiol. *4* : 541 - 554, 1977.
[Chl, chloroplast.]

31244 - **PEARSON, C.J., DAWBIN, K.W., MULDOON, D.K., CAMPBELL, L.C.** : Growth and quality of tropical forages in a temperate climate. - Aust. J. exp. Agr. anim.
Husb. *17* : 991 - 994, 1977. [Dry-matter production.]

31245 - **PEARSON, C.J., DERRICK, G.A.** : Thermal adaptation of *Pennisetum* : Leaf photosynthesis and photosynthate translocation. - Aust. J. Plant Physiol. *4* : 763
- 769, 1977.

31246 - **PEAVEY, D.G., STEUP, M., GIBBS, M.** : Characterization of starch breakdown in
intact spinach chloroplast. - Plant Physiol. *60* : 305 - 308, 1977.

31247 - **PEAVEY, D.G., STEUP, M., GIBBS, M.** : Characterization of starch breakdown in
the intact spinach chloroplasts. - In : COOMBS, J. (ed.) : 4th International
Congress on Photosynthesis. P. 288. UKISES, London 1977.

31248 - **PEAVEY, D.G., STEUP, M., GIBBS, M.** : Starch degradation in isolated spinach
chloroplasts. - In : MITSUI, A., MIYACHI, S., SAN PIETRO, A., TAMURA, S. (ed.):

Biological Solar Energy Conversion. Pp. 197 - 201. Academic Press, New York - San Francisco - London 1977.

31249 - PECHENOV, V.A. : Intensivnost' fotosinteza i ispol'zovanie fotosinteticheski aktivnoĭ radiatsii sakharnoĭ svekloĭ v svyazi s pitaniem. [Photosynthetic rate and utilization of photosynthetically active radiation of sugar beet in connection with nutrition.] - Fiziol. Biokhim. kul't. Rast. *9* : 53 - 57, 1977. [In R, ab : E.]

31250 - PEET, M.M., BRAVO, A., WALLACE, D.H., OZBUN, J.L. : Photosynthesis, stomatal resistance, and enzyme activities in relation to yield of field-grown dry bean varieties. - Crop Sci. *17* : 287 - 293, 1977.

31251 - PEET, M.M., OZBUN, J.L., WALLACE, D.H. : Physiological and anatomical effects of growth temperature on *Phaseolus vulgaris* L. cultivars. - J. exp. Bot. *28* : 57 - 69, 1977.

31252 - PEGELOW, E.J. Jr., BUXTON, D.R., BRIGGS, R.E., MURAMOTO, H., GENSLER, W.G. : Canopy photosynthesis and transpiration of cotton as affected by leaf type. - Crop Sci. *17* : 1 - 4, 1977.

31253 - PEISACH, J., ORME-JOHNSON, N.R., MIMS, W.B., ORME-JOHNSON, W.H. : Linear electric field effect and nuclear modulation studies of ferredoxins and high potential iron-sulfur proteins. - J. biol. Chem. *252* : 5643 - 5650, 1977.

31254 - PEISER, G.D., YANG, S.F. : Chlorophyll destruction by the bisulfite-oxygen system. - Plant Physiol. *60* : 277 - 281, 1977.

31255 - PEISKER, M. : Transpiration and CO_2 uptake at varying stomatal aperture. - In : UNGER, K. (ed.) : Biophysikalische Analyse Pflanzlicher Systeme. Pp. 151 - 153. VEB Gustav Fischer Verlag, Jena 1977.

31256 - PEISKER, M., APEL, P. : Influence of oxygen on photosynthesis and photorespiration in leaves of *Triticum aestivum* L. 3. Response of CO_2 gas exchange to oxygen at various temperatures. - Photosynthetica *11* : 29 - 37, 1977.

31257 - Peking Institute of Botany, 6th Laboratory, Academia sinica : [Preparation of chlorophyll *d*.] - Acta bot. sin. *19* : 222 - 225, 1977. [In Chin., ab : E.]

31258 - PELKONEN, P., HARI, P., LUUKKANEN, O. : Decrease of CO_2 exchange in Scots pine after naturally occurring or artificial low temperatures. - Can. J. Forest Res. *7* : 462 - 468, 1977.

31259 - PENHALE, P.A. : Macrophyte-epiphyte biomass and productivity in an eelgrass (*Zostera marina* L.) community. - J. exp. mar. Biol. Ecol. *26* : 211 - 224, 1977.

31260 - PENHALE, P.A., SMITH, W.O.,Jr. : Excretion of dissolved organic carbon by eelgrass (*Zostera marina*) and its epiphytes. - Limnol. Oceanogr. *22* : 400 - 407, 1977.

31261 - PENNING de VRIES, F.W.T. : Simulation der Assimilation und Transpiration der Pflanzendecke nach grundlegenden Gesetzen. - In : UNGER, K. (ed.) : Biophysikalische Analyse Pflanzlicher Systeme. Pp. 107 - 114. VEB Gustav Fischer Verlag, Jena 1977.

31262 - PENNING de VRIES, F.W.T., van LAAR, H.H. : Substrate utilization in germinating seeds. - In : LANDSBERG, J.J., CUTTING, C.V.(ed.) : Environmental Effects on Crop Physiology. Pp. 217 - 228. Academic Press, London - New York - San Francisco 1977. [Dry-matter accumulation.]

31263 - PEOPLES, M.B., DALLING, M.J. : Degradation of ribulose-1,5-bisphosphate carboxylase by proteolytic enzymes from crude extracts of wheat leaves. - Plant Physiol. *59* (6, Suppl.) : 44, 1977.

31264 - PEOPLES, T.R., KOCH, D.W. : The influence of potassium stress on photosynthesis in alfalfa. - In : COOMBS, J. (ed.) : 4th International Congress on Photosynthesis. P. 292. UKISES, London 1977.

31265 - PEPPER, G.E., PEARCE, R.B., MOCK, J.J. : Leaf orientation and yield of maize. - Crop Sci. *17* : 883 - 886, 1977. [Growth analysis.]

31266 - PERCHOROWICZ, J.T., GIBBS, M. : Light saturation, photorespiration, and the effect of oxygen concentration on photosynthesis in developing maize. - Plant Physiol. *59* (6, Suppl.) : 64, 1977.

31267 - PEREIRA, A.S.R. : Determination of leaf area coefficient in sunflower. - Neth. J. agr. Sci. *25* : 238 - 242, 1977.

31268 - PEREIRA, J.S., KOZLOWSKI, T.T. : Water relations and drought resistance of young *Pinus banksiana* and *P. resinosa* plantation trees. - Can. J. Forest Res. *7* : 132 - 137, 1977. [Resistances.]

31269 - PEREIRA, J.S., KOZLOWSKI, T.T. : Influence of light intensity, temperature, and leaf area on stomatal aperture and water potential of woody plants. - Can. J. Forest Res. *7* : 145 - 153, 1977. [Resistances.]

31270 - PEREIRA, J.S., KOZLOWSKI, T.T. : Variations among woody angiosperms in response to flooding. - Physiol. Plant. *41* : 184 - 192, 1977. [Resistances.]

31271 - PERKOWITZ, S., BEAN, B.L. : Far infrared absorption of chlorophyll in solution. - J. chem. Phys. *66* : 2231 - 2232, 1977.

31272 - PERRIER, A., ITIER, B., JAUSSELY, B. : Étude de la photosynthèse en plein champ. - In : MOYSE, A. (ed.) : Les Processus de la Production Végétale Primaire. Pp. 113 - 136. Gauthier-Villars, Paris 1977.

31273 - PERRY, L.J.,Jr., COMPTON, W.A. : Serial measures of dry matter accumulation and forage quality of leaves, stalks, and ears of three corn hybrids. - Agron. J. *69* : 751 - 755, 1977.

31274 - PERSANOV, V.M., GOGOTOV, I.N., GINS, V.K., MUKHIN, E.N. : Fotovydelenie vodoroda khloroplastami raznykh rasteniĭ. [Photoevolution of hydrogen by chloroplasts of different plants.] - Fiziol. Rast. *24* : 699 - 703, 1977. [In R, ab : E.]

31275 - PESCHEK, G.A. : Anoxygenic photosynthetic CO_2 fixation in the prokaryotic alga *Anacystis nidulans*. - In : COOMBS, J. (ed.) : 4th International Congress on Photosynthesis. P. 294. UKISES, London 1977.

31276 - PÉTERFI, S., MARTON, A., ŞTIRBAN, M., BERCEA, V. : The culture of some green filamentous algae under various conditions of light and nutritive media. I. Biomass increase and quantitative variation of pigments and proteins. - Rev. roum. Biol., Sér. Biol. vég. *22* : 49 - 58, 1977.

31277 - PETERSON, D.M., HOUSLEY, T.L., SCHRADER, L.E. : Long distance translocation of sucrose, serine, leucine, lysine, and CO_2 assimilates. II. Oats. - Plant Physiol. *59* : 221 - 224, 1977.

31278 - PETERSON, R.B., FRIBERG, E.E., BURRIS, R.H. : Diurnal variation in N_2 fixation and photosynthesis by aquatic blue-green algae. - Plant Physiol. *59* : 74 - 80, 1977.

31279 - PETINOV, N.S., IVANOV, V.P., KIRILLINA, V.I., GOLOVATYĬ, V.G., KALIMULINA, Kh.K. : Dinamika nakopleniya sukhogo veshchestva i plasticheskikh soedineniĭ v rasteniyakh yachmenya v zavisimosti ot vlazhnosti pochvy i urovnya mineral'nogo pitaniya. [Dynamics of accumulation of dry matter and plastic compounds in barley plants as a function of soil humidity and level of mineral nutrition.] - Fiziol. Rast. *24* : 593 - 600, 1977. [Growth analysis; in R, ab : E.]

31280 - PETKOVA, R.A., ZEINALOV, Yu. : On the analysis of oxygen induction phenomena in photosynthetizing systems. II. Some kinetic features of O_2 evolution in *Scenedesmus* and *Chlorella* cells. - Stud. biophys. *66* : 113 - 120, 1977.

31281 - PETR, J., LIPAVSKÝ, J. : Kapacita fotosyntézy a obsah chlorofylů u tří odrůd hrachu (*Pisum sativum* L.).[Photosynthetic capacity and chlorophyll content of three cultivars of pea (*Pisum sativum* L.).] - Rostlinná Výroba (Praha) *23* : 1177 - 1182, 1977. [In Czech, ab : E, G, R.]

31282 - PETROV, V.E., LOSKUTNIKOV, A.I. : Reparatsiya ėnergeticheskogo obmena assimiliruyushcheĭ kletki posle deĭstviya ėkstremal'noĭ temperatury. [Reparation of energy metabolism of assimilating cell after the effect of extreme temperature.] - Fiziol. Biokhim. kul't. Rast. *9* : 202 - 209, 1977. [In R, ab : E.]

31283 - PETROV, V.E., SEĬFULLINA,N.Kh., LOSKUTNIKOV, A.I. : Ėnergetika assimiliruyush-
chikh kletok i fotosintez 4. Rol' azota i fosfora v formirovanii vnutrikle-
tochnogo ėnergeticheskogo balansa assimiliruyushcheĭ kletkĭ. [Energetics of
assimilating cells and photosynthesis 4. The role of nitrogen and phosphorus in formation of intracellular energetic balance in assimilating cell.] -
Bot. Zh. *62* : 105 - 111, 1977. [In R.]

31284 - PETROVIC, J., MAREK, J., HRAŠKA, Š. : Vplyv gama-ožiarenia na niektoré kvan-
titatívne znaky a ultraštruktúra chloroplastov pšenice. [Effects of gamma-
rays on some quantitative indices and the structure of wheat chloroplasts.] -
Biolôgia (Bratislava) *32* : 33 - 42, 1977. [In Slovak, ab : E, R.]

31285 - PETRYA, V., SPIRESKU, I. : Vliyanie pestitsid Bromofos i Metakhlor na nekoto-
rye fiziologicheskie protsessy vodoros1i *Chlorella*. [Effect of pesticides
Bromofos and Metachlore on some physiological processes of the alga *Chlorella*.]-
Rev. roum. Biol., Sér. Biol. vég. *22* : 173 - 177, 1977. [Ps, Chl; in R, ab :
F.]

31286 - PETTEI, M.J., YUDD, A.P., NAKANISHI, K., HENSELMAN, R., STOECKENIUS, W. :
Identification of retinal isomers isolated from bacteriorhodopsin. - Biochemistry *16* : 1955 - 1959, 1977.

31287 - PETTY, K.M., JACKSON, J.B., DUTTON, P.L. : Kinetics and stoichiometry of proton binding in *Rhodopseudomonas sphaeroides* chromatophores. - FEBS Lett. *84* :
299 - 303, 1977.

31288 - PETTY, K.M., JACKSON, J.B., DUTTON, P.L. : Microsecond binding of two protons
per electron by *Rhodopseudomonas sphaeroides* chromatophores. - In : COOMBS, J.
(ed.) : 4th International Congress on Photosynthesis. P. 295. UKISES, London
1977.

31289 - PFAU, J. : Dose response, Weber law adherence and discrimination sensitivity
of the chromatophore movement of the brown alga *Dictyota dichotoma*. - Z.
Pflanzenphysiol. *81* : 17 - 25, 1977.

31290 - PFENNIG, N. : Phototrophic green and purple bacteria: a comparative, syste-
matic survey. - Annu. Rev. Microbiol. *31* : 275 - 290, 1977.

31291 - PFISTER, K., LICHTENTHALER, H.K. : Halogenated naphthoquinones as inhibitors
of photosynthetic electron flow. - In : COOMBS, J. (ed.) : 4th International
Congress on Photosynthesis. P. 296. UKISES, London 1977.

31292 - PFLÜGER, R., CASSIER, A. : Influence of monovalent cations on photosynthetic
CO_2 fixation. - In : Fertilizer Use and Production of Carbohydrate + Lipids.
Pp. 95 - 100. Int. Potash Inst., Worblaufen-Bern 1977.

31293 - PHAM THI, A.T., VIEIRA DA SILVA, J.B. : Action des déficits hydriques sur la
photosynthèse et sur la respiration des feuilles du cotonnier. - In : MOYSE,
A. (ed.) : Les Processus de la Production Végétale Primaire. Pp. 183 - 202.
Gauthier-Villars, Paris 1977.

31294 - PHAM THI, A.T., VIEIRA DA SILVA, J. : Sur la nature biochimique des corps
intrachloroplastiques chez le Cottonier. - Compt. rend. Acad. Sci. Paris,
Sér. D *284* : 2107 - 2109, 1977.

31295 - PHILOSOPH, S., BINDER, A., GROMET-ELHANAN, Z. : Coupling factor ATPase-complex
of *Rhodospirillum rubrum*. Purification and properties of a reconstitutively
active single subunit. - J. biol. Chem. *252* : 8747 - 8752, 1977.

31296 - PHILOSOPH, S., GROMET-ELHANAN, Z. : Coupling factor ATPase-complex of *Rhodospirillum rubrum* : Isolation of a four subunit soluble protein. - In : COOMBS,
J. (ed.) : 4th International Congress on Photosynthesis. P. 297. UKISES, London 1977.

31297 - PHUNG NHU HUNG, S., HOULIER, B., MOYSE, A. : Light-induced EPR signal I and
signal II in wheat etioplasts greened under intermittent light without or with
subsequent continuous light. - In : COOMBS, J. (ed.) : 4th International Congress on Photosynthesis. Pp. 297 - 298. UKISES, London 1977.

31298 - PICAUD, A. : Energy transfer between chlorophyll antennae in a Photosystem I
deficient mutant of *Chlamydomonas reinhardtii*. - In : COOMBS, J. (ed.) : 4th
International Congress on Photosynthesis. Pp. 298 - 299. UKISES, London 1977.

31299 - **PICOREL, R., del VALLE-TASCON, S., RAMIREZ, J.M.** : Isolation of a photosynthetic strain of *Rhodospirillum rubrum* with an altered reaction center. - Arch. Biochem. Biophys. *181* : 665 - 670, 1977.

31300 - **PIERRE, M.** : Action du SO_2 sur le métabolisme intermédiaire. II. Effet de doses subnécrotiques de SO_2 sur les enzymes de feuilles de Haricot. - Physiol. vég. *15* : 195 - 205, 1977.

31301 - **PILARSKI, J.** : Effect of changes in light intensity on net photosynthesis in *Nuphar luteum* (L.) SM. leaves. - Bull. Acad. pol. Sci., Sér. Sci. biol. *25* : 415 - 421, 1977.

31302 - **PILARSKI, J.** : Changes in net photosynthesis of *Nuphar luteum* (L.) SM. leaves on reduced light intensity. - Bull. Acad. pol. Sci., Sér. Sci. biol. *25* : 803 - 811, 1977.

31303 - **PILL, W.G., LANBETH, V.N.** : Effects of NH_4 and NO_3 nutrition with and without pH adjustment on tomato growth, ion composition, and water relations. - J. amer. Soc. hort. Sci. *102* : 78 - 81, 1977. [Stomatal resistance.]

31304 - **PIMENTEL, D.** : Energy budgets in natural and agricultural ecosystems. - In : BUVET, R., ALLEN, M.J., MASSUÉ, J.-P. (ed.) : Living Systems as Energy Converters. Pp. 299 - 306. North-Holland Publ. Co., Amsterdam - New York - Oxford 1977.

31305 - **PINTÉR, L., NÉMETH, J., PINTÉR, Z.** : A levélfelület változásának hatása a kukorica (*Zea mays* L.) szemtermésére. [Effect of changes in leaf surface on grain yield in maize.] - Növénytermelés *26* : 21 - 27, 1977. [Dry-matter accumulation; in Hung., ab : E.]

31306 - **PINTHUS, M.J., NERSON, H., MILLET, E., BAR SHALOM, I.** : Differentiation of the wheat spike and its function as yield factor. - Isr. J. Bot. *26* : 47, 1977.

31307 - **PÎRVU, C.** : Studiul producției primare a macrofitelor din mlaştinile Bîlbîitoarea, Stegardin şi Făget. [Primary production of macrophytes on Bîlbîitoarea, Stegardin and Făget marshes.] - Hidrobiologia *15* : 87 - 96, 1977. [In Roum., ab : G.]

31308 - **PIVOVAROVA, T.A., GORLENKO, V.M.** : Tonkoe stroenie *Chloroflexus aurantiacus* var. *mesophilus* (nom. prof.), vyrosshikh na svetu a aérobnykh i anaérobnykh usloviyakh. [Fine structure of *Chloroflexus aurantiacus* var. *mesophilus* (nom. prof.) grown in the light under aerobic and anaerobic conditions.] - Mikrobiologiya *46* : 329 - 334, 1977. [In R, ab: E.]

31309 - **PLAKUNOVA, V.G.** : Fotogeneratsiya dvukhvektornogo transmembrannogo gradienta protonov v kletkakh *Halobacterium halobium* R_1. [Photogeneration of two-vector transmembrane gradient of protons in the cells of *Halobacterium halobium* R_1.] - Biofizika *22* : 944 - 946, 1977. [In R, ab : E.]

31310 - **PLAKUNOVA, V.G.** : Énergeticheskaya kharakteristika stadii fotozavisimogo transporta ^{14}C-alanina v kletki *Halobacterium halobium* R_1. [Energy requirement of stages of photodependent ^{14}C-alanine transport into *Halobacterium halobium* R_1 cells.] - Mikrobiologiya *44* : 1116 - 1118, 1977. [In R, ab : E.]

31311 - **PLANTE-CUNY, M.-R.** : Pigments photosynthétiques et production primaire du microbenthos d'une lagune tropicale, la lagune Ébrié (Abidjan, Côte d'Ivoire). - Cah. ORSTOM , Sér. Océanographie *15* : 3 - 25, 1977.

31312 - **PLANTE-CUNY, M.R.** : Répartition à la surface et au sein du sédiment de la chlorophylle *a* et des phéopigments de quelques substrats meubles tropicaux immergés. - J. Rech. Océanogr. *2* (2) : 1 - 11, 1977.

31313 - **PLATT, S.G., BASSHAM, J.A.** : Separation of ^{14}C-labelled glycolate pathway metabolites from higher plant photosynthate. - J. Chromatogr. *133* : 396 - 401, 1977.

31314 - **PLATT, S.G., PLAUT, Z., BASSHAM, J.A.** : Steady-state photosynthesis in alfalfa leaflets. Effects of carbon dioxide concentration. - Plant Physiol. *60* : 230 - 234, 1977.

31315 - **PLATT, S.G., PLAUT, Z., BASSHAM, J.A.** : Ammonia regulation of carbon metabolism in photosynthesizing leaf discs. - Plant Physiol. *60* : 739 - 742, 1977.

31316 - **PLATT, T., DENMAN, K.L., JASSBY, A.D.** : Modeling the productivity of phytoplankton. - In : GOLDBERG, E.D., McCAVE, I.N., O'BRIEN, J.J., STEELE, J.H.
(ed.) : The Sea: Ideas and Observations on Progress in the Study of the Seas.
Vol. 6. Marine Modeling. Pp. 807 - 856. J. Wiley & Sons, New York 1977.

31317 - **PLATT, T., SUBBA RAO, D.V., DENMAN, K.L.** : Quantitative stimulation of phytoplankton productivity by deep water admixture in a coastal inlet. - Estuar.
coast. mar. Sci. *5* : 567 - 573, 1977. [Chl.]

31318 - **PLATT-ALOIA, K.A., THOMSON, W.W.** : Chloroplast development in young sesame
plants. - New Phytol. *78* : 599 - 605, 1977.

31319 - **PLAUT, Z., HEUER, B.** . Activity of ribulose diphosphate carboxylase of plants
grown under saline conditions and water stress. - In : COOMBS, J. (ed.) :
4th International Congress on Photosynthesis. P. 300. UKISES, London 1977.

31320 - **PLAUT, Z., LENDZIAN, K., BASSHAM, J.A.** : Nitrite reduction in reconstituted
and whole spinach chloroplasts during carbon dioxide reduction. - Plant Physiol. *59* : 184 - 188, 1977.

31321 - **PLAUT, Z., ZIESLIN, N.** : The effect of canopy wetting on plant water status,
CO_2 fixation, ion content and growth rate of "Baccara" roses. - Physiol. Plant.
39 : 317 - 322, 1977.

31322 - **PLESNICAR, M.** : Sulphur dioxide inhibition of photophosphorylation in isolated pea chloroplasts. - In : COOMBS, J. (ed.) : 4th International Congress
on Photosynthesis. P. 299. UKISES, London 1977.

31323 - **PLESSER, A., WEISSENBÖCK, G.** : Untersuchungen zur Lokalisation von Flavonoi-
den in Plastiden IV. Flavonoidgehalt intakter Chloroplasten aus *Avena sativa*
L. - Z. Pflanzenphysiol. *81* : 425 - 437, 1977.

31324 - **PLOTNIKOVA, A.N., KIRSHUNOVA, L.V.** : Osobennosti biosinteza adenozinfosfatov
pri zelenenii list'ev fasoli na svetu raznogo kachestva. [Peculiarities of
adenosine phosphates biosynthesis with greening of bean leaves in the light
of different quality.] - Fiziol. Biokhim. kul't. Rast. *9* : 497 - 500, 1977.
[In R, ab : E.]

31325 - **POINCELOT, R.P.** : Isolation of envelope membranes from bundle sheath chloroplasts of maize. - Plant Physiol. *60* : 767 - 770, 1977.

*31326 - **POKORNY, K.S., GOLD, K.** : Two morphological types of particulate inclusions
in marine dinoflagellates. - J. Phycol. *9* : 218 - 224, 1973. [Ps.]

*31327 - **POKROVSKAYA, T.N.** : O fotosinteze makrofitov v ozerakh. [Photosynthesis of
macrophytes in lakes.] - In : Tipologiya Ozernogo Nakopleniya Organicheskogo
Veshchestva. Pp. 35 - 45. Nauka, Moskva 1976. [In R.]

*31328 - **POKROVSKAYA, T.N.** : K tipologii ozer-nakopiteleĬ organicheskogo veshchestva.
[Typology of lakes - producers of organic matter.] - In : Tipologiya Ozernogo
Nakopleniya Organicheskogo Veshchestva. Pp. 46 - 89. Nauka, Moskva 1976.
[Ps; In R.]

31329 - **POLESCU-IONĂŞESCU, L.** : Quelques processus physiologiques suivis pendant le
cycle de développement d'une culture synchrone de *Chlorella coelastroides*. -
Rev. roum. Biol., Sér. Biol. vég. *22* : 133 - 140, 1977. [Chl, Car.]

31330 - **POLONSKIĬ, V.I., LISOVSKIĬ, G.M.** : Sostoyanie pigmentnogo apparata pshenitsy
pri vysokikh intensivnost'yakh FAR v svetokul'ture. [The state of wheat pigment apparatus at high intensities of PhAR in a light culture.] - Fiziol.
Rast. *24* : 1159 - 1164, 1977. [In R, ab : E.]

31331 - **POLONSKIĬ, V.I., LISOVSKIĬ, G.M., TRUBACHEV, I.N.** : Produktivnost' i biokhimicheskiĬ sostav pshenitsy pri vysokoĬ intensivnosti FAR v svetokul'ture.
[Productivity and biochemical composition of wheat at high PhAR irradiance.
in light culture.] - Fiziol. Rast. *24* : 718 - 724, 1977. [In R, ab : E.]

31332 - **POLUEKTOV, R.A.** : Mathematische Modelle der Pflanzenproduktivität. - In :
UNGER, K. (ed.) : Biophysikalische Analyse pflanzlicher Systeme. Pp. 31 - 37.
VEB Gustav Fischer Verlag, Jena 1977.

31333 - POPOV, V.I., KAUROV, B.S., GAVRILOV, A.G., TAGEEVA, S.V., RUBIN, L.B. : Primenenie metoda zamorazhivaniya-skalyvaniya dlya analiza povrezhdeniĭ membran khloroplastov, vyzyvaemykh izlucheniem lazerov. [Use of the freeze-etching method for analysis of destruction of chloroplast membranes induced by laser radiation.] - In : II Vsesoyuznyĭ Simpozium. Kriogennye Metody v Élektronnoĭ Mikroskopii. Tezisy Dokladov. Pp. 13 - 18. Pushchino 1977. [In R.]

31334 - POPOV, V.I., TAGEEVA, S.V. : Vozmozhnosti issledovaniya strukturno-konformatsionnykh izmeneniĭ fotosinteticheskikh membran metodom zamorazhivaniya-skalyvaniya. [Possibilities of studying the structural and conformational changes in photosynthetic membranes by the freeze etching method.] - In : II Vsesoyuznyĭ Simpozium. Kriogennye Metody v Élektronnoĭ Mikroskopii. Tezisy Dokladov. Pp. 9 - 13. Pushchino 1977. [In R.]

31335 - POPOV, V.I., TAGEEVA, S.V., KAUROV, B.S., GAVRILOV, A.G., RUBIN, A.B., RUBIN, L.B. : Effect of ruby laser irradiation on the ultrastructure of pea chloroplast membranes. - Photosynthetica 11 : 76 - 80, 1977.

31336 - POPOV, V.I., TAGEEVA, S.V., LADYGIN, V.G. : Nadmolekulyarnaya organizatsiya membran khloroplastov u mutantov Chlamydomonas reinhardi s neaktivnymi fotosistemami. [Supermolecular organization of chloroplast membranes in Chlamydomonas reinhardi mutants with inactive photosystems.] - Izv. Akad. Nauk SSSR, Ser. biol. 1977 : 856 - 868, 1977. [In R, ab : E.]

31337 - POPOVA, L., VAKLINOVA, S. : Influence of ferredoxin on the activity of NADPH$_2$-malatedehydrogenase in maize. - Dokl. bolg. Akad. Nauk 30 : 579 - 582, 1977.

31338 - POPOVA-STAEVSKA, L. : Photosynthesis in vitro under stationary conditions of CO$_2$ fixation. - Dokl. bolg. Akad. Nauk 30 : 901 - 904, 1977.

31339 - POPOVA-STAEVSKA, L. : Photosynthetic CO$_2$ fixation in reconstructed chloroplast system under steady-state photosynthesis. - Dokl. bolg. Akad. Nauk 30 : 1059 - 1062, 1977.

31340 - PORTER, E.M., BARTELS, P.G. : Use of single leaf cells to study mode of action of SAN 6706 on soybean and cotton. - Weed Sci. 25 : 60 - 65, 1977. [Ps.]

31341 - PORTER, G. : Living systems as energy converters. - In : BUVET, R., ALLEN, M. J., MASSUÉ, J.-P. (ed.) : Living Systems as Energy Converters. Pp. 1 - 4. North-Holland Publ. Co., Amsterdam - New York - Oxford 1977.

31342 - PORTER, G., SYNOWIEC, J.A., TREDWELL, C.J. : Intensity effects on the fluorescence of in vivo chlorophyll. - Biochim. biophys. Acta 459 : 329 - 336, 1977.

31343 - PORTIS, A.R. Jr., CHON, C.J., MOSBACH, A., HELDT, H.W. : Fructose- and sedoheptulosebisphosphatase. The sites of a possible control of CO$_2$ fixation by light-dependent changes of the stromal Mg^{2+} concentration. - Biochim. biophys. Acta 461 : 313 - 325, 1977.

31344 - POSNER, H.B., POSNER, R.S., GOWER, R.A. : Effects of DCMU on long-day flowering of Lemna perpusilla 6746 and photosynthetic mutant strain 1073. - Plant Cell Physiol. 18 : 1301 - 1307, 1977.

31345 - POSSINGHAM, J.V., ROSE, R.J. : Studies of the synthesis, location and segregation of chloroplast DNA in spinach. - In : BOGORAD, L., WEIL, J.H. (ed.) : Acides Nucléiques et Synthèse des Protéines chez les Végétaux. Coll. Int. CNRS No. 261. Pp. 85 - 91. Édit. CNRS, Paris 1977.

31346 - POTASEK, M.J., HOPFIELD, J.J. : The nature of electron transport in photosynthesis: Experimental test of the electron tunneling model. - In : COOMBS, J. (ed.) : 4th International Congress on Photosynthesis. P. 301. UKISES, London 1977.

31347 - POTTER, J.R. : Monitoring photosynthesis to measure translocation of bentazon in common cocklebur. - Weed Sci. 25 : 241 - 246, 1977.

31348 - POTTER, J.R., JONES, J.W. : Leaf area partitioning as an important factor in growth. - Plant Physiol. 59 : 10 - 14, 1977. [Growth analysis.]

31349 - POWLS, R., BRITTON, G. : A series of mutant strains of Scenedesmus obliquus with abnormal carotenoid compositions. - Arch. Microbiol. 113 : 275 - 280, 1977.

31350 - **POWLS, R., BRITTON, G.** : The roles of isomers of phytoene, phytofluene and
ζ-carotene in carotenoid biosynthesis by a mutant strain of *Scenedesmus obliquus*. - Arch. Microbiol. *115* : 175 - 179, 1977.

31351 - **POWLS, R., O'BRIEN, M.J., WOODROW, S., EASTERBY, J.S.** : Conversion of the
NADH-dependent glyceraldehyde 3-phosphate dehydrogenase of *Scenedesmus obliquus* into an NADPH-dependent form. - In : COOMBS, J. (ed.) : 4th International
Congress on Photosynthesis. P. 301. UKISES, London 1977.

31352 - **POZSÁR, B.I.** : A glükolaldehid-képződés három útja és kapcsolata a fotoszintézissel. [Three ways of glycolaldehyde production, and its connection with
photosynthesis.] - Bot. Közlem. *64* : 113 - 116, 1977. [In Hung., ab : E.]

31353 - **POZSÁR, B.I.** : Does photosynthetic glycolaldehyde formation occur at an endogenous level and what is its intensity? - Acta agron. Acad. Sci. hung. *26* :
207 - 208, 1977.

31354 - **PRADEL, J.** : Fluorescence induction in *Rhodopseudomonas sphaeroides* during
repigmentation. - In : COOMBS, J. (ed.) : 4th International Congress on Photosynthesis. P. 302. UKISES, London 1977.

31355 - **PRADEL, J., CLÉMENT-MÉTRAL, J.** : Cell division and photosynthetic apparatus
construction in cell envelope deficient "Phofil" mutant of *Rhodopseudomonas
sphaeroides*. - J. gen. Microbiol. *102* : 365 - 374, 1977.

31356 - **PRASAD, U., SINGHAL, G.S., MOHANTY, P.** : Effect of protons and cations on chloroplast membranes as visualized by the bound ANS fluorescence. - Biophys.
Struct. Mech. *3* : 259 - 274, 1977.

31357 - **PRÉCSÉNYI, I.** : Relationships between growth characteristics of maize hybrids
and sugar beet varieties. - Acta bot. Acad. Sci. hung. *23* : 361 - 366, 1977.

31358 - **PRÉCSÉNYI, I., BOGNÁR, C., CZIMBER, Gy., VIRÁGH, K.** : A Beta poly M/102 és
a Kawemono cukorrépa fajták növekedés-analízise. [Growth analysis of Beta
poly M/102 and Kawemono sugar beets.] - Növénytermelés *26* : 355 - 366, 1977.
[In Hung., ab : E.]

31359 - **PRÉCSÉNYI, I., CZIMBER, Gy., CSALA, G.** : Light energy transformation in maize
hybrids. - Acta agron. Acad. Sci. hung. *26* : 135 - 140, 1977.

31360 - **PREISS, J.** : Regulation of α1,4 glucan synthesis in photosynthetic systems. -
In : COOMBS, J. (ed.) : 4th International Congress on Photosynthesis. Pp.
302 - 303. UKISES, London 1977.

31361 - **PREŤOVÁ, A.** : Pigments in young embryos of *Linum usitatissimum* L. - Photosynthetica *11* : 217 - 219, 1977.

31362 - **PRÉZELIN, B.B., MEESON, B.W., SWEENEY, B.M.** : Characterization of photosynthetic rhythms in marine dinoflagellates I. Pigmentation, photosynthetic
capacity and respiration. - Plant Physiol. *60* : 384 - 387, 1977.

31363 - **PRÉZELIN, B.B., SWEENEY, B.M.** : Characterization of photosynthetic rhythms
in marine dinoflagellates. II. Photosynthesis-irradiance curves and *in vivo*
chlorophyll *a* fluorescence. - Plant Physiol. *60* : 388 - 392, 1977.

31364 - **PRÉZELIN, B.B., SWEENEY, B.M.** : Photosynthetic light adaptation in *Gonyaulax
polyedra*. - J. Phycol. *13* (Suppl.) : 55, 1977.

31365 - **PRICE, C.A., ZIELINSKI, R.E.** : Synthesis of cytochromes *b*559 and *f* by isolated chloroplasts from spinach. - In : COOMBS, J. (ed.) : 4th International
Congress on Photosynthesis. P. 303. UKISES, London 1977.

31366 - **PRIESTLEY, D.A., WOOLHOUSE, H.W.** : Properties of the chloroplast envelope
from primary leaves of *Phaseolus vulgaris*. - In : COOMBS, J. (ed.) : 4th
International Congress on Photosynthesis. P. 304. UKISES, London 1977.

31367 - **PŘIKRYL, K.** : Závislost tvorby výnosu zrna ozimé pšenice na množství suché
hmoty a obsahu živin v období květu. [The dependence of the formation of
winter wheat grain yield on the amount of dry matter and nutrient content
in the flowering period.] - Rostlinná Výroba (Praha) *23* : 831 - 836, 1977.
[In Czech, ab : E, G, R.]

31368 - **PRINCE, R.C., DUTTON, P.L.** : Single and multiple turnover reactions in the ubiquinone-cytochrome b-c_2 oxidoreductase of *Rhodopseudomonas sphaeroides*. The physical chemistry of the major electron donor to cytochrome c_2 and its coupled reactions. - Biochim. biophys. Acta *462* : 731 - 747, 1977.

31369 - PRINCE, R.C., LEIGH, J.S.,Jr., DUTTON, P.L. : The origin of the light-induced spin-polarized triplet or "biradical". - In : OLSON, J.M., HIND, G. (ed.) : Chlorophyll-Proteins, Reaction Centers, and Photosynthetic Membranes. Pp. 368 - 369. Brookhaven nat. Lab., Upton 1977.

31370 - PRINCE, R.C., THORNBER, J.P. : A novel electron paramagnetic resonance signal associated with the "primary" electron acceptor in isolated photochemical reaction centers of *Rhodospirillum rubrum*. - FEBS Lett. *81* : 233 - 237, 1977.

31371 - PRINCE, R.C., TIEDE, D.M., THORNBER, J.P., DUTTON, P.L. : Spectroscopic properties of the intermediary electron carrier in the reaction center of *Rhodopseudomonas viridis*. Evidence for its interaction with the primary acceptor. - Biochim. biophys. Acta *462* : 467 - 490, 1977.

31372 - PRIOUL, J.L., CHARTIER, P. : Partitioning of transfer and carboxylation components of intracellular resistance to photosynthetic CO_2 fixation: A critical analysis of the methods used. - Ann. Bot. *41* : 789 - 800, 1977.

31373 - PROBST, B. : 7. Primary production. - In : RHEINHEIMER, G. (ed.) : Microbial Ecology of a Brackish Water Environment. Pp. 71 - 78. Springer-Verlag, Berlin - Heidelberg - New York 1977.

31374 - PROCHASKA, L.J., GROSS, E.L. : Cation-induced quenching of chlorophyll *a* fluorescence in Triton X-100 subchloroplast particles. - Arch. Biochem. Biophys. *181* : 147 - 154, 1977.

31375 - PROCHASKA, L.J., GROSS, E.L. : Evidence for the location of divalent cation binding sites on the chloroplast membrane. - J. Membrane Biol. *36* : 13 - 32, 1977. [Ps, Chl.]

31376 - PROCHÁZKA, S., PEŠKA, J. : Distribuce asimilátů u vybraných odrůd pšenice ozimé v období tvorby zrna. [The distribution of assimilates in winter wheat varieties in the period of grain filling.] - In : Produkce Biomasy a Tvorba Výnosů Polních Plodin. Vol.1. Pp. 103 - 104. Česká Vědeckotechnická Společnost Zemědělská, Praha 1977. [In Czech, ab : E, G, R.]

31377 - PROCHÁZKA, S., SVOBODA, J., KVAPILOVÁ, M. : Distribuce asimilátů u sóji v období tvorby generativních orgánů. [Distribution of photosynthates in soybean plants during the period of formation of generative organs.] - Rostlinná Výroba *23* : 715 - 721, 1977. [In Czech, ab : E, G, R.]

31378 - PROCTOR, M.C.F. : Evidence on the carbon nutrition of moss sporophytes from $^{14}CO_2$ uptake and the subsequent movement of labelled assimilate. - J. Bryol. *9* : 375 - 386, 1977.

31379 - PROKHORCHIK, R.A. : Vliyanie abstsizovoĭ kisloty na fotokhimicheskuyu aktivnost' khloroplastov lyupina. [Effect of abscisic acid on photochemical activity of lupine chloroplasts.] - Dokl. Akad. Nauk belorus. SSR *21* : 644 - 646, 1977. [In R.]

*31380 - PRONINA, N.B. : Sravnitel'naya kharakteristika adenilatkinazy a pterinbelkovykh kompleksov iz khloroplastov gorokha. [Comparative characteristics of adenylate kinase and pterin-protein complexes from pea chloroplasts.] - Nauch. Dokl. vyssh. Shkoly, biol. Nauki *19*(1) : 94 - 100, 1976. [In R.]

31381 - PRONINA, N.B., LADONIN, V.F. : Ob adenozindifosfataznoĭ aktivnosti adenilatkinazy iz khloroplastov yachmenya i gorokha. [Adenosine diphosphatase activity of adenylate kinase from barley and pea chloroplasts.] - Nauch. Dokl. vyssh. Shkoly, biol. Nauki *20*(1) : 96 - 100, 1977. [In R.]

31382 - PROPP, M.V. : Exchange of energy, nitrogen and phosphorus between water, bottom and ice in a near-shore ecosystem of the Sea of Japan. - Helgol. wiss. Meeresunters. *30* : 598 - 610, 1977. [Ps, Chl.]

31383 - PROTICH, N. : Skorost na rastezha i .otmiraneto na listata pri shest vida zhitni trevi. [Growth rate and dying off of leaves in six cereal species.] - Rasteniev. Nauki *14*(5) : 90 - 97, 1977. [In Bulg., ab : E, R.]

31384 - PROTSENKO, D.F., EMCHUK, V.G. : Ustoĭchivost' pigmentov pshenitsy k razrushe-
niyu pri podkormke solyami azota i sakharozoĭ. [Resistance of wheat pigments
to destruction under nitrogen salts and sucrose dressing.] - Fiziol. Biokhim.
kul't. Rast. 9 : 571 - 575, 1977. [In R, ab : E.]

31385 - PROTSENKO, D.F., EMCHUK, V.G., KOMARENKO, N.I. : Ustoĭchivost' pigmentnogo
kompleksa i aktivnost' khlorofillazy kak pokazatel' zimostoĭkosti pshenitsy.
[Stability of pigment complex and activity of chlorophyllase as an index of
wheat winter hardiness.] - Fiziol. Biokhim. kul't. Rast. 9 : 266 - 272, 1977.
[In R, ab E.]

*31386 - PUCHKOVA, N.N., GORLENKO, V.M. : Novye korichnevye khlorobakterii *Prostheco-
chloris phaeoasteroidea* nov. sp. [A new brown chlorobacterium *Prosthecochloris phaeoasteroidea* sp. nov.] - Mikrobiologiya 45 : 655 - 660, 1976. [Chl;
in R, ab : E.]

*31387 - PUCHKOVA, N.N., GORLENKO, V.M., PIVOVAROVA, T.A. : Sravnitel'noe izuchenie
tonkogo stroeniya vibrioidnykh zelenykh serobakteriĭ. [Comparative study of
fine structure of vibrioid green sulphur bacteria.] - Mikrobiologiya 44 : 108
- 114, 1975. [Chromatophore; in R, ab : E.]

31388 - PUCKETT, K.J., TOMASSINI, F.D., NIEBOER, E., RICHARDSON, D.H.S. : Potassium
efflux by lichen thalli following exposure to aqueous sulphur dioxide. - New
Phytol. 79 : 135 - 146, 1977. [Ps.]

31389 - PULICH, W.,Jr. : Cytochrome c_{548} in *Nostoc* sp. (*Cyanophyceae*) : an electron
acceptor from reduced NADP in the dark. - J. Phycol. 13 : 40 - 45, 1977.

31390 - PULLES, M.P.J., Van GORKOM, H.J. : Two types of system 2 reaction centers in
Tris-washed chloroplasts. - In : COOMBS, J. (ed.) : 4th International Congress
on Photosynthesis. Pp. 304 - 305. UKISES, London 1977.

31391 - PULLIN, C.A., EVANS, E.H. : Photosynthetic reaction centres from a blue-green
alga. - In : COOMBS, J. (ed.) : 4th International Congress on Photosynthesis.
P. 300. UKISES, London 1977.

31392 - PÜMPEL, B. : Bestandesstruktur, Phytomassevorrat und Produktion verschiedener
Pflanzengesellschaften im Glocknergebiet. - In : CERNUSCA, A. (ed.) : Alpine
Grasheide Hohe Tauern. Veröffentl. Österr. MaB-Hochgebirgsprogrammes Hohe
Tauern. Band 1. Pp. 83 - 101. Universitätsverlag Wagner, Innsbruck 1977.

31393 - PUNNETT, T. : Multiple pathways for photosynthetic CO_2 fixation in vascular
plants. - In : COOMBS, J. (ed.) : 4th International Congress on Photosynthesis.
P. 305. UKISES, London 1977.

*31394 - PUROHIT, A.N., TREGUNNA, E.B. : Effects of carbon dioxide on the growth of
Douglas-fir seedlings. - Indian J. Plant Physiol. 19 : 164 - 170, 1976.

31395 - PUROHIT, (Km.) M., MALL, L.P., DUBEY, P.S. : Herbicidal pollution-chlorophyll
content as an index of residual toxicity. - Curr. Sci. 46 : 157 - 158, 1977.

*31396 - PYRINA, I.L., BASHKATOVA, E.L., SIGAREVA, L.E. : Pervichnaya produktsiya fi-
toplanktona v melkovodnoĭ zone Rybinskogo vodokhranilishcha v 1971 - 1972 gg.
[Primary production of phytoplankton in the shallow water zone of the Rybinsk
reservoir in 1971-72.] - Tr. Inst. Biol. vnutr. Vod Akad. Nauk SSSR 33 (Gidro-
biologicheskiĭ Rezhim Pribrezhnykh Melkovodiĭ Verkhnevolzhskikh Vodokhranilishch) : 106 - 132, 1976. [In R.]

*31397 - PYRINA, I.L., ELIZAROVA, V.A., NIKOLAEV, I.I. : Soderzhanie fotosinteti ches-
kikh pigmentov v fitoplanktone Onezhskogo ozera i ikh znachenie dlya otsenki
urovnya produktivnosti ètogo vodoema. [Content of photosynthetic pigments
in phytoplankton of Onega Lake and their significance for determining productivity of this reservoir.] - In : Mikrobiologiya i Pervichnaya Produktsiya
Onezhskogo Ozera. Pp. 108 - 122. Nauka, Leningrad 1973. [In R.]

31398 - QUANDT, L., GOTTSCHALK, G., ZIEGLER, H., STICHLER, W. : Isotope discrimination by photosynthetic bacteria. - FEMS Microbiol. Lett. 1 : 125 - 128, 1977.

31399 - QUAST, P. : Verteilung von Trockensubstanz, Kohlenhydraten und Gibberellin
bei Tomaten-, Kartoffel- und Auberginenpflanzen ohne und mit Früchten bzw.
Knollen. - Gartenbauwissenschaft 42 : 97 - 105, 1977.

31400 - **QUEBEDEAUX, B., CHOLLET, R.** : Comparative growth analyses of *Panicum* species with differing rates of photorespiration. - Plant Physiol. *59* : 42 - 44, 1977.

31401 - **QUINTANILHA, A.T., PACKER, L.** : Surface potential changes on energization of mitoplasts and chloroplasts. - In : PACKER, L., PAPAGEORGIOU, G.C., TREBST, A. (ed.) : Bioenergetics of Membranes. Pp. 55 - 60. Elsevier/North Holland Biomedical Press, Amsterdam - Oxford - New York 1977.

31402 - **RABOY, B., PADAN, E.** : Energization of α methyl glucoside transport in a cyanobacterium (blue-green alga), *Plectonema boryanum*, possessing intracytoplasmic photosynthetic lamella. - In : COOMBS, J. (ed.) : 4th International Congress on Photosynthesis. P. 306. UKISES, London 1977.

31403 - **RACHKOVSKAYA, M.M., KIM, L.O.** : Mikroėlementy i gazoustoĭchivost' rasteniĭ. [Trace elements and gas resistance of plants.] - In : Fiziologo-Biokhimicheskie i Ėkologicheskie Aspekty Ustoĭchivosti Rasteniĭ k Neblagopriyatnym Faktoram Vneshneĭ Sredy. Pp. 204 - 207. Akad. Nauk SSSR, Sib. Otd., Sib. Inst. Fiziol. Biokhim. Rast., Irkutsk 1977. [Chl; in R.]

31404 - **RACHKOVSKAYA, M.M., KIM, L.O., KAĬDALOVA, E.C., LYBINA, L.M., FRAĬMAN, I.Ya., KISLYUK, L.V.** : Vliyanie tekhnosfery na fiziologo-biokhimicheskie protsessy rasteniĭ. [Effect of technical environment on physiological and biochemical processes of plants.] - In : Problemy Okhrany Okruzhayushcheĭ Sredy Regiona s Intensivno Razvivayushcheĭsya Promyshlennost'yu. Pp. 70 -72. Kemerovo 1977. [Chl; in R.]

31405 - **RADCHENKO, M.Ï.** : Khemotaksonomichne vyvchennya pigmentiv u vydiv *Chlamydomonas* EHR. 1. Yakisnyĭ sklad i kil'kisnyĭ vmist pigmentiv u *Chlamydomonas* spp. v optymal'nykh umovakh seredovyshcha. [Chemotaxonomic study of pigments of *Chlamydomonas* EHR. species. 1. Qualitative composition and quantitative content of pigments in *Chlamydomonas* spp. under optimal conditions of medium.] - Ukr. bot. Zh. *34*: 367 - 371, 1977. [In Ukr., ab : E.]

31406 - **RADCHENKO, M.Ï.** : Khemotaksonomichne vyvchennya pigmentiv u vydiv *Chlamydomonas* EHR. 2. Yakisnyĭ sklad i kil'kisnyĭ vmist pigmentiv u *Chlamydomonas* ssp. v ekstremal'nykh umovakh seredovyshcha. [Chemotaxonomic study of pigments in *Chlamydomonas* EHR. species. 2. Qualitative composition and quantitative content of pigments in *Chlamydomonas* spp. under extremal conditions of the medium.] - Ukr. bot. Zh. *34* : 596 - 603, 672, 1977. [In Ukr., ab : E, R.]

31407 - **RADENOVIĆ, Č., FIDLER, D., LIVADA, M., VUČINIĆ, Ž., PENČIĆ, M.** : Decomposition of delayed light emission curve from an intact maize (*Zea mays* L.) leaf. - In : COOMBS, J. (ed.) : 4th International Congress on Photosynthesis. P. 307. UKISES, London 1977.

31408 - **RADEVA, V.** : Prouchvane v"rkhu vodniya i khranitelniya rezhim na ednogodishna lyutserna. [Water and nutrient regime of one-year alfalfa.] - Rasteniev. Nauki *14*(1) : 119 - 126, 1977. [Dry-matter accumulation; in Bulg., ab : E, R.]

31409 - **RADMER, R.** : The oxygen cycle : Apparent K_m values of CO_2 and O_2. - In : COOMBS, J. (ed.) : 4th International Congress on Photosynthesis. P. 308. UKISES, London 1977.

31410 - **RADMER, R., CHENIAE, G.** : Mechanisms of oxygen evolution. - In : BARBER, J. (ed.) : Primary Processes of Photosynthesis. Pp. 303 - 348. Elsevier, Amsterdam - New York - Oxford 1977.

31411 - **RADMER, R., KOK, B.** : Photosynthesis : Limited yields, unlimited dreams. - BioScience *27* : 599 - 605, 1977.

31412 - **RADMER, R.J., KOK, B.** : Light conversion efficiency in photosynthesis. - In : TREBST, A., AVRON, M. (ed.) : Photosynthesis I. (Encycl. Plant Physiol. N.S. Vol. 5.) Pp. 125 - 134. Springer-Verlag, Berlin - Heidelberg - New York 1977.

*31413 - **RADUNZ, A.** : Localization of the tri- and digalactosyl diglyceride in the thylakoid membrane with serological methods. - Z. Naturforsch. *31C* : 589 - 593, 1976.

31414 - **RADUNZ, A.** : Distribution of lipids and proteins in the outer surface of the thylakoid membrane. An investigation with specific antisera. - In : COOMBS, J.

(ed.) : 4th International Congress on Photosynthesis. Pp. 307 - 308. UKISES, London 1977.

31415 - RADWAY, J., ROSNER, D., GREENBAUM, J., HAYNES, L., MITSUI, A. : Hydrogen producing tropical marine photosynthetic microorganisms: isolation and morphology. - Plant Physiol. *59* (6, Suppl.) : 20, 1977.

31416 - RAGHAVENDRA, A.S., DAS, V.S.R. : Antitranspirant activity of inhibitors of cyclic photophosphorylation. - J. exp. Bot. *28* : 480 - 483, 1977.

31417 - RAGHAVENDRA, A.S., DAS, V.S.R. : Effects of light quality on photosynthetic carbon metabolism in C₄ and C₃ plants: Rapid movements of photosynthetic intermediates between mesophyll and bundle sheaths cells. - J. exp. Bot. *28* : 1169 - 1179, 1977.

31418 - RAGHAVENDRA, A.S., DAS, V.S.R. : Endogenous photophosphorylation by mesophyll and bundle sheath chloroplasts from *Setaria italica* and *Amaranthus paniculatus*. - Ann. Bot. *41* : 667 - 669, 1977.

31419 - RAGHAVENDRA, A.S., DAS, V.S.R. : Purification and properties of phosphoenolpyruvate and ribulose diphosphate carboxylases from C₄ and C₃ plants. - Z. Pflanzenphysiol. *82* : 315 - 321, 1977.

31420 - RAGHAVENDRA, A.S., DAS, V.S.R. : Comparative studies on C₄ and C₃ photosynthetic systems: Effect of metabolic inhibitors and biochemical intermediates on carbon metabolism. - Z. Pflanzenphysiol. *85* : 9 - 16, 1977.

31421 - RAGHAVENDRA, A.S., DAS, V.S.R. : Light-enhanced dark ¹⁴CO₂ fixation by leaves in relation to the C₄ dicarboxylic acid pathway of photosynthesis. - Aust. J. Plant Physiol. *4* : 833 - 841, 1977.

31422 - RAGHI-ATRI, F. : Zur Erfassung der Planktonbiomasse in stark eutrophen Gewässern im Uferbereich. - Z. Wasser- Abwasser-Forsch. *10* : 19 - 20, 1977.

31423 - RAI, A.K. : Induction of phycoerythrin in *Anabaena ambigua* RAO and its strains. - Experientia *33* : 595, 1977.

31424 - RAILTON, I.D. : 16,17-dihydro 16,17-dihydroxy gibberellin A₉ : A metabolite of [³H]-gibberellin Ag in chloroplast sonicates from *Pisum sativum* var. "Alaska". - Z. Pflanzenphysiol. *81* : 323 - 329, 1977.

31425 - RAILTON, I.D. : Gibberellin metabolism in chloroplasts of *Pisum sativum* L. var. Alaska. - S. afr. J. Sci. *73* : 22 - 23, 1977.

31426 - RAKHIMBERDIEVA, M.G., BUKHOV, N.G., KARAPETYAN, N.V. : Kharakteristika fotosistemy I iz tilakoidov grany i mezhgrannykh lamell khloroplastov. [Characteristics of photosystem I from grana thylakoids and stroma lamellae.] - Biokhimiya *42* : 1864 - 1871, 1977. [In R, ab : E.]

31427 - RAKHIMBERDIEVA, M.G., KARAPETYAN, N.V., KRASNOVSKIĬ, A.A. : Nizkotemperaturnaya induktsiya fluorestsentsii zeleneyushchikh list'ev fasoli. [Low-temperature induction of fluorescence of greening bean leaves.] - Dokl. Akad. Nauk SSSR *237* : 224 - 227, 1977. [In R.]

31428 - RAMADHAS, V., SUBRAMANIAN, B.R., VENUGOPALAN, V.K. : Significance of nannoplankton in primary production in Porto Novo waters. - Mahasagar *8* : 171 - 181, 1977. [Ps.]

31429 - RAMUS, J. : The significance of thallus anatomy to photon capture and productivity. - J. Phycol. *13* (Suppl.) : 57, 1977.

31430 - RAMUS, J., LEMONS, F., ZIMMERMAN, C. : Adaptation of light-harvesting pigments to downwelling light and the consequent photosynthetic performance of the eulittoral rockweeds *Ascophyllum nodosum* and *Fucus vesiculosus*. - Mar. Biol. *42* : 293 - 303, 1977.

31431 - RANCHYALIS, V. : Khlorofil'nye mutatsii pri vzaimodeĭstvii ėtilenimina s nekotorymi detergentami i kompleksonami. [Chlorophyll mutations as a result of interaction between ethylenimine and some detergents and complexing agents.] - Genetika *13* : 1446 - 1454, 1977. [In R, ab : E.]

31432 - RAND, R.H. : Gaseous diffusion in the leaf interior. - Trans. ASAE *20* : 701 - 704, 1977.

31433 - RANDALL, D.D., NELSON, C.J., ASAY, K.H. : Ribulose bisphosphate carboxylase. Altered genetic expression in tall fescue. - Plant Physiol. *59* : 38 - 41, 1977.

31434 - RANJEVA, R., ALIBERT, G., BOUDET, A.M. : Métabolisme des composés phénoliques chez le *Petunia* V. Utilisation de la phénylalanine par des chloroplastes isolés. - Plant Sci. Lett. *10* : 225 - 234, 1977.

31435 - RANJEVA, R., ALIBERT, G., BOUDET, A.M. : Métabolisme des composés phénoliques chez le *Petunia* VI. Intervention des chloroplastes dans la biosynthèse de la naringenine et de l'acide chlorogénique. - Plant Sci. Lett. *10* : 235 - 242, 1977.

31436 - RAO, C.K. : A chlorophyll-deficiency factor in the natural populations of *Tephrosia purpurea* (L.) PERS. - Proc. Indian Acad. Sci. *B 85* : 43 - 47, 1977.

31437 - RAO, K.K., DENNIS, G., REEVES, S.G., HALL, D.O. : Hydrogenases from photosynthetic bacteria and blue-green algae. - In : COOMBS, J. (ed.) : 4th International Congress on Photosynthesis. P. 309. UKISES, London 1977.

31438 - RAO, V.S.K., BRAND, J.J., MYERS, J. : Cold shock syndrome in *Anacystis nidulans*. - Plant Physiol.*59* : 965 - 969, 1977. [Ps.]

31439 - RAPER, C.D.,Jr., PARSONS, L.R., PATTERSON, D.T., KRAMER, P.J. : Relationship between growth and nitrogen accumulation for vegetative cotton and soybean plants. - Bot. Gaz. *138* : 129 - 137, 1977.

31440 - RASCHKE, K. : The osmotic motor of stomatal movement. - In : JUNGREIS, A.M., HODGES, T.K., KLEINZELLER, A., SCHULTZ, S.G. (ed.) : Water Relations in Membrane Transport in Plants and Animals. Pp. 47 - 53. Academic Press, New York - San Francisco - London 1977. [Ps.]

31441 - RASCHKE, K. : The stomatal turgor mechanism and its responses to CO_2 and abscisic acid: Observations and a hypothesis. - In : MARRÈ, E., CIFERRI, O. (ed.) : Regulation of Cell Membrane Activities in Plants. Pp. 173 - 183. Elsevier/North Holland Biomedical Press, Amsterdam 1977. [Ps.]

31442 - RASCHKE, K., DITTRICH, P. : [^{14}C] carbon-dioxide fixation by isolated leaf epidermes with stomata closed or open. - Planta *134* : 69 - 75, 1977.

31443 - RASCIO, N., MARIANI COLOMBO, P., ORSENIGO, M. : Maize chloroplast ontogenesis under continuous illumination. - Caryologia *30* : 497 - 498, 1977.

31444 - RASKIN, V.I. : Migratsiya ènergii mezhdu molekulami pigmentov v protsesse fotovosstanovleniya protokhlorofillida. [Energy migration among pigment molecules during photoreduction of protochlorophyllide.] - Dokl. Akad. Nauk belorus. SSR *21* : 272 - 275, 1977. [In R.]

31445 - RASQUAIN, A. : Aggregation properties of protochlorophyll pigments, *in vitro*. - In : COOMBS, J. (ed.) : 4th International Congress on Photosynthesis. P. 310. UKISES, London 1977.

31446 - RASQUAIN, A., HOUSSIER, C., SIRONVAL, C. : The dimerization of protochlorophyll pigments in non-polar solvents. - Biochim. biophys. Acta *462* : 622 - 641, 1977.

31447 - RATHNAM, C.K.M. : Relative contribution of mesophyll and bundle sheath cells to net CO_2 uptake by leaves of C_4 plants. - Plant Physiol. *59* (6, Suppl.) : 66, 1977.

31448 - RATHNAM, C.K.M. : Evidence for reassimilation of CO_2 released from C_4 acids by PEP carboxylase in leaves of C_4 plants. - FEBS Lett. *82* : 288 - 292, 1977.

31449 - RATHNAM, C.K.M., EDWARDS, G.E. : C_4-dicarboxylic acid metabolism in bundle-sheath chloroplasts, mitochondria and strands of *Echinochloa borumensis* HACK; a phosphoenolpyruvate-carboxykinase type C_4 species. - Planta *133* : 135 - 144, 1977.

31450 - RATHNAM, C.K.M., EDWARDS, G.E. : Use of inhibitors to distinguish between C_4 acid decarboxylation mechanisms in bundle sheath cells of C_4 plants. - Plant Physiol. *59* (6, Suppl.) : 66, 1977.

31451 - RATHNAM, C.K.M., EDWARDS, G.E. : C_4 acid decarboxylation and CO_2 donation to photosynthesis in bundle sheath strands and chloroplasts from species representing three groups of C_4 plants. - Arch. Biochem. Biophys. *182* : 1 - 13, 1977.

31452 - **RATHNAM, C.K.M., EDWARDS, G.E.** : Use of inhibitors to distinguish between C_4 acid decarboxylation mechanisms in bundle sheath cells of C_4 plants. - Plant Cell Physiol. *18* : 963 - 968, 1977.

31453 - **RATHNAM, C.K.M., ZILINSKAS, B.A.** : Reversal of 3-(3,4-dichlorophenyl)-1,1-dimethylurea inhibition of carbon dioxide fixation in spinach chloroplasts and protoplasts by dicarboxylic acids. - Plant Physiol. *60* : 51 - 53, 1977.

31454 - **RATINEN, H.** : Fluorescence spectra from dark- and light-grown pea leaves. - Rep. Univ. Oulu, Dep. Phys. *57* : 1 - 27, 1977.

31455 - **RATYNI, A.I., PYT'EVA, N.F., RUBIN, A.B.** : Opredelenie parametrov èlektron-transportnoĭ sistemy v khromatoforakh bakteriĭ *Ectothiorhodospira shaposhnikovii*. [Determining the parameters of the electron transport system in chromatophores of the bacterium *Ectothiorhodospira shaposhnikovii*.]-Nauch. Dokl. vyssh. Shkoly, biol. Nauki *20* (8) : 36 - 43, 1977. [In R.]

31456 - **RAVEN, J.A.** : ATP synthesis coupled to nitrate photoreduction in the alga *Hydrodictyon africanum*. - J. exp. Bot. *28* : 314 - 319, 1977. [Ps.]

31457 - **RAVEN, J.A., GLIDEWELL, S.M.** : C4 characteristics of chlorophyte photosynthesis. - In : COOMBS, J. (ed.) : 4th International Congress on Photosynthesis. P. 311. UKISES, London 1977.

*31458 - **RAVEN, J.A., SMITH, F.A.** : The evolution of chemiosmotic energy coupling. - J. theor. Biol. *57* : 301 - 312, 1976.

31459 - **RAVEN, J.A., SMITH, F.A.** : Characteristics, functions and regulation of active proton extrusion. - In : MARRÈ, E., CIFERRI, O. (ed.) : Regulation of Cell Membrane Activities in Plants. Pp. 25 - 40. North-Holland Publ. Comp., Amsterdam - Oxford - New York 1977.

31460 - **RAVEN, J.A., SMITH, F.A.** : "Sun" and "shade" species of green algae: relation to cell size and environment. - Photosynthetica *11* : 48 - 55, 1977.

31461 - **RAVIZZINI, R.A., ANDREO, C.S., VALLEJOS, R.H.** : Peptide alkaloids as inhibitors of photophosphorylation in spinach chloroplasts. - Plant Cell Physiol. *18* : 701 - 706, 1977.

31462 - **RAWLINGS, J., WHERLAND, S., GRAY, H.B.** : Electron transfer reactivity of spinach ferredoxin. - J. amer. chem. Soc. *99* : 1968 - 1971, 1977.

31463 - **RAWSON, H.M., BAGGA, A.K., BREMNER, P.M.:** Aspects of adaptation by wheat and barley to soil moisture deficits. - Aust. J. Plant Physiol. *4* : 389 - 401, 1977. [Production.]

31464 - **RAWSON, H.M., BEGG, J.E., WOODWARD, R.G.** : The effect of atmospheric humidity on photosynthesis, transpiration and water use efficiency of leaves of several plant species. - Planta *134* : 5 - 10, 1977.

31465 - **RAY, T.B., BLACK, C.C.** : The role of oxalacetate decarboxylation during photosynthesis in the C_4 species *Panicum maximum*. - Plant Physiol. *59* (6, Suppl.) : 64, 1977.

31466 - **RAY, T.B., BLACK, C.C.** : Oxaloacetate as the source of carbon dioxide for photosynthesis in bundle sheath cells of the C_4 species *Panicum maximum*. - Plant Physiol. *60* : 193 - 196, 1977.

*31467 - **REBELLO, A., WAGENER, K.** : Evaluation of ^{12}C and ^{13}C data on atmospheric CO_2 on the basis of a diffusion model for oceanic mixing. - In : NRIAGU, J.O. (ed.) : Environmental Biogeochemistry. Vol.1. Pp. 13 - 23. Ann Arbor Science Publishers, Ann Arbor 1976.

31468 - **REDLINGER, T.E., McDANIEL, R.G.** : Light-mediated oxygen uptake measured in wheat etioplasts. - Plant Physiol. *60* : 452 - 456, 1977.

31469 - **REED, A.J., HAGEMAN, R.H.** : Rates of nitrite assimilation by intact chloroplasts as affected by electron transport rates and supply of carbon skeletons. - Plant Physiol. *59* (6, Suppl.) : 127, 1977.

*31470 - REED, K.L., HAMERLY, E.R., DINGER, B.E., JARVIS, P.G. : An analytical model for field mesurement of photosynthesis. - J. appl. Ecol. *13* : 925 - 942, 1976.

31471 - REED, M.L. : The intracellular location of carbonic anhydrase (carbonate dehydratase) in C_3 leaf tissue. - Proc. aust. biochem. Soc. *10* : 14, 1977.

31472 - REED, M.L., GRAHAM, D. : Carbon dioxide and the regulation of photosynthesis: activities of photosynthetic enzymes and carbonate dehydratase (carbonic anhydrase) in *Chlorella* after growth or adaptation in different carbon dioxide concentrations. - Aust. J. Plant Physiol. *4* : 87 - 98, 1977.

31473 - REEVES, S.G., RAO, K.K., HALL, D.O. : Hydrogen production by biocatalytic photolysis of water using a chloroplast-hydrogenase system. - In : COOMBS, J. (ed.) : 4th International Congress on Photosynthesis. P. 312. UKISES, London 1977.

31474 - REIBACH, P.H., BENEDICT, C.R. : Fractionation of stable carbon isotopes by phosphoenolpyruvate carboxylase from C_4 plants. - Plant Physiol. *26* : 564 - 568, 1977.

31475 - REICH, R., SEWE, K.-U. : The effect of molecular polarization on the electrochromism of carotenoids - I. The influence of a carboxylic group. - Photochem. Photobiol. *26* : 11 - 17, 1977.

31476 - REICHENBÄCHER, D., RICHTER, J., SPAAR, D. : Unterschiede in der elektrophoretischen Beweglichkeit von Fraktion-I-Protein bei Gramineen und ihre mögliche Nutzung in Genetik und Züchtungsforschung. - Biochem. Physiol. Pflanzen *171* : 299 - 306, 1977.

31477 - REID, C.P.P., MEXAL, J.G. : Water stress effects on root exudation by lodgepole pine. - Soil Biol. Biochem. *9* : 417 - 421, 1977. [Photosynthates.]

31478 - REISS-HUSSON, F., RIVAS, E. : Preparation and properties of liposomes containing *Rhodopseudomonas sphaeroides* reaction centers. - In : COOMBS, J. (ed.) : 4th International Congress on Photosynthesis. P. 313. UKISES, London 1977.

31479 - REMENNIKOV, V.G., SAMUILOV, V.D. : Fotoindutsirovannoe pogloshchenie kisloroda khromatoforami i subkhromatofornymi pigment-belkovymi kompleksami *Rhodospirillum rubrum*. [Light-induced oxygen uptake by chromatophores and subchromatophore pigment-protein complexes.] - Biokhimiya *42* : 1997 - 2004, 1977. [In R, ab : E.]

31480 - REMY, R., HOARAU, J., LECLERC, J.C. : Electrophoretic and spectrophotometric studies of chlorophyll-protein complexes from tobacco chloroplasts. Isolation of a light harvesting pigment protein complex with a molecular weight of 70,000. - Photochem. Photobiol. *26* : 151 - 158, 1977.

31481 - REMY, R., HOARAU, J., LECLERC, J.-C. : Spectrophotometric study of chlorophyll-protein complexes from different photosynthesizing organisms and the heterogeneity of photoinduced spectral changes near 700 nm. - In : COOMBS, J. (ed.) : 4th International Congress on Photosynthesis. Pp. 414 - 415. UKISES, London 1977.

31482 - RENAUDIN, S., CAPDEPON, M. : Association of the endoplasmic reticulum and the plastids in *Tozzia alpina* L. scale leaves. - J. Ultrastructure Res. *61* : 303 - 308, 1977.

31483 - RENGER, G. : The use of trypsin as a structurally selective modifier of the thylakoid membrane. - In : PACKER, L., PAPAGEORGIOU, G.C., TREBST, A. (ed.) : Bioenergetics of Membranes. Pp. 339 - 350. Elsevier/North-Holland Biomedical Press, Amsterdam - Oxford - New York 1977.

31484 - RENGER, G. : A model for the molecular mechanism of photosynthetic oxygen evolution. - FEBS Lett. *81* : 223 - 228, 1977.

31485 - RENGER, G. : Photobiophysik. Photosynthese. - In : HOPPE, W., LOHMANN, W., MARKL, H., ZIEGLER, H. (ed.) : Biophysik. Ein Lehrbuch. Pp. 415 - 441. Springer-Verlag, Berlin - Heidelberg - New York 1977.

31486 - RENGER, G. : Über die strukturelle und funktionelle Organisation des System-II-Elektronentransportes in der Photosynthese. - Hoppe-Seyler's Z. physiol. Chem. *358* : 1380, 1977.

31487 - **RENGER, G., GLÄSER, M.** : Control of the dark reduction of photooxidized chlo-
rophyll-a_{II} by the inner thylakoid proton concentration. - In : **COOMBS, J.**
(ed.) : 4th International Congress on Photosynthesis. Pp. 313 - 314. UKISES,
London 1977.

31488 - **RENGER, G., GLÄSER, M., BUCHWALD, H.E.** : The control of the reduction kinetics in the dark of photooxidized chlorophyll a_{II}^+ by the inner thylakoid proton concentration. - Biochim. biophys. Acta *461* : 392 - 402, 1977.

31489 - **RENGER, G., SCHMID, G.H.** : On the correlation between the amplitude of the
electrochromic absorption changes and the number of bulk pigments. - Z. Naturforsch. *32C* : 963 - 967, 1977.

31490 - **RENGER, G., WOLFF, C.** : Further evidence for dissipative energy migration
via triplet states in photosynthesis. The protective mechanism of carotenoids
in *Rhodopseudomonas spheroides* chromatophores. - Biochim. biophys. Acta *460* :
47 - 57, 1977.

31491 - **RENTHAL, R.** : Apparent pK of photolabile proton binding to bacteriorhodopsin.
- Biochem. biophys. Res. Commun. *77* : 155 - 161, 1977.

31492 - **REPKA, J.** : Vplyv minerálnych živín na fotosyntézu, dýchanie a rast rastlín.
[Effect of mineral nutrition on photosynthesis, respiration and growth in
plants.] - Rostlinná Výroba (Praha) *23* : 733 - 740, 1977. [In Slovak, ab : E,
G, R.]

31493 - **REPKA, J.** : Vplyv minerálnej výživy na fotosynteticko-respiračnú aktivitu listov a produkciu hmoty. [Effect of mineral nutrients on the photosynthetic and
respiratory activity of leaves and on mass production.] - In : Produkce Biomasy a Tvorba Výnosu Polních Plodin. Vol. 1. Pp. 101 - 102. Česká vědeckotechnická Společnost zemědělská, Praha 1977. [In Slovak, ab : E, G, R.]

31494 - **REPKA, J.** : Fyziologicko-genetické a chemické faktory produktivity rastlín
(ciele a úlohy seminára). [Physiological, genetic and chemical factors of
plant productivity (aims and tasks of the seminar).] - In : **REPKA, J.** (ed.) :
Zborník Referátov zo Seminára Fyziologicko-Genetické a Chemické Faktory Produktivity Rastlín. Pp. 10 - 21. Vysoká Škola poľnohospodárska, Nitra 1977.
[Ps; in Slovak.]

31495 - **REPKA, J., GÁBORČÍK, N.** : Účinok pôsobenia dusíka a kyseliny giberelovej na
produkciu reznačky laločnatej (*Dactylis glomerata*). [The effect of nitrogen
and gibberellic acid on the production of *Dactylis glomerata*).] - In : **REPKA,
J.** (ed.) : Zborník Referátov zo Seminára Fyziologicko-Genetické a Chemické
Faktory Produktivity Rastlín. Pp. 111 - 116. Vysoká Škola poľnohospodárska,
Nitra 1977. [Chl, growth analysis; in Slovak.]

31496 - **REUTHER, W.** : Citrus. - In : **ALVIM, P.de T., KOZLOWSKI, T.T.** (ed.) : Ecophysiology of Tropical Crops. Pp. 409 - 439. Academic Press, New York - San Francisco - London 1977. [Dry-matter production.]

31497 - **REYNOLDS, J.A., STOECKENIUS, W.** : Molecular weight of bacteriorhodopsin solubilized in Triton X-100. - Proc. nat. Acad. Sci. USA *74* : 2803 - 2804, 1977.

31498 - **REYNOLDS, K.C., EDWARDS, K.** : A short-focus telescope for ground cover estimation. - Ecology *58* : 939 - 941, 1977. [Point quadrat sampling.]

31499 - **RHEE, C., BRIGGS, W.R.** : Some reponses of *Chondrus crispus* to light. I. Pigmentation changes in the natural habitat. - Bot. Gaz. *138* : 123 - 128, 1977.

31500 - **RHOADES, F.M.** : Growth rates of the lichen *Lobaria oregana* as determined from
sequential photographs. - Can. J. Bot. *55* : 2226 - 2233, 1977.

31501 - **RICHARDSON, S.G., SALISBURY, F.B.** : Plant responses to the light penetrating
snow. - Ecology *58* : 1152 - 1158, 1977. [Photometer.]

31502 - **RICKETTS, T.R.** : Changes in average cell concentrations of various constitu-
ents during synchronous division of *Platymonas striata* BUTCHER (*Prasinophy-
ceae*).- J. exp. Bot. *28* : 1278 - 1288, 1977.

31403 - **RIDLEY, S.M.** : Interaction of chloroplasts with inhibitors. Induction of chlorosis by diuron during prolonged illumination *in vitro*. - Plant Physiol. *59* :
724 - 732, 1977.

31504 - RIDLEY, S.M., RIDLEY, J. : The inhibition of carotenoid synthesis and its
effects on the conversion of etioplasts to chloroplasts. - In : COOMBS, J.
(ed.) : 4th International Congress on Photosynthesis. P. 315. UKISES, London
1977.

31505 - RIED, A., HESSENBERG, B., METZLER, H., ZIEGLER, R. : Distribution of excitation energy among Photosystem I and Photosystem II in red algae. I. Action
spectra of light reactions I and II. - Biochim. biophys. Acta 459 : 175 -
186, 1977.

31606 - RIED, A., REINHARDT, B. : Distribution of excitation energy between Photosystem I and Photosystem II in red algae. II. Kinetics of the transition between
state 1 and state 2. - Biochim. biophys. Acta 460 : 25 - 35, 1977.

31507 - RIED, A., REINHARDT, B. : Energy requirement of the induction of a transition
between state 2 and state 1 in red algae. - In : COOMBS, J. (ed.) : 4th International Congress on Photosynthesis. P. 316. UKISES, London 1977.

31508 - RIEMENSCHNEIDER, V.L., GILBERT, G.E. : Interception of solar radiation by
three deciduous forest communities in Neotoma, a valley in Southcentral Ohio.
- Ohio J. Sci. 77 : 231 - 235, 1977.

31509 - RIES, S.K., WERT, V. : Growth responses of rice seedlings to triacontanol in
light and dark. - Planta 135 : 77 - 82, 1977. [Growth analysis.]

31510 - RIGAU, M.C. : Relationship between production characteristics and chlorophyll
a content in Hedysarum coronarium L., Dyctylis glomerata L., and Festuca ela-
tior ssp. arundinacea (SCHREB.) HACK. - Photosynthetica 11 : 88 - 89, 1977.

31511 - RIJGERSBERG, C.P., AMESZ, J. : Fluorescence and absorbance changes in chloroplasts between 90 and 5 K. - In : COOMBS, J. (ed.) : 4th International Congress on Photosynthesis. Pp. 316 - 317. UKISES, London 1977.

31512 - ROARK, B., QUISENBERRY, J.E. : Environmental and genetic components of stomatal behavior in two genotypes of upland cotton. - Plant Physiol. 59 : 354 -
356, 1977. [Stomatal resistance.]

31513 - ROBARTS, R.D., SOUTHALL, G.C. : Nutrient limitation of phytoplankton growth
in seven tropical man-made lakes, with special reference to Lake McIlwaine,
Rhodesia. - Arch. Hydrobiol. 79 : 1 - 35, 1977.

31514 - ROBBINS, J.V. : The effects of total carbon supply, irradiance pH, temperature,
and salinity on short term photosynthesis of Palmaria palmata (L.) GREV. -
J. Phycol. 13 (Suppl.) : 58, 1977.

31515 - ROBELIN, M., MAUGET, J.C. : Comportement photosynthétique du Noyer (Juglans
regia). Premières observations sur un couvert continu obtenu à partir d'un
semis haute densité. - Ann. agron. 28 : 583 - 597, 1977.

31516 - ROBERTS, J. : The use of tree-cutting techniques in the study of the water
relations of mature Pinus sylvestris L. I. The technique and survey of the
results. - J. exp. Bot. 28 : 751 - 767, 1977. [Stomatal resistance.]

31517 - ROBERTS, S.W., MILLER, P.C. : Interception of solar radiation as affected by
canopy organization in two mediterranean shrubs. - Oecol. Plant. 12 : 273 -
290, 1977.

31518 - ROBINSON, H.H., YOCUM, C.F. : Photogeneration of reduced quinone catalysts of
photosystem I cyclic photophosphorylation. - Plant Physiol. 59 (6, Suppl.) :
23, 1977.

31519 - ROBINSON, J.M., GIBBS, M. : Glutathione reductase in spinach chloroplasts. -
Plant Physiol. 59 (6, Suppl.) : 129, 1977. [Ps.]

31520 - ROBINSON, J.M., GIBBS, M., COTLER, D.N. : Influence of pH upon the Warburg
effect in isolated intact spinach chloroplasts 1. Carbon dioxide photoassimilation and glycolate synthesis. - Plant Physiol. 59 : 530 - 534, 1977.

31521 - ROBINSON, S.J., YOCUM, C.F. : Site-specific electron transport reactions in
intact cells of a blue-green alga. - Plant Physiol. 59 (6, Suppl.) : 130,
1977.

31522 - ROBINSON, S.J., YOCUM, C.F., IKUMA, H., HAYASHI, F. : Inhibition of chloroplast electron transport reactions by trifluralin and diallate. - Plant Physiol. *60* : 840 - 844, 1977.

31523 - ROBINSON, S.P., WISKICH, J.T. : Uptake of ATP analogs by isolated pea chloroplasts and their effect on CO_2 fixation and electron transport. - Biochim. biophys. Acta *461* : 131 - 140, 1977.

31524 - ROBINSON, S.P., WISKICH, J.T. : p-chloromercuriphenyl sulphonic acid as a specific inhibitor of the phosphate transporter in isolated chloroplasts. - FEBS Lett. *78*: 203 - 206, 1977.

31525 - ROBINSON, S.P., WISKICH, J.T. : Pyrophosphate inhibition of carbon dioxide fixation in isolated pea chloroplasts by uptake in exchange for endogenous adenine nucleotides. - Plant Physiol. *59* : 422 - 427, 1977.

31526 - ROBINSON, S.P., WISKICH, J.T. : Inhibition of CO_2 fixation by adenosine 5'-diphosphate and the role of phosphate transport in isolated pea chloroplasts. - Arch. Biochem. Biophys. *184* : 546 - 554, 1977.

31527 - ROBSON, M., PARSONS, A.J. : Effect of nitrogen deficiency on photosynthesis, respiration and dry matter production of a small closed community of ryegrass. - In : COOMBS, J. (ed.) : 4th International Congress on Photosynthesis. Pp. 317 - 318. UKISES, London 1977.

31528 - RODSKJER, N., DAHLSTEDT, L. : Measurements of solar radiation in winter wheat. - Swed. J. agr. Res. *7* : 85 - 88, 1977.

31529 - ROGERS, L.J., FITZGERALD, M.P., HUSAIN, A., HUTBER, G.N. : Properties of the flavodoxins from a cyanobacterium and a red alga. - In : COOMBS, J. (ed.) : 4th International Congress on Photosynthesis. P. 319. UKISES, London 1977.

31530 - ROGERS, R.A., DUNN, J.H., NELSON, C.J. : Photosynthesis and cold hardening in zoysia and bermudagrass. - Crop Sci. *17* : 727 - 732, 1977.

31531 - ROMANENKO, T.I., IZMEST'EVA, L.R. : Soderzhanie khlorofilla i feopigmentov v vodakh Ust'-Ilimskogo vodokhranilishcha v avguste 1976 g. [Contents of chlorophyll and phaeopigments in waters of the Ust'-Ilimsk reservoir in August 1976.] - In : Biologicheskie Issledovaniya Vodoemov Vostochnoĭ Sibiri. Pp. 11 - 13. Irkutsk 1977. [In R.]

31532 - ROMANENKO, V.I., PERES EĬRIS, M. : K opredeleniyu chislennosti zhivykh vodorosleĭ s pomoshch'yu 14C. [Determination of viable algae numbers using 14C.] - Gidrobiol. Zh. *13* (6) : 87 - 90, 1977. [In R.]

31533 - ROMANOVA, E.Yu. : Tsiklicheskoe fotofosforilirovanie v khloroplastakh yachmenya v zavisimosti ot urovnya azotnogo pitaniya. [Cyclic photophosphorylation in barley chloroplasts depending on the level of nitrogen nutrition.] - Fiziol. Biokhim. kul't. Rast. *9* : 473 - 478, 1977. [In R, ab : E.]

31534 - ROMIJN, J.C., AMESZ, J. : Purification and photochemical properties of reaction centers of *Chromatium vinosum*. Evidence for the photoreduction of a naphthoquinone. - Biochem. biophys. Acta *461* : 327 - 338, 1977.

31535 - ROMNEY, E.M., WALLACE, A., KAAZ, H., HALE, V.Q., CHILDRESS, J.D. : Effect of shrubs on redistribution of mineral nutrients in zones near roots in the Mojave desert. - In : MARSHALL, J.K. (ed.) : The Belowground Ecosystem: A Synthesis of Plant-Associated Processes. Pp. 303 - 310. Colorado State Univ., Fort Collins 1977. [Leaf:plant ratio.]

31536 - ROOK, D.A., SWANSON, R.H., CRANSWICK, A.M. : Reaction of radiata pine to drought. - N. Zeal. Dep. Sci. Ind. Res. Inform. Ser. *126* (Proc. Soil Plant Water Symp.) : 55 - 68, 1977. [Ps, resistances.]

31537 - ROSE, C.W., MILLINGTON, R.J. : Mechanisms and models of plant growth. - In : RUSSELL, J.S., GREACEN, E.L. (ed.) : Soil Factors in Crop Production in a Semi-arid Environment. Pp. 224 - 240. University of Queensland Press, St. Lucia 1977. [Ps.]

*31538 - ROSENBERG, A. : Light-dependent stoichometry of galactosyl diglyceride and chlorophyll accretion during light-induced chloroplast membrane synthesis in *Euglena*. - Biochem. biophys. Res. Commun. *73* : 972 - 977, 1976.

31539 - ROSENTHAL, W.D., KANEMASU, E.T., RANEY, R.J., STONE, L.R. : Evaluation of an evapotranspiration model for corn. - Agron. J. *69* : 461 - 464, 1977. [Dry-matter accumulation.]

31540 - ROSENTRETER, R., AHMADJIAN, V. : Effect of ozone on the lichen *Cladonia arbuscula* and the *Trebouxia* phycobiont of *Cladina stellaris*. - Bryologist *80* : 600 - 605, 1977. [Chl.]

31541 - ROSHCHINA, V.D., ROSHCHINA, V.V. : Nekotorye aspekty mekhanizmov dvizheniya khloroplastov v toke tsitoplazmy. [Some aspects of mechanisms involved in the movement of chloroplasts in the flow of cytoplasm.] - Fiziol. Rast. *24* : 261 - 266, 1977. [In R, ab : E.]

31542 - ROSHCHINA, V.V., AKULOVA, E.A. : Light-induced absorption changes at 590 nm in pea etio-chloroplasts associated with plastocyanin. - Plant Sci. Lett. *9* : 253 - 258, 1977.

31543 - ROSINSKI, J., REED, R.L., KEE, S.-C. : Immunological characterization of protochlorophyllide holochrome from *Phaseolus*. - Plant Physiol. *59* (6, Suppl.): 8, 1977.

31544 - ROSKO, J.J., RACHLIN, J.W. : The effect of cadmium, copper, mercury, zinc and lead on cell division, growth, and chlorophyll *a* content of the chlorophyte *Chlorella vulgaris*. - Bull. Torrey bot. Club *104* : 226 - 233, 1977.

31545 - ROSNER, A., REISFELD, A., JAKOB, K.M., GRESSEL, J., EDELMAN, M. : Shifts in the RNA & protein metabolisms of *Spirodela* (duckweed). - In : BOGORAD, L., WEIL, J.H. (ed.) : Acides Nucléiques et Synthèse des Protéines chez les Végétaux. Coll. Int. CNRS No. 261. Pp. 561 - 568. Edit. CNRS, Paris 1977. [Chloroplast.]

31546 - ROSS, J. : Radiation conditions in the plant stand. - In : UNGER, K. (ed.) : Biophysikalische Analyse pflanzlicher Systeme. Pp. 115 - 119. VEB Gustav Fischer Verlag, Jena 1977.

31547 - ROSS, Yu.K. : Svetovoľ faktor produktivnosti. [Light factor in productivity.] - Jtogi Nauki Tekh., Ser. Fiziol. Rast. *3* : 55 - 89, 1977. [In R.]

31548 - ROTT, R., AVI-DOR, Y. : Effect of ionophoric compounds on aqueous suspensions of purple membrane. - FEBS Lett. *81* : 267 - 270, 1977.

31549 - ROTTENBERG, H. : Proton and ion transport across the thylakoid membranes. - In : TREBST, A., AVRON, M. (ed.) : Photosynthesis I. (Encycl. Plant Physiol. N.S. Vol.5.) Pp. 338 - 349. Springer-Verlag, Berlin - Heidelberg - New York 1977.

31550 - ROUHANI, I., KHOSH-KHUI, M. : Variations in photosynthetic rates of fourteen *Coleus* cultivars. - Plant Physiol. *59* : 114 - 115, 1977.

31551 - ROUX, E., FALUDI-DÁNIEL, Á. : Flash-induced 515 nm absorbance change in chloroplasts with various granum contents. - Biochim. biophys. Acta *461* : 25 - 30, 1977.

31552 - ROY, H., TERENNA, B., CHEONG, L.C. : Small subunit RUBPCase made on soluble pea polysomes. - Plant Physiol. *59* (6, Suppl.) : 9, 1977.

31553 - ROY, H., TERENNA, B., CHEONG, L.C. : Synthesis of the small subunit of ribulose-1,5-bisphosphate carboxylase by soluble fraction polyribosomes of pea leaves. - Plant Physiol. *60* : 532 - 537, 1977.

31554 - ROZHKO, I.I. : Ob izmenenii fotosinteticheskogo metabolizma pri ingibirovanii netsiklicheskogo fotofosforilirovaniya. [Change in photosynthetic metabolism with inhibition of non-cyclic photophosphorylation.] - Fiziol. Biokhim. kul't. Rast. *9* : 517 - 519, 1977. [In R, ab : E.]

*31555 - ROZOVA, T.L. : Avtokhtonnoe organicheskoe veshchestvo v vodoemakh-okhladitelyakh GRÉS No. 3. [Autochtonous organic matter in water reservoirs cooling GRES No.3.] - In : Tipologiya Ozernogo Nakopleniya Organicheskogo Veshchestva. Pp. 90 - 107. Nauka, Moskva 1976. [Ps; in R.]

31556 - RUBIN, A.B. : On the mechanism for primary change separation in the reaction centres of photosynthetic organisms. - In : COOMBS, J. (ed.) : 4th International Congress on Photosynthesis. P. 318. UKISES, London 1977.

31557 - RUBIN, B.A., SHEVYRËVA, V.V., MERZLYAK, M.N., VORONKOV, L.A. : Vysshie zhir-
         nye kisloty khloroplastov khlopchatnika pri zabolevanii vertitsilleznym vil-
         tom. [Higher fatty acids of cotton chloroplasts during infection with *Verti-
         cillium dahliae.*] - Fiziol. Rast. *24* : 1060 - 1066, 1977. [In R, ab : E.]

31558 - RUDICH, J., KALMAR, D., GEIZENBERG, C., HAREL, S. : Low water tensions in de-
         fined stages of processing tomato plants and their effects on yield and qua-
         lity. - J. hort. Sci. *52* : 391 - 399, 1977.

31559 - RUDOLPH, H., KABSCH, U., SCHMIDT-STOHN, G. : Änderungen des Chloroplastenpig-
         ment-Spiegels bei *Sphagnum magellanicum* im Verlauf der Synthese von sphagno-
         rubin und anderer membranochromer Pigmente. - Z. Pflanzenphysiol. *82* : 107 -
         116, 1977.

31560 - RUMBERG, B. : Field changes. - In : TREBST, A., AVRON, M. (ed.) : Photosyn-
         thesis I. (Encycl. Plant Physiol. N.S. Vol.5.) Pp. 405 - 415. Springer-Verlag,
         Berlin - Heidelberg - New York 1977. [Ps.]

31561 - RUMBERG, B., BUCHHOLZ, J., SCHNECKE, W. : Determination of the ATPase-coupled
         proton to ATP stoichiometry in isolated chloroplasts. - In : COOMBS, J. (ed.)
         : 4th International Congress on Photosynthesis. P. 320. UKISES, London 1977.

31562 - RUMBERG, B., RATHENOW, M. : Investigation of the proton to electron ratio in
         isolated chloroplasts. - In : COOMBS, J. (ed.) : 4th International Congress
         on Photosynthesis. Pp. 320 - 321. UKISES, London 1977.

31563 - RUMI, C.P., CARPINETTI, R.M. : Effect of sunlight on the development of *Tro-
         paeolum majus* : 1. Leaf growth. - Fyton *35* : 137 - 143, 1977.

31564 - RURAINSKI, H.J., MADER, G. : Regulation of the Hill reaction by cations and
         its abolishment by uncouplers. - Biochim. biophys. Acta *461* : 489 - 499,
         1977.

31565 - RUUGE, Ė.K., SUBCHINSKI, V.K., TIKHONOV, A.N. : Issledovanie ėlektronnogo
         transporta v fotosinteticheskikh sistemakh vysshikh rasteniĭ metodom ĖPR. V.
         Vzaimodeĭstvie paramagnitnogo zonda I(12,3) s membranami khloroplastov bo-
         bov. [Investigation of electron transport in photosynthetic systems of higher
         plants by the EPR method. V. Interaction of paramagnetic probe I(12,3) with
         the membranes of broad bean chloroplasts.] - Biofizika *22* : 840 - 845, 1977.
         [In R, ab : E.]

31566 - RUUGE, Ė.K., SUBCHINSKI, V.K., TIKHONOV, A.N. : Issledovanie struktury mem-
         bran khloroplastov vysshikh rasteniĭ pri pomoshchi paramagnitnykh zondov.
         [The structure of higher plant chloroplasts membranes as studied by paramag-
         netic probes.] - Mol. Biol. (Moskva) *11* : 646 - 655, 1977. [In R, ab : E.]

31567 - RYBIN, I.A., OBOLONSKIĬ, V.V., MIKHEEVA, S.A. : O dvukh komponentakh sveto-
         indutsirovannoĭ ėlektroreaktsii list'ev vysshikh rasteniĭ. [Two components
         of the light-induced electroreaction in the leaves of higher plants.] - Fizi-
         ol. Rast. *24* : 1165 - 1173, 1977. [In R, ab : E.]

31568 - RYBKINA, G.V., BIGLOVA, S.G., LEBEDEVA, L.A. : Sravnitel'noe prizhiznennoe
         izuchenie vozrastnykh osobennosteĭ vodoobmena kletok i khloroplastov. [Com-
         parative determinations of ontogenetic changes in water regime in cells and
         chloroplasts *in vivo*.] - Uchen. Zap. kazan. gos. ped. Inst. *180* (Faktory Sre-
         dy i Rastenie) : 69 - 77, 1977. [In R.]

31569 - RYBKINA, G.V., LEBEDEVA, L.A., BIGLOVA, S.G. : K voprosu o prizhiznennom izu-
         chenii vodnogo rezhima kletok i khloroplastov s pomoshch'yu mikrofotografii.
         [Determination of water regime in cells and chloroplasts *in vivo* using micro-
         photographs.] - Uchen. Zap. kazan. gos. ped. Inst. *180* (Faktory Sredy i Raste-
         nie) : 59 - 68, 1977. [In R.]

31570 - RYLE, G.J.A., POWELL, C.E., GORDON, A.J. : Respiration and the energy costs
         of symbiotic nitrogen fixation in grain and herbage legumes. - In : COOMBS, J.
         (ed.) : 4th International Congress on Photosynthesis. Pp. 321 - 322. UKISES,
         London 1977. [Ps.]

31571 - RYLSKA, T. : The mechanism of primary processes of photosynthesis photosensi-
         tization. 1. The hypothesis of biphotonic excitation of pigment molecules. -
         In : COOMBS, J. (ed.) : 4th International Congress on Photosynthesis. Pp. 322
         - 323. UKISES, London 1977.

31572 - **RYMAN, R.G., POWELL, C.E., RYLE, G.J.A.** : A comparison of three methods of measuring ^{14}C incorporated in plant material. - Int. J. appl. Radiat. Isotop. *28* : 436 - 439, 1977.

31573 - **SADEWASSER, D.A., DILLEY, R.A.** : The role of plastoquinone in photosynthetic water oxidation. - Plant Physiol. *59* (6,Suppl.) : 90, 1977.

31574 - **SADOWSKI, R., SYKUT, A.** : Effect of gibberellic acid GA_3 on the content of carotenoids in seedlings of *Phaseolus vulgaris* L. - Bull. Acad. pol. Sci., Sér. Sci. biol. *25* : 281 - 285, 1977.

31575 - **SAGROMSKY, H.** : Untersuchungen zur physiologischen Bedeutung von Chlorophyll *b*, eine Literaturübersicht. - Kulturpflanze *25* : 279 - 296, 1977.

31576 - **SAHL, H.G., TRÜPER, H.G.** : Enzymes of CO_2 fixation in *Chromatiaceae*. - FEMS Microbiol. Lett. *2* : 129 - 132, 1977.

31577 - **SAJI, T., BARD, A.J.** : Electrogenerated chemiluminescence. 29. The electrochemistry and chemiluminescence of chlorophyll *a* in *N,N*-dimethylformamide solutions. - J. amer. chem. Soc. *99* : 2235 - 2240, 1977.

31578 - **SAKA, H.** : [Changes in RuDP carboxylase activity of rice leaf blade at various growing stages.] - Jap. J. Crop Sci. *46* : 164 - 170, 1977. [In Jap., ab : E.]

31579 - **SAKHAROVA, O.V., KORZH, B.V.** : Pigmenty inbrednykh i gibridnykh form kukuruzy i ikh svyaz' s protsessom rosta i urozhaĭnost'yu. [Pigments of inbred and hybrid forms of maize and their relation to process of growth and yielding capacity.] - In : NASYROV, Yu.S. (ed.) : Genetika Fotosinteza. Pp. 250 - 256. Donish, Dushanbe 1977. [In R.]

31580 - **SAKURAI, H.** : A continuous-flow, density-gradient electrophoresis apparatus, and its application to concentration of proteins. - Plant Cell Physiol. *1977* (Spec. Issue 3 - Photosynthetic Organelles. Structure and Function) : 179 - 186, 1977. [ATPase.]

31581 - **SALAMA, F.M.** : Photosynthesis and formation of the native chlorophyll forms in light harvesting complexes of wheat chloroplasts under physiological drought (salinity). - In : COOMBS, J. (ed.) : 4th International Congress on Photosynthesis. P. 323a. UKISES, London 1977.

31582 - **SALAMON, Z.** : Influence of DCMU on slow fluorescence yield changes observed in synchronous cultures of *Chlorella pyrenoidosa*. - Bull. Acad. pol. Sci., Sér. Sci. biol. *25* : 653 - 658, 1977.

31583 - **SALARES, V.R., YOUNG, N.M., CAREY, P.R., BERNSTEIN, H.J.** : Excited state (exciton) interactions in polyene aggregates. Resonance Raman and absorption spectroscopic evidence. - J. Raman Spectrosc. *6* : 282 - 288, 1977. [Car.]

31584 - **SALATENKO, V.N.** : Fotosintez i rol' list'ev raznykh yarusov v formirovanii urozhaya i kachestva semyan kleshcheviny v posevakh. [Photosynthesis and role of leaves of different stages in formation of yield and quality of castor-oil plant seeds in crops.] - Fiziol. Biokhim. kul't. Rast. *9* : 86 - 92, 1977. [In R, ab : E.]

31585 - **SALATENKO, V.N.** : Fotosintez i produktivnost' kleshcheviny pri oroshenii. [Photosynthesis and productivity of castor-oil plant with irrigation.] - Fiziol. Biokhim. kul't. Rast. *9* : 424 - 431, 1977. [In R, ab : E.]

31586 - **SALE, P.J.M.** : Net carbon exchange rates of field-grown crops in relation to irradiance and dry weight accumulation. - Aust. J. Plant Physiol. *4* : 555 - 569, 1977.

31587 - **SALIH, F.A.** : Influence of plant age on the rates of photosynthesis and respiration, and chlorophyll content of four isogenic lines of barley (*Hordeum vulgare* L.). - Photosynthetica *11* : 207 - 210, 1977.

31588 - **SALIN, M.L., HOMANN, P.H.** : Photosystem I enriched subchloroplast particles from granaless chloroplasts. - Photosynthetica *11* : 5 - 10, 1977.

31589 - **SALISBURY, J.L., FLOYD, G.L.** : Molecular, enzymatic and ultrastructural characterization of the pyrenoid of *Micromonas squamata*, a scaly green monad. - Plant Physiol. *59* (6,Suppl.) : 111, 1977.

31590 - SALLAL, A-K.J., CODD, G.A. : An immunological and electrophoretic comparison of the thylakoids from vegetative cells and heterocysts of *Anabaena cylindrica*. - Brit. phycol. J. *12* : 163 - 169, 1977.

31591 - SALZER, J., BÜTTNER, R. : Untersuchungen über die Nettophotosynthese bei Apfelsorten 2. Mitt. Komponenten der Primärproduktion bei 11 Apfelsorten. - Arch. Züchtungsforsch. *7* : 159 - 168, 1977.

31592 - SAMBO, E.Y., MOORBY, J., MILTHORPE, F.L. : Photosynthesis and respiration of developing soybean pods. - Aust. J. Plant Physiol. *4* : 713 - 721, 1977.

31593 - SAMIEV, Kh.S., MARFINA, K.G. : Aktivnost' RNKazy khloroplastov khlopchatnika pri vodnom defitsite. [Activity of chloroplast RNase in cotton under water deficit.] - Biokhimiya *42* : 1361 - 1365, 1977. [In R, ab : E.]

31594 - SAMOSHINA, N.M., NOVIKOVA, N.A., NEKPASOV, L.I. : Spektral'nye svoĭstva khlorofilla,vklyuchennogo v poliakrilamidnyĭ gel'. [Spectral properties of chlorophyll embedded in polyacrylamide gel.] - Vest. mosk. Univ., Ser. 2 - Khimiya *18* (2) : 157 - 160, 1977. [In R, ab : E.]

31595 - SAMPSON, E.J., JONES, B.M.G. : The productivity of *Salix glauca* L. in Arctic Norway. - Ann. Bot. *41* : 155 - 161, 1977.

31596 - SAMSUDDIN, Z., IMPENS, I. : Relationship between photosynthetic rates and latex production in some *Hevea brasiliensis* MUELL. ARG. cultivars. - In : COOMBS, J. (ed.) : 4th International Congress on Photosynthesis. P. 323. UKISES, London 1977.

31597 - SAMUELSSON, G., ÖQUIST, G. : A method for studying photosynthetic capacities of unicellular algae based on *in vivo* chlorophyll fluorescence. - Physiol. Plant. *40* : 315 - 319, 1977.

31598 - SAMUILOV, V.D., REMENNIKOV, V.G. : Photooxidase activity of *Rhodospirillum rubrum* chromatophores. - In : COOMBS, J. (ed.) : 4th International Congress on Photosynthesis. P. 324. UKISES, London 1977.

31599 - SANADZE, G.A., DZOTSENIDZE, Ts.G., TARKHNISHVILI, G.M., PKHACHIASHVILI, S.Sh.: Izmeneniya udel'noĭ aktivnosti $^{14}CO_2$ i izoprena na kompensatsionnom punkte. [Changes in specific activity of $^{14}CO_2$ and isoprene on the compensation point.] - Dokl. Akad. Nauk SSSR *233* : 1226 - 1227, 1977. [In R.]

31600 - SANADZE, G.A., TEVZADZE, I.T., TARKHNISHVILI, G.M. : Izmenenie izotopnogo otnosheniya $C^{13}O_2/C^{12}O_2$ v atmosfere zamknutykh kamer pri fotosinteze. [A change in the $^{13}CO_2/^{12}CO_2$ isotope ratio in the atmosphere of closed chambers during photosynthesis.] - Fiziol. Rast. *24* : 646 - 647, 1977. [In R.]

31601 - SANAI, S., NAKAYAMA, M., OTA, Y. : [Plant growth-regulating activities of nicotinamide. I. Effect of nicotinamide on growth of rice seedlings.] - Jap. J. Crop Sci. *46* : 1 - 7, 1977. [Chl; in Jap., ab : E.]

31602 - SANAI, S., OTA, Y. : [Plant growth-regulating activities of nicotinamide. III. Effects of nicotinamide on the activities of nitrate reduction and photosynthesis.] - Jap. J. Crop Sci. *46* : 212 - 218, 1977. [In Jap., ab : E.]

31603 - SÁNCHEZ, C.S., DURAND-CHASTEL, H. : The utilization of *Spirulina* algae for industrial photosynthesis. - J. Phycol. *13* (Suppl.) : 60, 1977.

31604 - SANCHEZ, S.M. : The fine structure of the guard cells of *Helianthus annuus*. - Amer. J. Bot. *64* : 814 - 824, 1977. [Chloroplast.]

31605 - SANDERS, J.K.M. : N.M.R. spectral change as a probe of chlorophyll chemistry. - Chem. Soc. Rev. *6* : 467 - 487, 1977.

31606 - SANDERS, T.H., ASHLEY, D.A., BROWN, R.H. : Effects of partial defoliation on petiole phloem area, photosynthesis, and ^{14}C translocation in developing soybean leaves. - Crop Sci. *17* : 548 - 550, 1977.

31607 - SANDHU, B.S., HORTON, M.L. : Response of oats to water deficit. I. Physiological characteristics. - Agron. J. *69* : 357 - 360, 1977. [Ps.]

31608 - **SANDHU, B.S., HORTON, M.L.** : Response of oats to water deficit. II. Growth and yield characteristics. - Agron. J. *69* : 361 - 364, 1977. [Production.]

31609 - **SAND-JENSEN, K.** : Effect of epiphytes on eelgrass photosynthesis. - Aquat. Bot. *3* : 55 - 63, 1977.

31610 - **SANE, P.V.** : The topography of the thylakoid membrane of the chloroplast. - In : TREBST, A., AVRON, M. (ed.) : Photosynthesis I. (Encycl. Plant Physiol. N.S. Vol. 5.) Pp. 522 - 542. Springer-Verlag, Berlin - Heidelberg - New York 1977.

31611 - **SANE, P.V., BHAGWAT, A.S.** : Regulation of CO_2 fixation in maize (*Zea mays*). - In : COOMBS, J. (ed.) : 4th International Congress on Photosynthesis. P. 325. UKISES, London 1977.

31612 - **SANE, P.V., DESAI, T.S., TATAKE, V.G., GOVINDJEE** : On the origin of glow peaks in *Euglena* cells, spinach chloroplasts and subchloroplast fragments enriched in system I or II. - Photochem. Photobiol. *26* : 33 - 39, 1977.

31613 - **SAN JOSÉ, J.J.** : Gas exchange in *Paspalum repens* BERG. - Rev. Brasil. Biol. *37* : 525 - 533, 1977.

31614 - **SAN JOSÉ, J.J.** : Potencial hidrico e intercambio gaseoso de *Curatella americana* L. en la temporada seca de la sabana Trachypogon. [Water potential and gas exchange in *Curatella americana* L. during dry season in Trachypogon plain.] - Acta cient. Venez. *28* : 373 - 379, 1977. [Ps; in Span., ab : E.]

31615 - **SANKAWA, U.** : [Biosynthetic studies using ^{13}C as a tracer.] - Seikagaku *49* : 1238 - 1243, 1977. [Porphyrins; in Jap.]

31616 - **SANTARIUS, K.A., MILDE, H.** : Sugar compartmentation in frost-hardy and partially dehardened cabbage leaf cells. - Planta *136* : 163 - 166, 1977.

31617 - **SAPHON, S., CROFTS, A.R.** : On the release of protons associated with the photo-oxidation of water in spinach chloroplasts. - In : COOMBS, J. (ed.) : 4th International Congress on Photosynthesis. Pp. 325 - 326. UKISES, London 1977.

31618 - **SAPUNOV, V.V., TSVIRKO, M.P., SOLOV'EV, K.N.** : Tushenie tripletnykh sostoyaniĭ Fe(II)- i Fe(III)-porfirinami. [Quenching of triplet states with Fe(II)- and Fe(III)-porphyrines.] - Biofizika *22* : 766 - 770, 1977. [Chl; in R, ab : E.]

31619 - **SARIĆ, M.R.** : Metabolic activity of leaf parts of different colour in *Hedera canariensis* "Glorie de Marengo". - In : COOMBS, J. (ed.) : 4th International Congress on Photosynthesis. P. 369. UKISES, London 1977.

31620 - **SARIĆ, M.R., JELENIĆ, D., KEREČKI, B., KRSTIĆ, B.** : Effect of light of different wavelengths on total amino acid content in some plant species. - In : COOMBS, J. (ed.) : 4th International Congress on Photosynthesis. Pp. 326 - 327. UKISES, London 1977.

31621 - **SÁRVÁRI, É., LÁNG, F.** : Lincocin effect on the development of photosynthetic lamellae in bean (*Phaseolus vulgaris* L.). - In : COOMBS, J. (ed.) : 4th International Congress on Photosynthesis. P. 328. UKISES, London 1977.

31622 - **SARZHEVSKAYA, M.V., LOSEV, A.P.** : Vydelenie i identifikatsiya protokhlorofillov, poluchennykh okisleniem sootvetstvuyushchikh khlorofillov. [Isolation and identification of protochlorophylls obtained under oxidation of corresponding chlorophylls.] - Biokhimiya *42* : 2105 - 2109, 1977. [In R, ab : E.]

31623 - **SATOH, H., AIGA, I., OMURA, T.** : [Studies on character manifestation in chlorophyll mutants of rice II. Xantha mutant sensitive to low temperature.] - Sci. Bull. Fac. Agr. Kyushu Univ. *31* : 189 - 193, 1977. [In Jap., ab : E.]

31624 - **SATOH, J.** : [A denitrifying photosynthetic phototrophic bacterium.] - Kagaku to Seibutsu *15* : 498 - 499, 1977. [In Jap.]

31625 - **SATOH, K., BUTLER, W.L.** : Purification of a Photosystem II pigment complex. - In : COOMBS, J. (ed.) : 4th International Congress on Photosynthesis. P. 327. UKISES, London 1977.

31626 - **SATOH, K., YAMAGISHI, A., KATOH, S.** : Fluorescence induction in chloroplasts isolated from the green alga, *Bryopsis maxima*. II. Induction of cytochrome *f* photooxidation and the DPS transient in fluorescence induction. - Plant Cell Physiol. *1977* (Spec. Issue 3 - Photosynthetic Organelles. Structure and Function) : 75 - 86, 1977.

31627 - **SATOH, M., KRIEDEMANN, P.E., LOVEYS, B.R.** : Changes in photosynthetic activity and related processes following decapitation in mulberry trees. - Physiol. Plant. *41* : 203 - 210, 1977.

31628 - **SATOH, M., OHYAMA, K.** :[Studies on photosynthesis and translocation of photosynthate in mulberry tree. VI. Changes in amylase activity and in amount of carbohydrates in a storage organ after shoot pruding.]- Jap. J. Crop Sci. *46* : 499 - 503, 1977. [In Jap., ab : E.]

31629 - **SAUER, K., ACKER, S., MATHIS, P., VAN BEST, J.** : Optical studies of the back-reaction of $P-700^+$ with its electron acceptors in Photosystem-1 particles. Evidence for multiple acceptors. - In : **COOMBS, J.** (ed.) : 4th International Congress on Photosynthesis. Pp. 328 - 329. UKISES, London 1977.

31630 - **SAUER, K., ACKER, S., MATHIS, P., VAN BEST, J.A.** : Optical studies of photosystem I particles: Evidence for the presence of multiple electron acceptors. - In : **PACKER, L., PAPAGEORGIOU, G.C., TREBST, A.** (ed.) : Bioenergetics of Membranes. Pp. 351 - 359. Elsevier/North-Holland Biomedical Press, Amsterdam - Oxford - New York 1977.

31631 - **SAUER, K., BREWINGTON, G.T.** : Fluorescence lifetimes in chloroplasts, subchloroplast particles and *Chlorella* using single photon counting. - In : **COOMBS, J.** (ed.) : 4th International Congress on Photosynthesis. Pp. 329 - 330. UKISES, London 1977.

31632 - **SAUER, K., BREWINGTON, G.T.** : Fluorescence lifetimes of chloroplasts, subchloroplast particles and *Chlorella* using single photon counting. - Rep. LBL-6179: 1 - 13, 1977.

31633 - **SAUGIER, B.** : Micrometeorology on crops and grasslands. - In : **LANDSBERG, J. J., CUTTING, C.V.** (ed.) : Environmental Effects on Crop Physiology. Pp. 39 - 55. Academic Press, London - New York - San Francisco 1977. [Ps.]

31634 - **SAUTER, J.J.** : Electron microscopical localization of adenosine triphosphatase and β-glycerophosphatase in sieve cells of *Pinus nigra* var. *austriaca* (HOESS) BADOUX. - Z. Pflanzenphysiol. *81* : 438 - 458, 1977.

31635 - **SAWYER, D.T., BODINI, M.E., WILLIS, L.A., RIECHEL, T.L., MAGERS, K.D.** : Electrochemical and spectroscopic studies of manganese (II, III, IV) complexes as models for the photosynthetic oxygen-evolution reaction. - Adv. Chem. Ser. *162* (Bioinorg. Chem-2, Symp.) : 330 - 349, 1977.

31636 - **SAYRE, R.T., KENNEDY, R.A.** : Ecotypic differences in the C_3 and C_4 photosynthetic activity in *Mollugo verticillata*, a C_3-C_4 intermediate. - Planta *134* : 257 - 262, 1977.

31637 - **SAYRE, R.T., KENNEDY, R.A.** : Comparative photosynthetic carbon metabolism of four populations of *Mollugo verticillata*, a C_3-C_4 intermediate. - Plant Physiol. *59* (6, Suppl.) : 43, 1977.

31638 - **SCHAAFSMA, T.J., VAN DER BENT, S.J., KLEIBEUKER, J.F., KOOYMAN, R.P.H., GOEDHEER, J.C.** : Triplet states in algae. - In : **OLSON, J.M., HIND, G.** (ed.) : Chlorophyll-Proteins, Reaction Centers, and Photosynthetic Membranes. P. 369. Brookhaven nat. Lab., Upton 1977.

31639 - **SCHAEDLE, M., BASSHAM, J.A.** : Chloroplast glutathione reductase. - Plant Physiol. *59* : 1011 - 1012, 1977.

31640 - **SCHAEFFER, G.W.** : Culture and morphogenetic response of a lethal, chlorophyll-deficient mutant of tobacco to hormones, amino acids and sucrose. - In Vitro *13* : 31 - 35, 1977.

31641 - **SCHÄFER, G., HEBER, U., HELDT, H.W.** : Glucose transport into spinach chloroplasts. - Plant Physiol. *60* : 286 - 289, 1977.

31642 - **SCHAFFERNICHT, H., JUNGE, W.** : On the mutual orientation of pigments in Photosystem I particles. - In : **COOMBS, J.** (ed.) : 4th International Congress on Photosynthesis. Pp. 330 - 331. UKISES, London 1977.

31643 - **SCHANTZ, R., BAR-NUN, S., OHAD, I.** : Preparation of antibodies against specific chloroplast membrane polypeptides associated with the formation of Photosystems I and II in *Chlamydomonas reinhardi* y-1. - Plant Physiol. *59* : 167 - 172, 1977.

31644 - SCHANTZ, R., PELLEGRINI, M. : Correlations between mitochondrial and plastid
structure in *Euglena gracilis*. - In : BOGORAD, L., WEIL, J.H. (ed.) : Acides
Nucléiques et Synthèse des Protéines chez les Végétaux. Coll. Int. CNRS, No.
261. Pp. 159 - 167. Édit. CNRS, Paris 1977.

31645 - SCHAPENDONK, A.H.C.M., VREDENBERG, W.J : Salt-induced absorbance changes of
P-515 in broken chloroplasts. - Biochim. biophys. Acta *462* : 613 - 621, 1977.

31646 - SCHAPENDONK, A.H.C.M., VREDENBERG, W.J. : Voltage calibration of the P515
band shift in broken chloroplasts. - In : COOMBS, J. (ed.) : 4th International Congress on Photosynthesis. P. 331. UKISES, London 1977.

31647 - SCHÄTZLER, H.P., KÜHN, W. : Growth studies on plant plots by gamma scanning.
- Int. J. appl. Radiat. Isotop. *28* : 645 - 652, 1977. [Biomass production.]

31648 - SCHAUBERGER, C.W., WILDMAN, R.B. : Accumulation of aldrin and dieldrin by
blue-green algae and related effects on photosynthetic pigments. - Bull. environ. Contamination Toxicol. *17* : 534 - 541, 1977.

31649 - SCHEER, H., KATZ, J.J., NORRIS, J.R. : Proton-electron hyperfine coupling
constants of the chlorophyll *a* cation radical by ENDOR spectroscopy. - J. amer.
chem. Soc. *99* : 1372 - 1381, 1977.

31650 - SCHENK, H.E.A. : Inwieweit können biochemische Untersuchungen der Endocyano-
sen zur Klärung der Plastiden-Entstehung beitragen? - Arch. Protistenk. *119* :
274 - 300, 1977.

31651 - SCHENK, H.E.A. : Zur osmotischen Beeinflussung der Photosynthese von *Cyanophora paradoxa* KORSCH. durch Saccharose. - Arch. Microbiol. *114* : 261 - 266,
1977.

31652 - SCHIDLOWSKI, M., EICHMAN, R. : Evolution of the terrestrial oxygen budget. -
In : PONNAMPERUMA, C. (ed.) : Chemical Evolution of the Early Precambrian.
Pp. 87 - 99. Academic Press, New York - San Francisco - London 1977. [Ps.]

*31653 - SCHIDLOWSKI, M., EICHMANN, R., JUNGE, C.E. : Precambrian sedimentary carbona-
tes : Carbon and oxygen isotope geochemistry and implications for the terrestrial oxygen budget. - Precambrian Res. *2* : 1 - 69, 1975.

31654 - SCHIEDER, O. : Attempts in regeneration of mesophyll protoplasts of haploid
and diploid wild type lines, and those of chlorophyll-deficient strains from
different *Solanaceae*. - Z. Pflanzenphysiol. *84* : 275 - 281, 1977.

31655 - SCHIEDER, O. : Hybridisation experiments with protoplasts from chlorophyll-
-deficient mutants of some Solanaceous species. - Planta *137* : 253 - 257,
1977.

31656 - SCHIEMANN, J., WOLLGIEHN, R., PARTHIER, B. : Isolation of a transcription-
-active RNA polymerase-DNA complex from *Euglena* chloroplasts. - Biochem. Physiol. Pflanzen *171* : 474 - 478, 1977.

31657 - SCHINAS, S., ROWELL, D.L. : Lime-induced chlorosis. - J. Soil Sci. *28* : 351 -
368, 1977.

31658 - SCHLESINGER, W.H. : Carbon balance in terrestrial detritus. - Annu. Rev. Ecol.
Syst. *8* : 51 - 81, 1977.

31659 - SCHLESINGER, W.H., CHABOT, B.F. : The use of water and minerals by evergreen
and deciduous shrubs in Okefenokee Swamp. - Bot. Gaz. *138* : 490 - 497, 1977.
[Stomatal resistance.]

31660 - SCHLOSS, J.V., HARTMAN, F.C. : Reaction of ribulosebisphosphate carboxylase
from *Rhodospirillum rubrum* with the potential affinity label 3-bromo-1,4-di-
hydroxy-2-butanone 1,4-bisphosphate. - Biochem. biophys. Res. Commun. *75* :
320 - 328, 1977.

31661 - SCHLOSS, J.V., HARTMAN, F.C. : Inactivation of ribulosebisphosphate carboxy-
lase/oxygenase from spinach with the affinity label *N*-bromoacetylethanolamine
phosphate. - Biochem. biophys. Res. Commun. *77* : 230 - 236, 1977.

31662 - SCHMID, G.H., KOENIG, F., RADUNZ, A., MENKE, W. : The properties of a thylakoid membrane polypeptide with molecular weight 11,000 and its relationship
to Photosystem II. - In : OLSON, J.M., HIND, G. (ed.) : Chlorophyll-Proteins,
Reaction Centers, and Photosynthetic Membranes. P. 360. Brookhaven nat. Lab.,
Upton 1977.

31663 - **SCHMID, G.H., LEHMANN-KIRK, U.** : Photooxidation reactions of diphenylcarba-
zide and their DCMU-sensitivity in thylakoids of the blue-green alga *Oscillatoria chalybea*. - Arch. Microbiol. *115* : 265 - 269, 1977.

*31664 - **SCHMID, G.H., MENKE, W., KOENIG, F., RADUNZ, A.** : Inhibition of electron
transport on the oxygen-evolving side of photosystem II by an antiserum to
a polypeptide isolated from the thylakoid membrane. - Z. Naturforsch. *31 C* :
304 - 311, 1976.

31665 - **SCHMID, G.H., RADUNZ, A., MENKE, W.** : Localization and function of plastocyanin and cytochrome f in the thylakoid membrane of higher plant chloroplasts. - In : COOMBS, J. (ed.) : 4th International Congress on Photosynthesis. P. 332. UKISES, London 1977.

*31666 - **SCHMID, G.H., RENGER, G., GLÄSER, M., KOENIG, F., RADUNZ, A., MENKE, W.** :
Effect of an antiserum to a thylakoid membrane polypeptide on the primary
photoreaction of photosystem II. - Z. Naturforsch. *31 C* : 594 - 600, 1976.

31667 - **SCHMID, R., JAGENDORF, A.T., HULKOWER, S.** : Arginine modifiers as energy
transfer inhibitors in photophosphorylation. - Biochim. biophys. Acta *462* :
177 - 186, 1977.

31668 - **SCHMIDT, G.W., MATLIN, K.S., CHUA, N.H.** : Rapid procedure for selective enrichment of photosynthetic electron transport mutants. - Proc. nat. Acad.
Sci. USA *74* : 610 - 614, 1977.

31669 - **SCHMIDT, H.-W., HAMPP, R.** : Regulation of membrane properties of mitochondria
and plastids during chloroplast development. II. Action of phytochrome in
a cell-free system. - Z. Pflanzenphysiologie *82* : 428 - 434, 1977.

31670 - **SCHMIDT, L.** : Phytomassevorrat und Nettoprimärproduktivität alpiner Zwerg-
strauchbestände. - Oecol. Plant. *12* : 195 - 213, 1977.

31671 - **SCHMIDT, U.D., PARADIES, H.H.** : The structure of the ε-subunit from the chloroplast coupling factor (CF$_1$) studied by means of small angle X-ray scattering and inelastic light scattering. - Biochem. biophys. Res. Commun. *78* :
383 - 392, 1977.

31672 - **SCHMIDT, U.D., PARADIES, H.H.** : The structure of the δ-subunit from chloroplast coupling factor (CF$_1$) in solution. - Biochem. biophys. Res. Commun.
78 : 1043 - 1052, 1977.

31673 - **SCHMIDT-STOHN, G.** : Änderungen der Plastidenpigmente bei *Sphagnum magellani-
cum* BRID.in Abhängigkeit von Standort, Verfärbungsgrad und Alter. - Z. Pflanzenphysiol. *81* : 289 - 303, 1977.

31674 - **SCHMITT, J.M., CROUSE, E.J., DRIESEL, A.J., HERRMANN, R.G., BOHNERT, H.-J.** :
Studies on the organization of chloroplast DNAs. - In : COOMBS, J. (ed.) :
4th International Congress on Photosynthesis. P. 333. UKISES, London 1977.

31675 - **SCHMITZ, K.** : Long distance transport of assimilates in *Laminariales*. -
J. Phycol. *13* (Suppl.) : 61, 1977.

31676 - **SCHMITZ, K., KREMER, B.P.** : Carbon fixation and analysis of assimilates in
a coral-dinoflagellate symbiosis. - Mar. Biol. *42* : 305 - 313, 1977.

31677 - **SCHNARRENBERGER, C., BURKHARD, C.** : *In-vitro* association of chloroplasts and
peroxisomes from spinach (*Spinacia oleracea* L.) leaves. - In : COOMBS, J.
(ed.) : 4th International Congress on Photosynthesis. Pp. 333 - 334. UKISES,
London 1977.

31678 - **SCHNARRENBERGER, C., BURKHARD, C.** : *In-vitro* interaction between chloroplasts
and peroxisomes as controlled by inorganic phosphate. - Planta *134* : 109 -
114, 1977.

31679 - **SCHNEIDER, K., FRISCHKNECHT, K.** : The exogenous inorganic carbon source in
algal photosynthesis measured under steady state conditions with *Anacystis
nidulans*. - In : COOMBS, J. (ed.) : 4th International Congress on Photosynthesis. Pp. 334 - 335. UKISES, London 1977.

31680 - **SCHNEIDER, M.M., HAMPP, R., ZIEGLER, H.** : Envelope permeability to possible
precursors of carotenoid biosynthesis during chloroplast-chromoplast transformation. - Plant Physiol. *60* : 518 - 520, 1977.

31681 - SCHOBERT, B. : The influence of water stress on the metabolism of diatoms. II. Proline accumulation under different conditions of stress and light. - Z. Pflanzenphysiol. *85* : 451 - 461, 1977. [Ps.]

31682 - SCHOCH, S., LEMPERT, U., RÜDIGER, W. : Über die letzten Stufen der Chlorophyll-Biosynthese. Zwischenprodukte zwischen Chlorophyllid und phytolhaltigem Chlorophyll. - Z. Pflanzenphysiol. *83* : 427 - 436, 1977.

31683 - SCHÖNFELD, M., NEUMANN, J. : Mode of action of uncouplers and energy transfer inhibitors as shown by effects on the proton conductance of the thylakoid membrane. - In : COOMBS, J. (ed.) : 4th International Congress on Photosynthesis. Pp. 335 - 336. UKISES, London 1977.

31684 - SCHÖNFELD, M., NEUMANN, J. : Proton conductance of the thylakoid membrane : modulation by light. - FEBS Lett. *73* : 51 - 54, 1977.

31685 - SCHOPFER, P., BAJRACHARYA, D., BERGFELD, R., FALK, H. : Light-mediated organelle transformation: The transformation of glyoxysomes into peroxisomes by phytochrome. - In : NOVER, L., MOTHES, K. (ed.) : Cell Differentiation in Microorganisms, Plants and Animals. Pp. 625 - 639. North-Holland, Amsterdam - New York - Oxford 1977.

31686 - SCHREIBER, U., ARMOND, P.A. : Heat-induced changes in chlorophyll fluorescence and related heat-damage at the pigment level. - Carnegie Inst. Year Book *76* : 341 - 346, 1977.

31687 - SCHREIBER, U., AVRON, M. : ATP-induced chlorophyll luminescence in isolated spinach chloroplasts. - Carnegie Inst. Year Book *76* : 236 - 239, 1977.

31688 - SCHREIBER, U., AVRON, M. : ATP-induced chlorophyll luminescence in isolated spinach chloroplasts. - FEBS Lett. *82* : 159 - 162, 1977.

31689 - SCHREIBER, U., AVRON, M. : The involvement of proton gradients and the Q-R electron transfer in ATP-induced reverse electron flow. - In : COOMBS, J. (ed.) : 4th International Congress on Photosynthesis. P. 337. UKISES, London 1977.

31690 - SCHREIBER, U., BERRY, J.A. : Heat-induced chlorophyll fluorescence changes in intact leaves correlated with damage of the photosynthetic apparatus. - Carnegie Inst. Year Book *76* : 323 - 327, 1977.

31691 - SCHREIBER, U., BERRY, J.A. : Heat-induced changes of chlorophyll fluorescence in intact leaves correlated with damage of the photosynthetic apparatus. - Planta *136* : 233 - 238, 1977.

31692 - SCHREIBER, U., FINK, R., VIDAVER, W.: Fluorescence induction in whole leaves: Differentiation between the two leaf sides and adaptation to different light regimes. - Planta *133* : 121 - 129, 1977.

31693 - SCHULDINER, S. : Acid base ATP synthesis in chloroplasts. - In : TREBST, A., AVRON, M. (ed.) : Photosynthesis I. (Encycl. Plant Physiol. N.S. Vol.5.) Pp. 416 - 422. Springer-Verlag, Berlin - Heidelberg - New York 1977.

31694 - SCHULZ, D. : Chromoplasten bei Mossen. - Z. Pflanzenphysiol. *81* : 85 - 88, 1977. [Chloroplast.]

31695 - SCHULZE, E.-D., FUCHS, M.I., FUCHS, M. : Spacial distribution of photosynthetic capacity and performance in a mountain spruce forest of Northern Germany. I. Biomass distribution and daily CO_2 uptake in different crown layers. - Oecologia *29* : 43 - 61, 1977.

31696 - SCHULZE, E.-D., FUCHS, M., FUCHS, M.I. : Spacial distribution of photosynthetic capacity and performance in a mountain spruce forest of Northern Germany III. The significance of the evergreen habit. - Oecologia *30* : 239 - 248, 1977.

*31697 - SCHULZE, E.-D., ZIEGLER, H., STICHLER, W. : Environmental control of Crassulacean Acid Metabolism in *Welwitschia mirabilis* HOOK. FIL. in its range of natural distribution in the Namib desert. - Oecologia *24* : 323 - 334, 1976.

31698 - SCHÜRMANN, P., BUCHANAN, B.B. : Regulation of fructose 1,6-bis-phosphatase and sedoheptulose 1,7-bis-phosphatase in chloroplasts. - Experientia *33* : 799, 1977.

31699 - SCHÜRMANN, P., WOLOSIUK, R.A., BUCHANAN, B.B., BREAZEALE, V.D., McKINNEY, D. W. : Chloroplast thioredoxin: Properties and function in light-induced enzyme regulation in chloroplasts. - In : COOMBS, J. (ed.) : 4th International Congress on Photosynthesis. Pp. 338. UKISES, London 1977.

31700 - SCHURR, J.M., RUCHTI, J. : Dynamics of O_2 and CO_2 exchange, photosynthesis, and respiration in rivers from time-delayed correlations with ideal sunlight. - Limnol. Oceanogr. 22 : 208 - 225, 1977.

31701 - SCHUSTER, A., BOLHAR-NORDENKAMPF, H.R. : Photosynthesis and atrazine effect as influenced by changing light intensities, CO_2 tensions and temperatures. - In : COOMBS, J. (ed.) : 4th International Congress on Photosynthesis. P. 339. UKISES, London 1977.

31702 - SCHWARZ, G., SCHÄLIKE, W., SCHULMEISTER, T. : Beziehungen zwischen Purin- und Energiestoffwechsel. - In : UNGER, K. (ed.) : Biophysikalische Analyse pflanzlicher Systeme. Pp. 201 - 210. VEB Gustav Fischer Verlag, Jena 1977. [Ps.]

31703 - SCHWENN, J.D. : Assimilatory sulfate reduction by chloroplasts : The regulatory influence of adenosine-mono- and -diphosphate. - In : COOMBS, J. (ed.) : 4th International Congress on Photosynthesis. P. 340. UKISES, London 1977.

31704 - SCOTT, B.D., JITTS, H.R. : Photosynthesis of phytoplankton and zooxanthellae on a coral reef. - Mar. Biol. 41 : 307 - 315, 1977.

31705 - SCOTT, N.S. : Chloroplast DNA synthesis in Euglena gracilis. - In : BOGORAD, L., WEIL, J.H. (ed.) : Acides Nucléiques et Synthèse des Protéines chez les Végétaux. Coll. Int. CNRS. No. 261. Pp. 147 - 151. Édit. CNRS, Paris 1977. [Chl.]

31706 - SCOTT, W.R., DOUGHERTY, C.T., LANGER, R.H.M. : Development and yield components of high-yielding wheat crops. - New Zeal. J. agr. Res. 20 : 205 - 212, 1977.

31707 - SEARLE, G.F.W., BARBER, J., HARRIS, L., PORTER, G., TREDWELL, C.J. : Picosecond laser study of fluorescence lifetimes in spinach chloroplast phytosystem I and photosystem II preparations. - Biochim. biophys. Acta 459 : 390 - 401, 1977.

31708 - SEARLE, G.F.W., BARBER, J., MILLS, J.D. : 9-Amino-acridine as a probe of the electrical double layer associated with the chloroplast thylakoid membranes. - Biochim. biophys. Acta 461 : 413 - 425, 1977.

31709 - SEARLE, G.F.W., BARBER, J., PORTER, G., TREDWELL, C.J. : A study of energy transfer in Porphyridium cruentum using picosecond laser technique. - In : COOMBS, J. (ed.) : 4th International Congress on Photosynthesis. Pp. 340 - 341. UKISES, London 1977.

31710 - SEDIYAMA, G.C., PRUITT, W.O. : Estudo do microclima e dos perfis de umidade e dioxido de carbono no interior e acima do dossel vegetativo da cultura do sorgo (Sorghum vulgare PERS.). [Study of the microclimate and the humidity and carbon dioxide profiles in the interior and upper part of the vegetative canopy of the sorghum culture (Sorghum vulgare PERS.).] - Rev. Ceres 24 : 563 - 570, 1977. [In Port., ab : E.]

31711 - SEELEY, E.J., KAMMERECK, R. : Carbon fluxes in apple trees: Use of a closed system to study the effect of a mild cold stress on "Golden Delicious". - J. amer. Soc. hort. Sci. 102 : 282 - 286, 1977.

31712 - SEELEY, E.J., KAMMERECK, R. : Carbon flux in apple trees : The effects of temperature and light intensity on photosynthetic rates. - J. amer. Soc. hort. Sci. 102 : 731 - 733, 1977.

31713 - SEELY, G.R. : Chlorophyll in model systems : Clues to the role of chlorophyll in photosynthesis. - In : BARBER, J. (ed.) : Primary Processes of Photosynthesis. Pp. 1 -53. Elsevier, Amsterdam - New York - Oxford 1977.

31714 - SELIGER, H.H. : Environmental photobiology. - In : SMITH, K.C. (ed.) : The Science of Photobiology. Pp. 143 - 173. Plenum Press, New York - London 1977. [Chl.]

31715 - SELLDÉN, G. : The appearance and development of photophosphorylation and proton uptake in greening barley seedlings. - In : COOMBS, J. (ed.) : 4th International Congress on Photosynthesis. Pp. 341 - 342. UKISES, London 1977.

31716 - SELMAN, B.R. : The influence of the internal buffering capacity and pH on the yield of ATP in the dark with ferricyanide pre-treated spinach thylakoid membranes. - In : COOMBS, J. (ed.) : 4th International Congress on Photosynthesis. Pp. 342 - 343. UKISES, London 1977.

31717 - SELMAN, B.R., ORT, D.R. : Oxidation-reduction coupled phosphorylation in the dark with isolated spinach chloroplasts. - Biochim. biophys. Acta 460 : 101 - 112, 1977.

31718 - SELMAN, B.R., SMITH, D.D., DILLEY, R.A. : Localization of plastocyanin in chloroplast membranes determined by chemical modification reagents. - In : COOMBS, J. (ed.) : 4th International Congress on Photosynthesis. P. 336. UKISES, London 1977.

31719 - SELSTAM, E. : Photodecomposition of monogalactosyldiglyceride in etioplasts. - In : COOMBS, J. (ed.) : 4th International Congress on Photosynthesis. Pp. 343 - 344. UKISES, London 1977.

31720 - SEMENENKO, V.E., AVRAMOVA, S., GEORGIEV, D., PRONINA, N.A. : Sravnitel'noe izuchenie aktivnosti i lokalizatsii karboangidrazy v kletkakh Chlorella i Scenedesmus. [Comparison of activity and localization of carbonic anhydrase in the cells of Chlorella and Scenedesmus.] - Fiziol. Rast. 24 : 1055 - 1059, 1977. [In R, ab : E.]

31721 - SEMENOVA, G.A., LADYGIN, V.G., TAGEEVA, S.V. : Élektronno-mikroskopicheskoe izuchenie khloroplastov v zigotakh Chlamydomonas reinhardi. [Electron microscopic study of chloroplasts in zygotes of Chlamydomonas reinhardi.] - In : NASYROV, Yu.S. (ed.) : Genetika Fotosinteza. Pp. 150 - 154. Donish, Dushanbe 1977. [In R.]

31722 - SEMENOVA, G.A., LADYGIN, V.G., TAGEEVA, S.V. : Ul'trastrukturnaya organizatsiya membrannoĭ sistemy khloroplastov mutantov Chlamydomonas reinhardi s neaktivnymi fotosistemami. [Ultrastructural organization of the chloroplast membrane system in mutants of Chlamydomonas reinhardi with inactive photosystems.] - Fiziol. Rast. 24 : 18 - 22, 1977. [In R, ab : E.]

31723 - SENGER, H. : Changes in the sensitivity to inhibitors of photosynthesis during the life cycle of unicellular algae. - Plant Cell Physiol. 1977 (Spec. Issue 3 - Photosynthetic Organelles. Structure and Function) : 229 - 238, 1977.

31724 - SENGER, H., BISHOP, N.I. : Comparison of the mechanisms of oxygen evolution and photohydrogen production during the life cycle of synchronized cells of Scenedesmus obliquus. - Plant Physiol. 59 (6, Suppl.) : 130, 1977.

31725 - SENGER, H., MELL, V. : Photochemical activities, pigment distribution and photosynthetic unit size of chloroplast particles isolated from synchronized cells of Scenedesmus obliquus. - In : COOMBS, J. (ed.) : 4th International Congress on Photosynthesis. P. 344. UKISES, London 1977.

31726 - SENGER, H., MELL, V. : Preparation of photosynthetically active particles from synchronized cultures of unicellular algae. - In : PRESCOTT, D.M. (ed.) : Methods in Cell Biology. Vol. 15. Pp. 201 - 219. Academic Press, New York - San Francisco - London 1977.

31727 - SENSER, M., BECK, E. : On the mechanisms of frost injury and frost hardening of spruce chloroplasts. - Planta 137 : 195 - 201, 1977.

31728 - SENSER, M., BECK, E. : Seasonal changes of photosynthetic activities and frost resistance of spruce chloroplasts. - In : COOMBS, J. (ed.) : 4th International Congress on Photosynthesis. P. 345. UKISES, London 1977.

31729 - SEPASKHAH, A.R. : Estimation of individual and total leaf areas of safflowers. - Agron. J. 69 : 783 - 785, 1977.

31730 - SEREBRYAKOVA, L.T., ZORIN, N.A., GOGOTOV, I.N. : Ochistka i svoĭstva svyazannoĭ s khromatoforami gidrogenazy fototrofnoĭ bakterii Thiocapsa roseopersicina. [Purification and properties of phototrophic bacteria Thiocapsa roseopersicina hydrogenase bound with chromatophores.] - Biokhimiya 42 : 740 - 745, 1977. [In R, ab : E.]

31731 - SERVAITES, J., OGREN, W. : Chemical inhibition of the glycolate pathway in soybean leaf cells. - Plant Physiol. 59 (6, Suppl.) : 42, 1977.

31732 - SERVAITES, J.C., OGREN, W.L. : Rapid isolation of mesophyll cells from leaves of soybean for photosynthetic studies. - Plant Physiol. 59 : 587 - 590, 1977.

31733 - SERVAITES, J.C., OGREN, W.L. : Chemical inhibition of the glycolate pathway in soybean leaf cells. - Plant Physiol. 60 : 461 - 466, 1977.

31734 - SERVAITES, J.C., OGREN, W.L. : pH dependence of photosynthesis and photorespiration in soybean leaf cells. - Plant Physiol. 60 : 693 - 696, 1977.

31735 - ŠESTÁK, Z. : Photosynthetic characteristics during ontogenesis of leaves. 1. Chlorophylls. - Photosynthetica 11 : 367 - 448, 1977.

31736 - ŠESTÁK, Z. : Photosynthetic characteristics during ontogenesis of leaves. 2. Photosystems, components of electron transport chain, and photophosphorylation. - Photosynthetica 11 : 449 - 474, 1977.

31737 - ŠESTÁK, Z., ZIMA, J., STRNADOVÁ, H. : Ontogenetic changes in the internal limitations to bean-leaf photosynthesis. 2. Activities of photosystems 1 and 2 and non-cyclic photophosphorylation and their dependence on photon flux density. - Photosynthetica 11 : 282 - 290, 1977.

31738 - ŠETLÍK, I., ŠETLÍKOVÁ, E. : The measurement of photosynthetic electron transport on permeabilized cells of chlorococcal algae in the course of their cell cycle. - In : COOMBS, J. (ed.) : 4th International Congress on Photosynthesis. P. 346. UKISES, London 1977.

31739 - SEVERI, A., LAUDI, G., FORNASIERO, R.B. : Ultrastructural researches on the plastids of parasitic plants. VI. Scales of Cytisus hypocistis. - Caryologia 30 : 500, 1977.

31740 - SEWE, K.-U., REICH, R. : Carotenoid-chlorophyll complexes as origin of electrochromic absorption-changes in the membranes of photosynthesis. - In : COOMBS, J. (ed.) : 4th International Congress on Photosynthesis. P. 347. UKISES, London 1977.

31741 - SEWE, K.-U., REICH, R. : Influence of the chlorophylls on the electrochromism of carotenoids in the membranes of photosynthesis. - FEBS Lett. 80 : 30 - 34, 1977.

31742 - SEWE, K.-U., REICH, R. : The effect of molecular polarization on the electrochromism of carotenoids. II. Lutein-chlorophyll complexes: The origin of the field-indicating absorption-change at 520 nm in the membranes of photosynthesis. - Z. Naturforsch. 32C : 161 - 171, 1977.

31743 - SEYER, P., LESCURE, A.M. : Inhibition of chloroplast differentiation by the thymidine analogue, 5-bromo-2' deoxyuridine in cultured tobacco cells. - Cell Differentiation 6 : 65 - 74, 1977.

31744 - SEYER, P., LESCURE, A.M. : Effect of 5-bromodeoxyuridine (BrdUrd) on chloroplast differentiation in tobacco cell cultures. - In : BOGORAD, L., WEIL, J. H. (ed.) : Acides Nucléiques et Synthèse des Protéines chez les Végétaux. Coll. Int. CNRS. No. 261. Pp. 467 - 471. Édit. CNRS, Paris 1977.

31745 - SEYFER, J.R., WILHM, J. : Variation with stream order in species composition, diversity, biomass, and chlorophyll of periphyton in Otter Creek, Oklahoma. - Southwest. Naturalist 22 : 455 - 467, 1977.

31746 - SHAHAK, Y., SIDERER, Y., AVRON, M. : Acid-base induced reverse electron flow in chloroplasts. - In : PACKER, L., PAPAGEORGIOU, G.C., TREBST, A. (ed.) : Bioenergetics of Membranes. Pp. 405 - 414. Elsevier/North-Holland Biomedical Press, Amsterdam - Oxford - New York 1977.

31747 - SHAHAK, Y., SIDERER, Y., AVRON, M. : Reverse electron flow induced luminescence triggered by acid-base transition of chloroplasts. - Plant Cell Physiol. 1977 (Spec. Issue 3 - Photosynthetic Organelles. Structure and Function) : 115 - 127, 1977.

31748 - SHANKARNARAYAN, K.A., DABADGHAO, P.M., KUMAR, R., RAI, P. : Effect of defoliation management and manuring on dry matter yields and quality in Sehima nervosum, Cenchrus ciliaris and Cenchrus setigerus. - Ann. Arid Zone 16 : 441 - 454, 1977. [Biomass.]

31749 - SHAR, A.O., MULLIGAN, H.F. : Simulated seasonal marine mining impacts on plankton. - Int. Rev. ges. Hydrobiol. *62* : 505 - 510, 1977.

31750 - SHARMA, D.P., FERREE, D.C., HARTMAN, F.O. : Effect of some soil-applied herbicides on net photosynthesis and growth of apple trees. - HortScience *12* : 153 - 154, 1977.

31751 - SHARMA, D.P., FERREE, D.C., HARTMAN, F.O. : Multiple applications of dicofol and dodine sprays on net photosynthesis of apple leaves. - HortScience *12* : 154 - 155, 1977.

31752 - SHARMA, R.P., SUTAR, R.S. : Improvement in yield and yield attributes through mutation breeding in barley. - J. nucl. agr. Biol. *6* (4) : 114 - 118, 1977. [Ps.]

31753 - SHATILOV, I.S. : Pochvennye faktory fotosinteticheskoĭ deyatel'nosti i produktivnosti i printsipy polucheniya planiruemykh urozhaev. [Soil factors of photosynthetic activity and productivity and principles of reaching the planned yields.] - Itogi Nauki Tekh., Ser. Fiziol. Rast. *3* : 126 - 134, 1977. [In R.]

31754 - SHAVIT, N. : Bound nucleotides and conformational changes in photophosphorylation. - In : TREBST, A., AVRON, M. (ed.) : Photosynthesis I. (Encycl. Plant Physiol. N.S. Vol. 5.) Pp. 350 - 357. Springer-Verlag, Berlin - Heidelberg - New York 1977.

31755 - SHAVIT, N., LIEN, S., SAN PIETRO, A. : On the role of membrane-bound ADP and ATP in photophosphorylation in chloroplast membranes. - FEBS Lett. *73* : 55 - 58, 1977.

31756 - SHAW, M.A., RICHARDS, W.R. : Isolation of a bacteriochlorophyll-protein complex from *Rhodopseudomonas sphaeroides* which may be a thylakoid precursor. - In : OLSON, J.M., HIND, G. (ed.) : Chlorophyll-Proteins, Reaction Centers, and Photosynthetic Membranes. P. 366. Brookhaven nat. Lab., Upton 1977.

31757 - SHEATH, R.G., HELLEBUST, J.A., SAWA, T. : Changes in plastid structure, pigmentation and photosynthesis of the conchocelis stage of *Porphyra leucosticta (Rhodophyta, Bangiophyceae)* in response to low light and darkness. - Phycologia *16* : 265 - 276, 1977.

*31758 - SHEATH, R.G., MUNAWAR, M., HELLEBUST, J.A. : Phytoplankton biomass composition and primary productivity during the ice-free period in a tundra pond. - In : Proceedings of the Circumpolar Conference on Northern Ecology. Pp. III--21 - III-31. Nat. Res. Council Canada, Ottawa 1975.

31759 - SHEEHY, J.E. : Microclimate, canopy structure and photosynthesis in canopies of three contrasting temperate forage grasses. III. Canopy photosynthesis, individual leaf photosynthesis and the distribution of current assimilate. - Ann. Bot. *41* : 593 - 604, 1977.

31760 - SHEEHY, J.E., COOK, D. : Irradiance distribution and carbon-dioxide flux in forage grass canopies. - Ann. Bot. *41* : 1017 - 1029, 1977.

31761 - SHEEHY, J.E., PEACOCK, J.M. : Microclimate, canopy structure and photosynthesis in canopies of three contrasting temperate forage grasses I. Canopy structure and growth. - Ann. Bot. *41* : 567 - 578, 1977.

31762 - SHEEHY, J.E., WINDRAM, A., PEACOCK, J.M. : Microclimate, canopy structure and photosynthesis in canopies of three contrasting temperate forage grasses II. Microclimate and canopy structure. - Ann. Bot. *41* : 579 - 592, 1977.

31763 - SHEIKHOLESLAM, S.N., CURRIER, H.B. : Phloem pressure differences and ^{14}C-assimilate translocation in *Ecballium elaterium*. - Plant Physiol. *59* : 376 - 380, 1977.

31764 - SHEIKHOLESLAM, S.N., CURRIER, H.B. : Effect of water stress on turgor differences and ^{14}C-assimilate movement in phloem of *Ecballium elaterium*. - Plant Physiol. *59* : 381 - 383, 1977.

31765 - SHEKHAR, V.C., IRITANI, W.M. : A study of methods of incorporation and source of ^{14}C on distribution of radioactivity in potato plants. - Amer. Potato J. *54* : 195 - 201, 1977.

31766 - SHELDON, R.B., BOYLEN, C.W. : Maximum depth inhabited by aquatic vascular plants. - Amer. Midland Naturalist *97* : 248 - 254, 1977. [Light profile.]

31767 - SHELEF, G. : Solar energy conversion *via* algal wastewater treatment and protein production. - In : COOMBS, J. (ed.) : 4th International Congress on Photosynthesis. P. 350. UKISES, London 1977.

31768 - SHEN-MILLER, J., GAWLIK, S.R. : Effects of indoleacetic acid on the quantity of mitochondria, microbodies, and plastids in the apical and expanding cells of dark-grown oat coleoptiles. - Plant Physiol. *60* : 323 - 328, 1977.

31769 - SHERIFF, D.W. : The effect of humidity on water uptake by, and viscous flow resistance of, excised leaves of a number of species : Physiological and anatomical observations. - J. exp. Bot. *28* : 1399 - 1407, 1977.

31770 - SHERIFF, D.W., KAYE, P.E. : The response of diffusive conductance in wilted and unwilted *Atriplex hastata* L. leaves to humidity. - Z. Pflanzenphysiol. *83* : 463 - 466, 1977.

31771 - SHERIFF, D.W., KAYE, P.E. : Responses of diffusive conductance to humidity in a drought avoiding and a drought resistant (in terms of stomatal response) legume. - Ann. Bot. *41* : 653 - 655, 1977.

31772 - SHERMAN, L.A., CUNNINGHAM, J. : Isolation and characterization of temperature--sensitive, high-fluorescence mutations of the blue-green alga, *Synechococcus cedrorum*. - Plant Sci. Lett. *8* : 319 - 326,1977.

31773 - SHERMAN, W.V., CAPLAN, S.R. : Chromophore mobility in bacteriorhodopsin. - Nature *265*: 273 - 274, 1977.

31774 - SHEVCHENKO, Zh.P., LUK'YANOVA, E.N. : Vliyanie virusa polosatoĭ mozaiki na anatomicheskoe stroenie i biokhimicheskie protsessy list'ev ozimoĭ pshenitsy. [Effect of streak mosaic virus on the anatomical structure and biochemical processes of winter wheat leaves.] - Nauch. Tr. ukr. sel'skokhoz. Akad. *159* : 49 - 51, 1977. [Chl, Car; in R.]

31775 - SHIBATA, H., OCHIAI, H. : Purification and properties of δ-aminolevulinic acid dehydratase from radish cotyledons. - Plant Cell Physiol. *18* : 421 - 429, 1977.

31776 - SHIBATA, K. : Thermoluminescence from the O_2-evolving system in leaves and algal cells under development of photosynthetic apparatus. - In : COOMBS, J. (ed.) : 4th International Congress on Photosynthesis. P. 348. UKISES, London 1977.

31777 - SHIEH, Y.-J. : Effect of planting density on community photosynthesis and on yielding components of rice plants. - Bot. Bull. Acad. sin. *18* : 153 - 168, 1977.

31778 - SHIHIRA-ISHIKAWA, I. : De- and re-generation of chloroplast in *Euglena* and *Chlorella*. - J. Phycol. *13* (Suppl.) : 63, 1977.

31779 - SHIHIRA-ISHIKAWA, I., OSAFUNE, T., EHARA, T., OHKURO, I., HASE, E. : An early light-independent phase of chloroplast development in dark-grown cells of *Euglena gracilis* Z. I. Dependence of the plastid development on previous culture conditions. - Plant Cell Physiol. *1977* (Spec. Issue 3 - Photosynthetic Organelles. Structure and Function) : 445 - 454, 1977.

31780 - SHIMIZU, M., NISHIMURA, M. : Formation of electrical field accompanying temperature jump in isolated spinach chloroplasts. - Biochim. biophys. Acta *459* : 412 - 417, 1977.

31781 - SHIMOKAWA, K., SAKANOSHITA, A. : [Ethylene-induced chloroplast senescence of Satsuma mandarin (*Citrus unshiu* MARC.)] - Bull. Fac. Agr. Miyazaki Univ. *24* : 27 - 33, 1977. [In Jap., ab : E.]

31782 - SHIMOKAWA, K., SHIMADA, S. : [Chlorophyll degrading enzyme in Satsuma mandarin (*Citrus unshiu* MARC.) fruit.] - Bull. Fac. Agr. Miyazaki Univ. *24* : 309 - 317, 1977. [In Jap., ab : E.]

31783 - SHIN, M., SUKENOBU, M., OSHINO, R., KITAZUME, Y. : Two plant-type ferredoxins from a blue-green alga, *Nostoc verrucosum*. - Biochim. biophys. Acta *460* : 85 - 93, 1977.

*31784 - SHINDE, P.A. : Effect of *Corynebacterium insidiosum* on fresh weight and chlorophyll content of lucerne seedlings. - J. Maharashtra agr. Univ. *1* : 160, 1976.

31785 - SHIOZAWA, J.A. : The *P*700-chlorophyll *a*-protein : further characterization of the complex isolated from higher plants. - In : OLSON, J.M., HIND, G. (ed.) : Chlorophyll-Proteins, Reaction Centers, and Photosynthetic Membranes. Pp. 361 - 362. Brookhaven nat. Lab., Upton 1977.

*31786 - SHIOZAWA, J.A., THORNBER, J.P. : Further characterization of the *P*700-chlorophyll *a*-protein isolated from higher plants. - Plant Physiol. *57* (Suppl.) : 95, 1976.

*31787 - SHIPMAN, L.L., COTTON, T.M., NORRIS, J.R., KATZ, J.J. : An analysis of the visible absorption spectrum of chlorophyll *a* monomer, dimer, and oligomers in solution. - J. amer. chem. Soc. *98* : 8222 - 8230, 1976.

31788 - SHIPMAN, L.L., COTTON, T.M., NORRIS, J.R., KATZ, J.J. : Vibronic exciton theory: toward a better understanding of the visible absorption spectrum of antenna chlorophyll. - In : OLSON, J.M., HIND, G. (ed.) : Chlorophyll-Proteins, Reaction Centers, and Photosynthetic Membranes. Pp. 369 - 370. Brookhaven nat. Lab., Upton 1977.

31789 - SHIPMAN, R.H., FAN, L.T., KAO, I.C. : Single-cell protein production by photosynthetic bacteria. - Adv. appl. Microbiol. *21* : 161 - 183, 1977.

31790 - SHIROYA, M. : Translocation of organic substances in sunflower I. Downward translocation of ^{14}C-sucrose. - Plant Cell Physiol. *18* : 633 - 639, 1977.

31791 - SHIRYAEV, A.I., ZAICHKIN, È.I., MANUIL'SKAYA, S.V., MIKHNO, A.I., OSTROVSKAYA, L.K. : Struktura model'nykh analogov fotosinteticheskikh membran. [Structure of model analogues of photosynthetic membranes.] - In : II Vsesoyuznyĭ Simpozium. Kriogennye Metody v Èlektronnoĭ Mikroskopii. Tezisy Dokladov. Pp. 22 - 25. Pushchino 1977. [In R.]

31792 - SHISHIDO, Y., HORI, Y. : Studies on translocation and distribution of photosynthetic assimilates in tomato plants. II. Distribution pattern as affected by phyllotaxis. - Tohoku J. agr. Res. *28* : 82 - 95, 1977.

31793 - SHIVE, J.B.,Jr., BROWN, K.W. : Quaking and gas exchange in leaves of cottonwood (*Populus deltoides*, MICHX.). - Plant Physiol. *59* (6, Suppl.) : 61, 1977.

31794 - SHKUROPATOV, A.Ya., KURBANOV, K.B., STOLOVITSKIĬ, Yu.M., EVSTIGNEEV, V.B. : Fotokhimicheskie i fotoèlektronnye svoĭstva komponentov fotosinteticheskogo apparata. II. Statsionarnaya fotoprovodimost' v lamellyarnykh sistemakh khlorofill-aktseptor i khlorofill + belok-aktseptor. [Photochemical and photoelectronic properties of components of the photosynthetic apparatus. II. Stationary photoconductivity in lamellar systems of chlorophyll-acceptor and chlorophyll + protein-acceptor.] - Biofizika *22* : 407 - 412, 1977. [In R, ab : E.]

31795 - SHLYK, A.A., PRUDNIKOVA, I.V., SAVCHENKO, G.E., AVERINA, N.G., KOSTYUK, N.N., KAMYSHENKO, L.K., VLASENOK, L.I., GAPONENKO, V.I., BALEVA, E.F., PARAMONOVA, T.K., LOSITSKAYA, T.V., VEZITSKIĬ, A.Yu. : Svyaz' biosinteza khlorofilla s sintezom belkov i RNK v zeleneyushchikh i zelenykh list'yakh. [Relation of chlorophyll biosynthesis to protein and RNA synthesis in greening and green leaves.] - In : NASYROV, Yu.S. (ed.) : Genetika Fotosinteza. Pp. 51 - 60. Donish, Dushanbe 1977. [In R.]

31796 - SHLYK, A.A., ULASÈNAK, L.I., VRUBEL', S.V., GRONSKAYA, N.I., BAŬRYNA, G.U. : Suadnosiny pakazchykaŭ metabalizmu khlarafilu ŭ fragmentakh khlaraplastaŭ pry ikh mekhanichnaĭ dèzintègratsii. [Correlation of chlorophyll metabolic indexes in fragments of chloroplasts during their mechanical disintegration.] - Vestsi Akad. Navuk belorus. SSR, Ser. biyal. Navuk *1977* (5) : 66 - 73, 139, 1977. [In Belorus., ab : R.]

31797 - SHOAF, W.T., LIUM, B.W. : The quantitative determination of chlorophylls *a* and *b* from freshwater algae without interference from degradation products. - J. Res. US geol. Survey *5* : 263 - 264, 1977.

31798 - SHOMER-ILAN, A., GALUN, M., WAISEL, Y. : Seasonal variations in carbon isotope ratios in lichens. - Isr. J. Bot. *26* : 46, 1977.

31799 - SHOSHAN, V., SHAVIT, N. : Characterization and activity of a four subunit coupling factor in photophosphorylation. - In : COOMBS, J. (ed.) : 4th International Congress on Photosynthesis. P. 349. UKISES, London 1977.

*31800 - SHUKANOV, A.S. : Vliyanie vozbuditelya lozhnoĭ muchnistoĭ rosy na urozhaĭ korneplodov i semyan sakharnoĭ svekly. [Effect of *Pseudoperonospora* on yield of sugar beet bulbs and seeds.] - In : KAKHNOVICH, L.V. (ed.) : Optimizatsiya Fotosinteticheskogo Apparata Vozdeĭstviem Razlichnykh Faktorov. Pp. 153 - 164, 175 - 176. Izd.belorus. gos. Univ., Minsk 1976. [Chl; in R.]

31801 - SHUL'GIN, I.A., NICHIPOROVICH, A.A., KLIMOV, S.V., MUREĬ, I.A. : K strukturnoi organizatsii lista kak optiko-fotosinteziruyushchei sistemy. [Structural organization of the leaf as an optical photosynthetic system.] - Fiziol. Rast. *24* : 684 - 690, 1977. [In R, ab : E.]

31802 - SHUMILOVA, A.A., STEPANOVA, A.M. : Vliyanie ėkzogennogo glitsina na uglekislotnyi obmen list'ev makhorki i kukuruzy. [The effect of exogenous glycine on carbon dioxide exchange in makhorka and maize leaves.] - Vest. leningr. Univ., Biol. *1977* (1) : 112 - 117, 1977. [In R, ab : E.]

31803 - SHUTTLEWORTH, W.J. : Comments on 'Resistance of a partially wet canopy: Whose equation fails?'. - Boundary-Layer Meteorol. *12* : 385 - 386, 1977.

31804 - SHUTTLEWORTH, W.J. : The exchange of wind-driven fog and mist between vegetation and the atmosphere. - Boundary-Layer Meteorol. *12* : 463 - 489, 1977. [Resistances.]

31805 - SHUVALOV, V.A., ASADOV, A.A., KRAKHMALEVA, I.N. : Exciton interaction between the pigment molecules in reaction centers of *Rhodopseudomonas viridis*. In : COOMBS, J. (ed.) : 4th International Congress on Photosynthesis. P. 351. UKISES, London 1977.

31806 - SICHER, R., JENSEN, R.G. : Chloroplast ribulose-1,5-bisphosphate carboxylase: effects of CO_2, Mg^{2+} and RuBP. - Plant Physiol. *59* (6, Suppl.) : 42, 1977.

31807 - SIDDELL, S.G., ELLIS, R.J. : Protein synthesis by etioplasts. - Biochem. Soc. Trans. *5* : 98 - 102, 1977.

31808 - SIDERER, Y., MALKIN, S. : On the role of manganese in Photosystem II. - In : COOMBS, J. (ed.) : 4th International Congress on Photosynthesis. P. 351. UKISES, London 1977.

31809 - SIDERER, Y., MALKIN, S., POUPKO, R., LUZ, Z. : Electron spin resonance and photoreaction of Mn(II) in lettuce chloroplasts. - Arch. Biochem. Biophys. *179* : 174 - 182, 1977.

31810 - SIEFERMANN-HARMS, D. : The xanthophyll cycle in higher plants. - In : TEVINI, M., LICHTENTHALER, H.K. (ed.) : Lipids and Lipid Polymers in Higher Plants. Pp. 218 - 230. Springer-Verlag, Berlin - Heidelberg - New York 1977.

31811 - SIEFERMANN-HARMS, D. : On the mobility of violaxanthin in chloroplasts. - In : COOMBS, J. (ed.) : 4th International Congress on Photosynthesis. P. 352. UKISES, London 1977.

31812 - SIEGELMAN, H.W., KYCIA, J.H., HAXO, F.T. : Peridinin-chlorophyll *a*-protein of dinoflagellate algae. - In : OLSON, J.M., HIND, G. (ed.) : Chlorophyll-Proteins, Reaction Centers, and Photosynthetic Membranes. Pp. 162 - 169. Brookhaven nat. Lab., Upton 1977.

31813 - SIEGENTHALER, P.-A., DEPÉRY, F. : Aging of the photosynthetic apparatus VI. Changes in pH dependence of ΔpH, thylakoid internal pH and proton uptake and relationships to electron transport. - Plant Cell Physiol. *18* : 1047 - 1055, 1977.

31814 - SIEGENTHALER, P.A., MVE AKAMBA, L., RAWYLER, A., NOVAK-HOFER, I. : Deterioration processes in chloroplasts: Control of structure and functions by unsaturated fatty acids. - In : COOMBS, J. (ed.) : 4th International Congress on Photosynthesis. P. 353. UKISES, London 1977.

31815 - SIEGENTHALER, P.-A., NOVAK-HOFER, I. : Thylakoid membrane structure: Two-dimensional separation of proteins by isoelectric focusing and electrophoresis in sodium dodecylsulphate. - In : PACKER, L., PAPAGEORGIOU, G.C., TREBST, A.

(ed.) : Bioenergetics of Membranes. Pp. 269 - 286. Elsevier/North-Holland Biomedical Press, Amsterdam - Oxford - New York 1977.

31816 - SIEGENTHALER, P.-A., RAWYLER, A. : Aging of photosynthetic apparatus. V. Change in pH dependence of electron transport and relationships to endogenous free fatty acids. - Plant Sci. Lett. *9* : 265 - 273, 1977.

31817 - SIEGHARDT, H. : Untersuchungen zur Energienutzung von *Phragmites communis* TRIN. - Arch. Hydrobiol. *79* : 172 - 181, 1977.

31818 - SIGALAT, C., de KOUCHKOVSKY, Y. : Continuous measurement of photosynthetic phosphorylation kinetics in chloroplasts with the luminescent method luciferine-luciferase. - In : COOMBS, J. (ed.) : 4th International Congress on Photosynthesis. P. 354. UKISES, London 1977.

31819 - SIGGEL, U. : New results on the absorption changes of the plastoquinone system in spinach chloroplasts. - In : COOMBS, J. (ed.) : 4th International Congress on Photosynthesis. P. 355. UKISES, London 1977.

31820 - SIGGEL, U., KHANNA, R., RENGER, G., GOVINDJEE : Investigation of the absorption changes of the plastoquinone system in broken chloroplasts. The effect of bicarbonate-depletion. - Biochim. biophys. Acta *462* : 196 - 207, 1977.

31821 - SILAEVA, A.M., OSTROVSKAYA, L.K., GAMAYUNOVA, M.S., GRIGORA, M.Yu., KOLEGOVA, L.E., MOSHKOV, D.A. : Issledovanie fotosinteticheskikh membran khloroplastov kukuruzy metodom krioskalyvaniya. [Study of photosynthetic membranes of maize chloroplasts by the freeze-etching method.] - In : II Vsesoyuznyĭ Simpozium. Kriogennye Metody v Ėlektronnoĭ Mikroskopii. Tezisy Dokladov. Pp. 40 - 44. Pushchino 1977. [In R.]

31822 - SILSBURY, J.H., FUKAI, S. : Effects of sowing time and sowing density on the growth of subterranean clover at Adelaide. - Aust. J. agr. Res. *28* : 427 - 440, 1977.

31823 - SILVIUS, J.E., JOHNSON, R.R., PETERS, D.B. : Effect of water stress on carbon assimilation and distribution in soybean plants at different stages of development. - Crop Sci. *17* : 713 - 716, 1977.

31824 - SIMIONESCU, C.I., SIMIONESCU, B.C., MORA, R., LEANCĂ, M., IOANID, E. : Synthèse abiotique des porphyrines : obtention de composés du type porphyrinique par la décharge de haute fréquence dans un melange de méthane, d'ammoniac et de vapeur d'eau. - Compt. rend. Acad. Sci. Paris, Sér. C *284* : 743 - 746, 1977. [Chl phylogenesis.]

31825 - ŠIMON, J. : Tvorba a produkce biomasy cukrovky v závlaze na lehké půdě. [Formation and production of phytomass in sugarbeet under irrigation on light soil.] - In : Produkce Biomasy a Tvorba Výnosu Polních Plodin. Vol. 2. Pp. 69 - 77. Česká vědeckotechnická Společnost zemědělská, Praha 1977. [Growth analysis; in Czech, ab : E, G, R.]

31826 - ŠIMON, J. : Tvorba biomasy, struktura a výše výnosu bobu obecného (*Vicia faba* L.) na lehkých půdách v závlaze. [Biomass production, structure and level of yield of broad-bean (*Vicia faba* L.) on light-textured irrigated soils.] - Rostlinná Výroba (Praha) *23* : 1169 - 1176, 1977. [In Czech, ab : E, G, R.]

31827 - SIMON, J.C. : Étude des influences agronomiques des brise-vent dans les périmètres irriqués du Centre-Ouest de l'Argentine. I. - Effets des brise-vent sur la croissance et le développement d'une culture type : la vigne. - Ann. agron. *28* : 75 - 93, 1977. [Dry-matter production.]

31828 - SIMONIS, W., LEE KADEN, J. : Photosystem I activity and aminoacid uptake in *Anacystis nidulans*. - In : COOMBS, J. (ed.) : 4th International Congress on Photosynthesis. P. 356. UKISES, London 1977.

31829 - SIMONOVA, E.I., KUDINOVA, L.I., NOVIKOVA, N.S. : Regulyatsiya biosinteza khlorofilla svetom raznogo spektral'nogo sostava. [Regulation of chlorophyll biosynthesis by light of different spectral composition.] - Fiziol. Rast. *24* : 1154 - 1158, 1977. [In R, ab : E.]

31830 - SIMPSON, D.J. : Wild-type and mutant thylakoid structure in barley plastids analysed by freeze-fracturing. - In : COOMBS, J. (ed.) : 4th International Congress on Photosynthesis. P. 357. UKISES, London 1977.

31831 - SIMPSON, D.J., BAQAR, M.R., LEE, T.H. : Chemical regulation of plastid deve-
         lopment III. Effect of light and CPTA on chromoplast ultrastructure and ca-
         rotenoids of *Capsicum anuum*. - Z. Pflanzenphysiol. *82* : 189 - 209, 1977.

31832 - SIMPSON, D.J., BAQAR, M.R., LEE, T.H. : Chromoplast ultrastructure of *Capsi-
         cum* carotenoid mutants. I. Ultrastructure and carotenoid composition of a new
         mutant. - Z. Pflanzenphysiol. *83* : 293 - 308, 1977.

31833 - SIMPSON, D.J., BAQAR, M.R., LEE, T.H. : Fine structure and carotenoid compo-
         sition of the fibrillar chromoplasts of *Asparagus officinalis* L. - Ann. Bot.
         41 : 1101 - 1108, 1977. [Chl.]

31834 - SIMPSON, D.J., LEE, T.H. : Chromoplast ultrastructure of *Capsicum* carotenoid
         mutants. II. Effect of light and CPTA. - Z. Pflanzenphysiol. *83* : 309 - 325,
         1977.

31835 - SINCLAIR, J., GARLAND, S., ARNASON, T., HOPE, P., GRANVILLE, M. : Polychlori-
         nated biphenyls and their effects on photosynthesis and respiration. - Can.
         J. Bot. *55* : 2679 - 2684, 1977.

31836 - SINCLAIR, T.R., GOUDRIAAN, J., de WIT, C.T. : Mesophyll resistance and CO_2
         compensation concentration in leaf photosynthesis models. - Photosynthetica
         11 : 56 - 65, 1977.

31837 - SINESHCHEKOV, O.A., LITVIN, F.F. : Induktsiya fluorestsentsii odinochnogo
         khloroplasta v tseloĭ kletke zelenoĭ vodorosli *Haematococcus pluvialis*. [The
         fluorescence induction of a single chloroplast in an intact cell of the green
         alga *Haematococcus pluvialis*.] - Biofizika 22 : 58 - 63, 1977. [In R, ab : E.]

31838 - SINESHCHEKOV, V.A., LITVIN, F.F. : Luminescence of bacteriorhodopsin from
         Halobacterium halobium and its connection with the photochemical conversions
         of the chromophore. - Biochim. biophys. Acta *462* : 450 - 466, 1977.

31839 - SINGH, K.P., SINGH, R.P. : The chlorophyll content of sun- and shade-leaves
         of common trees growing at Varanasi, India. - Indian J. Ecol. *4* : 46 - 54,
         1977.

31840 - SINGH, L., SRIVASTAVA, M.P., SINGH, R.P., TIWARI, A.S. : Leaf number and area
         and their association with yield in pigeon pea. - Indian J. Genet. Plant
         Breed. *37* : 450 - 452, 1977.

31841 - SINGH, L.B. : Mango. - In : ALVIM, P. de T., KOZLOWSKI, T.T. (ed.) : Ecophy-
         siology of Tropical Crops. Pp. 479 - 487. Academic Press, New York - San
         Francisco - London 1977. [Production.]

31842 - SINGH, R.P., RAMAKRISHNA, Y.S. : Yield and moisture utilization patterns of
         dryland crops grown on conserved şoil moisture. - Ann. arid Zone *16* : 257 -
         262, 1977. [Dry-matter accumulation.]

31843 - SIPOŞ, G., PĂLTINEANU, R., AVRIGEANU, G., IARCA, P., PĂTRĂŞCOIU, G., CIOTEA,
         V. : Regimul de irigare şi consumul de apă la bumbac în condiţiile Cîmpiei
         Române. [Irrigation regime and water consumption at cotton cultivated in the
         Romanian Plain.] - An. Inst. Cercetări Pentru Cereale Plante Tehnice - Fundu-
         lea *42* : 315 - 321, 1977. [Dry-matter accumulation; in Roum., ab : E, R.]

31844 - SIREVÅG, R., BUCHANAN, B.B., BERRY, J.A., TROUGHTON, J.H. : Mechanisms of CO_2
         fixation in bacterial photosynthesis studied by the carbon isotope fractiona-
         tion technique. - Arch. Microbiol. *112* : 35 - 38, 1977.

31845 - SIREVÅG, R., ORMEROD, J.G. : Synthesis, storage and degradation of polygluco-
         se in *Chlorobium thiosulfatophilum*. - Arch. Microbiol. *111* : 239 - 244, 1977.
         [Chl.]

31846 - SIRONVAL, C. : Early photobiochemical activities in etiolated leaves illumi-
         nated for the first time. - In : COOMBS, J. (ed.) : 4th International Cong-
         ress on Photosynthesis. P. 358. UKISES, London 1977.

31847 - SIRONVAL, C., BONOTTO, S., PAQUES, M., DUJARDIN, E. : Temporary periodic lo-
         calization of the chloroplasts in the form of transverse bands, in the stalk
         of *Acetabularia mediterranea*. - In : WOODCOCK, C.L.F. (ed.) : Progress in
         Acetabularia Research. Pp. 207 - 217. Academic Press, New York - San Francisco -
         London 1977.

31848 - **SIROTENKO, O.D.** : Ein nichtstationäres Modell für den Wasser- und Wärmehaushalt und für die Ertragsleistung eines Pflanzenbestandes. - In : **UNGER, K.** (ed.) : Biophysikalische Analyse pflanzlicher Systeme. Pp. 38 - 50. VEB Gustav Fischer Verlag, Jena 1977.

31849 - **SIROTENKO, O.D., GORBACHEV, V.A.** : Dinamicheskie modeli produktivnosti agrotsenozov i problemy modelirovaniya protsessov energo- i massoobmena v sisteme "pochva - rastenie - atmosfera". [Dynamic models of productivity of agrocoenoses and problems of modeling processes of energy and mass exchange in the system "soil - plant - atmosphere".]-Itogi Nauki Tech., Ser. Fiziol. Rast. *3* : 90 - 107, 1977. [In R.]

*31850 - **SISLER, E.C.** : Ethylene analogs : effect of some unsaturated sulfides (thioethers) on tobacco leaves. - Tobacco *178* (2) : 29 - 34, 1976. Tobacco Sci. *20* : 6 - 10, 1976. [Chl.]

*31851 - **SISLER, E.C.** : Photobleaching of tobacco leaves. - Tobacco *178* (6) : 55 - 59, 1976. Tobacco Sci. *20* : 35 - 39, 1976. [Chl,Car.]

*31852 - **SISLER, E.C., PIAN, A.** : Effect of ethylene and cyclic olefins on tobacco leaves. - Tobacco *175* (10) : 27 - 31, 1973. Tobacco Sci. *17* : 68 - 72, 1973. [Chl.]

31853 - **SISSON, W.B., CALDWELL, M.M.** : Atmospheric ozone depletion : Reduction of photosynthesis and growth of a sensitive higher plant exposed to enhanced u.v.-B radiation. - J. exp. Bot. *28* : 691 - 705, 1977.

31854 - **SITTE, P.** : Functional organization of biomembranes. - In : **TEVINI, M., LICHTENTHALER, H.K.** (ed.) : Lipids and Lipid Polymers in Higher Plants. Pp. 1 - 28. Springer-Verlag, Berlin - Heidelberg - New York 1977. [Thylakoids.]

31855 - **SIVTSEV, M.V., DONDO, V.V.** : Korrelyatsiya dinamiki soderzhaniya khlorofilla i aktivnosti khlorofillazy v list'yakh rasteniĭ. [Correlation of chlorophyll dynamics and chlorophyllase activity in plant leaves.] - Izv. Akad. Nauk SSSR, Ser. biol. *77* (2) : 186 - 193, 1977. [In R, ab : E.]

31856 - **SKOKUT, T.A., WU, J.H., DANIEL, R.S.** : Retardation of ultraviolet light accelerated chlorosis by visible light or by benzyladenine in *Nicotiana glutinosa* leaves : changes in amino acids content and chloroplast ultrastructure. - Photochem. Photobiol. *25* : 109 - 118, 1977.

31857 - **SLABAS, A.R., EVANS, M.C.W.** : EPR signals in chloroplasts responding to illumination sequence of four flashes. - Nature *270* : 169 - 171, 1977.

31858 - **SLABAS, A.R., EVANS, M.C.W.** : Photosystem II e.p.r. signals which respond to flash number. - In : **COOMBS, J.** (ed.) : 4th International Congress on Photosynthesis. P. 359. UKISES, London 1977.

31859 - **SLABBERS, P.J.** : Surface roughness of crops and potential evapotranspiration. - J. Hydrol. *34* : 181 - 191, 1977. [Resistances.]

31860 - **SLATER, E.C.** : Biological membranes as energy transducers. - In : **BUVET, R., ALLEN, M.J., MASSUÉ, J.-P.** (ed.) : Living Systems as Energy Converters. Pp. 221 - 227. North-Holland Publ. Co., Amsterdam - New York - Oxford 1977. [Ps.]

31861 - **SLATYER, R.O.** : Altitudinal variation in the photosynthetic characteristics of snow gum, *Eucalyptus pauciflora* SIEB ex SPRENG. III Temperature response of material grown in contrasting thermal environments. - Aust. J. Plant Physiol. *4* : 301 - 312, 1977.

31862 - **SLATYER, R.O.** : Altitudinal variation in the photosynthetic characteristics of snow gum, *Eucalyptus pauciflora* SIEB. ex SPRENG. IV. Temperature response of four populations grown at different temperatures. - Aust. J. Plant Physiol. *4* : 583 - 594, 1977.

31863 - **SLATYER, R.O.** : Altitudinal variation in the photosynthetic characteristics of snow gum, *Eucalyptus pauciflora* SIEB ex SPRENG. VI. Comparison of field and phytotron responses to growth temperature. - Aust. J. Plant Physiol. *4* : 901 - 916, 1977.

31864 - **SLATYER, R.O., FERRAR, P.J.** : Altitudinal variation in the photosynthetic characteristics of snow gum, *Eucalyptus pauciflora* SIEB. ex SPRENG. II.

Effects of growth temperature under controlled conditions. - Aust. J. Plant
Physiol. *4* : 289 - 299, 1977.

31865 - **SLATYER, R.O., FERRAR, P.J.** : Altitudinal variation in the photosynthetic
characteristics of snow gum, *Eucalyptus pauciflora* SIEB. ex SPRENG. V. Rate
of acclimation to an altered growth environment. - Aust. J. Plant Physiol.
4 : 595 - 609, 1977.

31866 - **SLATYER, R.O., MORROW, P.A.** : Altitudinal variation in the photosynthetic
characteristics of snow gum, *Eucalyptus pauciflora* SIEB ex SPRENG. I. Seasonal changes under field conditions in the Snowy Mountains area of south-eastern Australia. - Aust. J. Bot. *25* : 1 - 20, 1977.

31867 - **SLAVÍK, B.** : An analogue model for the estimation of the transmesophyllar diffusive resistance for CO_2 in a photosynthesizing amphistomatous anisolateral
leaf. - In : UNGER, K. (ed.) : Biophysikalische Analyse pflanzlicher Systeme.
Pp. 146 - 150. VEB Gustav Fischer Verlag, Jena 1977.

31868 - **SLAVÍK, L.** : Posouzení vlivu řízeného vodního režimu na tvorbu výnosu ozimé
pšenice. [Appreciation of the coordinated moisture supply influence upon the
farming of winter wheat yield.] - In : Produkce Biomasy a Tvorba Výnosů Polních Plodin. Vol. 2. Pp. 52 - 64. Česká vědeckotechnická Společnost zemědělská, Praha 1977. [Dry-matter accumulation; in Czech, ab : E, G, R.]

31869 - **SLAWYK, G., COLLOS, Y., AUCLAIR, J.-C.** : The use of the ^{13}C and ^{15}N isotopes
for the simultaneous measurement of carbon and nitrogen turnover rates in marine phytoplankton. - Limnol. Oceanogr. *22* : 925 - 932, 1977.

31870 - **SLEPER, D.A., NELSON, C.J., ASAY, K.H.** : Diallel and path coefficient analysis of tall fescue (*Festuca arundinacea*) regrowth under controlled conditions.
- Can. J. Genet. Cytol. *19* : 557 - 564, 1977. [Ps.]

31871 - **SLOOTEN, L.** : Effects of preillumination on coupling factor activities in
Rhodospirillum rubrum chromatophores. - In : COOMBS, J. (ed.) : 4th International Congress on Photosynthesis. Pp. 359 - 360. UKISES, London 1977.

31872 - **SLOVACEK, R., HIND, G.** : Regulation of photosynthesis by cyclic electron
transport. - Plant Physiol. *59* (6, Suppl.) : 23, 1977.

31873 - **SLOVACEK, R.E., HIND, G.** : Influence of antimycin A and uncouplers on anaerobic photosynthesis in isolated chloroplasts. - Plant Physiol. *60* : 538 - 542,
1977.

31874 - **SLUITERS-SCHOLTEN, C.M.T., MOLL, W.A.W., STEGWEE, D.** : Ferredoxin and ferre-
doxin-NADP-oxidoreductase in leaves of *Phaseolus vulgaris* L. - Planta *133* :
289 - 294, 1977.

31875 - **SLUKHAЇ, S.I., TKACHUK, E.S., KIRICHENKO, V.P., KOLOMIETS, N.G., GRINCHUK,
M.A.** : Vliyanie azotnogo pitaniya na vodnyї rezhim ozimoї pshenitsy. [Effect
of nitrogen nutrition on water regime of winter wheat.] - Fiziol. Biokhim.
kul't. Rast. *9* : 122 - 128, 1977. [Dry-matter accumulation; in R, ab : E.]

31876 - **SMILLIE, R.M.** : Regulation of development and activity of chloroplast membranes by temperature. - Proc. aust. biochem. Soc. *10* : Q9, 1977.

31877 - **SMILLIE, R.M., NIELSEN, N.C., HENNINGSEN, K.W., WETTSTEIN, D. von** : Development of photochemical activity in chloroplast membranes. I. Studies with mutants of barley grown under a single environment. - Aust. J. Plant Physiol.
4 : 415 - 438, 1977.

31878 - **SMILLIE, R.M., NIELSEN, N.C., HENNINGSEN, K.W., WETTSTEIN, D. von** : Development of photochemical activity in chloroplast membranes. II. Studies with a
mutant of barley grown under different environments. - Aust. J. Plant Physiol. *4* : 439 - 449, 1977.

*31879 - **SMITH, A.P.** : Stratification of temperate and tropical forests. - Amer. Naturalist *107* : 671 - 683, 1973. [Ps.]

31880 - **SMITH, B.B., REBEIZ, C.A.** : Chloroplast biogenesis : Detection of Mg-proto-
porphyrin chelatase *in vitro*. - Arch. Biochem. Biophys. *180* : 178 - 185, 1977.

31881 - **SMITH, B.B., REBEIZ, C.A.** : Spectrofluorometric determination of Mg-protopor-
phyrin monoester and longer wavelength metalloporphyrins in the presence of

Zn-protoporphyrin. - Photochem. Photobiol. *26* : 527 - 532, 1977. [Chl.]

*31882 - SMITH, D.C. : Symbiosis and the biology of lichenised fungi. - In : JENNINGS, D.H., LEE, D.L. (ed.) : Symbiosis. Symposia of the Society for Experimental Biology. No. 29. Pp. 373 - 405. Cambridge Univ. Press, New York - Cambridge 1975. [Ps.]

31883 - SMITH, D.D., SELMAN, B.R., VOEGELI, K.K., JOHNSON, G., DILLEY, R.A. : Chloroplast membrane sidedness. Location of plastocyanin determined by chemical modifiers. - Biochim. biophys. Acta *459* : 468 - 482, 1977.

31884 - SMITH, J.A., OLIVER, R.E., BERRY, J.K. : A comparison of two photographic techniques for estimating foliage angle distribution. - Aust. J. Bot. *25* : 545 - 553, 1977.

31885 - SMITH, P.R., NEALES, T.F. : Analysis of effects of virus infection on the photosynthetic properties of peach leaves. - Aust. J. Plant Physiol. *4* : 723 - 732, 1977.

*31886 - SMITH, R.C., CALKINS, J. : The use of the Robertson meter to measure the penetration of solar middle-ultraviolet radiation (UV-B) into natural waters. - Limnol. Oceanogr. *21* : 746 - 749, 1976.

31887 - SMITH, W.K., NOBEL, P.S. : Influences of seasonal changes in leaf morphology on water-use efficiency for three desert broadleaf shrubs. - Ecology *58* : 1033 - 1043, 1977.

31888 - SMITH, W.K., NOBEL, P.S. : Temperature and water relations for sun and shade leaves of a desert broadleaf, *Hyptis emoryi*. - J. exp. Bot. *28* : 169 - 183, 1977.

*31889 - SMITH, W.O. Jr. : The extracellular release of glycolic acid by a marine diatom. - J. Phycol. *10* : 30 - 33, 1974. [Photosynthates.]

31890 - SMITH, W.O. Jr. : The respiration of photosynthetic carbon in eutrophic areas of the ocean. - J. mar. Res. *35* : 557 - 565, 1977.

31891 - SMITH, W.O. Jr., BARBER, R.T., HUNTSMAN, S.A. : Primary production off the coast of northwest Africa : excretion of dissolved organic matter and its heterotrophic uptake. - Deep-Sea Res. *24* : 35 - 47, 1977.

31892 - SNOZZI, M. : Isolierung und Charakterisierung von Reaktionszentren aus *Rhodospirillum rubrum*. - Ber. deut. bot. Ges. *90* : 485 - 492, 1977.

31893 - SNOZZI, M., BRUNI, A., BACHOFEN, R., CUENDET, P. : Phospholipid content of reaction centers and light harvesting complexes from *Rhodospirillum rubrum*. - In : COOMBS, J. (ed.) : 4th International Congress on Photosynthesis. Pp. 360 - 361. UKISES, London 1977.

31894 - SNYDER, F.W., CARLSON, G.E., SILVIUS, J.E. : A genetically controlled photosynthate-partitioning system in sugarbeet. - Plant Physiol. *59* (6, Suppl.) : 98, 1977.

31895 - SOFFE, R.W., LENTON, J.R., MILFORD, G.F.J. : Effects of photoperiod on some vegetable species. - Ann. appl. Biol. *85* : 411 - 415, 1977. [Growth analysis.]

31896 - SOFIELD, I., EVANS, L.T., COOK, M.G., WARDLAW, I.F. : Factors influencing the rate and duration of grain filling in wheat. - Aust. J. Plant Physiol. *4* : 785 - 797, 1977. [Ps.]

31897 - SOKOLOVE, P.M., MARSHO, T.V. : Slow fluorescence quenching of type A chloroplasts. Relationship to electron-flow with CO_2 as acceptor. - FEBS Lett. *75* : 28 - 32, 1977.

31898 - SOKOLOVE, P.M., MARSHO, T.V. : Slow fluorescence quenching of type A chloroplasts. Resolution into two components. - Biochim. biophys. Acta *459* : 27 - 35, 1977.

31899 - SOLÁROVÁ, J., VÁCLAVÍK, J., POSPÍŠILOVÁ, J. : Leaf conductance and gas exchange through adaxial and abaxial surfaces in water stressed primary bean leaves. - Biol. Plant. *19* : 59 - 64, 1977.

31900 - SOLOV'EV, A.A., CHUGUNOV, V.A., EROKHIN, Yu.E. : Vydelenie i kharakteristika pigment-belkovykh kompleksov iz zelenoĭ serobakterii *Chlorobium limicola* forma *thiosulfatophilum*. [Isolation and characterization of pigment-protein complexes from green sulphur bacterium *Chlorobium limicola* form *thiosulfatophilum*.] - Biokhimiya *42* : 1697 - 1703, 1977. [In R, ab : E.]

31901 - SONG, P.-S. : [Molecular topography of solar energy harvesting pigments in marine algae.]-Hwahak Kwa Kongop Ui Chinbo *17* : 236 - 240, 1977. [In Korean.]

31902 - SONG, P.-S., CHIN, C.-A., YAMAZAKI, I., BABA, H. : Excited states of photobiological receptors. II. Chlorophylls, phytochrome, and stentorin. - Int. J. Quantum Chem., Quantum Biol. Symp. *4* : 305 - 315, 1977.

31903 - SOONG, T.-S.T., HAGEMAN, R.H. : Changes in leaf proteolytic enzymes as a function of water deficit in corn (*Zea mays* L.). - Plant Physiol. *59* (6, Suppl.) : 53, 1977. [Chl.]

31904 - SÖRENSEN, L., HALLDAL, P. : Comparative analyses of action spectra of glycolate excretion ("photorespiration") and photosynthesis in blue-green alga *Anacystis nidulans*. - Photochem. Photobiol. *26* : 511 - 518, 1977.

31905 - SOROKA, V.P. : Matematicheskiǐ metod opredeleniya ploshchadi list'ev konopli. [Mathematical method for leaf area determination in hemp.] - Nauch. Dokl. vyssh. Shkoly, biol. Nauki *20* (1) : 135 - 137, 1977. [In R.]

31906 - SOROKIN, E.M. : Dve formy zavisimosti intensivnosti zamedlennoǐ fluorestsentsii khlorofilla *a* v kletke ot intensivnosti vozbuzhdayushchego sveta. [Two types of dependence of the delayed fluorescence intensity of chlorophyll *a* in the cell upon the intensity of exciting light.] - Izv. Akad. Nauk SSSR, Ser. biol. *1977* : 388 - 395, 1977. [In R, ab : E.]

31907 - SOSA, F.M., CARDENAS, J. : NADPH as electron donor for nitrate reduction in *Chlamydomonas reinhardi*. - Z. Pflanzenphysiol. *85* : 171 - 175, 1977.

31908 - SOSINSKA, A., MALESZEWSKI, S., KACPERSKA-PALACZ, A. : Carbon photosynthetic metabolism in leaves of cold-treated rape plants. - Z. Pflanzenphysiol. *83* : 285 - 291, 1977.

31909 - SOSNOWSKA, J. : Qualitative and quantitative studies on phytoplankton and chlorophyll content in the pelagic water of Lake Żarnowieckie in 1974. - Acta Soc. Bot. Pol. *46* : 403 - 422, 1977.

31910 - SOUNDARARAJAN, M., ROCHER, J.P. : Carbohydrate metabolism and photosynthetic adaptation in two species of ryegrass - *Lolium multiflorum* and *Lolium perenne*. - In : COOMBS, J. (ed.) : 4th International Congress on Photosynthesis. P. 358. UKISES, London 1977.

31911 - SOURNIA, A. : Analyse et bilan de la production primaire dans les récifs coralliens. - Ann. Inst. océanogr. (Paris) *53* : 47 - 73, 1977.

31912 - SOURNIA, A. : Notes on primary productivity of coastal waters in the Gulf of Elat (Red Sea). - Int. Rev. ges. Hydrobiol. *62* : 813 - 819, 1977.

31913 - SPALDING, M.H., EDWARDS, G.E. : Isolation of photosynthetically active cells from *Sedum telephium*, a CAM plant. - Plant Physiol. *59* (6, Suppl.) : 115, 1977.

31914 - SPENCER, K.G., TOGASAKI, R.K. : Growth on glycolate and the glycolate : DCPIP oxidoreductase in *Chlamydomonas reinhardtii*. - Plant Physiol. *59* (6, Suppl.) : 66, 1977.

31915 - SPERK, G., TUPPY, H. : Differences between adenosine triphosphatases from monocotylous and dicotylous plants. - Plant Physiol. *59* : 155 - 157, 1977.

31916 - SPIERTZ, J.H.J. : The influence of temperature and light intensity on grain growth in relation to the carbohydrate and nitrogen economy of the wheat plant. - Neth. J. agr. Sci. *25* : 182 - 197, 1977. [Ps.]

31917 - SPIESS, H. : Analysis of the chloroplast ribosomal proteins from *Chlamydomonas reinhardii*, streptomycin-resistant and dependent mutants by two-dimensional gel electrophoresis. - Plant Sci. Lett. *10* : 103 - 113, 1977.

31918 - SPILLER, H., BÖGER, P. : Photosynthetic nitrite reduction by dithioerythritol and the effect of nitrite on electron transport in isolated chloroplasts. - Photochem. Photobiol. *26* : 397 - 402, 1977.

31919 - SPILLER, H., BÖGER, P. : Photosynthetic electron transport activity of the cyanobacterium *Nostoc muscorum*. - In : COOMBS, J. (ed.) : 4th International Congress on Photosynthesis. P. 362. UKISES, London 1977.

31920 - SPITTLEHOUSE, D.L., RIPLEY, E.A. : Carbon dioxide concentrations over a native grassland in Saskatchewan. - Tellus *29* : 54 - 65, 1977.

31921 - SPRENT, J.I., BRADFORD, A.M. : Nitrogen fixation in field beans (*Vicia faba*) as affected by population density, shading and its relationship with soil moisture. - J. agr. Sci. *88* : 303 - 310, 1977. [Dry-matter accumulation.]

31922 - SPRENT, J.I., BRADFORD, A.M., NORTON, C. : Seasonal growth patterns in field beans (*Vicia faba*) as affected by population density, shading and its relationship with soil moisture. - J. agr. Sci. *88* : 293 - 301, 1977. [Dry-matter accumulation.]

31923 - SPREY, B. : Lamellae-bound inclusions in isolated spinach chloroplasts. Ultrastructure and isolation. - Z. Pflanzenphysiol. *83* : 159 - 179, 1977.

31924 - SPREY, B., LAMBERT, C. : Lamellae-bound inclusions in isolated spinach chloroplasts. 2. Identification and composition. - Z. Pflanzenphysiol. *83* : 227 - 247, 1977.

31925 - SQUIRE, G.R. : Seasonal changes in photosynthesis of tea (*Camellia sinensis* L.) - J. appl. Ecol. *14* : 303 - 316, 1977.

31926 - SRIVASTAVA, G.C., DESHMUKH, P.S., TOMAR, D.P.S. : Significance of leaf orientation and bract in seed yield in sunflower. - Indian J. Plant Physiol. *20* : 151 - 156, 1977.

31927 - SRIVASTAVA, H.S., CHAUHAN, J.S. : Seed germination, seedling growth and nitrogen and pigment concentration in dodder as affected by inorganic nitrogen. - Z. Pflanzenphysiol. *84* : 391 - 398, 1977.

31928 - STABENAU, H. : Einfluß der CO_2-Konzentration auf Wachstum und Stoffwechsel von *Chlorogonium elongatum*. - Biochem. Physiol. Pflanz. *171* : 449 - 454, 1977.

31929 - STADNICHUK, I.N., LITVIN, F.F. : Razlozhenie na komponenty spektrov pogloshcheniya i fluorestsentsii zelenykh bakterii. [Computer curve analysis of bacteriochlorophyll absorption and fluorescence spectra in green bacteria.] - Biofizika *22* : 170 - 174, 1977. [In R, ab : E.]

31930 - STAEHELIN, L.A., ARMOND, P.A., MILLER, K.R. : Chloroplast membrane organization at the supramolecular level and its functional implications. - In : OLSON, J.M., HIND, G. (ed.) : Chlorophyll-Proteins, Reaction Centers, and Photosynthetic Membranes. Pp. 278 - 315. Brookhaven nat. Lab., Upton 1977.

31931 - STAFFORD, H.A., BROWN, M.A. : Photochemical dimerization of ferulic acid by chloroplasts from *Sorghum*. - Plant Physiol. *59* : 94 - 96, 1977.

31932 - STANEV, V.P. : Sravnitelni izsledvannya na fotosintetichnata deinost na sl"nchogleda (C_3-tip) i tsarevitsata (C_4-tip) rasteniya. [Comparative study of photosynthetic activity of sunflower (C_3-type) and maize (C_4-type) plants.] - Fiziol. Rast. (Sofia) *3* (3) : 5 - 13, 1977. [In Bulg., ab : E, R.]

31933 - STANHILL, G. : Allometric growth studies on the carrot crop I. Effects of plant development and cultivar. - Ann. Bot. *41* : 533 - 540, 1977.

31934 - STANHILL, G. : Allometric growth studies of the carrot crop II. Effects of cultural practices and climatic environment. - Ann. Bot. *41* : 541 - 552, 1977.

31935 - STANHILL, G. : Quantifying weather-crop relations. - In : LANDSBERG, J.J., CUTTING, C.V. (ed.) : Environmental Effects on Crop Physiology. Pp. 23 - 37. Academic Press, London - New York - San Francisco 1977.

31936 - STANHILL, G., FUCHS, M. : The relative flux density of photosynthetically active radiation. - J. appl. Ecol. *14* : 317 - 322, 1977.

31937 - STANIER, R.Y., COHEN-BAZIRE, G. : Phototropic prokaryotes : the cyanobacteria. - Annu. Rev. Microbiol. *31* : 225 - 274, 1977. [Ps apparatus.]

31938 - STANKOVIČ, Ž.S. : Starch formation in isolated pea chloroplasts. - In : COOMBS, J. (ed.) : 4th International Congress on Photosynthesis. P. 363. UKISES, London 1977.

31939 - STANKOVIC, Z.S., WALKER, D.A. : Photosynthesis by isolated pea chloroplasts.
Some effects of adenylates and inorganic pyrophosphate. - Plant Physiol. *59* :
428 - 432, 1977.

31940 - STARCK, Z., KARWOWSKA, R. : Gibberellins as factors affecting photosynthesis
and assimilates. - In : KUDREV, T., IVANOVA, I., KARANOV, E. (ed.) : Plant
Growth Regulators. Pp. 183 - 186. Publ. House Bulgar. Acad. Sci., Sofia 1977.

31941 - STARCK, Z., STRADOWSKA, M. : Pattern of growth and ^{14}C-assimilates distributions in relation to photosynthesis in radish plants treated with growth substances. - Acta Soc. Bot. Pol. *46* : 617 - 627, 1977.

31942 - STEELE, J.A., UCHYTIL, T.F., DURBIN, R.D. : The binding of tentoxin to a tryptic digest of chloroplast coupling factor 1. - Biochim. biophys. Acta *459* :
347 - 350, 1977.

31943 - STEFFAN, H., ULLEMEYER, H. : Eine Gaswechselküvette für radioaktive Pflanzen-
versuche zum Selbstbauen. - Angew. Bot. *51* : 287 - 292, 1977.

31944 - STEIGER, H.M., BECK, E., BECK, R. : Oxygen concentration in isolated chloroplasts during photosynthesis. - Plant Physiol. *60* : 903 - 906, 1977.

31945 - STEIGER, H.M., BECK, E., BECK, R. : The oxygen pressure in isolated intact
and broken chloroplasts during photosynthesis. - In : COOMBS, J. (ed.) :
4th International Congress on Photosynthesis. Pp. 363 - 364. UKISES, London
1977.

31946 - STEINHÜBEL, G. : Rozdiely v intenzite spotreby CO_2 vyvolané odlišným spôso-
bom dopestovania semenáčikov a sadeníc. [Different intensity of CO_2 consumption in seedlings and planting stock induced by different methods of cultivation.] - Biológia (Bratislava) *32* : 255 - 264, 1977. [In Slovak, ab : E.]

31947 - STEINHÜBEL, G. : Metodické osobitosti pri štúdiu fotosyntézy drevín. [Special
methodology in the study of photosynthesis of woody plants.] - In : HUZULÁK,
J., MASAROVIČOVÁ, E. (ed.) : Fotosyntéza a Vodný Režim Drevín. Pp. 142 - 166.
Modra-Piesky 1977. [In Slovak, ab : E, R.]

31948 - STEMLER, A. : Membrane bound bicarbonate and Photosystem II. - In : COOMBS, J.
(ed.) : 4th International Congress on Photosynthesis. P. 361. UKISES, London
1977.

31949 - STEMLER, A. : Photosynthetic electron flow from photosystem II requires membrane-bound bicarbonate. - Carnegie Inst. Year Book *76* : 217 - 219, 1977.

31950 - STEMLER, A. : The binding of bicarbonate ions to washed chloroplast grana. -
Biochim. biophys. Acta *460* : 511 - 522, 1977.

31951 - STEPANOVA, A.M., CHESNOKOV, V.A. : Proiskhozhdenie khloroplastov. [Origin of
chloroplasts.] - In : NASYROV, Yu.S. (ed.) : Genetika Fotosinteza. Pp. 21 -
27. Donish, Dushanbe 1977. [In R.]

31952 - STEPHEN, R.C., McDONALD, R.C., KELSON, A. : Comparison of dry matter production from spring-sown cereals sown on different dates. - New. Zeal. J. exp.
Agr. *5* : 59 - 62, 1977.

31953 - STEPHENSON, R.A., WILSON, G.L. : Patterns of assimilate distribution in soybeans at maturity. I The influence of reproductive developmental stage and
leaf position. - Aust. J. agr. Res. *28* : 203 - 209, 1977.

31954 - STEPHENSON, R.A., WILSON, G.L. : Patterns of assimilate distribution in soybeans at maturity. II. The time course of changes in ^{14}C distribution in pods
and stem sections. - Aust. J. agr. Res. *28* : 395 - 400, 1977.

31955 - STEPONKUS, P.L., GARBER, M.P., MYERS, S.P., LINEBERGER, R.D. : Effects of
cold acclimation and freezing on structure and function of chloroplast thylakoids. - Cryobiology *14* : 303 - 321, 1977.

31956 - STERNE, R.E., KAUFMANN, M.R., ZENTMYER, G.A. : Environmental effects on transpiration and leaf water potential in avocado. - Physiol. Plant. *41* : 1 - 6,
1977. [Resistances.]

31957 - STEUP, M., SSYMANK, V., WINKLER, U., GLOCK, H. : Photoregulation of transfer
and 5S ribosomal RNA synthesis in *Chlorella*. - Planta *137* : 139 - 144, 1977.
[Ps.]

31958 - STEVENS, F.J., PANKRATZ, H.S., UFFEN, R.L. : Demonstration of two 3,3-diami-
nobenzidine oxidation reactions associated with photosynthetic membranes in
anaerobic light-grown *Rhodospirillum rubrum*. - J. Histochem. Cytochem. *25* :
1264 - 1268, 1977.

31959 - STEWART, A., LARKUM, A.W.D. : Effect of the plastoquinone antagonist dibro-
mothymoquinone on electron transport reactions in chloroplast preparations. -
Aust. J. Plant Physiol. *4* : 253 - 261, 1977.

31960 - STEWART, A.C., BENDALL, D.S. : Photosynthetic electron transport in a cell-
-free preparation from a thermophilic blue-green alga, *Phormidium laminosum*.
- In : COOMBS, J. (ed.) : 4th International Congress on Photosynthesis.
Pp. 364 - 365. UKISES, London 1977.

31961 - STEWART, I. : Provitamin A and carotenoid content of citrus juices. - J. agr.
Food Chem. *25* : 1132 - 1137, 1977.

31962 - STEWART, J.B. : Evaporation from the wet canopy of a pine forest. - Water
Resour. Res. *13* : 915 - 921, 1977. [Resistances.]

31963 - STEWART, R., AUCHTERLONIE, C.C., CODD, G.A. : Studies on the subunit structure of ribulose-1,5-diphosphate carboxylase from the blue-green alga *Micro-
cystis aeruginosa*. - Planta *136* : 61 - 64, 1977.

31964 - STEWART, W.D.P., ROWELL, P., CODD, G.A., APTE, S.K. : Inter-relations of
photosynthesis and N_2-fixation in photosynthetic prokaryotes. - In :
COOMBS, J. (ed.) : 4th International Congress on Photosynthesis. P. 365.
UKISES, London 1977.

31965 - STEWART, W.D.P., SINADA, F., CHRISTOFI, N., DAFT, M.J. : Primary production
and microbial activity in Scottish fresh water habitats. - In : SKINNER, F.A.,
SHEWAN, J.M. (ed.) : Aquatic Microbiology. Pp. 31 - 54. Academic Press,
London - New York 1977. [Chl.]

31966 - STIGTER, C.J. : Water vapour pressure within a maize crop. - Arch. Meteorol.
Geophys. Bioklimatol., Ser. B, *24* : 349 - 359, 1977.

31967 - STIGTER, C.J., GOUDRIAAN, J., BOTTEMANNE, F.A., BIRNIE, J., LENGKEEK, J.G.,
SIBMA, L. : Experimental evaluation of a crop climate simulation model for
Indian corn (*Zea mays* L.). - Agr. Meteorol. *18* : 163 - 186, 1977. [Ps.]

31968 - STILES, W. : Enclosure method for measuring photosynthesis, respiration and
transpiration of crops in the field. - Grassl. Res. Inst. (Hurley) tech. Rep.
18 : 1 - 42, 1977.

31969 - STILLWELL, E.F., BRYANT, J.P. : Diphenyl mercury effects on cell division,
calcification and net photosynthesis in coccolithophorid *Cricosphaera carte-
rae*. - Sci. Biol. J. *3* : 251 - 258, 1977.

31970 - STILLWELL, W. : On the origin of photophosphorylation. - J. theor. Biol. *65*
: 479 - 497, 1977.

31971 - STILLWELL, W., TIEN, H.T. : Acidification,bleaching and oxygen consumption
with chlorophyll-containing lipid microvesicles. - Biochim. biophys. Acta
461 : 239 - 252, 1977.

31972 - STILLWELL, W., TIEN, H.T. : Protection of chlorophyll by phospholipids from
photooxidation. - Biochem. biophys. Res. Commun. *76* : 232 - 238, 1977.

31973 - STOCKING, C.R., NIVISON, H., LIN, C.H. : Intracellular movement of polypeptides during chloroplast development. - In : COOMBS, J. (ed.) : 4th International Congress on Photosynthesis. P. 366. UKISES, London 1977.

31974 - STOEV, K., IVANTCHEV, V. : Données nouvelles sur le problème de la transло-
cation descendante et ascendante des produits de la photosynthèse de la vigne.
- Vitis *16* : 253 - 262, 1977.

*31975 - STOLBOVA, A.V., FAM TKHAN' KHO, KVITKO, K.V. : Gruppa stsepleniya mutatsiĭ,
narushayushchikh funktsii khloroplasta u khlamidomonady. [The linkage groups
of mutations disturbing the chloroplast functions in *Chlamydomonas*.] - In :
Issledovaniya po Genetike. Vol. 6. Pp. 54 - 60. Izd. leningrad. Univ., Leningrad 1976. [In R.]

31976 - STOLLER, M.L., MALKIN, R., KNAFF, D.B. : Oxidation-reduction properties of photosynthetic nitrite reductase. - FEBS Lett. *81* : 271 - 274, 1977.

31977 - STOUT, R., CLELAND, R.E. : Effects of fusicoccin on the activity of a key pH-stat enzyme, PEP carboxylase. - Plant Physiol. *59* (6, Suppl.) : 16, 1977.

31978 - STOY, V. : Trockensubstanzproduktion und Assimilateinlagerung in das Getreidekorn. - Z. Pflanzenern. Bodenk. *140* : 35 - 50, 1977. [Ps.]

31979 - STRASSER, R.J., BUTLER, W.L. : Energy distribution in the photosynthetic apparatus during development. - In : COOMBS, J. (ed.) : 4th International Congress on Photosynthesis. Pp. 366 - 367. UKISES, London 1977.

31980 - STRASSER, R.J., BUTLER, W.L. : Energy transfer and the distribution of excitation energy in the photosynthetic apparatus of spinach chloroplasts. - Biochim. biophys. Acta *460* : 230 - 238, 1977.

31981 - STRASSER, R.J., BUTLER, W.L. : Fluorescence emission spectra of Photosystem I, Photosystem II and the light-harvesting chlorophyll *a/b* complex of higher plants. - Biochim. biophys. Acta *462* : 307 - 313, 1977.

31982 - STRASSER, R.J., BUTLER, W.L. : The yield of energy transfer and the spectral distribution of excitation energy in the photochemical apparatus of flashed bean leaves. - Biochim. biophys. Acta *462* : 295 - 306, 1977.

*31983 - STREBEYKO, P. : Problem wody i światła w produkcji roślinnej. [The problem of water and light in plant production.] - Biul. warzywniczy *13* (Suppl.) : 83 - 91, 1972. [Ps; in Pol., ab : E, R.]

*31984 - STREBEYKO, P. : Wymaganie świetlne rośliny. [Light requirements of plants.] - Postępy Nauk roln. *1976* (3) : 3 - 20, 1976. [Ps; in Pol.]

31985 - STROTMANN, H., BICKEL-SANDKÖTTER, S. : Energy-dependent exchange of adenine nucleotides on chloroplast coupling factor (CF_1). - Biochim. biophys. Acta *460* : 126 - 135, 1977.

31986 - STROTMANN, H., BICKEL-SANDKÖTTER, S., EDELMANN, K., SCHLIMME, E., BOOS, K.S., LÜSTORFF, J. : Studies on the light adenine nucleotide binding site of chloroplast coupling factor (CF_1). - In : van DAM, K., van GELDER, B.F. (ed.) : Structure and Function of Energy-Transducing Membranes. Pp. 307 - 317. Elsevier, Amsterdam 1977.

31987 - STROTMANN, H., BICKEL-SANDKÖTTER, S., LÜSTORFF, J., BOOS, K.S., SCHLIMME, E., ECKSTEIN, F. : Photophosphorylation and CF_1 binding of ADP analogues. - In : COOMBS, J. (ed.) : 4th International Congress on Photosynthesis. Pp. 367 - 368. UKISES, London 1977.

*31988 - STULL, E.A., AMEZAGA, E. de, GOLDMAN, C.R. : The contribution of individual species of algae to primary productivity of Castle Lake, California. - Verh. int. Verein. Limnol. *18* : 1776 - 1783, 1973.

31989 - SUBRAMANIAN, B.R., RAMADHAS, V., VENUGOPALAN, V.K. : Nannoplankton production in Vellar estuary. - Curr. Sci. *46* : 212 - 215, 1977.

B31990 - SUDACHKOVA, N.E. : Metabolizm Khvoĭnykh i Formirovanie Drevesiny. [Metabolism of Conifers and Wood Formation.] - Nauka, Sibirskoe Otdelenie, Novosibirsk 1977. [Chl, photosynthates; in R.]

31991 - SUD'INA, E.G., DOVBYSH, E.F., GOLOD, M.G., DONTSOVA, I.G., BAĬDULOVA-BABKO, T.Yu. : Khlorofillaza v izolirovannykh list'yakh tabaka. [Chlorophyllase in isolated tobacco leaves.] - Fiziol. Biokhim. kul't. Rast. *9* : 418 - 423, 1977. [In R, ab : E.]

31992 - SUGIYAMA, K.-I., MURATA, N. : Analyses of absorption and fluorescence spectra of water-soluble chlorophyll proteins, pigment system II particles and chlorophyll *a* in diethylether solution by the curve-fitting method. - Biochim. biophys. Acta *503* : 107 - 119, 1977.

31993 - SUGIYAMA, T. : [Biochemical aspects of C_4 photosynthetic CO_2-fixation. Crassulacean acid metabolism.] - Tampakushitsu Kakusan Koso *22* : 366 - 373, 1977. [In Jap.]

31994 - SUMAYAO, C.R., KANEMASU, E.T., HODGES, T. : Soil moisture effects on transpi-
         ration and net carbon dioxide exchange of sorghum. - Agr. Meteorol. *18* : 401
         - 408, 1977.

31995 - SUMIDA, S. : [Mechanism of photosynthesis inhibition by herbicides.] - Kagaku
         to Seibutsu *15* : 404 - 410, 1977. [In Jap.]

31996 - SUMMERFIELD, R.J., DART, P.J., HUXLEY, P.A., EAGLESHAM, A.R.J., MINCHIN, F.R.,
         DAY, J.M. : Nitrogen nutrition of cowpea (*Vigna unguiculata*). I. Effects of
         applied nitrogen and symbiotic nitrogen fixation on growth and seed yield. -
         Exp. Agr. *13* : 129 - 142, 1977. [Dry-matter production.]

*31997 - SUMPER, M., REITMEIER, H., OESTERHELT, D. : Zur Biosynthese der Purpurmembran
         von Halobakterien. - Angew. Chem. *15* : 203 - 210, 1976.

*31998 - SUNDBERG, I., TILBERG, J.-E., BJÖRKLUND, G. : Effects of 2,5-dibromo-3-methyl-
         6-isopropyl-*p*-benzoquinone (DBMIB) on light induced and glycolytic phosphate
         uptake in *Scenedesmus*. - Physiol. Plant. *38* : 141 - 146, 1976. [Ps.]

31999 - SUNDBOM, E., BJÖRN, L.O. : Phytoluminography : Imaging plants by delayed light
         emission. - Physiol. Plant. *40* : 39 - 41, 1977.

32000 - SUNDBOM, E., SAMUELSSON, G., ÖQUIST, G. : Temperature induced changes of chlo-
         rophyll a fluorescence from leaves, isolated chloroplasts and algae. - In :
         COOMBS, J. (ed.) : 4th International Congress on Photosynthesis. P. 368. UKI-
         SES, London 1977.

32001 - SUNDQVIST, C. : Phytochrome regulated protochlorophyllide accumulation in dark
         grown leaves treated with δ-aminolevulinic acid. - In : COOMBS, J. (ed.) :
         4th International Congress on Photosynthesis. P. 370. UKISES, London 1977.

32002 - SUSALLA, A.A., MAHLBERG, P.G. : Young plastids in shoot apical cells of a ge-
         netic albino strain of *Nicotiana (Solanaceae)*. - Can. J. Bot. *55* : 1429 -
         1433, 1977.

B32003 - SUTCLIFFE, J. : Plants and Temperature. (Studies in Biology. Vol. 86.) - Edward
         Arnold, London 1977. [Ps.]

B32004 - SUTCLIFFE, J.F., PATE, J.S. (ed.) : The Physiology of the Garden Pea. (Expe-
         rimental Botany: An International Series of Monographs. Vol.12.) - Academic
         Press, London - New York - San Francisco 1977. [Ps.]

32005 - SUZUKI, K. : Photochemical reaction systems of photosynthesis in *Phytolacca
         americana*. II. Isolation and characterization of the photosystem particles. -
         Plant Cell Physiol. *1977* (Spec. Issue 3 - Photosynthetic Organelles. Structure
         and Function) : 415 - 425, 1977.

32006 - SUZUKI, M., NAKAMURA, K. :[On dry matter production and efficiency for solar
         energy utilization by rice population in warmer region of Japan.]- Jap. J.
         Crop Sci. *46* : 530 - 536, 1977. [In Jap., ab : E.]

32007 - SUZUKI, R., FUJITA, Y. : Carotenoid photobleaching induced by the action of
         photosynthetic reaction center II : DCMU-sensitivity. - Plant Cell Physiol.
         18 : 625 - 631, 1977.

32008 - ŠVACHULA, V., KOHOUT, V. : Porovnání metabolismu cukrové řepy a merlíku bílé-
         ho v diferencovaných vláhových podmínkách. [Comparison of metabolism of sugar
         beet and dungweed under differentiated moisture conditions.] - Rostlinná Vý-
         roba (Praha) *23* : 77 - 88, 1977. [Chl; in Czech, ab : E, R.]

32009 - ŠVACHULA, V., ŠVACHULOVÁ, J. : Die Erhöhung der Peroxidasenaktivität, des
         Chlorophyllgehaltes und der Zuckerproduktion in der Zuckerrübe in bewässerter
         Fruchtfolge. - Z. Acker- Pflanzenbau *145* : 142 - 153, 1977.

32010 - ŠVACHULA, V., TORNIKIDU, J., ZAHRADNÍČEK, J. : Vliv vědecky řízené závlahy
         cukrovky na její technologickou jakost a posklizňový metabolismus. [Influence
         of scientific control of irrigation of sugar beet on its technological quality
         and on metabolism after harvesting.]-Listy cukrovarnické (Praha) *93* : 35 - 41,
         1977. [Production; in Czech, ab : E, G, R.]

32011 - ŠVIHRA, J. : Fotosyntetická variabilita listov niektorých druhov ovocných dre-
         vín. [Photosynthetic variability of leaves of some fruit tree species.] - In :

HUZULÁK, J., MASAROVIČOVÁ, E. (ed.) : Fotosyntéza a Vodný Režim Drevín. Pp. 175 - 184. Modra-Piesky 1977. [In Slovak, ab : E,R.]

32012 - SVOBODA, J. : Ecology and primary production of raised beach communities, Truelove Lowland. - In : BLISS, L.C. (ed.) : Truelove Lowland, Devon Island, Canada : A High Arctic Ecosystem. Pp. 185 - 216. Univ. Alberta Press, Edmonton 1977.

32013 - SWADER, J.A. : Autoxidation of tricine by 2-anthraquinonesulfonic acid and light. - Photosynthetica 11 : 327 - 329, 1977.

32014 - SWYSEN, C., SYMONS, M. : The light-induced carotenoid absorbance changes in Rhodopseudomonas sphaeroides : A calibration of the bandshifts versus the membrane potential. - In : COOMBS, J. (ed.) : 4th International Congress on Photosynthesis. Pp. 370 - 371. UKISES, London 1977.

32015 - SYBER, A.Yu., MOLDAU, Kh.A. : Apparatura dlya opredeleniya soprotivleniya ust'its i vodosoderzhaniya lista s razdel'nym konditsionirovaniem rasteniya i otdel'nogo lista. [An apparatus with separate conditioning of the plant and the leaf for the determination of stomatal resistance and leaf water content.] - Fiziol. Rast. 24 : 1301 - 1307, 1977.

32016 - SYKUT, A. : Influence of mineral medium components - calcium, phosphorus and potassium - on carotenoid biosynthesis in the alga Scenedesmus acuminatus. - Acta Soc. Bot. Pol. 46 : 339 - 346, 1977.

32017 - SYMONS, M., SWYSEN, C., SYBESMA, C. : An investigation into the different populations of carotenoids in chromatophores from Rhodopseudomonas sphaeroides and capsulata. - In : PACKER, L., PAPAGEORGIOU, G.C., TREBST, A. (ed.) : Bioenergetics of Membranes. Pp. 477 - 483. Elsevier/North-Holland Biomedical Press, Amsterdam - Oxford - New York 1977.

32018 - SYMONS, M., SWYSEN, C., SYBESMA, C. : The light-induced carotenoid absorbance changes in Rhodopseudomonas sphaeroides. An analysis and interpretation of the band shifts. - Biochim. biophys. Acta 462 : 706 - 717, 1977.

32019 - SYMONS, M., SYBESMA, C. : Light-induced carotenoid absorbance changes in chromatophores from Rhodopseudomonas sphaeroides G1C : Analysis and interpretation of the bandshifts. - In : COOMBS, J. (ed.) : 4th International Congress on Photosynthesis. Pp. 371 - 372. UKISES, London 1977.

32020 - SYVERTSEN, J.P., CUNNINGHAM, G.L. : Rate of leaf production and senescence and effect of leaf age on net gas exchange in creosotebush. - Photosynthetica 11 : 161 - 166, 1977.

32021 - SZALAY, L. : Research on photosynthetic pigments in the Institute of Biophysics, University of Szeged. - Acta Univ. Iodz., Ser. II, 16 : 15 - 33, 1977.

32022 - SZANIAWSKI, R., ŻELAWSKI, W., WIERZBICKI, B. : Wymiana gazowa i gospodarka wodna. [Gas exchange and water relations.] - In : BIAŁOBOK, S. (ed.) : Swierk Pospolity - Picea abies (L.) KARST. (Nasze Drzewa Lesne - Monografie Popularnonaukowe). Pp. 131 - 152. Państwowe Wydawnictwo Naukowe, Warszawa 1977. [In Pol., ab : E.]

32023 - SZANIAWSKI, R.K. : Diurnal and seasonal patterns of gas exchange in Scots pine (Pinus silvestris L.) seedlings grown under laboratory conditions. - In : HUZULÁK, J., MASAROVIČOVÁ, E. (ed.) : Fotosyntéza a Vodný Režim Drevín. Pp. 251 - 258. Modra-Piesky 1977.

32024 - SZAREK, S.R. : Osmoregulation of malic acid accumulation in leaf tissue. - Plant Physiol. 59 (6, Suppl.) : 116, 1977.

32025 - SZAREK, S.R., TING, I.P. : The occurrence of crassulacean acid metabolism among plants. - Photosynthetica 11 : 330 - 342, 1977.

32026 - SZAREK, S.R., WOODHOUSE, R.M. : Ecophysiological studies of Sonoran desert plants. II. Seasonal photosynthesis patterns and primary production of Ambrosia deltoidea and Olneya tesota. - Oecologia 28 : 365 - 375, 1977.

32027 - SZAREK, S.R., WOODHOUSE, R.M. : Gas exchange estimates of primary productivity for Sonoran desert plants. - Plant Physiol. 59 (6, Suppl.) : 98, 1977.

32028 - SZEWCZYK, E., RYCZKOWSKI, M. : Changes in the chlorophyll content in the developing embryo (dicotyledonus plants). - Bull. Acad. pol. Sci., Sér. Sci. biol. *25* : 679 - 683, 1977.

32029 - TAGEEVA, S.V., POPOV, V.I. : Relation between structural arrangement of chloroplast membranes and their functional state. - In : COOMBS, J. (ed.) : 4th International Congress on Photosynthesis. P.373. UKISES, London 1977.

*32030 - TAGEEVA, S.V., SEMYENOVA, G.A., LADYGIN, V.I. : Peculiarities of ultrafine structure of pigment mutants *Chlamydomonas reinhardi*. - In : Fourth European Regional Conference on Electron Microscopy. Abstracts. Pp. 391 - 392. Rome 1968.

32031 - TAGUCHI, S. : Photosynthesis of nannoplankton and netplankton at the subsurface chlorophyll maximum layer in the tropical North Pacific Ocean. - J. Phycol. *13* (Suppl.) : 66, 1977.

32032 - TAGUCHI, S., PLATT, T. : Assimilation of $^{14}CO_2$ in the dark compared to phytoplankton production in a small coastal inlet. - Estuar. coast. mar. Sci. *5* : 679 - 684, 1977.

32033 - TAI, E.A. : Banana. - In : ALVIM, P. de T., KOZLOWSKI, T.T. (ed.) : Ecophysiology of Tropical Crops. Pp. 441 - 460. Academic Press, New York - San Francisco - London 1977. [Production.]

32034 - TAI, P.Y.P., HAMMONS, R.O., MATLOCK, R.S. : Genetic relationships among three chlorophyll-deficient mutants in peanut, *Arachis hypogaea* L. - Theor. appl. Genet. *50* : 35 - 40, 1977.

32035 - TAKABE, T. : Ribulose-1,5-bisphosphate carboxylase from the blue-green alga, *Anabaena cylindrica*. - Agr. biol. Chem. *41* : 2255 - 2260, 1977.

32036 - TAKABE, T., AKAZAWA, T. : A comparative study on the effect of O_2 on photosynthetic carbon metabolism by *Chlorobium thiosulfatophilum* and *Chromatium vinosum*. - Plant Cell Physiol. *18* : 753 - 765, 1977.

32037 - TAKAGI, S., MATSUGAMI, M. : [Effect of carotenoid on leaf lipoxygenase activity. II. Studies on the inhibition of lipoxygenase activity by lutein.] - J. agr. chem. Soc. Japan *51* : 489 - 495, 1977. [In Jap., ab : E.]

32038 - TAKAGI, S., MATSUGAMI, M., MORITOKI, T. : [Effect of carotenoid on leaf lipoxygenase activity I. Purification of spinach chloroplast lipoxygenase and the inhibition by lutein to its activity.] - Sci. Rep. Fac. Agr. Okayama Univ. *49* : 35 - 44, 1977. [In Jap., ab : E.]

32039 - TAKAHAMA, U., NISHIMURA, M. : Light-induced chemiluminescence of luminol in spinach chloroplast fragments : Reaction of O_2^- with electron transfer components. - Plant Cell Physiol. *18* : 1139 - 1148, 1977.

32040 - TAKAHAMA, U., SHIMIZU, M., NISHIMURA, M. : Temperature-jump-induced release of H^+ from chloroplasts : A relationship between the release of H^+ and reverse electron transfer . - Plant Cell Physiol. *1977* (Spec. Issue 3 - Photosynthetic Organelles. Structure and Function) : 149 - 156, 1977.

32041 - TAKAHASHI, K., INABA, T., KAJIWARA, T. : Distribution of ^{14}C assimilated from $[^{14}C]O_2$ in cucumber leaves infected with downy mildew. - Physiol. Plant Pathol. *11* : 255 - 259, 1977.

32042 - TAKAHASHI, M., ASADA, K. : Manganese binding to sodium cyanide-treated chloroplasts : Effects of light and redox-potentials on the binding. - Plant Cell Physiol. *18* : 807 - 814, 1977.

32043 - TAKAHASHI, M., NORTON, A.B. : Seasonal changes in microbial biomass in the Fraser River estuary, British Columbia. - Arch. Hydrobiol. *79* : 133 - 143, 1977. [Chl.]

32044 - TAKAHASHI, M., SEIBERT, D.L., THOMAS, W.H. : Occasional blooms of phytoplankton during summer in Saanich Inlet, B.C., Canada. - Deep-Sea Res. *24* : 775 - 780, 1977. [Chl.]

32045 - TAKAMIYA, K.-I., DUTTON, P.L. : The influence of transmembrane potentials of the redox equilibrium between cytochrome c_2 and the reaction center in *Rhodopseudomonas sphaeroides* chromatophores. - FEBS Lett. *80* : 279 - 284, 1977.

32046 - TAKEDA, G., IWAKI, H., TAKAYANAGI, S. : [Ecological studies on the photosynthesis of winter cereals IV. Model simulation of dry matter growth of six-rowed barley.] - Jap. J. Crop Sci. *46* : 178 - 192, 1977. [In Jap., ab : E.]

32047 - TAKEDA, T., AGATA, W., HAKOYAMA, S., TANAKA, H. : [Studies on weed vegetation in non-cultivated paddy fields. II. The relation between the ecological distribution of gramineous C_3- and C_4-weeds and the soil moisture condition in non-cultivated paddy fields.] - Jap. J. Crop Sci. *46* : 558 - 568, 1977. [Ps; in Jap., ab : E.]

32048 - TAKEMOTO, J., HUANG KAO, M.Y.C. : Effects of incident light levels on photosynthetic membrane polypeptide composition and assembly in *Rhodopseudomonas sphaeroides*. - J. Bacteriol. *129* : 1102 - 1109, 1977.

32049 - TAMAS, I.A., OBERLANDER, R., BECKER, J., SHERMAN, D. : Oxidation of indole-acetic acid by chloroplast preparations. - Plant Physiol. *59* (6, Suppl.) : 11, 1977.

*32050 - TAMURA, K., NARUSAWA, T., KANAZAWA, T. :[Steady state levels of ATP in *Chlorella* cells and transitional changes of ATP level accompanied with environmental changes.]- Sagami Joshi Daigaku Kiyo [J. Sagami Women's Univ.] *40* : 27 - 32, 1976. [In Jap.]

32051 - TAN, C.S., BLACK, T.A., NNYAMAH, J.U. : Characteristics of stomatal diffusion resistance in a Douglas fir forest exposed to soil water deficits. - Can. J. Forest Res. *7* : 595 - 604, 1977.

32052 - TAN, W.K., TAN, G.Y., WALTON, P.D. : Canopy characters and their relationship to spring productivity in *Bromus inermis* LEYSS. - Crop Sci. *17* : 7 - 10, 1977.

32053 - TANDEAU de MARSAC, N. : Occurrence and nature of chromatic adaptation in *Cyanobacteria*. - J. Bacteriol. *130* : 82 - 91, 1977.

32054 - TANDEAU de MARSAC, N., COHEN-BAZIRE, G. : Molecular composition of cyanobacterial phycobilisomes. - Proc. nat. Acad. Sci. USA *74* : 1635 - 1639, 1977.

32055 - ŢÂRA, G., BERCEA, V., ŞTIRBAN, M. : Activitatea fiziologică a frunzelor clorozate la vita de vie. [Physiological activity of chlorosis-affected grapevine leaves.] - Stud. Cercet. Biol., Ser. Biol. Veg. *29* (1) : 67 - 71, 1977. [Ps, Chl; in Roum.]

B32056 - TARCHEVSKIĬ, I.A. : Osnovy Fotosinteza. [Fundamentals of Photosynthesis.] - Vysshaya Shkola, Moskva 1977.

32057 - TARCHEVSKIĬ, I.A., CHIKOV, V.I., IVANOVA, A.P., SULEĬMANOVA, A.Yu., ANDRIANOVA, Yu.E. : Osobennosti fotosinteza i ottoka assimilyatov u razlichnykh sortov yarovoĭ pshenitsy. [Characteristics of photosynthesis and photosynthate efflux in various cultivars of spring wheat.] - In : NASYROV, Y.S. (ed.) : Genetika Fotosinteza. Pp. 234 - 237. Donish, Dushanbe 1977. [In R.]

32058 - TARGON, P.G., DOBROVOL'SKAYA, M.G., GRUBAYA, Zh.F., MAKOVETSKAYA, R.V. : Vodouderzhivayushchaya sposobnost' nekotorykh drevesnykh porod, introdutsirovannykh v Moldavii, i ee zavisimost' ot obmena veshchestv. [Water holding capacity of some tree species introduced in Moldavia and dependence of it on metabolism.] - Nauch. Dokl. vyssh. Shkoly, biol. Nauki *20* (12) : 103 - 107, 1977. [Chl; in R.]

32059 - TARILA, A.G.I., ORMROD, D.P., ADEDIPE, N.O. : Effects of phosphorus nutrition and light intensity on growth and development of the cowpea (*Vigna unguiculata* L.).- Ann. Bot. *41* : 75 - 83, 1977.

32060 - TAYLOR, J.A., MACKENDER, R.O. : Envelopes from developing plastids of light--grown *Avena sativa* L. leaves. - In : COOMBS, J. (ed.) : 4th International Congress on Photosynthesis. P. 374. UKISES, London 1977.

32061 - TAYLOR, J.A., MACKENDER, R.O. : Plastid development in the first leaf of *Avena sativa* L. - Plant Physiol. *59* (6, Suppl.) : 10, 1977.

32062 - **TAYO, T.O.** : Comparative analysis of the growth, development and yield of three soya-bean varieties (*Glycine max* L.). - J. agr. Sci. *88* : 151 - 157, 1977. [Production.]

32063 - **TCHAN, Y.T., CHIOU, A.C.M., NEW, P.B., FUNNELL, G.R.** : The photobioluminometer, an instrument for the study of ecological factors affecting photosynthesis. - Microbial Ecol. *3* : 327 - 332, 1977.

*32064 - **TeBEEST, D.O., DURBIN, R.D., KUNTZ, J.E.** : Stomatal resistance of red oak seedlings infected by *Ceratocystis fagacearum*. - Phytopathology *66* : 1295 - 1297, 1976.

32065 - **TEERI, J.A., PATTERSON, D.T., ALBERTE, R.S., CASTLEBERRY, R.M.** : Changes in the photosynthetic apparatus of maize in response to simulated natural temperature fluctuations. - Plant Physiol. *60* : 370 - 373, 1977.

32066 - **TEH, K.H., SWANSON, C.A.** : Sulfur dioxide inhibition on translocation and photosynthesis in *Phaseolus vulgaris* L. - Plant Physiol. *59* (6, Suppl.) : 123, 1977.

32067 - **TEKALE, N.S., JOSHI, R.N.** : Studies on stability of carotenoid pigments in lucerne vegetation and juice. - Indian J. Nutr. Diet. *14* : 161 - 166, 1977.

32068 - **TELFER, A., BARBER, J.** : Dual action of ionophore A23187 on intact chloroplasts. - In : COOMBS, J. (ed.) : 4th International Congress on Photosynthesis. Pp. 375 - 376. UKISES, London 1977.

32069 - **TEL-OR, E., CAMMACK, R., RAO, K.K., ROGERS, L.J., STEWART, W.D.P., HALL, D.O.:** Comparative immunochemistry of bacterial, algal and plant ferredoxins. - Biochim. biophys. Acta *490* : 120 - 131, 1977.

32070 - **TEL-OR, E., LUIJK, L., PACKER, L.** : Hydrogenase in heterocystous cyanobacteria. - In : COOMBS, J. (ed.) : 4th International Congress on Photosynthesis. Pp. 374 - 375. UKISES, London 1977.

32071 - **TEL-OR, E., MALKIN, S.** : The photochemical and fluorescence properties of whole cells, spheroplasts and spheroplast particles from the blue-green alga *Phormidium luridum*. - Biochim. biophys. Acta *459* : 157 - 174, 1977.

32072 - **TEL-OR, E., STEWART, W.D.P.** : Photosynthetic components and activities of nitrogen-fixing isolated heterocysts of *Anabaena cylindrica*. - Proc. roy. Soc. Lond. Ser. B *198* : 61 - 86, 1977.

32073 - **TENHUNEN, J.D., WEBER, J.A., FILIPEK, L.H., GATES, D.M.** : Development of a photosynthesis model with an emphasis on ecological applications. III. Carbon dioxide and oxygen dependencies. - Oecologia *30* : 189 - 207, 1977.

32074 - **TERAO, J., MATSUSHITA, S.** : Structures of monohydroperoxides produced from chlorophyll sensitized photooxidation of methyl linoleate. - Agr. biol. Chem. (Tokyo) *41* : 2467 - 2468, 1977.

32075 - **TEREKHOVA, I.V., BELYAEVA, E.V., DOMAN, N.G.** : Izuchenie produktov fotosinteza khlorelly s pomoshch'yu prizhiznennogo ělektroforeza. [Products of photosynthesis of *Chlorella* studied by *intravitam* electrophoresis.] - Fiziol. Rast. *24* : 1088 - 1093, 1977. [In R, ab : E.]

32076 - **TERMAN, G.L., NOGGLE, J.C., HUNT, C.M.** : Growth rate - nutrient concentration relationships during early growth of corn as affected by applied N, P, and K. - Soil Sci. Soc. Amer. J. *41* : 363 - 368, 1977.

32077 - **TERPSTRA, W.** : Possible relationship between chlorophyllase and other enzymes in photosynthetic membranes. - In : COOMBS, J. (ed.) : 4th International Congress on Photosynthesis. P. 376. UKISES, London 1977.

32078 - **TERPSTRA, W.** : A study of properties and activity of chlorophyllase in photosynthetic membranes. - Z. Pflanzenphysiol. *85* : 139 - 146, 1977.

32079 - **TERRY, N.** : Photosynthesis, growth, and role of chloride. - Plant Physiol. *60* : 69 - 75, 1977.

32080 - **TERRY, N.** : Is chloride really required for photosynthesis? - In : COOMBS, J. (ed.) : 4th International Congress on Photosynthesis. P. 377. UKISES, London 1977.

32081 - **TERSKOV, I.A., KHARUK, V.I., SPIROV, V.V.** : Issledovanie khlorofillsoderzhashchikh tkanei metodom mikrofotometricheskogo skanirovaniya. [Chlorophyll-containing tissues studied by microphotometric scanning.] - Fiziol. Rast. *24* : 10 - 17, 1977. [In R, ab : E.]

32082 - **TETLEY, R.M., BISHOP, N.I.** : Oxygen stimulation of H_2 production in a blue-green algae. - Plant Physiol. *59* (6, Suppl.) : 130, 1977.

32083 - **TETT, P., KELLY, M.G., HORNBERGER, G.M.** : Estimation of chlorophyll a and pheophytin a in methanol. - Limnol. Oceanogr. *22* : 579 - 580, 1977.

32084 - **TEVINI, M.** : Light, function, and lipids during plastid development. - In : TEVINI, M., LICHTENTHALER, H.K. (ed.) : Lipids and Lipid Polymers in Higher Plants. Pp. 121 - 145. Springer-Verlag, Berlin - Heidelberg - New York 1977.

32085 - **TEVINI, M., FREY, R.** : Lipid metabolism and function of developing and senescing spinach chloroplasts. - In : COOMBS, J. (ed.) : 4th International Congress on Photosynthesis. P. 378. UKISES, London 1977.

32086 - **TEVINI, M., HERM, K., LEONHARDT, H.-D.** : Lipids and function of etiochloroplasts after ultraviolet, blue and red light-illumination. - Biochem. Soc. Trans. *5* : 95 - 98, 1977. [Ps, Chl.]

32087 - **TEVZADZE, I.T., TARKHNISHVILI, G.M., SANADZE, G.A.** : Vliyanie kisloroda na izmenenie izotopnogo otnosheniya $C^{13}O_2/C^{12}O_2$ v atmosfere zamknutykh kamer pri fotosinteze. [The effect of oxygen on a change in the $^{13}CO_2/^{12}CO_2$ isotope ratio in the atmosphere of closed chambers during photosynthesis.] - Fiziol. Rast. *24* : 1073 - 1075, 1977. [In R.]

32088 - **TEW, J.** : The effect of light intensity, oxygen and nitrate concentration on the rate of $^{14}CO_2$ in the light and dark from C_3 and C_4 photosynthetic plant leaves. - In : COOMBS, J. (ed.) : 4th International Congress on Photosynthesis. P. 372. UKISES, London 1977.

32089 - **TEZUKA, Y.** : The effect of nutrient concentration on the standing crop of *Scenedesmus obliquus* grown under continuous culture condition. - Rikusuigaku Zasshi [Jap. J. Limnol.] *38* (3) : 90 - 93, 1977. [Chl.]

* 32090 - **THEODÓRSSON, P., BJARNASON, J.Ö.** : The acid-bubbling method for primary productivity measurements modified and tested. - Limnol. Oceanogr. *20* : 1018 - 1019, 1975.

*32091 - **THERING, D.B., WOHLER, J.R.** : Studies on the productivity of a periphyton community. - Proc. Pennsylvania Acad. Sci. *48* : 61 - 64, 1974. [Ps.]

32092 - **THIMANN, K.V., TETLEY, R.M., KRIVAK, B.M.** : Metabolism of oat leaves during senescence. V. Senescence in light. - Plant Physiol. *59*: 448 - 454, 1977. [Chl.]

32093 - **THINH, L.V., GRIFFITHS, D.J.** : Studies of the relationship between the ascidian *Diplosoma virens* and its associated microscopic algae. I. Photosynthetic characteristics of the algae. - Aust. J. mar. Freshwater Res. *28* : 673 - 681, 1977.

32094 - **THOFELT, L.** : Thermographic field studies on leaf temperature. - Oikos *29* : 180 - 185, 1977.

32095 - **THOMANN, R.V.** : Comparison of lake phytoplankton models and loading plots. - Limnol. Oceanogr. *22* : 370 - 373, 1977.

32096 - **THOMAS, D.A., REBELLA, C., CHARTIER, P.** : An analysis of the vertically reflected radiation from a maize crop as a possible means of determining its biomass and water content. - Agr. Meteorol. *18* : 101 - 114, 1977.

32097 - **THOMAS, H.** : Ultrastructure, polypeptide composition and photochemical activity of chloroplasts during foliar senescence of a non-yellowing mutant genotype of *Festuca pratensis* HUDS. - Planta *137* : 53 - 60, 1977.

32098 - **THOMAS, H., STODDART, J.L.** : Biochemistry of leaf senescence in grasses. - Ann. appl. Biol. *85* : 461 - 463, 1977.

32099 - **THOMAS, J.B., KLEINEN HAMMANS, J.W., VERWER, W.** : On the quantitative relationship of chlorophyll b and the chlorophyll a form c_a685 in the light-harvesting pigment-protein complex of chloroplasts. - In : COOMBS, J. (ed.) : 4th International Congress on Photosynthesis. Pp. 378 - 379. UKISES, London 1977.

32100 - THOMAS, J.B., KLEINEN HAMMANS, J.W., VERWER, W. : On the relative extractability of chlorophyll $a685$ and chlorophyll b complexes by two detergents. - Acta bot. neerl.26 : 321 - 325, 1977.

32101 - THOMAS, J.B., KLEINEN HAMMANS, J.W., VERWER, W. : On the quantitative relationship between chlorophyll b and the chlorophyll a from C_a685 in the light-harvesting pigment-protein complex of chloroplasts. - Acta bot. neerl. 26 : 377 - 383,1977.

32102 - THOMAS, J.C. : Modification of phycobiliprotein content and Photosystem II alteration in relation to nitrogen privation in two oscillatoriacean species. - In : COOMBS, J. (ed.) : 4th International Congress on Photosynthesis. Pp. 379 - 380. UKISES, London 1977.

32103 - THOMAS, J.R., GAUSMAN, H.W. : Leaf reflectance vs. leaf chlorophyll and carotenoid concentrations for eight crops. - Agron. J. 69 : 799 - 802, 1977.

32104 - THOMAS, S., BIRD, I.F., CORNELIUS, M.J., KEYS, A.J. : Photosynthesis, photorespiration and productivity: Measurement of photorespiration in field crops. - In : COOMBS, J. (ed.) : 4th International Congress on Photosynthesis. Pp. 380 - 381. UKISES, London 1977.

32105 - THOMAS, T.A. : An automated procedure for the determination of soluble carbohydrates in herbage. - J. Sci. Food Agr. 28 : 632 - 642, 1977.

32106 - THOMAS, T.H. : Hormonal aspects of senescence in green vegetable crops. - Ann. appl. Bot. 85 : 421 - 424, 1977.

32107 - THOMAS, W.H., HOLM-HANSEN, O., SEIBERT, D. : Effects of pollutants on natural marine phytoplankton: The CEPEX experience. - J. Phycol. 13 (Suppl.) : 67, 1977.

32108 - THOMAS, W.H., HOLM-HANSEN, O., SEIBERT, D.L.R., AZAM, F., HODSON, R., TAKAHASHI, M. : Effects of copper on phytoplankton standing crop and productivity: controlled ecosystem pollution experiment. - Bull. mar. Sci. 27 : 34 - 43, 1977. [Chl.]

32109 - THOMPSON, A., VOGEL, J., LEE, R.E. : Carbon dioxide uptake in relation to a plastid inclusion body in the succulent *Kalanchoë pinnata* PERSOON. - J. exp. Bot. 28 : 1037 - 1041, 1977.

32110 - THOMPSON, D.R., HINCKLEY, T.M. : A simulation of water relations of white oak based on soil moisture and atmospheric evaporative demand. - Can. J. Forest Res. 7 : 400 - 409, 1977. [Resistances.]

32111 - THOMPSON, D.R., HINCKLEY, T.M. : Effect of vertical and temporal variations in stand microclimate and soil moisture on water status of several species in an oak-hickory forest. - Amer. Midl. Naturalist 97 : 373 - 380, 1977. [Resistances.]

32112 - THORNBER, J.P., ALBERTE, R.S. : The organization of chlorophyll *in vivo*. - In : TREBST, A., AVRON, M. (ed.) : Photosynthesis I. (Encycl. Plant Physiol. N.S. Vol.5.) Pp. 574 - 582. Springer-Verlag, Berlin - Heidelberg - New York 1977.

32113 - THORNBER, J.P., ALBERTE, R.S., HUNTER, F.A., SHIOZAWA, J.A., KAN, K.-S. : The organization of chlorophyll in the plant photosynthetic unit. - In : OLSON, J.M., HIND, G. (ed.) : Chlorophyll-Proteins, Reaction Centers, and Photosynthetic Membranes. Pp. 132 - 148. Brookhaven nat. Lab., Upton 1977.

32114 - THORNBER, J.P., DUTTON, P.L., FAJER, J., FORMAN, A., OLSON, J.M., PRINCE, R.C. : Isolated photochemical reaction centers from bacteriochlorophyll b-containing organisms. - In : COOMBS, J. (ed.) : 4th International Congress on Photosynthesis. P. 382. UKISES, London 1977.

32115 - THORNE, G.N., THOMAS, S.M., PEARMAN, I. : Effect of nitrogen on photosynthesis and respiration in crops of winter wheat. - In : Produkce Biomasy a Tvorba Výnosů Polních Plodin. Vol.1. Pp. 91 - 100. Česká vědeckotechnická Společnost zemědělská, Praha 1977.

32116 - THORNE, S.W., NEWCOMB, E.H., OSMOND, C.B. : Identification of chlorophyll b in extracts of prokaryotic algae by fluorescence spectroscopy. - Proc. nat. Acad. Sci. USA 74 : 575 - 578, 1977.

32117 - THORNLEY, J.H.M. : Interpretation of shoot : root relationships. - Ann. Bot.
        41 : 461 - 464, 1977. [Dry-matter accumulation.]

32118 - THORNLEY, J.H.M. : Growth, maintenance and respiration: a re-interpretation.
        - Ann. Bot. *41* : 1191 - 1203, 1977.

32119 - THORNLEY, J.H.M. : Modelling as a tool in plant physiological research. - In:
        LANDSBERG, J.J., CUTTING, C.V. (ed.) : Environmental Effects on Crop Physio-
        logy. Pp. 339 - 350. Academic Press, London - New York - San Francisco 1977.
        [Ps.]

32120 - THORPE, M.R., BUTLER, D.R. : Heat transfer coefficients for leaves on orchard
        apple trees. - Boundary-Layer Meteorol. *12* : 61 - 73, 1977. [Resistances.]

32121 - THORPE, N., MILTHORPE, F.L. : Stomatal metabolism: CO_2 fixation and respirat-
        ion. - Aust. J. Plant Physiol. *4* : 611 - 621, 1977.

32122 - THROWER, S.L. : Translocation into mature leaves - the effect on growth pat-
        tern.-New Phytol. *78* : 361 - 364, 1977. [Photosynthates.]

*32123 - THURNAUER, M.C., KATZ, J.J., NORRIS, J.R. : The triplet state in bacterial
        photosynthesis: Possible mechanisms of the primary photo-act. - Proc. nat.
        Acad. Sci. USA *72* : 3270 - 3274, 1975.

32124 - THURNAUER, M.C., KATZ, J.J., NORRIS, J.R. : Electronic properties of singlet
        chlorophyll as revealed by a study of the triplet state. - In : OLSON, J.M.,
        HIND, G. (ed.) : Chlorophyll-Proteins, Reaction Centers, and Photosynthetic
        Membranes. P. 369. Brookhaven nat. Lab., Upton 1977.

32125 - TIBONI, O., PARISI, B., CIFERRI, O. : The cellular site of synthesis chloro-
        plast elongation factors. - In : BOGORAD, L., WEIL, J.H. (ed.) : Acides Nuclé-
        iques et Synthèse des Protéines chez les Végétaux. Coll. Int. CNRS. No. 261.
        Pp. 345 - 349. Édit. CNRS, Paris 1977.

32126 - TICHÁ, I., ČATSKÝ, J. : Ontogenetic changes in the internal limitations to
        bean-leaf photosynthesis. 3. Leaf mesophyll structure and intracellular con-
        ductance for carbon dioxide transfer. - Photosynthetica *11* : 361 - 366, 1977.

32127 - TIEDE, D.M., PRINCE, R.C., DUTTON, P.L. : Electron paramagnetic resonance and
        optical properties of an electron-carrier intermediate in reaction centers of
        Chromatium vinosum. - In : OLSON, J.M., HIND, G. (ed.) : Chlorophyll-Proteins,
        Reaction Centers, and Photosynthetic Membranes. P. 368. Brookhaven nat. Lab.,
        Upton 1977.

32128 - TIEFERT, M.A., ROY, H., MOUDRIANAKIS, E.N. : Binding of adenine nucleotides
        and pyrophosphate by the purified coupling factor of photophosphorylation. -
        Biochemistry *16* : 2396 - 2404, 1977..

32129 - TIEFERT, M.A., ROY, H., MOUDRIANAKIS, E.N. : Conversion of bound adenine nuc-
        leotides by the purified coupling factor of photophosphorylation.-Biochemistry
        16 : 2404 - 2409, 1977.

32130 - TIEMANN, R., RENGER, G., GRÄBER, P., WITT, H.T. : On the coupling of vectori-
        al transport of electrons and protons at the plastoquinone pool in photosynthe-
        sis. - In : COOMBS, J. (ed.) : 4th International Congress on Photosynthesis.
        P. 381a. UKISES, London 1977.

32131 - TIEN, H.T. : Bilayer lipid membranes (BLM): An experimental model for the thy-
        lakoid membrane. - In : COOMBS, J. (ed.) : 4th International Congress on Pho-
        tosynthesis. P. 381. UKISES, London 1977.

32132 - TIEN, H.T. : Photoelectric bilayer lipid membrane : a model for the thylakoid
        membrane. - In : OLSON, J.M., HIND, G. (ed.) : Chlorophyll-Proteins, Reaction
        Centers, and Photosynthetic Membranes. Pp. 105 - 131. Brookhaven nat. Lab.,
        Upton 1977.

*32133 - TIKHOMIROV, A.A., ZOLOTUKHIN, I.G., SID'KO, F.Ya. : Vliyanie svetovykh rezhi-
        mov na produktivnost' i kachestvo urozhaya redisa. [Effect of illuminance on
        the productivity and quality of radish crops.] - Fiziol. Rast. *23* : 502 - 507,
        1976. [Ps, Chl; in R, ab : E.]

32134 - TIKHONOV, A.N., RUUGE, È.K., SUBCHINSKII, V.K. : Issledovanie élektronnogo trans-
        porta v fotosinteticheskikh sistemakh vysshikh rastenil metodom ÈPR. IV. Vliya-

nie dvukhvalentnykh kationov na fotoindutsirovannye okislitel'no-vosstanovi-tel'nye prevrashcheniya P700. [Investigation of electron transport in photo-synthetic systems of higher plants by the EPR method. IV. Effect of bivalent cations on photoinduced redox transformation of P700.] - Biofizika 22 : 833 - 839, 1977. [In R, ab : E.]

32135 - TILLBERG, J.-E., GIERSCH, C., HEBER, U. : CO_2 reduction by intact chloroplasts under a diminished proton gradient. - Biochim. biophys. Acta 461 : 31 - 47, 1977.

*32136 - TILZER, M.M., GOLDMAN, C.R., RICHARDS, R.C., WRIGLEY, R.C. : Influence of se-diment inflow on phytoplankton primary productivity in Lake Tahoe (California--Nevada). - Int. Rev. ges. Hydrobiol. 61 : 169 - 182, 1976.

32137 - TILZER, M.M., HILLBRICHT-ILKOWSKA, A., KOWALCZEWSKI, A., SPODNIEWSKA, I., TURCZYŃSKA, J. : Diel phytoplankton periodicity in Mikołajskie Lake, Poland, as determined by different methods in parallel. - Int. Rev. ges. Hydrobiol. 62 : 279 - 289, 1977. [Chl.]

32138 - TILZER, M.M., PAERL, H.W., GOLDMAN, C.R. : Sustained viability of aphotic phy-toplankton in Lake Tahoe (California-Nevada). - Limnol. Oceanogr. 22 : 84 - 91, 1977. [Chl.]

32139 - TING, I.P., HANSCOM, Z.,III. : Induction of acid metabolism in *Portulacaria afra*. - Plant Physiol. 59 : 511 - 514, 1977. [Ps.]

32140 - TISCHER, W., STROTMANN, H. : Relationship between inhibitor binding by chloro-plasts and inhibition of photosynthetic electron transport. - In : COOMBS, J. (ed.) : 4th International Congress on Photosynthesis. P. 383. UKISES, London 1977.

32141 - TISCHER, W., STROTMANN, H. : Relationship between inhibitor binding by chloro-plasts and inhibition of photosynthetic electron transport. - Biochim. biophys. Acta 460 : 113 - 125, 1977.

32142.- TITLYANOV, É.A., LI, B.D. : Amplituda fenotipicheskikh variatsiĭ fotosinteti-cheskikh pigmentov soderzhaniya u morskikh zelenykh vodorosleĭ i adaptatsiya vida k intensivnosti svetovogo potoka. [Amplitude of phenotypic variations of photosynthetic pigments contents in marine green algae and species adaptation to irradiance.]- In : NASYROV, Yu.S. (ed.) : Genetika Fotosinteza. Pp. 278 - 284. Donish, Dushanbe 1977. [In R.]

32143 - TITLYANOV, É.A., ZVALINSKIĬ, V.I., YADYKIN, A.A., LI, B.D., CHERNOVA, S.I. : Adaptatsiya benticheskikh rasteniĭ k svetu. II. Ustoĭchivost' k fotodestruktsii pigmentnogo apparata morskoĭ zelenoĭ vodorosli *Ulva fenestrata*. [Adaptation of benthic plants to light. II. Resistance of the green attached alga *Ulva fenes-trata* to high illuminance.] - Biol. Morya 1977 (2) : 3 - 10, 1977. [In R, ab : E.]

32144 - TITOV, A.F., OLIMPIENKO, G.S. : Kolichestvennaya svyaz' mezhdu chislom élek-troforeticheskikh variantov peroksidazy i chastotoĭ estestvennykh khlorofill'-nykh mutatsiĭ. [Quantitative relationship between the number of electrophore-tic variants of peroxidase and the frequency of natural chlorophyll mutations.] - Genetika 13 : 1165 - 1167, 1977. [In R, ab : E.]

32145 - TODOROVA-TRIFONOVA, A., ĬORDANOV, I.T. : Vliyanie na khloramfenikola i tetra-tsiklina v"rkhu fotofosforilirashchata aktivnost na *Scenedesmus acutus* MEYEN. [Effect of chloramphenicol and tetracycline on *Scenedesmus acutus* MEYEN phos-phorylating activity.] - Fiziol. Rast. (Sofia) 3 (1) : 64 - 68, 1977. [In Bulg., ab : E, R.]

32146 - TOMASSINI, F.D., LAVOIE, P., PUCKETT, K.J., NIEBOER, E., RICHARDSON, D.H.S. : The effect of time of exposure to sulphur dioxide on potassium loss from and photosynthesis in the lichen, *Cladina rangiferina* (L.) HARM. - New Phytol. 79 : 147 - 155, 1977.

*32147 - TOMBESI, L. : Rapporti acqua-terreno evapotraspirazione potenziale, bilanci energetici ed idrologici nell' ambiente climatico di Roma. [Water-soil relat-ions, potential evapotranspiration, energy and water balances in climate of Rome.] - Agrochimica 20 : 101 - 172, 1976. [Ps; In Ital., ab : E, F, G, Span.]

*32148 - TOMKIEWICZ, M., KLEIN, M.P. : Photooxidation of chlorophyll *b* by quinones studied by chemically-induced dynamic nuclear polarization. - Proc. nat. Acad.
Sci. USA *70* : 143 - 146, 1973.

32149 - TONN, S.J., GOGEL, G.E., LOACH, P.A. : Isolation and characterization of an
organic solvent soluble polypeptide component from photoreceptor complexes
of *Rhodospirillum rubrum*. - Biochemistry *16* : 877 - 885, 1977.

B32150 - TOOMING, Kh.G. : Solnechnaya Radiatsiya i Formirovanie Urozhaya. [Solar Radiation and Yield Formation.] - Gidrometeoizdat, Leningrad 1977. [In R.]

32151 - TÖRMÄLÄ, T. : Effects of mowing and ploughing on the primary production and
flora and fauna of a reserved field in Central Finland. - Acta Agr. scand.
27 : 253 - 264, 1977.

32152 - TORSSELL, B.W.R., McPHERSON, H.G. : An improved model for simulating the penetration, propagation and absorption of radiation within plant canopies. -
Aust. J. Ecol. *2* : 245 - 256, 1977.

32153 - TOWNSLEY, P.M. : The development of secondary by-products from plant cell
cultures. - Development Ind. Microbiol. *18* : 619 - 625, 1977. [Car.]

32154 - TOYAMA, M., TAKEUCHI, Y. : Studies on the photosynthesis and respiration of
some horticultural crop plants grown in the sand field. I. Variations in the
CO_2 concentration of the atmosphere of melon plants (*Cucumis melo* L.) grown
in plastic tunnels. - J. jap. Soc. hort. Sci. *45* : 369 - 374, 1977.

32155 - TOYOSHIMA, Y., MORINO, M., MOTOKI, H., SUKIGARA, M. : Photo-oxidation of water in phospholipid bilayer membranes containing chlorophyll *a*. - Nature *265* :
187 - 189, 1977.

32156 - TRAVIS, A.J., MANSFIELD, T.A. : Studies of malate formation in "isolated"
guard cells. - New Phytol. *78* : 541 - 546, 1977.

32157 - TREBST, A. : Electron transfer in photosynthetic systems : Energy conservation in photosynthetic electron transport of chloroplasts. - In : BUVET, R.,
ALLEN, M.J., MASSUÉ, J.-P. (ed.) : Living Systems as Energy Converters. Pp.
209 - 220. North-Holland Publ. Co., Amsterdam - New York - Oxford 1977.

32158 - TREBST, A. : Transmembrane electron transport and energy conservation in chloroplasts. - In : PACKER, L., PAPAGEORGIOU, G.C., TREBST, A. (ed.) : Bioenergetics of Membranes. Pp. 389 - 394. Elsevier/North-Holland Biomedical Press,
Amsterdam - Oxford - New York 1977.

B32159 - TREBST, A., AVRON, M. (ed.) : Photosynthesis I. (Encyclopedia of Plant Physiology. N.S. Vol.5.)- Springer-Verlag, Berlin - Heidelberg - New York 1977.

32160 - TREBST, A., AVRON, M. : Introduction. - In : TREBST, A., AVRON, M. (ed.) :
Photosynthesis I. (Encycl.Plant Physiol. N.S. Vol. 5.) Pp. 1 - 4. Springer-
Verlag, Berlin - Heidelberg - New York 1977. [Ps, chloroplast.]

32161 - TREBST, A., REIMER, S. : Reversal of the inhibition of photosynthetic electron
flow in chloroplasts by an internal TMPD bypass. - Plant Cell Physiol. *1977* :
(Spec. Issue 3 - Photosynthetic Organelles. Structure and Function) : 201 -
209, 1977.

32162 - TREDWELL, C.J., PORTER, G., SYNOWIEC, J.A. : Picosecond time-resolved fluorescence of *Chlorella pyrenoidosa*. - In : COOMBS, J. (ed.) : 4th International
Congress on Photosynthesis. Pp. 383 - 384. UKISES, London 1977.

32163 - TREGUBENKO, M.Ya., FILIPPOV, G.L., VISHNEVSKIĬ, N.V., MAKSIMOVA, L.A. : Fizio-
logicheskaya reaktsiya razlichnykh gibridov kukuruzy na zasukhu. [Physiologi-
cal reactions of different maize hybrids to drought.] - Sel'skokhoz. Biol.
12 : 412 - 418, 1977. [Ps; in R, ab : E.]

*32164 - TRENBATH, B.R. : Application of a growth model to problems of the productivi-
ty and stability of mixed stands. - In : Proceedings of the 12th International
Grassland Congress. Vol.1. Pp. 546 - 558. Moskva 1974.

32165 - TRENBATH, B.R., HARTLEY, P.R., MACPHERSON, D.K. : Plant distribution and individual plant photosynthesis. - In: COOMBS,J.(ed.): 4th International Congress
on Photosynthesis. Pp. 384 - 385. UKISES, London 1977.

32166 - TRENKENSHU, A.P., SID'KO, F.Ya. : Fotobiosintez khlorelly pri preryvistom osveshchenii s razlichnymi sootnosheniyami sveto-temnovykh periodov. [Photobio-
synthesis of *Chlorella* in flashing light with different light-dark period ratios.] - Izv. sib. Otd. Akad. Nauk SSSR, Ser. biol. Nauk *1977* (3) : 86 - 90,
1977. [In R, ab : E.]

32167 - TRISSL, H.-W., MONTAL, M. : Electrical demonstration of rapid light-induced
conformational changes in bacteriorhodopsin. - Nature *266* : 655 - 657, 1977.

32168 - TROSPER, T., BENSON, D.L. : Spectral properties of an improved *Rhodopseudomonas viridis* reaction-center preparation. - In : OLSON, J.M., HIND, G. (ed.) :
Chlorophyll-Proteins, Reaction Centers, and Photosynthetic Membranes. Pp. 367
- 368. Brookhaven nat. Lab., Upton 1977.

32169 - TROSPER, T.L., BENSON, D.L., THORNBER, J.P. : Isolation and spectral characteristics of the photochemical reaction center of *Rhodopseudomonas viridis*. -
Biochim. biophys. Acta *460* : 318 - 330, 1977.

32170 - TROUGHTON, A. : The effect of phosphorus nutrition upon the growth and morphology of young plants of *Lolium perenne* L. - Ann. Bot. *41* : 85 - 92, 1977.
[Production.]

32171 - TROUGHTON, A. : The rate of growth and partitioning of assimilates in young
grass plants: A mathematical model. - Ann. Bot. *41* : 553 - 565, 1977.

32172 - TROUGHTON, J.H., CURRIE, B.G., CHANG, F.H. : Relations between light level,
sucrose concentration, and translocation of carbon 11 in *Zea mays* leaves. -
Plant Physiol. *59* : 808 - 820, 1977.

32173 - TROXLER, R.F., BROWN, S.B. : Mechanism of bile pigment synthesis in plants. -
Plant Physiol. *59* (6, Suppl.): 8, 1977.

32174 - TSCHUMI, P.A., ZBÄREN, D., ZBÄREN, J. : An improved oxygen method for measuring primary production in lakes. - Schweiz. Z. Hydrol. *39* : 306 - 313, 1977.

32175 - TSEL'NIKER, Yu.L. : Regulyatsiya protsessov gazoobmena CO_2 i morfogeneza u
sazhentsev lesnykh derev'ev pri zatenenii. [Regulation of carbon dioxide gas
exchange and morphogenesis in young forest trees during shading.] - Fiziol.
Rast. *24* : 57 - 64, 1977. [In R, ab : E.]

32176 - TSENOVA, E.N. : Effect of chloramphenicol and cycloheximide on glyceraldehy-
de-3-phosphate dehydrogenase activity in greening barley seedlings during nitrate and ammonium assimilation. - Dokl. bolg. Akad. Nauk *30* : 591 - 594,
1977.

32177 - TSENOVA, M. : Fotosintetichna asimilyatsiya na ^{14}C pri izolirani khloroplasti
v pris"stvie na NO_3^-, SO_4^- i NH_4^+. [Photosynthetic assimilation of ^{14}C by isolated chloroplasts in the presence of nitrate, sulfate, and ammonium ions.] -
Fiziol. Rast. (Sofia) *3* (3) : 14 - 21, 1977. [In Bulg., ab : E, R.]

32178 - TSENOVA, M.P. : Vliyanie na kaliev feritsianid v"rkhu fosforilirashchata aktivnost na izolirani khloroplasti. [Effect of potassium ferricyanide on the
photophosphorylating activity of isolated chloroplasts.] - Fiziol. Rast. (Sofia) *3* (1) : 11 - 17, 1977. [In Bulg., ab : E, R.]

32179 - TSENOVA, M.P. : Vliyanie na indolilotsetnata kiselina v"rkhu fotosintetichnata aktivnost pri izolirani khloroplasti. [Effect of indole-3-acetic acid on
the photosynthetic activity of isolated chloroplasts.] - Fiziol. Rast. (Sofia)
3 (2) : 26 - 32, 1977. [In Bulg., ab : E, R.]

32180 - TSOGLIN, L.N., BAKULIN, V.A. : Vliyanie vozrastnoĭ struktury populyatsiĭ na
fiziologicheskie parametry i produktivnost' kul'tur mikrovodorosleĭ. [The
effect of age structure of populations on physiological parameters and productivity of microalgal cultures.] - Fiziol. Rast. *24* : 1295 - 1300, 1977.
[In R, ab : E.]

32181 - TSUNO, Y. :[Some characteristics of the photosynthesis of the potato plant.]-
Bull. Fac. Agr., Tottori Univ. *29* : 89 - 95, 1977. [In Jap., ab : E.]

32182 - TSUNO, Y. : [Dry matter production of potato plants with special reference
to photosynthesis and respiration.] - Bull. Fac. Agr., Tottori Univ. *29* :
96 - 102, 1977. [In Jap., ab : E.]

32183 - TSUNO, Y., SATO, T., TOMOSAWA, N. : [An analysis of dry matter production
on heterosis in grain sorghum.] - Bull. Sand Dune Res. Inst., Tottori Univ.
16 : 27 - 34, 1977. [Ps; in Jap., ab : E.]

32184 - TSVYLEV, O.P., TKACHENKO, V.N. : O vozmozhnosti fotokhemilyuminestsentnoy
otsenki produktsionnoy sposobnosti fitoplanktona. [Possibility of photochemi-
luminescent evaluation of the production ability of phytoplankton.] - Okeanologiya *17* : 883 - 889, 1977. [In R, ab : E.]

32185 - TUBA, Z. : Examination of the vertical pigment structure in an oak forest
(*Quercetum petreae-cerris*). - Acta bot. Acad. Sci. hung. *23* : 413 - 426,
1977.

32186 - TUCKER, C.J. : Spectral estimation of grass canopy variables. - Remote Sensing Environ. *6* : 11 - 26, 1977. [Chl, Car, biomass.]

32187 - TUCKER, C.J. : Asymptotic nature of grass canopy spectral reflectance. -
Appl. Optics *16* : 1151 - 1156, 1977.

32188 - TUCKER, C.J., GARRATT, M.W. : Leaf optical system modeled as a stochastic
process. - Appl. Optics *16* : 635 - 642, 1977.

32189 - TUCKER, C.J., MILLER, L.D. : Soil spectra contributions to grass canopy spectral reflectance. - Photogram. Eng. remote Sensing *43* : 721 - 726, 1977.

32190 - TUGARINOV, V.V. : Izuchenie kharaktera mutabil'nosti svetochuvstvitel'nykh
shtammov *Chlamydomonas reinhardi*. [Character of mutability of light-sensitive
strains of *Chlamydomonas reinhardi*.] - In : NASYROV, Yu.S. (ed.) : Genetika
Fotosinteza. Pp. 144 - 149. Donish, Dushanbe 1977. [Chl, Car; in R.]

32191 - TUGNAWAT, R.K., SONI, A.K. : Estimation of leaf area index. - Sci. Cult. *43* :
488 - 489, 1977.

32192 - TUQUET, C., GUILLOT-SALOMON, T., DE LUBAC, M., SIGNOL, M. : Granum formation
and the presence of phosphatidylglycerol containing *trans*-Δ_3-hexadecenoic
acid. - Plant Sci. Lett. *8* : 59 - 64, 1977.

*32193 - TURNER, J.B., FRIEDMANN, E.I. : Fine structure of capitular filaments in the
coenocytic green alga *Penicillus*. - J. Phycol. *10* : 125 - 134, 1974. [Chloro-
plast.]

32194 - TURNER, N.C., HEICHEL, G.H. : Stomatal development and seasonal changes in
diffusive resistance of primary and regrowth foliage of red oak (*Quercus rub-
ra* L.) and red maple (*Acer rubrum* L.). - New Phytol. *78* : 71 - 81, 1977.

*32195 - TUTTLE, R.C., LOEBLICH, A.R., III. : The discovery of genetic recombination
in the dinoflagellate *Crypthecodinium cohnii*. - J. Phycol. *10* (Suppl.) : 16,
1974. [Car.]

32196 - TYREE, M.T., CAMERON, S.I. : A new technique for measuring oscillatory and
diurnal changes in leaf thickness. - Can. J. Forest Res. *7* : 540 - 544, 1977.

32197 - TYSHKEVICH, G.L. : Biologo-fiziologicheskie osobennosti raznykh vidov i ėko-
tipov buka. [Biological and physiological characteristics of different beech
species and ecotypes.] - Nauch. Dokl. vyssh. Shkoly, biol. Nauki *20* (7) :
102 - 108, 1977. [Ps; in R.]

32198 - TYSZKIEWICZ, E., POPOVIC, R., ROUX, E. : 1) Relationship between P_{700} redox
state and proton translocation; 2) action of reduced mediators on photophosphorylation and two steps phosphorylation, studied with dark grown *Pinus nigra* cotyledons. - In : COOMBS, J. (ed.) : 4th International Congress on Photosynthesis. P. 385. UKISES, London 1977.

32199 - TYSZKIEWICZ, E., POPOVIC, R., ROUX, E. : Relationship between the redox-state
of *P*-700 and photosystem I-mediated proton translocation studied with chloroplasts from dark-grown *Pinus nigra* seedlings. - FEBS Lett. *81* : 65 - 68, 1977.

32200 - UDOVENKO, G.V., GONCHAROVA, Ė.A. : Peredvizhenie ^{14}C-assimilyatov v list'ya
i plody pri zasukhe i zasolenii. [Transport of ^{14}C-photosynthates into leaves
and fruits during drought and salinization.] - Fiziol. Rast. *24* : 901 - 905,
1977. [In R, ab : E.]

32201 - UEDA, J., NAKAGAWA, S. : [Effect of low light intensity on early summer defoliation of grapevines.] - J. Jap. Soc. hort. Sci. 46 : 158 - 168, 1977.
[Ps; in Jap., ab : E.]

32202 - UHEDA, E., KURAISHI, S. : Increase of cytokinin activity in detached etiolated
cotyledons of squash after illumination. - Plant Cell Physiol. 18 : 481 - 483,
1977. [Chl.]

32203 - UHLMANN, D., FRITZSCHE, I. : Die planktische Primärproduktion bei extrem hoher organischer Belastung - ein Beitrag zur Diskussion "Trophie/Saprobie". -
Ergeb. Limnol. 9 : 169 - 175, 1977.

32204 - UHRIG, H., TEVINI, M. : Lipid composition and photochemical activities of
phospholipase C-treated spinach chloroplasts. - In : COOMBS, J. (ed.) : 4th
International Congress on Photosynthesis. P. 386. UKISES, London 1977.

32205 - UL-HAQUE, M.I., GALLON, J.R., CHAPLIN, A.E. : The intermediary metabolism of
the unicellular blue-green alga Gloeocapsa sp. LB795. - Biochem. Soc. Trans.
5 : 1484 - 1486, 1977. [Photosynthates.]

32206 - ULLRICH, W.R., EISELE, R. : Relations between nitrate uptake and nitrate reduction
in Ankistrodesmus braunii. - In : Echanges Ioniques Transmembranaires chez
les Végétaux. Coll. Int. CNRS. No. 258. Pp. 307 - 315. CNRS, Paris 1977. [Ps.]

32207 - ULUBEKOVA, M.V., AKSENOV, V.P. : Kislorodnyĭ obmen kletok Scenedesmus obliquus
v dlinnovolnovoĭ oblasti spektra (λ>700 nm). [Oxygen exchange by Scenedesmus
obliquus in the long-wavelength region of the spectrum (λ>700 nm).] - Fiziol.
Rast. 24 : 5 - 9, 1977. [In R, ab : E.]

32208 - UMRIKHINA, A.V., BYSTROVA, M.I., BUBLICHENKO, N.V., MAL'GOSHEVA, I.N., KRAS-
NOVSKIĬ, A.A. : Obrazovanie svobodnykh radikalov pri fotookislenii bakterio-
viridina v monomernom i agregirovannom sostoyanii. [Free radical formation during photooxidation of bacterioviridin in monomeric and aggregated state.] -
Biofizika 22 : 780 - 788, 1977. [In R, ab : E.]

B32209 - UNGER, K. (ed.) : Biophysikalische Analyse Pflanzlicher Systeme. - VEB Gustav
Fischer Verlag, Jena 1977. [Ps.]

32210 - UNGER, K. : Energieumsatz in Ökosystemen. - In : UNGER, K. (ed.) : Biophysikalische Analyse Pflanzlicher Systeme. Pp. 290 - 301. VEB Gustav Fischer Verlag, Jena 1977.

32211 - UNGER, K., CLAUS, S. : Zur Modellierung der Biomassenproduktion von Kulturpflanzen. - In : UNGER, K. (ed.) : Biophysikalische Analyse pflanzlicher
Systeme. Pp. 85 - 95. VEB Gustav Fischer Verlag, Jena 1977.

32212 - UPADHYAYA, S.D., SINGH, V.P. : Effect of N, P, K on the pigment system in Tra-
gus biflorus SCHULT. - Sci. Cult. 43 : 353 - 354, 1977.

32213 - UPHAUS, R.A., BLAKE, M.I., KOSTKA, A.G., KATZ, J.J. : Automated growth chambers for production of isotopically substituted higher plants. - Photosynthetica 11 : 314 - 321, 1977.

32214 - UPHAUS, R.A., COTTON, T.M., KATZ, J.J. : An electron spin resonance study of
the photoactive chlorophyll-protein complex of Chenopodium album. - In :
OLSON, J.M., HIND, G. (ed.) : Chlorophyll-Proteins, Reaction Centers, and Photosynthetic Membranes. P. 359. Brookhaven nat. Lab., Upton 1977.

*32215 - URBACH, D., SUCHANKA, M., URBACH, W. : Effect of substituted pyridazinone her-
bicides and of difunone (EMD-IT 5914) on carotenoid biosynthesis in green algae. - Z. Naturforsch. 31C : 652 - 655, 1976.

32216 - URBACH, W. : Eukaryotic algae. - In : TREBST, A., AVRON, M. (ed.) : Photosynthesis I. (Encycl. Plant Physiol. N.S. Vol.5.) Pp. 603 - 624. Springer-Verlag,
Berlin - Heidelberg - New York 1977. [Ps, Chl.]

32217 - URBACH, W., KAISER, W. : Endogenous cyclic and pseudo-cyclic photophosphorylation in isolated intact chloroplasts. - In : COOMBS, J. (ed.) : 4th International Congress on Photosynthesis. P. 387. UKISES, London 1977.

32218 - URSINO, D.J., SCHEFSKI, H., LATOUR, P.W. : Translocation of photoassimilates
in gamma-irradiated soybean plants. - Environm. exp. Bot. 17 : 35 - 42, 1977.

32219 - URSINO, D.J., SCHEFSKI, H., MCCABE, J. : Radiation-induced changes in rates of photosynthetic CO_2 uptake in soybean plants. - Environm. exp. Bot. *17* : 27 - 34, 1977.

32220 - USHIJIMA, T., TAZAKI, T. : The influence of sulfur dioxide on the photosynthetic and transpiration rate in several higher plants. - In : KASUGA, S., SUZUKI, N., YAMADA, T.(ed.) : Proceedings of the 4th International Clean Air Congress. Pp. 84 - 87. Jap. Union Air Pollut. Prev. Assoc. 1977.

32221 - USMANOV, P.D. : K voprosu o geticheskom kontrole funktsionirovaniya khloroplastov. [Genetic control of chloroplast activity.] - Sel'skokhoz. Biol. *12* : 769 - 778, 1977. [In R, ab : E.]

32222 - USMANOV, P.D., ABDULLAEV, Kh.A., USMANOVA, O.V., SOKHIBNAZAROV, Sh. : Mutatsionnaya izmenchivost' khloroplastov. [Mutation variability of chloroplasts.] - In : NASYROV, Yu.S. (ed.) : Genetika Fotosinteza. Pp. 104 - 114. Donish, Dushanbe 1977. [In R.]

32223 - USUDA, H., MIYACHI, S. : Coupling of malate decarboxylation to CO_2 fixation and the reduction of 3-phosphoglyceric acid in corn bundle sheath cells. - Plant Cell Physiol. *18* : 1109 - 1120, 1977.

32224 - VACEK, K., NAUŠ, J., ŠVÁBOVÁ, M., VAVŘINEC, E., KAPLANOVÁ, M., HÁLA, J. : Fluorescence polarization spectra of chlorophyll *a* molecules in oriented polymer matrix. - Stud. biophys. *62* : 201 - 207, 1977.

32225 - VACEK, K., WONG, D., GOVINDJEE : Absorption and fluorescence properties of highly enriched reaction center particles of Photosystem I and of artificial systems. - Photochem. Photobiol. *26* : 269 - 276, 1977.

32226 - VÁCLAVÍK, J. : Distribution pattern of gas exchange in the area of maize leaf blades during the generative phase. - Biol. Plant. *19* : 457 - 461, 1977.

*32227 - VAĬNSHTEĬN, M.B. : Rezul'taty opredeleniya fotosinteza fitoplanktona radiouglerodnym metodom i mikrobiologicheskaya kharakteristika Kharbeĭskikh ozer. [Results of determining phytoplankton photosynthesis with the radiocarbon method and microbiological characteristics of Kharbeĭan lakes.] - In : Produktivnost' Ozer Vostochnoĭ Chasti Bol'shezemel'skoĭ Tundry. Pp. 76 - 79. Nauka, Leningrad 1976. [In R.]

32228 - VAKLINOVA, S., KAFALIEVA, D., MANOLOVA, N. : Photochemical activity of mutant forms of pea, lacking chlorophyll "b". - In : COOMBS, J. (ed.) : 4th International Congress on Photosynthesis. P. 387 a. UKISES, London 1977.

32229 - VALADON, L.R.G., MUMMERY, R.S. : Carotenoids of lilies and red pepper : Biogenesis of capsanthin and capsorubin. - Z. Pflanzenphysiol. *82* : 407 - 416, 1977.

32230 - VALANNE, N. : Effect of continuous light on CO_2 fixation, chlorophyll content, growth and chloroplast structure in *Ceratodon purpureus*. - Z. Pflanzenphysiol. *81* : 347 - 357, 1977.

32231 - VALANNE, N. : The combined effects of light intensity and continuous light on the CO_2 fixation, chlorophyll content and chloroplast structure of the protonema of *Ceratodon purpureus*. - Z. Pflanzenphysiol. *83* : 275 - 283, 1977.

32232 - VALCKE, R., VAN POUCKE, M., CLIJSTERS, H. : Effects of age and relative humidity on the greening of etiolated seedlings of barley (*Hordeum vulgare* var. Union). - In : COOMBS, J. (ed.) : 4th International Congress on Photosynthesis. P. 388. UKISES, London 1977.

32233 - VALLEJOS, R.H., ANDREO, C.S. : An essential arginyl residue in chloroplast coupling factor 1. - In : COOMBS, J. (ed.) : 4th International Congress on Photosynthesis. P. 389. UKISES, London 1977.

32234 - VALLEJOS, R.H., RAVIZZINI, R.A., ANDREO, C.S. : Sulphydryl groups in photosynthetic energy conservation. IV. Inhibition of the ATPase of chloroplast coupling factor 1 by sulphydryl reagents. - Biochim. biophys. Acta *459* : 20 - 26, 1977.

32235 - VALLEJOS, R.H., VIALE, A., ANDREO, C.S. : Essential role of an arginyl residue at the catalytic site(s) of chloroplast coupling factor. - FEBS Lett. *84* : 304 - 308, 1977.

32236 - VAN, T.K., GARRARD, L.A., WEST, S.H. : Effects of 298-nm radiation on photosynthetic reactions of leaf discs and chloroplast preparations of some crop species. - Environ. exp. Bot. *17* : 107 - 112, 1977.

32237 - VAN, T.K., HALLER, W.T., BOWES, G., GARRARD, L.A. : Effects of light quality on growth and chlorophyll composition in *Hydrilla*. - J. aquatic Plant Management *15* : 29 - 31, 1977.

32238 - VAN BEST, J.A., DUYSENS, L.N.M. : A one microsecond component of chlorophyll luminescence suggesting a primary acceptor of System II of photosynthesis different from Q. - Biochim. biophys. Acta *459* : 187 - 206, 1977.

32239 - VAN BEST, J.A., MATHIS, P. : Analysis of absorption changes at 820 nm due to oxidized primary donors of Photosystems 1 and 2. - In : **COOMBS, J.** (ed.) : 4th International Congress on Photosynthesis. P. 390. UKISES, London 1977.

32240 - VANDEN DRIESSCHE,T., CERF, E. : Effects of morphactins on the chloroplast ultrastructure. - J. Ultrastructure Res. *59* : 140 - 148, 1977.

*32241 - VANDEN DRIESSCHE,T., DUJARDIN, E., MAGNUSSON, A., SIRONVAL, C. :*Acetabularia mediterranea* : Circadian rhythms of photosynthesis and associated changes in molecular structure of the thylakoid membranes. - Internat. J. Chronobiol.*4* : 111 - 124, 1976.

32242 - VANDEN DRIESSCHE,T., LANNOYE, R., GLORY, M., WILLECOME, L. : Exogenous NADP and photosynthesis in *Acetabularia*. - In : **COOMBS, J.** (ed.) : 4th International Congress on Photosynthesis. Pp. 390 - 391. UKISES, London 1977.

32243 - VANDERLIP, R.L., ARKIN, G.F. : Simulating accumulation and distribution of dry matter in grain sorghum. - Agron. J. *69* : 917 - 923, 1977.

*32244 - VANDERMEULEN, D.L., GOVINDJEE : Interactions of fluorescent analogs of adenine nucleotides with coupling factor isolated from chloroplasts. - Biophys. J. *16* (2,Part 2) : 159, 1976.

32245 - VANDERMEULEN, D.L., GOVINDJEE : Binding of modified adenine nucleotides to isolated coupling factor from chloroplasts as measured by polarization of fluorescence. - Europe. J. Biochem. *78* : 585 - 598, 1977.

32246 - VAN GINKEL, G. : Action spectra of photophosphorylation catalyzed by PS-1 or PS-2 separately in spinach chloroplasts. - In : **COOMBS, J.** (ed.) : 4th International Congress on Photosynthesis. P. 391. UKISES, London 1977.

32247 - VAN GINKEL, G. : Ageing of isolated chloroplasts in a stabilizing medium: influence of different types of photophosphorylation. - Acta bot. neerl. *26* : 303 - 311, 1977.

32248 - VAN GINKEL, G. : Detailed action spectra of photophosphorylation catalyzed by photosystem I. - Acta bot. neerl. *26* : 313 - 319, 1977.

32249 - VAN GORKOM, H.J., PULLES, M.P.J. : Calibration of chlorophyll fluorescence quenching by Photosystem 2 reaction centers. - In : **COOMBS, J.** (ed.) : 4th International Congress on Photosynthesis. Pp. 393 - 394. UKISES, London 1977.

32250 - VAN GRONDELLE, R., DUYSENS, L.N.M., VAN DER WEL, J.A., VAN DER WAL, H.N. : Function and properties of a soluble *c*-type cytochrome *c*-551 in secondary photosynthetic electron transport in whole cells of *Chromatium vinosum* as studied with flash spectroscopy. - Biochim. biophys. Acta *461* : 188 - 201, 1977.

32251 - VAN GRONDELLE, R., HOLMES, N.G., RADEMAKER, H., DUYSENS, L.N.M. : The relationship between fluorescence yield and reaction centre triplet yield in bacterial photosynthesis. - In : **COOMBS, J.** (ed.) : 4th International Congress on Photosynthesis. P. 394. UKISES, London 1977.

32252 - VAN HOLSTEIJN, H.M.C., BEHBOUDIAN, M.H., BONGERS, H.C.M.L. : Water relations of lettuce. II. Effects of drought on gas exchange properties of two cultivars. - Sci. Hort. *7* : 19 - 26, 1977.

32253 - VAN HUYSTEE, R.B. : Porphyrin and peroxidase synthesis in cultured peanut cells. - Can. J. Bot. *55* : 1340 - 1344, 1977.

32254 - **VAN METTER, R.L.** : Excitation energy transfer in the light-harvesting chlorophyll *a/b* protein. - Biochim. biophys. Acta *462* : 642 - 658, 1977.

32255 - **VAN RENSEN, J.J.S.** : Effects of the herbicide 4,6-dinitro-*O*-cresol on electron transport and chlorophyll *a* fluorescence in isolated chloroplasts. - In : **COOMBS, J.** (ed.) : 4th International Congress on Photosynthesis. P. 392. UKISES, London 1977.

32256 - **VAN RENSEN, J.J.S., VAN DER VET, W., VAN VLIET, W.P.A.** : Inhibition and uncoupling of electron transport in isolated chloroplasts by the herbicide 4,6-dinitro-*o*-cresol. - Photochem. Photobiol. *25* : 579 - 583, 1977.

32257 - **VANSEVEREN, J.P., HERBAUTS, J.** : Index foliaire, paramètres foliaires et caractéristiques édaphiques stationnelles dans quelques peuplements forestiers de Lorraine belge. - Ann. Sci. forest. *34* : 215 - 229, 1977.

32258 - **VAN VALKENBURG, S.D., KARLANDER, E.P., PATTERSON, G.W., COLWELL, R.R.** : Features for classifying photosynthetic aerobic nanoplankton by numerical taxonomy. - Taxon *26* : 497 - 505, 1977.

32259 - **VARSHNEY, K.A., BAIJAL, B.D.** : Effect of salt-stress on chlorophyll contents of some grasses. - Indian J. Plant Physiol. *20* : 161 - 163, 1977.

32260 - **VASCONCELOS, A.C., TRIEMER, R.H.** : Synthesis and ultrastructural localization of coupling factor CF_1 in *Euglena* chloroplasts. - In : **COOMBS, J.** (ed.) : 4th International Congress on Photosynthesis. Pp. 392 - 393. UKISES, London 1977.

32261 - **VASEV, V.A.** : Produktivnost' fotosinteza dvukh prostykh mezhlineĭnykh gibridov kukuruzy i ikh roditel'skikh liniĭ. [Photosynthetic productivity of two simple interline maize hybrids and their parental lines.] - Sel'skokhoz. Biol. *12* : 934 - 937, 1977. [In R, ab : E.]

32262 - **VASILEVA, V., TSENOVA, E., FEDINA, I., V"LKOVA RADEVA, R., GUSHCHINA, L., VAKLINOVA, S.** : Vliyanie na khloramfenikola i tsiklokheksimida v"rkhu fotosinteza-ta i fotodishaneto v pozelenyavashchi prorast"tsi ot tsarevitsa pri asimilatsiya na nitratni i amonievi·ĭoni. I. Aktivnost na fosfoenolpiruvat-karboksilaza, ribulozodifosfat-karboksilaza i glitseraldekhid-3-fosfatdekhidrogenaza. [Effect od chloramphenicol and cycloheximide on photosynthesis and photorespiration of greening maize seedlings during nitrate and ammonium ion assimilation. I. Phosphoenolpyruvate carboxylase, ribulosebisphosphate carboxylase and glyceraldehyde 3-phosphate dehydrogenase activities.]-Fiziol.Rast.(Sofia) *3*(2): 13 - 25, 1977. [In Bulg., ab : E, R.]

32263 - **VASIL'EVA, V.D., KRYLOVA, I.L.** : Opredelenie nadzemnoĭ biomassy po nekotorym morfologicheskim pokazatelyam nadzemnykh organov. [Determination of aboveground biomass from some morphological parameters of shoot.] - Nauch. Dokl. vyssh. Shkoly, biol. Nauki *20* (3) : 96 - 102, 1977. [In R.]

32264 - **VATER, J., SALNIKOW, J., KLEINKAUF, A.** : A fluorimetric study of substrate and effector binding of D-ribulose-1,5-biphosphate carboxylase/oxygenase from spinach. - Biochem. biophys. Res. Commun. *74* : 1618 - 1625, 1977.

32265 - **VAVILOV, P.P., ZELENIN, G.G., KISELEV, V.N.** : Urozhaĭ i kachestvo klubneĭ rannego kartofelya pri razlichnykh rezhimakh orosheniya. [Yield and quality of potato tubers cultivated in different irrigation regimes.] - Izv. timiryazev. sel'skokhoz. Akad. (Moskva) *1977* (4) : 33 - 41, 1977. [Dry-matter accumulation; in R, ab : E.]

32266 - **VECHER, A.S., KALER, V.L., PREDKEL', K.I., ADAMCHIK, G.G., ARTEM'EV, P.N.** : Lokalizatsiya tsitokhromov b_6 and f in pigment-belkovykh kompleksakh khloroplastov. [Localization of cytochromes b_6 and f in pigment-protein chloroplast complexes.] - Dokl. Akad. Nauk belorus. SSR *21* : 1043 - 1046, 1056, 1977. [In R.]

32267 - **VECHER, A.S., RESHETNIKOV, V.N., KOVAL'CHUK, R.A., MAS'KO, A.A., BULKO, O.P., PREDKEL', K.I.** : Razlichiya kletochnykh yader i khloroplastov di- i tetraploidnoĭ rzhi. [Differences in cell nuclei and chloroplasts of di- and tetraploid rye.] - In : **NASYROV, Yu.S.** (ed.) : Genetika Fotosinteza. Pp. 133 - 136. Donish, Dushanbe 1977. [In R.]

32268 - **VEGTER, F.** : The closure of the Grevelingen estuary: its influence on phytoplankton primary production and nutrient content. - Hydrobiologia *52* : 67 - 71, 1977.

32269 - VELICHKO, I.M. : Produktsiya nekotorykh zelenykh nitchatykh vodorosleĭ v prirodnykh assotsiyatsiyakh. [Production of some green filamentous algae in natural associations.] - Gidrobiol. Zh. *13* (2) : 23 - 27, 1977. [Ps; in R.]

32270 - VELTHUYS, B.R. : Photosynthetic oxygen evolution from hydrogen peroxide. - In: COOMBS, J. (ed.) : 4th International Congress on Photosynthesis. P. 395. UKISES, London 1977.

32271 - VENATOR, C.R., HOWES, C.D., TELEK, L. : Chlorophyll and carotenoid contents of *Pinus caribaea* seedlings and inferences for adaptability. - Turrialba *27* : 169 - 173, 1977.

32272 - VENKATESH, C.S., THAPLIYAL, R.C. : Short note: natural chlorophyll mutants in a Himalayan pine. - Silvae Genet. *26* : 142, 1977.

32273 - VENNEWITZ, M.K., ZINGMARK, R. : Netplankton and nannoplankton productivity in a South Carolina estuary. - J. Phycol. *13* (Suppl.) : 70, 1977. [Chl.]

32274 - VENRICK, E.L., BEERS, J.R., HEINBOKEL, J.F. : Possible consequences of containing microplankton for physiological rate measurements. - J. exp. mar. Biol. Ecol. *26* : 55 - 76, 1977. [Chl.]

32275 - VERBELEN, J.P., De GREEF, J.A., MOEREELS, E. : Red light sensitive ATP-ase function in etiolated plants - microcalorimetric evidence. - Plant Physiol. *59* (6, Suppl.) : 49, 1977.

32276 - VERDIER, G. : Changes in polyadenylic acid containing RNAs during chloroplast development in *Euglena gracilis*. - In : COOMBS, J. (ed.) : 4th International Congress on Photosynthesis. P. 396. UKISES, London 1977.

32277 - VERMA, S.B., ROSENBERG, N.J. : Brown-Rosenberg resistance model of crop evapotranspiration modified tests in an irrigated sorghum field. - Agron. J. *69* : 332 - 335, 1977. [Resistances.]

32278 - VERMEERSCH, J., MONÉGER, R., LECHEVALLIER, D. : Sur le microdosage séparé des nucléotides pyridiniques réduits et oxydés dans les feuilles et les plantes de Blé et de Spirodèle. - Compt. rend. Acad. Sci. Paris, Sér. D *284* : 1529 - 1531, 1977.

32279 - VERMEGLIO, A. : Secondary electron transfer in reaction centers of *Rhodopseudomonas sphaeroides*. Out-of-phase periodicity of two for the formation of ubisemiquinone and fully reduced ubiquinone. - Biochim. biophys. Acta *459* : 516 - 524, 1977.

32280 - VERMEGLIO, A., BRETON, J., PAILLOTIN, G. : Relative orientation of different pigments in reaction centers from *Rhodopseudomonas sphaeroides*. - In : COOMBS, J. (ed.) : 4th International Congress on Photosynthesis. P. 395. UKISES, London 1977.

32281 - VERMEGLIO, A., CLAYTON, R.K. : Kinetics of electron transfer between the primary and the secondary electron acceptor in reaction centers from *Rhodopseudomonas sphaeroides*. - Biochim. biophys. Acta *461* : 159 - 165, 1977.

32282 - VERMEGLIO, A., CLAYTON, R.K. : Orientation of bacterial reaction-center chromophores : evidence for two absorption bands of the dimeric primary electron donor. - In : OLSON, J.M., HIND, G. (ed.) : Chlorophyll-Proteins, Reaction Centers, and Photosynthetic Membranes. P. 367. Brookhaven nat. Lab., Upton 1977.

32283 - VERNON, L.P., KLEIN, S.M. : Polypeptide components of a PS I reaction center complex from *Anabaena flos-aquae*. - In : COOMBS, J. (ed.) : 4th International Congress on Photosynthesis. P. 397. UKISES, London 1977.

32284 - VERSHININ, A.V. : Fiziologo-biokhimicheskie aspekty monogibridnogo geterozisa, poluchennogo na osnove khlorofil'nykh mutantov gorokha. Soobshchenie II. Pigmenty v khode individual'nogo razvitiya. [Physiological and biochemical aspects of monohybrid heterosis derived from pea chlorophyll mutants. Part II. Pigments in the course of individual development.] - Genetika *13* : 1153 - 1164, 1977. [In R, ab : E.]

32285 - VERWER, W., KLEINEN HAMMANS, J.W. : Intramembranous particles and chlorophyll complexes in *Chlamydononas* spec. - Acta bot. neerl. *26* : 299 - 302, 1977.

32286 - VERWER, W., KLEINEN HAMMANS, J.W., THOMAS, J.B., VERVERGAERT, P.H.J.T. : Intramembranous particles and chlorophyll complexes in chloroplasts. - Biochim. biophys. Acta *461* : 202 - 208, 1977.

*32287 - VESELOVSKIĬ, V.A., LESHCHINSKAYA, L.V., MARKAROVA, E.N., VESELOVA, T.V., TARU-SOV, B.N. : Vliyanie osveshchennosti list'ev khlopchatnika na teploustoĭchivost' fotosinteticheskogo apparata. [Effect of illumination of cotton leaves on heat resistance of the photosynthetic apparatus.] - Fiziol. Rast. *23* : 467 - 472, 1976. [In R, ab : E.]

32288 - VESK, M., JEFFREY, S.W. : Effect of blue-green light on photosynthetic pigments and chloroplast structure in unicellular marine algae from six classes. - J. Phycol. *13* : 280 - 288, 1977.

32289 - VETTER, J., HARASZTI, E. : Adatok a zárvatermők β-karotin- és klorofilltartalmáról. [Amount of β-carotene and chlorophylls in angiospermous plants.] - Bot. Közlem. *64* : 35 - 41, 1977. [In Hung., ab : E.]

32290 - VIDAL, J., GADAL, P., CAVALIE, G., CAILLIAU-COMMANAY, L. : NADH and NADPH dependent malate dehydrogenases of *Phaseolus vulgaris*. - Physiol. Plant. *39* : 190 - 195, 1977.

32291 - VIDAL, J., RIO, M.C., GADAL, P. : Etude de l'évolution de la malate deshydrogénase a NADP durant le verdissement des feuilles de *Phaseolus vulgaris* L. - Plant Sci. Lett. *8* : 243 - 249, 1977.

32292 - VIDOVIČ, J. : Fyziologická charakteristika ideotypu kukurice. [Physiological characteristics of the maize ideotype.] - In : REPKA, J. (ed.) : Zborník Referátov zo Seminára Fyziologicko-Genetické a Chemické Faktory Produktivity Rastlín. Pp. 55 - 76. Vysoká Škola poľnohospodárska, Nitra 1977. [Ps; in Slovak.

32293 - VIERKE, G. : Kinetics of the back reaction of Photosystem II in *Chlorella fusca* after extraction of membrane bound manganese by hydroxylamine. - In : COOMBS, J. (ed.) : 4th International Congress on Photosynthesis. Pp. 397 - 398. UKISES, London 1977.

32294 - VIETOR, D.M., ARIYANAYAGAM, R.P., MUSGRAVE, R.B. : Photosynthetic selection of *Zea mays* L. I. Plant age and leaf position effects and a relationship between leaf and canopy rates. - Crop Sci. *17* : 567 - 573, 1977.

32295 - VIGNAIS, P.M. : Energy compartmentation in the cell. - In : BUVET, R., ALLEN, M.J., MASSUÉ, J.-P. (ed.) : Living Systems as Energy Converters. Pp. 135 - 151. North-Holland Publ. Co., Amsterdam - New York - Oxford 1977. [Ps.]

32296 - VIGNES, D., CARLES, J. : Influence du vent sur l'activité photosynthétique et les échanges gazeux. - Oecol. Plant. *12* : 149 - 158, 1977.

32297 - VIIL, J., LAISK, A., OJA, V., PÄRNIK, T. : Enhancement of photosynthesis caused by oxygen under saturating irradiance and high CO_2 concentrations. - Photosynthetica *11* : 251 - 259, 1977.

32298 - VIRGIN, H. : The spectral response of light dependent chlorophyll *b* formation. - Physiol. Plant. *40* : 45 - 49, 1977.

32299 - VIRGIN, H.I. : The spectral response of light dependent chlorophyll *b* formation. - In : COOMBS, J. (ed.) : 4th International Congress on Photosynthesis. P. 398. UKISES, London 1977.

32300 - VISHNEVSKIĬ, N.V., MAKSIMOVA, L.A. : Soderzhanie khlorofilla v list'yakh i intensivnost' fotosinteza razlichnykh gibridov kukuruzy v usloviyakh orosheniya. [Chlorophyll content in the leaves and rate of photosynthesis in different maize hybrids under irrigation.] - Dokl. vsesoyuz. Akad. sel'skokhoz. Nauk *1977* (7) : 41 - 43, 1977. [In R.]

32301 - VISSER, J.W.M., RIJGERSBERG, C.P., GAST, P. : Photooxidation of chlorophyll in spinach chloroplasts between 10 and 180 K. - Biochim. biophys. Acta *460* : 36 - 46, 1977.

32302 - VLASYUK, P.A., LOBANOVA, Z.I., KLIMOVITSKAYA, Z.M. : Soderzhanie nukleinovykh
kislot v khloroplastakh rasteniĭ pri margantsevoĭ nedostatochnosti. [Content
of nucleic acids in plant chloroplasts under manganese deficiency.] - Fiziol.
Biokhim. kul't. Rast. 9 : 139 - 143, 1977. [In R, ab : E.]

32303 - VOEGELI, K.K., O'KEEFE, D., WHITMARSH, J., DILLEY, R.A. : Valinomycin inhibition of chloroplast electron transport at or near plastoquinone. - Arch. Biochem. Biophys. 183 : 333 - 339, 1977.

32304 - VOGEL, P. : Bestimmung des Xanthophyllgehaltes von Pflanzenölen. - Fette-Sei-
fen-Anstrichmittel 79 (3) : 97 - 103, 1977.

32305 - VOGELMANN, T.C., SCHEIBE, J. : Action spectra for chromatic adaptation in the
blue-green alga Fremyella diplosiphon DROUET. - Plant Physiol. 59 (6, Suppl.) :
48, 1977.

32306 - VOIGTLÄNDER, G., KÜHBAUCH, W., LANG, W. : Wanderung von ^{14}C-Assimilaten im
Ampfer (Rumex obtusifolius L.) und Wiesenknöterich (Polygonum bistorta L.).
- Landwirtsch. Forsch. 30 : 3 - 12, 1977.

*32307 - VOĬNOVSKAYA, K.K., DARKANBAEVA, G.T., MAĬCHEKINA, R.M., KOSHANOVA, K.Sh. :
Sostoyanie khlorofilla v list'yakh pshenitsy v svyazi s ontogenezom i uslovi-
yami vozdelyvaniya rasteniĭ. [State of chlorophyll in wheat leaves in relation to ontogenesis and conditions of plant cultivation.] - In : Fotosintez
i Produktivnost' Ozimoĭ Pshenitsy na Yugo-Vostoke Kazakhstana. Pp. 77 - 83,
132. Nauka kazakh. SSR, Alma-Ata 1976. [In R.]

32308 - VOITURIEZ, B., HERBLAND, A. : Production primaire, nitrate et nitrite dans
l'Atlantique tropical. II - Distribution du nitrate et production de nitrite.
- Cah. ORSTOM, Sér. Océanographie 15 : 57 - 65, 1977.

32309 - VOLKOVA, N.V., IVASHCHENKO, Ya.N., VASILENOK, L.I., MUSHKETIK, L.S., KANIVETS,
N.P., GRIDASOVA, V.I., YASNIKOV, A.A. : Vliyanie rostaktiviruyushchego veshchestva N-okisi 2,4-lutidina na fotofosforilirovanie i svetoindutsiruemyĭ
transport protona v khloroplaste. [Effect of growth-activating agent N-oxide-
2,4-lutidine on photophosphorylation and light-induced transport of proton
in chloroplasts.] - Fiziol. Biokhim. kul't. Rast. 9 : 527 - 531, 1977. [In
R, ab : E.]

*32310 - VOLODARSKIĬ, N.I., BYSTRYKH, E.E. : Funktsional'naya aktivnost' fotosinteti-
cheskogo apparata pri narushenii vodnogo rezhima podsolnechnika. [Functional
activity of the photosynthetic apparatus in the sunflower during disturbance
of the water state.] - Fiziol. Rast. 23 : 497 - 501, 1976. [In R, ab : E.]

32311 - VOLODARSKIĬ, N.I., BYSTRYKH, E.E., NIKOLAEVA, E.K. : Izmenenie fotosinteticheskoĭ deyatel'nosti lista v ontogeneze pshenitsy. [Transformation of the leaf
photosynthetic activity in wheat ontogenesis.] - Sel'skokhoz. Biol. 12 : 853
- 859, 1977. [In R, ab : E.]

32312 - VONG, N.Q., MURATA, Y. : Studies on the physiological characteristics of C_3
and C_4 crop species. I. The effects of air temperature on the apparent photosynthesis, dark respiration, and nutrient absorption of some crops. - Jap. J.
Crop Sci. 46 : 45 - 52, 1977.

32313 - VOROBEĬKOV, G.A., ANIKINA, R.D. : Vliyanie regulyatorov rosta na ustoĭchivost'
soi k pochvennomu zatopleniyu. [Effect of growth regulators on resistance of
soya to soil flooding.] - Fiziol. Rast. 24 : 1269 - 1275, 1977. [Chl; in R,
ab : E.]

32314 - VOROB'EVA, L.M., KRASNOVSKIĬ, A.A. : Vliyanie kisloroda na promezhutochnye
stadii obrazovaniya khlorofilla v ètiolirovannykh list'yakh. [Effect of oxygen on intermediate stages of chlorophyll formation in etiolated leaves.] -
Dokl. Akad. Nauk SSSR 232 : 225 - 228, 1977. [In R.]

32315 - VOSKRESENSKAYA, N.P., DROZDOVA, I.S., KRENDELEVA, T.E. : Effect of light quality on the organization of photosynthetic electron transport chain of pea
seedlings. - Plant Physiol. 59 : 151 - 154, 1977.

*32316 - VOSKRESENSKAYA, N.P., MAZHUL', M.M. : Svetozavisimoe izmenenie aktivnosti gli-
tseral'degid-3-fosfatdegidrogenazy i svyaz' ego s reaktsiyami fotosinteza v
list'yakh gorokha. [Light dependent changes in the activity of glyceraldehyde-

-3-phosphate dehydrogenase related to photosynthetic reactions in pea leaves.]
- Fiziol. Rast. *23* : 483 - 489, 1976. [In R, ab : E.]

32317 - **VOSKRESENSKAYA, N.P., POYARKOVA, N.M., KHODZHIEV, A., DROZDOVA, I.S.** : Regulya-
tornoe deĭstvie sinego sveta na aktivnost' karboksiliruyushchikh fermentov i
fermentov glikolatnogo puti u rasteniĭ bobov i kukuruzy. [Control action of blue
radiation on the activity of carboxylases and glycolate pathway enzymes in broad
bean and maize plants.] - In : NASYROV, Yu.S. (ed.) : Genetika Fotosinteza. Pp.
71 - 77. Donish, Dushanbe 1977. [In R.]

B32318 - **VOZNESENSKIĬ, V.L.** : Fotosintez Pustynnykh Rasteniĭ (Yugo-Vostochnye Karaku-
my). [Photosynthesis of Desert Plants (South-East Kara-kum).] - Nauka, Leningrad 1977. [In R.]

32319 - **VREDENBERG, W.J.** : Photo-electric responses and ion transfers at the membra-
nes of intact chloroplasts. - In : THELLIER, M., MONNIER, A., DEMARTY, M.,
DAINTY, J. (ed.) : Échanges Ioniques Transmembranaires chez les Végétaux.
Coll. CNRS 258. Pp. 583 - 590. Édit. CNRS, Paris 1977.

32320 - **VREDENBERG, W.J.** : Some electrophysiological aspects of energy coupling and
ion transport in intact chloroplasts. - Biochem. Soc. Trans. *5* : 499 - 503,
1977.

32321 - **VREDENBERG, W.J., SCHAPENDONK, A.H.C.M., TONK, W.J.M.** : Electrogenesis at chloroplast membranes: P515- and microelectrode measurements. - In : COOMBS, J.
(ed.) : 4th International Congress on Photosynthesis. Pp. 398 - 399. UKISES,
London 1977.

32322 - **VRKOČ, F.** : K dynamice růstu a produktivitě hlavních polních plodin. [Growth
dynamics and productivity of main field crops.] - In : Produkce Biomasy a
Tvorba Výnosů Polních Plodin. Vol.1. Pp. 13 - 23. Česká Vědeckotechnická Společnost Zemědělská, Praha 1977. [Growth analysis; In Czech, ab : E, G, R.]

32323 - **VUČINIČ, Ž., RADENOVIČ, Č., PENČIČ, M.** : The temperature dependence of the
millisecond component of delayed light emission fromaan intact maize (*Zea
mays* L.) leaf. - In : COOMBS, J. (ed.) : 4th International Congress on Photosynthesis. P. 399. UKISES, London 1977.

32324 - **VYAS, A.B., JOY, M.P.** : Studies on caloric contents of some edible and non-
-edible plants during postmonsoon season. - Comp. Physiol. Ecol. *2* : 17 - 19,
1977.

32325 - **VYAS, N.L., GARG, R.K., VYAS, L.N.** : Plant biomass and net production relations of *Dyospiros melanoxylon* ROXB. COR. at deciduous forest near Udaipur
(Rajasthan), India. - Biológia (Bratislava) *32* : 461 - 467, 1977.

32326 - **VYSKOT, B., NOVÁK, F.J.** : Habituation and organogenesis in callus cultures
of chlorophyll mutants of *Nicotiana tabacum* L. - Z. Pflanzenphysiol. *81* : 34
- 42, 1977.

32327 - **WAGENER, K.** : Recycling of excess carbon dioxide from fossil energy conversion
by plants. - In : BUVET, R., ALLEN, M.J., MASSUÉ, J.-P. (ed.) : Living Systems
as Energy Converters. Pp. 319 - 328. North-Holland Publ. Co., Amsterdam - New
York - Oxford 1977.

32328 - **WAGGONER, P.E.** : Simulation modelling of plant physiological processes to pre-
dict crop yields. - In : LANDSBERG, J.J., CUTTING, C.V. (ed.) : Environmental
Effects on Crop Physiology. Pp. 351 - 363. Academic Press, London - New York
- San Francisco 1977.

32329 - **WAGNER, E.** : Molecular basis of physiological rhythms. - In : JENNINGS, D.H.
(ed.) : Integration of Activity in the Higher Plant. Pp. 33 - 72. Cambridge
University Press, Cambridge - London - New York - Melbourne 1977. [Ps.]

32330 - **WAGNER, G., OESTERHELT, D.** : Light-dependent potassium transport and ATP synthesis in *Halobacterium halobium*. - In : COOMBS, J. (ed.) : 4th International
Congress on Photosynthesis. P. 400. UKISES, London 1977.

32331 - **WAGNER, R., JUNGE, W.** : Gated proton conductance across the thylakoid membrane through OPDM-modified coupling factor CF_1. - In : COOMBS, J. (ed.) : 4th International Congress on Photosynthesis. P. 401. UKISES, London 1977.

32332 - WAGNER, R., JUNGE, W. : Gated proton conduction *via* the coupling factor of photophosphorylation modified by *N,N-orthophenyldimaleimide*. - Biochim. biophys. Acta *462* : 259 - 272, 1977.

32333 - WAGNER, W., FOLLMANN, H. : A thioredoxin from green algae. - Biochem. biophys. Res. Commun. *77* : 1044 - 1051, 1977.

32334 - WALBOT, V. : Heavy metal impurities impair spectrophotometric assay of ribulose bisphosphate carboxylase activity. - Plant Physiol. *59* : 107 - 110, 1977.

32335 - WALBOT, V. : Use of silica sol step gradients to prepare bundle sheath and mesophyll chloroplasts from *Panicum maximum*. - Plant Physiol. *60* : 102 - 108, 1977.

32336 - WALBOT, V. : The dimorphic chloroplasts of the C_4 plant *Panicum maximum* contain identical genomes. - Cell *11* : 729 - 737, 1977. [Ps.]

32337 - WALK, R.-A., HOCK, B. : Glyoxysomal and mitochondrial malate dehydrogenase of watermelon (*Citrullus vulgaris*) cotyledons. II. Kinetic properties of the purified isoenzymes. - Planta *136* : 221 - 228, 1977.

32338 - WALK, R.-A., MICHAELI, S., HOCK, B. : Glyoxysomal and mitochondrial malate dehydrogenase of watermelon (*Citrullus vulgaris*) cotyledons. I. Molecular properties of the purified isoenzymes. - Planta *136* : 211 - 220, 1977.

32339 - WALKER, A.J., HO, L.C. : Carbon translocation in the tomato : carbon import and fruit growth. - Ann. Bot. *41* : 813 - 823, 1977.

32340 - WALKER, A.J., HO, L.C. : Carbon translocation in the tomato : Effects of fruit temperature on carbon metabolism and the rate of translocation. - Ann. Bot. *41* : 825 - 832, 1977.

32341 - WALKER, A.J., THORNLEY, J.H.M. : The tomato fruit : import, growth, respiration and carbon metabolism at different fruit sizes and temperatures. - Ann. Bot. *41* : 977 - 985, 1977.

32342 - WALKER, D.A. : *In vitro* photosynthesis. - In : CASTELLANI, A. (ed.) : Research in Photobiology. Pp. 153 - 167. Plenum Press, New York - London 1977.

32343 - WALKER, D.A., HEROLD, A. : Can the chloroplast support photosynthesis unaided? - Plant Cell Physiol. *1977* (Spec. Issue 3 - Photosynthetic Organelles. Structure and Function) : 295 - 310, 1977.

32344 - WALKER, N.A., SMITH, F.A. : Bicarbonate uptake and membrane currents in *Chara* photosynthesis. - In : COOMBS, J. (ed.) : 4th International Congress on Photosynthesis. P. 425. UKISES, London 1977.

32345 - WALKER, N.A., SMITH, F.A. : Circulating electric currents between acid and alkaline zones associated with HCO_3^- assimilation in *Chara*. - J. exp. Bot. *28* : 1190 - 1206, 1977.

32346 - WALL, J.D., JOHANSSON, B.C., GEST, H. : A pleiotropic mutant of *Rhodopseudomonas capsulata* defective in nitrogen metabolism. - Arch. Microbiol. *115* : 259 - 263, 1977. [Ps growth rates.]

32347 - WALLACE, A., BAMBERG, S.A., CHA, J.W., ROMNEY, E.M. : Partitioning of photosynthetically fixed ^{14}C in perrenial plants of the northern Mojave desert. - In : MARSHALL, J.K. (ed.) : The Belowground Ecosystem: A Synthesis of Plant-Associated Processes. Pp. 141 - 148. Colorado State University, Fort Collins 1977.

32348 - WALLENTINUS, I. : Productivity studies on baltic macroalgae. - J. Phycol. *13* (Suppl.) : 71, 1977. [Chl.]

32349 - WALNE, P.L., PAGNI, P.S. : Characterization of pigments from isolated stigmata of *Euglena gracilis* var. *bacillaris*. - J. Phycol. *13* (Suppl.) : 71, 1977.

*32350 - WALNE, P.L., PALISANO, J.R. : Observations on non-chlorophyllous cells of *Euglena gracilis* as revealed by fluorescence and electron microscopy. - J. Phycol. *10* (Suppl.) : 18, 1974.

32351 - **WALTER, G., AVERINA, N.G., MEISTER, A.** : Protochlorophyllid-Resynthese unter dem Einfluß von Kinetin. Spektrophotometrische Untersuchungen an Weizenkeimpflanzen (*Triticum aestivum* L.) *in vivo*. - Biochem. Physiol. Pflanzen *171* : 409 - 417, 1977.

32352 - **WALTON, D.C., GALSON, E., HARRISON, M.A.** : The relationship between stomatal resistance and abscisic-acid levels in leaves of water-stressed bean plants. - Planta *133* : 145 - 148, 1977.

32353 - **WALTON, D.W.H.** : Studies on *Acaena (Rosaceae)*. II. Leaf production and senescence in *Acaena magellanica* (LAM.) VAHL. - Br. Antarct. Surv. Bull. *45* : 93 - 100, 1977. [Ps.]

32354 - **WALZ, D.** : Change of lecithin aggregation due to valinomycin-lipid interaction, and its relevance to energy conversion. - In : **PACKER, L., PAPAGEORGIOU, G.C., TREBST, A.**(ed.) : Bioenergetics of Membranes. Pp. 485 - 494. Elsevier/North-Holland Biomedical Press, Amsterdam - Oxford - New York 1977.

32355 - **WALZ, D.** : Pigment containing lipid vesicles. III. Role of chlorophyll *a* as sensor for aggregational states of lecithin. - J. Membrane Biol. *31* : 31 - 64, 1977.

*32356 - **WANDERS, J.B.W.** : The role of benthic algae in the shallow reef of Curaçao (Netherlands Antilles). I : Primary productivity in the coral reef. - Aquat. Bot. *2* : 235 - 270, 1976. [Chl.]

32357 - **WANG, R.T., MYERS, J.** : Evidence for reverse energy transfer from chlorophyll to phycobilins in *Anacystis nidulans*. - In : **OLSON, J.M., HIND, G.** (ed.) : Chlorophyll-Proteins, Reaction Centers, and Photosynthetic Membranes. Pp. 363 - 364. Brookhaven nat. Lab., Upton 1977.

32358 - **WANG, R.T., MYERS, J.** : Reverse energy transfer from chlorophyll to phycobilin in *Anacystis nidulans*. - Plant Cell Physiol. *1977* (Spec. Issue 3 - Photosynthetic Organelles. Structure and Function) : 3 - 7, 1977.

32359 - **WANG,R.T.,STEVENS, C.L.R., MYERS, J.** : Action spectra for photoreactions I and II of photosynthesis in the blue-green alga *Anacystis nidulans*. - Photochem. Photobiol. *25* : 103 - 108, 1977.

32360 - **WANG, W.** : Photoconversion of protochlorophyllide in *Chlamydomonas reinhardtii*. - Plant Physiol. *59* (6, Suppl.) : 92, 1977.

32361 - **WANG, W., BOYNTON, J.E., GILLHAM, N.W.** : Genetic control of chlorophyll biosynthesis : Effect of increased δ-aminolevulinic acid synthesis on the phenotype of the *y-1* mutant of *Chlamydomonas*. - Mol. gen. Genet. *152* : 7 - 12, 1977.

*32362 - **WARD, H.B.** : Dark reduction of 2,6-dichlorophenolindophenol by *Anacystis nidulans*. - J. Phycol. *10* (Suppl.) : 5, 1974.

32363 - **WAREMBOURG, F.R., PAUL, E.A.** : Seasonal transfers of assimilated ^{14}C in grassland : plant production and turnover, soil and plant respiration. - Soil Biol. Biochem. *9* : 295 - 301, 1977.

32364 - **WARRINGTON, I.J., PEET, M., PATTERSON, D.T., BUNCE, J., HASLEMORE, R.M., HELLMERS, H.** : Growth and physiological responses of soybean under various thermoperiods. - Aust. J. Plant Physiol. *4* : 371 - 380, 1977. [Ps.]

32365 - **WASHITANI, I., SATO, S.** : Studies on the function of proplastids in the metabolism of *in vitro* cultured tobacco cells I. Localization of nitrite reductase and NADP-dependent glutamate dehydrogenase. - Plant Cell Physiol. *18* : 117 - 125, 1977.

32366 - **WASHITANI, I., SATO, S.** : Studies on the function of proplastids in the metabolism of *in vitro* cultured tobacco cells II. Glutamine synthetase/glutamate synthetase pathway. - Plant Cell Physiol. *18* : 505 - 512, 1977.

32367 - **WASIELEWSKI, M.R., SMITH, U.H., COPE, B.T., KATZ, J.J.** : A synthetic biomimetic model of special pair bacteriochlorophyll *a*. - J. amer. chem. Soc. *99* : 4172 - 4173, 1977.

32368 - **WATANABE, M.F.** : Phycoerythrin in the deeper water layer of a stratified eutrophic lake: an application of bile pigment in determining the standing crop of blue-green algae. - Int. Rev. ges. Hydrobiol. *62* : 549 - 556, 1977.

32369 - WATTS, W.R. : Field studies of stomatal conductance. - In : LANDSBERG. J.J., CUTTING, C.V. (ed.) : Environmental Effects on Crop Physiology. Pp. 173 - 196. Academic Press, London - New York - San Francisco 1977.

32370 - WEAVER, E.C., CORKER, G.A. : Electron paramagnetic resonance spectroscopy. - In : TREBST, A., AVRON, M. (ed.) : Photosynthesis I. (Encycl. Plant Physiol. N.S. Vol.5.) Pp. 168 - 178. Springer-Verlag, Berlin - Heidelberg - New York 1977. [Ps.]

32371 - WEAVING, G.S., FILSHIE, J. : A solarimeter utilizing silicon semiconductor diodes. - J. agr. Eng. Res. 22 : 113 - 126, 1977.

32372 - WEBB, W.L. : Seasonal allocation of photoassimilated carbon in Douglas fir seedlings. - Plant Physiol. 60 : 320 - 322, 1977.

32373 - WEBER, D.J., ANDERSEN, W.R., HESS, S., HANSEN, D.J., GUNASEKARAN, M. : Ribulose-1,5-bisphosphate carboxylase from plants adapted to extreme environments. - Plant Cell Physiol. 18 : 693 - 699, 1977.

32374 - WEBSTER, B.D., LEOPOLD, A.C. : The ultrastructure of dry and imbibed cotyledons of soybean. - Amer. J. Bot. 64 : 1286 - 1293, 1977. [Proplastids.]

32375 - WEBSTER, G.D., EDWARDS, P.A., JACKSON, J.B. : Interconversion of two kinetically distinct states of the membrane-bound and solubilized H^+-translocating ATPase from Rhodospirillum rubrum. - FEBS Lett. 76 : 29 - 35, 1977.

32376 - WEBSTER, G.D., JACKSON, J.B. : Effect of anions on the H^+-translocating ATPase from Rhodospirillum rubrum. - In : COOMBS, J. (ed.) : 4th International Congress on Photosynthesis. P. 402. UKISES, London 1977.

32377 - WEBSTER, T.R., JAGELS, R. : Morphology and development of aerial roots of Selaginella martensii grown in moist containers. - Can. J. Bot. 55 : 2149 - 2158, 1977. [Chloroplasts.]

32378 - WEGMANN, K. : Biochemische Anpassung von Pflanzen an Wassermangel. - Chemiker-Zeit. 101 : 169 - 173, 1977. [Ps.]

32379 - WEGMANN, K., KAMEKE, E. von : Properties and function of two manganese-containing proteins from Dunaliella chloroplasts. - In : COOMBS, J. (ed.) : 4th International Congress on Photosynthesis. P. 403. UKISES, London 1977.

32380 - WEIGEL, H.P. : On the distribution of particulate metals, chlorophyll and seston in the Baltic sea. - Mar. Biol. 44 : 217 - 222, 1977.

32381 - WEINSTEIN, J.D., CASTELFRANCO, P.A. : Protoporphyrin IX biosynthesis from glutamate in isolated greening chloroplasts. - Arch. Biochem. Biophys. 178 : 671 - 673, 1977.

32382 - WEINSTEIN, J.D., CASTELFRANCO, P.A. : Incorporation of ^{14}C-glutamate in 5-aminolevulinic acid (ALA) by isolated chloroplasts. - Plant Physiol. 59 (6, Suppl.) : 92, 1977.

*32383 - WEISE, G., AUERBACH, S., HORBACH, W., HORNIG, L., PRÜFER, P., SEIDEL, K. : IRGA-Messung des CO_2-Umsatzes des Phytobenthos als Voraussetzung seiner Nutzbarkeit als Indikations- und Kontrollsystem. - In : Acta IMEKO Vol. III. Measurement and Instrumentation. Pp. 173 - 184. Akadémiai Kiadó, Budapest 1973.

32384 - WEISS, M.A., McCARTY, R.E. : Cross-linking within a subunit of coupling factor 1 increases the proton permeability of spinach chloroplast thylakoids. - J. biol. Chem. 252 : 8007 - 8012, 1977.

32385 - WEISTROP, J.S., STERN, A.I. : Appearance of photochemical activity in isolated chloroplasts from far-red-illuminated leaves of Phaseolus vulgaris. - J. exp. Bot. 28 : 354 - 365, 1977.

32386 - WELLBURN, A.R. : Distribution of chloroplast coupling factor (CF_1) particles on plastid membranes during development. - In : COOMBS, J. (ed.) : 4th International Congress on Photosynthesis. Pp. 403 - 404. UKISES, London 1977.

32387 - WELLBURN, A.R. : Distribution of chloroplast coupling factor (CF_1) particles on plastid membranes during development. - Planta 135 : 191 - 198, 1977.

*32388 - WELLBURN, A.R., HAMPP, R. : Uptake of mevalonate and acetate during plastid development. - Biochem. J. *158* : 231 - 233, 1976.

32389 - WELLBURN, A.R., HAMPP, R. : Changes in plastid and mitochondrial envelopes during greening. - Biochem. Soc. Trans. *5* : 91 - 94, 1977.

32390 - WELLBURN, A.R., QUAIL, P.H., GUNNING, B.E.S. : Examination of ribosome-like particles in isolated prolamellar bodies. - Planta *134* : 45 - 52, 1977.

32391 - WENDT, C.W., ONKEN, A.B., WILKE, O.C., HARGROVE, R., BAUSCH, W., BARNES, L. : Effects of irrigation systems on the water requirements of sweet corn. - Soil Sci. Soc. Amer. J. *41* : 785 - 788, 1977. [Growth analysis.]

32392 - WESSELS, J.S.C. : Fragmentation. - In : TREBST, A., AVRON, M.(ed.) : Photosynthesis I. (Encycl. Plant Physiol. N.S. Vol. 5.) Pp. 563 - 573. Springer-Verlag, Berlin - Heidelberg - New York 1977. [Chloroplast.]

32393 - WESSELS, J.S.C., BORCHERT, M.T. : SDS-gel electrophoresis of the reaction-center containing complexes and the light-harvesting complex of spinach chloroplasts. - In : COOMBS, J. (ed.) : 4th International Congress on Photosynthesis. P. 405. UKISES, London 1977.

32394 - WESTMAN, W.E., ROGERS, R.W. : Biomass and structure of a subtropical eucalypt forest, North Stradbroke Island. - Aust. J. Bot. *25* : 171 - 191, 1977.

32395 - WESTRIN, H., JOHANSSON, G. : Specific extraction of intact chloroplasts using aqueous biphasic systems. - In : COOMBS, J. (ed.) : 4th International Congress on Photosynthesis. Pp. 404 - 405. UKISES, London 1977.

32396 - WETTERN, M., WEBER, A. : A new method for dry weight determination of marine planctonic diatoms. - Bot. mar. *20* : 537 - 539, 1977.

32397 - WETTSTEIN, D. von : Genetic control of chloroplast protein synthesis in higher plants. - In : COOMBS, J. (ed.) : 4th International Congress on Photosynthesis. P. 406. UKISES, London 1977.

32398 - WEZELMAN, B., GASSMAN, M.L., CASTELFRANCO, P.A. : 5-aminolevulinic acid oxidase from dark-grown barley shoots. - Plant Physiol. *59* (6, Suppl.) : 103, 1977.

32399 - WHATLEY, J.M. : Variations in the basic pathway of chloroplast development. - New Phytol. *78* : 407 - 420, 1977.

32400 - WHATLEY, J.M. : The effect of cotyledons on chloroplast development in primary leaves of *Phaseolus vulgaris*. - New Phytol. *79* : 55 - 60, 1977.

32401 - WHATLEY, J.M. : The fine structure of *Prochloron*. - New Phytol. *79* : 309 - 313, 1977.

32402 - WHEELER, H. : Increase with age sensitivity of oat leaves to victorin. - Phytopathology *67* : 859 - 861, 1977. [Chl.]

32403 - WHIGHAM, D., SIMPSON, R. : Growth, mortality, and biomass partitioning in freshwater tidal wetland populations of wild rice (*Zizania aquatica* var. *aquatica*). - Bull. Torrey bot. Club *104* : 347 - 351, 1977.

32304 - WHITE, E., PAYNE, G.W. : Chlorophyll production, in response to nutrient additions, by the algae in Lake Taupo water. - N.Z.J. mar. Freshwater Res. *11* : 501 - 507, 1977.

32405 - WHITE, R.C., JONES, I.D., GIBBS, E., BUTLER, L.S. : Estimation of copper pheophytins, chlorophylls, and pheophytins in mixtures in diethyl ether. - J. agr. Food Chem. *25* : 143 - 145, 1977.

*32406 - WHITMAN, W., TABITA, F.R. : Inhibition of D-ribulose 1,5-biphosphate carboxylase by pyridoxal 5'-phosphate. - Biochem. biophys. Res. Commun. *71* : 1034 - 1039, 1976.

32407 - WHITMARSH, J., CRAMER, W.A. : A pathway for the reduction of cytochrome b-559 by Photosystem II in spinach chloroplasts. - In : COOMBS, J. (ed.) : 4th International Congress on Photosynthesis. Pp. 406 - 407. UKISES, London 1977.

32408 - WHITMARSH, J., CRAMER, W.A. : Kinetics of the photoreduction of cytochrome b-559 by photosystem II in chloroplasts. - Biochim. biophys. Acta *460* : 280 - 289, 1977.

32409 - WHITTENBURY, R., DOW, C.S. : Morphogenesis and differentiation in *Rhodomicro-bium vanniellii* and other budding and prosthecate bacteria. - Bacteriol. Rev. *41* : 754 - 808, 1977. [Chl.]

B32410 - WHITTINGHAM, C.P. : Photosynthesis. - 2nd rev. Ed. Carolina Biology Readers. Vol. 9. Carolina Biological Supply Company, Burlington, N.C. 1977.

32411 - WIDART, M., DINANT, M., AGHION, J. : Chlorophylle *a* fixée à des globules de lipides et de protéines extraits du lait. Photosensibilisation de la réduc-tion du rouge de méthyle par l'ion ascorbate en milieux aqueux. - Physiol. vég. *15* : 705 - 710, 1977.

32412 - WIEBE, H.-J., LORENZ, H.-P. : Wirkung von Wechseltemperatur und lichtabhän-giger Temperaturregelung auf das Wachstum von Kopfsalat. - Gartenbauwissenschaft *42* : 42 - 45, 1977.

32413 - WIEBE, W.J., SMITH, D.F. : Direct measurement of dissolved organic carbon re-lease by phytoplankton and incorporation by microheterotrophs. - Mar. Biol. *42* : 213 - 223, 1977. [Ps.]

32414 - WIĘCKOWSKI, S., FICEK, S. : Effects of some inhibitors of protein synthesis on chloroplast fine structure, CO_2 fixation and the Hill reaction activity. - Acta Soc. Bot. Pol. *46* : 251 - 258, 1977.

32415 - WIELGOLASKI, F.E. : Primary production of alpine communities in Norway esti-mated by CO_2-exchange and harvesting techniques. - In : COOMBS, J. (ed.) : 4th International Congress on Photosynthesis. Pp. 407 - 408. UKISES, London 1977.

32416 - WIESSNER, W., DUBERTRET, G., MENDE, D. : *In vivo* regulation of the Photosys-tem II mediated electron transport. - In : COOMBS, J. (ed.) : 4th Internatio-nal Congress on Photosynthesis. P. 410. UKISES, London 1977.

32417 - WILD, A., FULDNER, K.-H. : The concentration of cytochrome *f* and *P*700 in chlorophyll-deficient mutants of *Chlorella fusca*. - Planta *136* : 281 - 281, 1977.

32418 - WILD, A., OBERWEIS, A.L., RÜHLE, W. : Wirkung der Insektizide Allethrin, Lin-dan und Jacutin-Fogetten-Sublimat auf den photosynthetischen Elektronentrans-port. - Z. Pflanzenphysiol. *82* : 161 - 172, 1977.

32419 - WILD, A., TROSTMANN, U. : Development of the photosynthetic apparatus during light-dependent greening of a *Chlorella*-mutant. - In : COOMBS, J. (ed.) : 4th International Congress on Photosynthesis. Pp. 408 - 409. UKISES, London 1977.

32420 - WILDMAN, S.G., JOPE, C.A. : Origin of chloroplasts the size of single grana in apical meristems. - Plant Cell Physiol. *1977* (Spec. Issue 3 - Photosynthe-tic Organelles. Structure and Function) : 385 - 401, 1977.

*32421 - WILDNER, G.F. : The greening process in *Euglena gracilis* I. The kinetics of appearance of chloroplast proteins and the effect of cycloheximide and chlo-ramphenicol on their synthesis. - Z. Naturforsch. *31 C* : 157 - 162, 1976.

32422 - WILDNER, G.F., HENKEL, J. : The influence of oxidants and reductants on the enzymatic properties of ribulose-1,5-bisphosphate carboxylase and its oxyge-nase activity. - In : COOMBS, J. (ed.) : 4th International Congress on Pho-tosynthesis. P. 411. UKISES, London 1977.

32423 - WILHM, J., DORRIS, T., SEYFER, J.R., McCLINTOCK, N. : Seasonal variation in plankton populations in the Arkansas River near the confluence of Red Rock Creek. - Southwest. Naturalist *22* : 411 - 420, 1977.

32424 - WILKINSON, R.E. : Zeaxanthin epoxidation inhibition by EPTC (S-ethyl dipro-pylthiocarbamate). - Bot. Gaz. *138* : 270 - 275, 1977.

32425 - WILLERT, D.J. von, CURDTS, E., WILLERT, K. von : Veränderung der PEP-Carbo-xylase während einer durch NaCl geförderten Ausbildung eines CAM bei *Mesem-bryanthemum crystallinum*. - Biochem. Physiol. Pflanz. *171* : 101 - 107, 1977.

32426 - WILLERT, D.J. von, THOMAS, D.A., LOBIN, W., CURDTS, E. : Ecophysiologic in-vestigations in the family of the *Mesembryanthemaceae*. Occurrence of a CAM and ion content. - Oecologia *29* : 67 - 76, 1977.

32427 - **WILLIAMS, A.M., MARINOS, N.G.** : Regulation of the movement of assimilate into ovules of *Pisum sativum* cv. Greenfeast : effect of pod temperature. - Aust. J. Plant Physiol. *4* : 515 - 521, 1977.

32428 - **WILLIAMS, G.J., KEMP, P.R., OUTLON, K.** : Anatomical, physiological, and biochemical evidence for the C_3 pathway in *Verbascum thapsus* L. - New Phytol. *79* : 489 - 492, 1977.

32429 - **WILLIAMS, L.E., KENNEDY, R.A.** : Carboxylase, decarboxylase and aminotransferase enzyme studies during leaf ontogeny in *Zea mays*. - Plant Physiol. *59* (6, Suppl.) : 65, 1977.

32430 - **WILLIAMS, L.E., KENNEDY, R.A.** : Relationship between early photosynthetic products, photorespiration, and stage of leaf development in *Zea mays*. - Z. Pflanzenphysiol. *81* : 314 - 322, 1977.

32431 - **WILLIAMS, R.H., HAYES, J.D.** : The breeding implications of studies on yield and its components in contrasting genotypes of spring barley. - Cereal Res. Commun. *5* : 113 - 118, 1977.

32432 - **WILLIAMS, W.P.** : The two photosystems and their interactions. - In : **BARBER, J.** (ed.) : Primary Processes of Photosynthesis. Pp. 99 - 147. Elsevier, Amsterdam - New York - Oxford 1977.

32433 - **WILLIAMS, W.P., NUTBEAM, A.R., SALAMON, Z.** : Relationship between slow fluorescence yield changes observed in CMU poisoned *Chlorella pyrenoidosa* and the state I/state II phenomenon. - In : **COOMBS, J.** (ed.) : 4th International Congress on Photosynthesis. P. 412. UKISES, London 1977.

32434 - **WILLIAMS, W.T., BOUNDY, C.A.P., MILLINGTON, A.J.** : The effect of sowing date on the growth and yield of three sorghum cultivars in the Ord River valley. II. The components of growth and yield. - Aust. J. agr. Res. *28* : 381 - 387, 1977.

32435 - **WILLMER, C.M., RUTTER, J.C.** : Guard cell malic acid metabolism during stomatal movements. - Nature *269* : 327 - 328, 1977.

32436 - **WILSON, L.A.** : Root crops. - In : **ALVIM, P. de T., KOZLOWSKI, T.T.** (ed.) : Ecophysiology of Tropical Crops. Pp. 187 - 236. Academic Press, New York - San Francisco - London 1977. [Ps.]

32437 - **WIMMER, M.J., ROSE, I.A.** : Mechanism for oxygen exchange in the chloroplast photophosphorylation system. - J. biol. Chem. *252* : 6769 - 6775, 1977.

32438 - **WINGET, G.D., KANNER, N., RACKER, E.** : Formation of ATP by the adenosine triphosphatase complex from spinach chloroplasts reconstituted together with bacteriorhodopsin. - Biochim. biophys. Acta *460* : 490 - 499, 1977.

32439 - **WINTER, K., KRAMER, D., TROUGHTON, J.H., CARD, K.A., FISCHER, K.** : C_4 pathway of photosynthesis in a member of *Polygonaceae* : *Calligonum persicum* (BOISS. & BUHSE) BOISS. - Z. Pflanzenphysiol. *81* : 341 - 346, 1977.

32440 - **WIRTH, E., KELLY, G.J., FISCHBECK, G., LATZKO, E.** : Enzyme activities and products of CO_2 fixation in various photosynthetic organs of wheat and oat. - Z. Pflanzenphysiol. *82* : 78 - 87, 1977.

32441 - **WITHERS, N.W., BRITTON, G., GOODWIN, T.W., ALBERTE, R.S., LEWIN, R.A., THORNBER, J.P.** : Carotenoid and chlorophyll-protein composition of *Prochloron* sp., a prokaryotic green alga. - In : **COOMBS, J.** (ed.) : 4th International Congress on Photosynthesis. Pp. 413 - 414. UKISES, London 1977.

32442 - **WITHERS, N.W., COX, E.R., TOMAS, R., HAXO, F.T.** : Pigments of the dinoflagellate *Peridinium balticum* and its photosynthetic endosymbiont. - J. Phycol. *13* : 354 - 358, 1977.

32443 - **WITT, H.T.** : On the role of the electrical field in photosynthesis. - In : **ROUX, E.** (ed.) : Electrical Phenomena at the Biological Membrane Level. Pp. 507 - 517. Elsevier, Amsterdam - New York - Oxford 1977.

32444 - **WITT, H.T.** : On the bioenergetics mechanism of photosynthesis. Results by pulse methods. - In : **BUVET, R., ALLEN, M.J., MASSUÉ, J.-P.** (ed.) : Living

Systems as Energy Converters. Pp. 185 - 197. North-Holland Publ. Co., Amsterdam - New York - Oxford 1977.

32445 - WITT, H.T., SCHLODDER, E., GRÄBER, P. : Conformational change, ATP generation and turnover rate of the chloroplast ATPase analyzed by energization with an external electric field. - In : PACKER, L., PAPAGEORGIOU, G.C., TREBST, A. (ed.) : Bioenergetics of Membranes. Pp. 447 - 457. Elsevier/North-Holland Biomedical Press, Amsterdam - Oxford - New York 1977.

32446 - WITT, U. : Auswirkungen der künstlichen Düngung eines Hochgebirgssees (Vorderer Finstertaler See, Kühtai, Tirol). - Arch. Hydrobiol. *81* : 211 - 232, 1977. [Primary production.]

32447 - WITZTUM, A., SHAPIRA, Z. : Exudation and chloroplast fragmentation as a result of ultraviolet irradiation in *Spirodela oligorhiza*. - Isr. J. Bot. *26* : 109 - 114, 1977.

32448 - WODZINSKI, R.S., LABEDA, D.P., ALEXANDER, M. : Toxicity of SO_2 and NO_x : Selective inhibition of blue-green algae by bisulfite and nitrite. - J. Air Pollut. Control Assoc. *27* : 891 - 893, 1977. [Ps.]

32449 - WOLEDGE, J. : Differences in photosynthesis between vegetative and reproductive swards of ryegrass. - In : COOMBS, J. (ed.) : 4th International Congress on Photosynthesis. Pp. 412 - 413. UKISES, London 1977.

32450 - WOLEDGE, J. : The effects of shading and cutting treatments on the photosynthetic rate of ryegrass leaves. - Ann. Bot. *41* : 1279 - 1286, 1977.

32451 - WOLF, F.T. : Effects of chemical agents in inhibition of chlorophyll synthesis and chloroplast development in higher plants.-Bot. Rev. *43* : 395 - 425, 1977.

32452 - WOLFF, B., SCHANTZ, R. : Studies on plastidial proteins in *Euglena gracilis*. - In : COOMBS, J. (ed.) : 4th International Congress on Photosynthesis. P. 409. UKISES, London 1977.

32453 - WOLLMAN, F.A. : Control of the redox state of the secondary electron acceptor of Photosystem II. - In : COOMBS, J. (ed.) : 4th International Congress on Photosynthesis. P. 414. UKISES, London 1977.

32454 - WOLOSIUK, R.A., BUCHANAN, B.B. : Thioredoxin and glutathione regulate photosynthesis in chloroplasts. - Nature *266* : 565 - 567, 1977.

32455 - WOLOSIUK, R.A., BUCHANAN, B.B., CRAWFORD, N.A. : Regulation of NADP-malate dehydrogenase by the light-actuated ferredoxin/thioredoxin system of chloroplasts. - FEBS Lett. *81* : 253 - 258, 1977.

32456 - WONG, D., JURSINIC, P., GOVINDJEE : Direct effects of mono and divalent cations on the activity of the reaction center of photosystem II in pea chloroplasts. - Plant Physiol. *59* (6, Suppl.) : 24, 1977.

32457 - WONG, J.H.H., BENEDICT, C.R. : Enzyme profile of CO_2 fixing pathways in castor bean endosperms. - Plant Physiol. *59* (6, Suppl.) : 43, 1977.

*32458 - WONG, S.C., HEW, C.S. : Diffusive resistance, titratable acidity, and CO_2 fixation in two tropical epiphytic ferns. - Amer. Fern J. *66* : 121 - 124, 1976.

32459 - WONG, W.W.L., BENEDICT, C.R., McGRATH, T., KOHEL, R.J. : The fractionation of stable carbon isotopes by RuDP carboxylase. - Plant Physiol. *59* (6, Suppl.) : 42, 1977.

32460 - WOO, K.C., BERRY, J.A., OSMOND, C.B., LORIMER, G.H. : Photorespiration and the metabolism of ammonia. - In : COOMBS, J. (ed.) : 4th International Congress on Photosynthesis. P. 415. UKISES, London 1977.

32461 - WOO, K.C., OSMOND, C.B. : Intramitochondrial localization and properties of glycine decarboxylation and serine synthesis in plant leaves. - Plant Physiol. *59* (6, Suppl.) : 43, 1977.

32462 - WOO, M.D., OSMOND, C.B. : Participation of leaf mitochondria in the photorespiratory carbon oxidation cycle : Glycine decarboxylation activity in leaf mitochondria from different species and its intra-mitochondrial location. - Plant Cell Physiol. *1977* (Spec. Issue 3 - Photosynthetic Organelles. Structure and Function) : 315 - 323, 1977.

32463 - WOOD, K.G.: Chemical enhancement of CO_2 flux across the air-water interface. - Arch. Hydrobiol. *79* : 103 - 110, 1977.

32464 - WOOD, P.M. : *C*-type cytochromes and plastocyanin in higher plant and algal photosynthesis. - In : COOMBS, J. (ed.) : 4th International Congress on Photosynthesis. Pp. 415 - 416. UKISES, London 1977.

32465 - WOOD, P.M. : The roles of *c*-type cytochromes in algal photosynthesis. Extraction from algae of a cytochrome similar to higher plant cytochrome *f*. - Europe. J. Biochem. *72* : 605 - 612, 1977.

32466 - WOODS, S.J., SWEARINGIN, M.L. : Influence of simulated early lodging upon soybean seed yield and its components. - Agron. J. *69* : 239 - 242, 1977.

32467 - WRAIGHT, C.A. : Electron acceptors of photosynthetic bacterial reaction centers. Direct observation of oscillatory behaviour suggesting two closely equivalent ubiquinones. - Biochim. biophys. Acta *459* : 525 - 531, 1977.

32468 - WRAIGHT, C.A. : The role of iron in the electron acceptor region of reaction centres of photosynthetic bacteria. - In : COOMBS, J. (ed.) : 4th International Congress on Photosynthesis. P. 416. UKISES, London 1977.

32469 - WRENCH, P., WRIGHT, L., BRADY, C.J., HINDE, R.W. : The source of carbon for proline synthesis in osmotically stressed artichoke tuber slices. - Aust. J. Plant Physiol. *4* : 703 - 711, 1977.

32470 - WRIGHT, S.J.L., STAINTHORPE, A.F., DOWNS, J.D. : Interactions of the herbicide propanil and a metabolite, 3,4-dichloroaniline, with blue-green algae. - Acta phytopathol. Acad. Sci. hung. *12* : 51 - 60, 1977. [Ps.]

32471 - WRISCHER, M. : Ultrastructural localization of photosystem I in plastids of senescent spinach leaves. - Acta bot. croat. *36* : 57 - 61, 1977.

32472 - WRÓBEL, D. : Second and fourth derivative absorption spectra of chlorophyll *a* and chlorophyll *c in vivo* and *in vitro*. - Photosynthetica *11* : 90 - 92, 1977.

32473 - WURTZ, E.A., BOYNTON, J.E., GILLHAM, N.W. : Perturbation of chloroplast DNA amounts and chloroplast gene transmission in *Chlamydomonas reinhardtii* by 5-fluorodeoxyuridine. - Proc. nat. Acad. Sci. USA *74* : 4552 - 4556, 1977.

32474 - WYDRZYNSKI, T., GOVINDJEE, MARKS, S.B., SCHMIDT, P.G., GUTOWSKY, H.S. : NMR studies on chloroplast membranes : frequency and temperature dependence on water proton relaxation rates. - Biophys. J. *17* : 198a, 1977.

32475 - WYDRZYNSKI, T., MARKS, S., GOVINDJEE : The role of manganese in O_2 evolution during photosynthesis. - Plant Physiol. *59* (6, Suppl.) : 24, 1977.

32476 - WYDRZYNSKI, T., MARKS, S.B., GOVINDJEE : The role of manganese in oxygen evolution. - In : COOMBS, J. (ed.) : 4th International Congress on Photosynthesis. P. 417. UKISES, London 1977.

32477 - WYN JONES, R.G., STOREY, R., POLLARD, A. : Ionic and osmotic regulation in plants particularly halophytes. - In : THELLIER, M., MONNIER, A., DEMARTY, M., DAINTY, J. (ed.) : Échanges Ioniques Transmembranaires chez les Végétaux. Pp. 537 - 544. Édit. CNRS, Paris 1977. [Ps.]

32478 - YABUKI, K. : [Growth environment and photosynthesis of plants. 5.] - Nogyo Oyobi Engei *52* : 23 - 26, 1977. [In Japan.]

32479 - YAGI, T. : Use of an enzymic electric cell and immobilized hydrogenase in the study of the biophotolysis of water to produce hydrogen. - In : MITSUI, A., MIYACHI, S., SAN PIETRO, A., TAMURA, S. (ed.) : Biological Solar Energy Conversion. Pp. 61 - 68. Academic Press, New York - San Francisco - London 1977. [Ps.]

32480 - YAGI, T., MUKOHATA, Y. : Effects of purine nucleotides on photosynthetic electron transport in isolated chloroplasts. - J. Bioenerg. Biomembranes *9* : 31 - 40, 1977.

32481 - YAKUBOVA, M.M., NAZAROVA, Z.A., KRENDELEVA, T.E. : Kharakteristika pervichnykh reaktsiĭ fotosinteza u mutanta *Arabidopsis thaliana* (L.). [Characteristics of primary reactions of photosynthesis in the mutant *Arabidopsis thalia-na* (L.).] - Nauch. Dokl. vyssh. Shkoly, biol. Nauki *20* (11) : 110 - 114, 1977. [In R.]

32482 - YAKUBOVA, M.M., RUBIN, A.B., KHRAMOVA, G.A., MATORIN, D.N. : O fotokhimiches-kikh protsessakh fotosinteza u mutantov *Gossypium hirsutum*. [Photochemical processes of photosynthesis in *Gossypium hirsutum* mutants.] - In : NASYROV, Yu. S. (ed.) : Genetika Fotosinteza. Pp. 195 - 200. Donish, Dushanbe 1977. [In R.]

32483 - YAKUSHKINA, N.I., DULIN, A.F. : Fotosinteticheskaya aktivnost' prorostkov yachmenya pri obrabotke kinetinom i gibberellinom. [Photosynthetic activity of barley seedlings treated with kinetin and gibberellin.] - Sel'skokhoz. Biol. *12* : 212 - 214, 1977. [In R, ab : E.]

32484 - YAKUSHKINA, N.I., STARIKOVA, V.T. : Vliyanie kumarina i gibberellina na ne-kotorye storony énergeticheskogo obmena prorostkov kukuruzy. [The effect of coumarin and gibberellin on energy metabolism in maize seedlings.] - Fiziol. Rast. *24* : 1211 - 1216, 1977. [Ps; in R, ab : E.]

32485 - YAMAMOTO, T., WATANABE, S. : Studies on leaf burn of pear trees. VI. Resis-tances and rates of transpiration of the leaves *in situ* of Bartlett pear in the late period of the rainy season. - Bull. Yamagata Univ., agr. Sci. *7* (4) : 451 - 462, 1977.

32486 - YAMAMOTO, Y., NISHIMURA, M. : Characteristics of light-induced 515-nm absor-bance change in spinach chloroplasts at lower temperatures I. Participation of structural changes of thylakoid membranes in light-induced 515-nm absorban-ce change. - Plant Cell Physiol. *18* : 55 - 66, 1977.

32487 - YAMAMOTO, Y., NISHIMURA, M. : Characteristics of light-induced 515-nm absor-bance change in spinach chloroplasts at lower temperatures. II. Relationship between 515-nm absorbance change and rapid H^+ uptake in chloroplasts after a short flash illumination. - Plant Cell Physiol. *18* : 293 - 301, 1977.

32488 - YAMAYA, T., OJIMA, K., OHIRA, K. : Studies on the greening of cultured soy-bean and *Ruta* cells. II. Photosynthetic activities of the cultured green cells. - Soil Sci. Plant Nutr. *23* : 59 - 66, 1977.

32489 - YASNIKOV, A.A. : Ion-radical mechanisms of photophosphorylation and light-de-pendent transport of protons in chloroplasts. - In : COOMBS, J. (ed.) : 4th International Congress on Photosynthesis. Pp. 417 - 418. UKISES, London 1977.

32490 - YASNIKOV, A.A., BERSHTEĬN, B.I., VOLKOVA, N.V., VASILENOK, L.I., VOLOVIK, O.I., ZAĬTSEVA, N.A., KANIVETS, N.P., MUSHKETIK, L.S., OKANENKO, A.S., OSTROVSKAYA, L.K., PETRENKO, S.S., POLISHCHUK, A.I., REĬNGARD, T.A., SEMENYUK, I.I. : Regulyatsiya vklyucheniya neorganicheskogo fosfata v fotofosforilirovanie (piruvatkinazoĭ i fosfatazoĭ).[Regulation of inorganic phosphate incorpora-tion in photophosphorylation (by pyruvate kinase and phosphatase).] - In : NASYROV, Yu.S. (ed.) : Genetika Fotosinteza. Pp. 216 - 221. Donish, Dushanbe 1977. [In R.]

32491 - YELENOSKY, G., GUY, C.L. : Carbohydrate accumulation in leaves and stems of "Valencia" orange at progressively colder temperatures. - Bot. Gaz. *138* : 13 - 17, 1977.

32492 - YEN, H.-C., MARRS, B. : Growth of *Rhodopseudomonas capsulata* under anaerobic dark conditions with dimethyl sulfoxide. - Arch. Biochem. Biophys. *181* : 411 - 418, 1977. [Ps.]

32493 - YENTSCH, C.M., YENTSCH, C.S., STRUBE, L.R. : Variations in ammonium enhance-ment, an indication of nitrogen deficiency in New England coastal phytoplank-ton populations. - J. mar. Res. *35* : 537 - 555, 1977. [Chl.]

32494 - YENTSCH, C.S. : On the contribution of plant physiology to the study of pri-mary production. - In : COOMBS, J. (ed.) : 4th International Congress on Pho-tosynthesis. P. 419. UKISES, London 1977. [Ps.]

32495 - YOCH, D.C., CARITHERS, R.P., ARNON, D.I. : Isolation and characterization of bound iron-sulfur proteins from bacterial photosynthetic membranes. I. Ferredoxins III and IV from *Rhodospirillum rubrum* chromatophores. - J. biol. Chem. *252* : 7453 - 7460, 1977.

32496 - YOCUM, C.F. : Energy conserving cyclic electron transport associated with
Photosystem II. - In : COOMBS, J. (ed.) : 4th International Congress on Photosynthesis. P. 418. UKISES, London 1977.

32497 - YOCUM, C.F. : Photophosphorylation associated with photosystem II. II. Effects
of electron donors, catalyst oxidation, and electron transport inhibitors
on photosystem II cyclic photophosphorylation. - Plant Physiol. 60 : 592 -
596, 1977.

32498 - YOCUM, C.F. : Photophosphorylation associated with photosystem II. III. Characterization of uncoupling, energy transfer inhibition, and proton uptake
reactions associated with photosystem II cyclic photophosphorylation. - Plant
Physiol. 60 : 597 - 601, 1977.

32499 - YOCUM, C.F., GUIKEMA, J.A. : Photophosphorylation associated with photosystem II. 1. Photosystem II cyclic photophosphorylation catalyzed by p-phenyle-
nediamine. - Plant Physiol. 56 : 33 - 37, 1977.

32500 - YOKOHAMA, Y., KAGEYAMA, A. : Pigment composition and photosynthetic capabili-
ties of deep-water green algae. - J. Phycol. 13 (Suppl.) : 76, 1977.

32501 - YOKOHAMA, Y., KAGEYAMA, A., IKAWA, T., SHIMURA, S. : A carotenoid characteristic of chlorophycean seaweeds living in deep costal waters. - Bot. Mar.
20 : 433 - 436, 1977.

32502 - YORDANOV, I., DILOVA, S., PETKOVA, R., ZEINALOV, Yu. : Post action of high
temperature on the formation of the photosynthetic apparatus in etiolated
bean (Phaseolus vulgaris L.) leaves. - In : COOMBS, J. (ed.) : 4th International Congress on Photosynthesis. P. 420. UKISES, London 1977.

32503 - YOSHIDA, K., YOSHIDA, M. : Relation between main shoot and tillers in corn
(Zea mays L.), with special reference to translocation of ^{14}C-assimilates. -
Jap. J. Crop Sci. 46 : 171 - 177, 1977. [In Jap., ab : E.]

32504 - YOSHIDA, S. : Rice. - In : ALVIM, P. de T., KOZLOWSKI, T.T. (ed.) : Ecophysiology of Tropical Crops. Pp. 57 - 87. Academic Press, New York - San Francisco - London 1977. [Production.]

32505 - YOSHIDA, S., HARA, T. : Effects of air temperature and light on grain filling
of an indica and a japonica rice (Oryza sativa L.) under controlled environmental conditions. - Soil Sci. Plant Nutr. 23 : 93 - 107, 1977. [Growth.]

32506 - YOUNG, R.H., GARNSEY, S.M. : Water uptake patterns in blighted citrus trees. -
J. amer. Soc. hort. Sci. 102 : 751 - 756, 1977. [Resistances.]

32507 - YOUNIS, H.M., WINGET, G.D. : CF_1-dependent restoration of energy-linked reactions reconstituted with a hydrophobic protein from spinach chloroplasts. -
Biochem. Biophys. Res. Comm. 77 : 168 - 174, 1977.

32508 - YOUNIS, H.M., WINGET, G.D., RACKER, E. : Requirement of the δ subunit of
chloroplast coupling factor 1 for photophosphorylation. - J. biol. Chem.
252 : 1814 - 1818, 1977.

32509 - YU, S.L., DIETRICH, W.E., Jr. : Effect of host homogenates of photosynthate
excretion by zoochlorellae of Hydra viridis. - Proc. Pennsylvania Acad. Sci.
51 : 137 - 138, 1977.

32510 - YU, W., PELLEGRINO, F., ALFANO, R.R. : Time-resolved fluorescence spectroscopy of spinach chloroplast. - Biochim. biophys. Acta 460 : 171 - 181, 1977.

*32511 - YUBISUI, T., TAKESHITA, M., YONEYAMA, Y. : Reactivation by glycerol and ethy-
lene glycol of inactivated δ-aminolevulinic acid synthetase of Rhodopseudomonas spheroides. - Experientia 32 : 859 - 860, 1976.

32512 - ZAĬTSEV, S.V., KOLBANOVSKAYA, E.Yu., VARFOLOMEEV, S.D. : Issledovanie zakono-
mernosteĭ fotovosstanovleniya zameshchennykh 4,4'-dipiridila s pomoshch'yu
izolirovannykh khloroplastov. [Relationships of photoreduction of substituted salts of 4,4'-dipyridyl by isolated chloroplasts.] - Biokhimiya 42 :
1069 - 1076, 1977. [In R, ab : E.]

32513 - ZAĬTSEV, S.V., PETUKHOV, S.A., VARFOLOMEEV, S.D. : Kinetics of inactivation and effects of stabilization of chloroplasts. - J. Solid-Phase Biochem. 2 : 123 - 130, 1977.

32514 - ZAĬTSEVA, T.A., VRUBLEVSKAYA, K.G., MANDEL', T.E. : Dinamika soderzhaniya adenozinfosfatov v zelenyushchikh ètiolirovannykh rasteniyakh pshenitsy v zavisimosti ot spektral'nogo sostava sveta. [Dynamics of contents of adenosine phosphates in greening etiolated wheat plants in dependence on spectral composition of light.] - Nauch. Dokl. vyssh. Shkoly, biol. Nauki 20 (6) : 31 - 34, 1977. [In R.]

B32515 - ZALENSKIĬ, O.V. : Èkologo-fiziologicheskie Aspekty Izucheniya Fotosinteza. [Eco-physiological Aspects of Studying Photosynthesis.] - Timiryazevskie Chteniya. Vol. 37. Pp. 1 - 57. Nauka, Leningrad 1977. [In R.]

32516 - ZAMSKI, E., TSIVION, Y. : Translocation in plants possessing supernumerary phloem. 1. ^{14}C-assimilates and auxin in internal phloem of tobacco (*Nicotiana tabacum* L.). - J. exp. Bot. 28 : 117 - 126, 1977.

32517 - ZAVODNIK, N. : Note on the effects of lead on oxygen production of several littoral seaweeds of the Adriatic sea. - Bot. Mar. 20 : 167 - 170, 1977.

32518 - ZDANOVSKI, B., KORYTSKA, A., BNIN'SKA, M., SOSNOVSKA, Ĭ., RADZEĬ, Ĭ., ZAKHVE-YA, Ĭ. : Izmeneniya v distrofnom ozere pod vliyaniem udobreniya. [Changes in dystrophic lake under the effect of fertilization.] - Gidrobiol. Zh. 13 (6) : 32 - 38, 1977. [Chl; in R, ab : E.]

32519 - ZEEVAART, J.A.D. : Translocation pattern in the short-day plant *Xanthium* in relation to long-day inhibition of flowering. - Plant Physiol. 59 (6, Suppl.): 63, 1977.

32520 - ZEEVAART, J.A.D., BREDE, J.M., CETAS, C.B. : Translocation patterns in *Xanthium* in relation to long day inhibition of flowering. - Plant Physiol. 60 : 747 - 753, 1977. [Photosynthates.]

32521 - ZEIGER, E., HEPLER, P. : Light and energy transduction in guard cells. - Plant Physiol. 59 (6, Suppl.) : 96, 1977. [Ps.]

32522 - ZEĬNALOV, Yu. : Vliyanie na povishenite temperaturi v"rkhu spektralnite kharakteristiki i funktsioniraneto na intaktni kletki ot zeleni vodorasli. I. Vliyanie na povishenite temperaturi v"rkhu absorbtsionnite spektri i fluorestsentnite svoĭstva. [Influence of higher temperatures on the spectral characteristics and the function of intact green algae cells. I. Influence of higher temperatures on absorption spectra and fluorescent properties.] - Fiziol. Rast. (Sofia) 3 (3) : 32 - 44, 1977. [In Bulg., ab : E, R.]

32523 - ZEĬNALOV, Yu. : Vliyanie na povishenite temperaturi v"rkhu spektralnite kharakteristiki i funktsioniraneto na intaktni kletki ot zeleni vodorasli. II. Vliyanie na povishenite temperaturi v"rkhu kislorodnata induktsiya. [Effect of high temperatures on the spectral characteristics and the functioning of intact green algae cells. II. Effect of high temperatures on oxygen induction.] - Fiziol. Rast. (Sofia) 3 (4) : 19 - 25, 1977. [In Bulg., ab : E, R.]

32524 - ZEINALOV, Yu. : On the analysis of oxygen induction phenomena in photosynthetizing systems. I. Kinetics of the O_2 evolution in green algae cells. - Stud. biophys. 65 : 227 - 238, 1977.

32525 - ZELAWSKI, W., ŁOTOCKI, A., WIERZBICKI, B., LAUDANSKI, Z. : Model of interrelations between photosynthetic productivity and some physiological and morphological features of Scots pine (*Pinus silvestris* L.) seddlings. - In : COOMBS, J. (ed.) : 4th International Congress on Photosynthesis. Pp. 420 - 421. UKISES, London 1977.

32526 - ZELEŇÁKOVÁ, E., POLEK, B. : Zmeny v obsahu chlorofylov a dusíkatých látok počas vegetačného obdobia v listoch marhule (*Prunus armeniaca* L.).[Changes in the content of chlorophyll and in nitrogen substances in apricot leaves during the vegetation period.] - In : HUZULÁK, J., MASAROVIČOVÁ, E. (ed.) : Fotosyntéza a Vodný Režim Drevín. Pp. 224 - 231. Modra-Piesky 1977. [In Slovak, ab : E, R.]

32527 - ZELITCH, I. : The future of photosynthesis. (A scientist who discovered why some plants are twice as efficient as others seeks new answers to the world food problem.) - Horticulture 55 : 48 - 54, 1977.

B32528 - ZELITCH, I. : Fotosynteza Fotooddychanie i Produktivność Roślin. [Photosynthesis, Photorespiration and Plant Productivity.] - Państwowe Wydawnictwo Rolnicze i Lesne, Warszawa 1977. [In Pol.]

32529 - ZELITCH, I., OLIVER, D.J., BERLYN, M.B. : Increasing photosynthetic carbon dioxide fixation by the biochemical and genetic regulation of photorespiration. - In : MITSUI, A., MIYACHI, S., SAN PIETRO, A., TAMURA, S. (ed.) : Biological Solar Energy Conversion. Pp. 231 - 242. Academic Press, New York - San Francisco - London 1977.

32530 - ZEMÁNEK, M. : Spolupůsobení vody a dusíku při tvorbě výnosu jarního ječmene. [Interaction of water and nitrogen in formation of spring barley crop.] - Rostlinná Výroba (Praha) 23 : 683 - 693, 1977.[In Czech, ab : E, G, R.]

32531 - ZEN'KEVICH, É.I., KOCHUBEEV, G.A., LOSEV, A.P., GURINOVICH, G.P. : Izuchenie énergeticheskogo vzaimodeľstviya mezhdu protokhlorofillom i khlorofillom v smeshannykh assotsiyatakh v zavisimosti ot kontsentratsii khlorofilla. [Energetic interaction between protochlorophyll and chlorophyll in mixed associates in dependence on chlorophyll concentration.] - Mol. Biol. (Moskva) 11 : 1039 - 1056, 1977. [In R, ab : E.]

32532 - ZEYNALOV, Y. : On the nature of the induction phenomena at oxygen evolution in intact cells of green algae. - Dokl. bolg. Akad. Nauk 30 : 1201 - 1204, 1977.

32533 - ZEYNALOV, Y. : Non-additiveness in the action of light at the photosynthesis of green plants. - Dokl. bolg. Akad. Nauk 30 : 1479 - 1482, 1977.

32534 - ZEYNALOV, Y. : The principle of non-additiveness in the action of light and the concept of two photosystems at the photosynthesis in green plants. - Dokl. bolg. Akad. Nauk 30 : 1641 - 1644, 1977.

32535 - ZIEGLER, I., HAMPP, R. : Control of $^{35}SO_4^-$ and $^{35}SO_3^-$ binding to chloroplast thylakoides by electron transport. - In : COOMBS, J. (ed.) : 4th International Congress on Photosynthesis. Pp. 421 - 422. UKISES, London 1977.

32536 - ZIEGLER, I., HAMPP, R. : Control of $^{35}SO_4^{2-}$ and $^{35}SO_3^{2-}$ incorporation into spinach chloroplasts during photosynthetic CO_2 fixation. - Planta 137 : 303 - 307, 1977.

32537 - ZIELINSKI, R.E., PRICE, C.A. : Synthesis of cytochrome b559 by isolated spinach chloroplasts. - Plant Physiol. 59 (6, Suppl.) : 8, 1977.

32538 - ŽILA, L. : Niektoré fyziologické procesy cukrovej repy pri rôznych metódach závlahového režimu. [Some physiological processes of sugar-beet irrigated according to different methods.] - Rostlinná Výroba (Praha) 23 : 695 - 703, 1977. [Growth analysis; in Slovak, ab : E, G, R.]

32539 - ŽILA, L., HUZULÁK, J., LEDEČOVÁ, B. : Metodický príspevok k meraniu charakteristík vodného režimu marhule a broskyne. [Methodical contribution to the measurement of the characteristics of water relations in apricot and peach.] - In : HUZULÁK, J., MASAROVIČOVÁ, E. (ed.) : Fotosyntéza a Vodný Režim Drevín. Pp. 115 - 122. Modra-Piesky 1977. [Resistances; in Slovak, ab : E, R.]

32540 - ZILINSKAS, B.A., ZIMMERMAN, B.K., GANTT, E. : Allophycocyanin forms from the blue-green alga Nostoc sp. - In : COOMBS, J. (ed.) : 4th International Congress on Photosynthesis. Pp. 422 - 423. UKISES, London 1977.

32541 - ZIMA, M. : Vzťah medzi obsahom vody a fotosyntézou listov topoľa. [The relationship between water content and photosynthesis of poplar leaves.] - In : HUZULÁK, J., MASAROVIČOVÁ, E. (ed.) : Fotosyntéza a Vodný Režim Drevín. Pp. 99 - 105. Modra-Piesky 1977. [In Slovak, ab : E, R.]

32542 - ZIMA, M. : Možnosti regulácie fyziologickej aktivity odnoží obilnín. [Possibilities to control physiological activity of grain crop tillers.] - In : REPKA, J. (ed.) : Zborník Referátov zo Seminára Fyziologicko-Genetické a Chemické Faktory Produktivity Rastlín. Pp. 152 - 159. Vysoká Škola poľnohospodárska, Nitra 1977. [Growth analysis; in Slovak.]

32543 - **ZIMMERMAN, U.D., KUCERA, C.L.** : Effects of composition changes on productivity and biomass relationships in tallgrass prairie. - Amer. Midland Naturalist
97 : 465 - 469, 1977. [Dry-matter accumulation.]

32544 - **ZINGMARK, R.G., WOHLGEMUTH, S., WAGNER, G.L., VENNEWITZ, M.K., REIS, R.R.,
HALL, M.O., EBELING, D.E., BROWN, D.C.** : Ecology of macroalgae in a temperate
salt marsh. I. Biomass and productivity of intertidal species. - J. Phycol.
13 (Suppl.) : 77, 1977.

*32545 - **ZOLOTOV, V.I., RAZUVAEV, A.I.** : Vliyanie kompleksa agrotekhnicheskikh priemov
na produktivnost' fotosinteza i urozhaĭ gibridov kukuruzy. [Effect of complex
of cultural practices on the net assimilation rates and yield of maize hybrids.] - Byul. vses. nauchno-issl. Inst. Kukuruzy *1975* (4/40) : 11 - 14,
1975. [In R.]

32546 - **ZRŮST, J., SMOLÍKOVÁ, A.** : Rozdíly v rychlosti fotosyntézy u kříženců a někte-
rých rodičovských odrůd brambor. [Differences in photosynthetic rate in potato hybrids and in some parental cultivars.] - Rostlinná Výroba (Praha) *23* :
723 - 732, 1977. [In Czech, ab : E, G, R.]

32547 - **ZSCHOCHE, W.C., TING, I.P.** : Microbody-malate dehydrogenase in plants with C_4-
photosynthesis. - Plant Sci. Lett. *9* : 103 - 106, 1977.

32548 - **ZSOLNAY, A.** : Hydrocarbon content and chlorophyll correlation in the waters
between Nova Scotia and the Gulf Stream. - Deep-Sea Res. *24* : 199 - 207,
1977.

32549 - **ZUERRER, H., BACHOFEN, R.** : 5-aminolevulinic acid as precursor in the biosynthesis of bacteriochlorophyll and bacteriopheophytin in *Rhodospirillum rubrum*.
- In : COOMBS, J. (ed.) : 4th International Congress on Photosynthesis. Pp.
423 - 424. UKISES, London 1977.

32550 - **ZÜRRER, H., SNOZZI, M., HANSELMANN, K., BACHOFEN, R.** : Localisation of the
subunits of the photosynthetic reaction centers in the chromatophore membrane of *Rhodospirillum rubrum*. - Biochim. biophys. Acta *460* : 273 - 279, 1977.

32551 - **ZURZYCKI, J.** : Specific effects of the high intensity of blue light on the
structure of chloroplasts. - In : COOMBS, J. (ed.) : 4th International Congress on Photosynthesis. P. 426. UKISES, London 1977.

32552 - **ZURZYCKI, J., GABRYŠ, H.** : Changes in light absorption by the chloroplast, related to its structural transformations. - Acta Soc. Bot. Pol. *46* : 369 -
380, 1977.

32553 - **ZURZYCKI, J., METZNER, H.** : Volume changes of chloroplasts *in vivo* at high
densities of blue and red radiation. - Photosynthetica *11* : 260 - 267, 1977.

32554 - **ZUTSHI, D.P., VASS, K.K.** : Estimates of phytoplankton production in Manasbal
Lake, Kashmir using carbon-14 method. - Trop. Ecol. *18* : 103 - 108, 1977.
[Ps.]

Authors' names are presented in the form in which they appear in the respective publication. The names from papers published in Cyrillic characters are transcribed as
shown on p.III of this volume. Alternative spellings and forms of the name of the
same author are usually cross-indexed. The numbers in *italics* refer to publications
in which the respective author acts as an editor.

A

COOMBS, J. (continued)
28719, 28724, 28729, 29738, 28742,
28745-6, 28753, 28755, 28759, 28771,
28779, 28784, 28789, 28794, 28798,
28802, 28812, 28816-7, 28822, 28828,
28833, 28837, 28845, 28856, 28860,
28862, 28877, 28884, 28887, 28889,
28893, 28901, 28903-4, 28912, 28924-
-5, 28929, 28942, 28946, 28958,
28965, 28999, 29002, 29017, 29032,
29037, 29043, 29051, 29059, 29062,
29067, 29075, 29087, 29105, 29125,
29138, 29140, 29143, 29148, 29160-1,
29165, 29165, 29174, 29180, 29187,
29192, 29194, 29200-1, 29219, 29225,
29230-1, 29260, 29266, 29268, 29271-
-3, 29275, 29278-9, 29286, 29294,
29305, 29307, 29310, 29318, 29322,
29332, 29341, 29348, 29353, 29359,
29376, 29392, 29406, 29412, 29424,
29436, 29438, 29445, 29453, 29463-4,
29472, 29503, 29514, 29519, 29529,
29541, 29552, 29555, 29557, 29563,
29568, 29579, 29584, 29586, 29598,
29602, 29604, 29608-10, 29614-5,
29631, 29633-4, 29637, 29641, 29664,
29666, 29668, 29682-3, 29700, 29713,
29719, 29728, 29736, 29745, 29751,
29756, 29760, 29786, 29800, 29805,
29808, 29810, 29830, 29832, 29834,
29838, 29845, 29857, 29870, 29875,
29880, 29884-5, 29888, 29892, 29894,
29901, 29908, 29910, 29912, 29921,
29925, 29934, 29942, 29946, 29961,
29969, 29978, 29980, 29987, 29991-2,
29997, 30001, 30004, 30006, 30012,
30016, 30027, 30050, 30094, 30116,
30119, 30122, 30127, 30134, 30144,
30153, 30159, 30168, 30174, 30176-7,
30190, 30210, 30213, 30229, 30240,
30246, 30251, 30254, 30257, 30266,
30273, 30288, 30307, 30318, 30320,
30322, 30330, 30340, 30362, 30372,
30382, 30385, 30402, 30411, 30414,
30424, 30429, 30444, 30467, 30498,
30514, 30518, 30526, 30528, 30528,
30538, 30542, 30552, 30561, 30569-
-70, 30574, 30580, 30584, 30590,
30595, 30607-9, 30626, 30631, 30640,
30644, 30656, 30662, 30670, 30673,
30684, 30690, 30693, 30695, 30701,
30709, 30717, 30732, 30740, 30744,
30750, 30759, 30765, 30769, 30778-
-9, 30794, 30798, 30806, 30812,
30823, 30826, 30828, 30838-9, 30845,
30848, 30853, 30855, 30859, 30867,
30887, 30903, 30913, 30922, 30935-8,
30940, 30948, 30955, 30963, 30965,
30977, 30980, 30986, 30999, 31003,
31008, 31026, 31038, 31043-5, 31049,
31051, 31078-9, 31088, 31094, 31096,
31098, 31102, 31118, 31137, 31143,
31145, 31149, 31150, 31157, 31176,
31179, 31182, 31185, 31194, 31205,

31220, 31247, 31264, 31275, 31288,
31291, 31296-8, 31319, 31322, 31346,
31351, 31354, 31360, 31365-6, 31390-
-1, 31393, 31402, 31407, 31409,
31414, 31437, 31445, 31457, 31473,
31478, 31481, 31487, 31504, 31507,
31511, 31527, 31529, 31556, 31561-
-2, 31570-1, 31581, 31596, 31598,
31617, 31619-21, 31625, 31629,
31631, 31642, 31646, 31665, 31674,
31677, 31679, 31683, 31689, 31699,
31701, 31703, 31709, 31715-6, 31718-
-9, 31725, 31728, 31738, 31740,
31767, 31776, 31799, 31808, 31811,
31814, 31818-9, 31828, 31830,
31846, 31858, 31871, 31893, 31910,
31919, 31938, 31945, 31948, 31960,
31964, 31973, 31979, 31987, 32000-1,
32014, 32019, 32060, 32068, 32070,
32077, 32080, 32085, 32088, 32099,
32102, 32104, 32114, 32130-1, 32140,
32162, 32165, 32204, 32217, 32228,
32232-3, 32239, 32242, 32246, 32249,
32251, 32255, 32260, 32270, 32276,
32280, 32283, 32293, 32299, 32321,
32323, 32330-1, 32344, 32376, 32379,
32386, 32393, 32395, 32397, 32407,
32415-6, 32419, 32422, 32433,
32441, 32449, 32452-3, 32460, 32464,
32468, 32475, 32489, 32494, 32496,
32502, 32525, 32535, 32540, 32549,
32551
COOPER, J.P.   29166, 31156
COOPER, S.   29034, 29388
COPE, B.T.   32367
COPE, F.W.   29167
COPONY, W.   29168
CORBU, S.   30087
CORKER, G.A.   32370
CORMACK, W.F.   30861
CORNELIUS, M.J.   28807, 30288-9, 32104
CORNIC, G.   29169
CORNIDES, I.   31187
CORRELL, D.L.   29466
COSSETTE. C.   29170
COSTA, F.   29048
COSTE, B.   29171
COSTES, C.   29172
COTLER, D.N.   31520
COTTON, T.M.   29173, 31787-8, 32214
COUCH, R.   29175
COUDERC, H.   29122
COUDRET, A.   29495
COUGHLAN, S.   29176-8
COUTTS, J.H.   31022
COWAN, I.R.   29179
COWLES, J.R.   31170
COX, E.R.   32442
COX, R.P.   29180
COYNE, P.I.   28803
CRĂCIUN, C.   30770
CRAIG, S.   29181
CRAMER, W.A.   29182-3, 32407-8
CRANE, F.L.   28693-6, 29184

FITZGERALD, M.P. 31529
FLEMMING, B.-U. 30316
FLETCHER, R.A. 31229
FLINN, A.M. 28613, 29509
FLINT, E.A. 29510
FLINT, R.W. 29511
FLORENZANO, G. 31203
FLOROV, R.J. 31014
FLOROVA, N.B. 31053
FLOYD, G.L. 29512, 31589
FLÜGGE, U.-I. 29513
FOCK, H. 29032, 29514, 30023, 30542-5
FOCKE, R. 29515
FOKHT, A.S. 28664-5
FOLKERS, K. 28696
FOLLMANN, H. 32333
FONDA, S.A. 28542
FONG, F. 29516-7
FONG, F.K. 29497, 29518-22
FONTES, J.-C. 18723
FORD, E.D. 30876
FORD, M. 29523
FORD, M.A. 28624
FORDHAM, R. 29524
FORK, D.C. 28634, 29525-9, 30971-4,
  31241
FORMAN, A. 29452, 29454, 32114
FORNASIERO, R.B. 31739
FORTI, G. 29530-3, 29599-600, 29634-5,
  30093
FOSTER, A. 29534-5
FOSTER, J.G. 29536
FOSTER, R.B. 29791
FOTT, J. 29293
FOULDS, W. 29537
FOURY, C. 29538
FOWLE, C.D. 28996
FOWLER, C.F. 29539-41
FOX, J.L. 29782
FOX, S.B. 29839
FOY, C.D. 29542
FOY, M.G. 30032
FOYER, C. 29808
FRACKOWIAK, D. 29543-5
FRADKIN, L.I. 29546-7
FRAGATA, M. 29548-9
FRAIMAN, I.Ya. 31404
FRANCKE, B. 28651
FRANK, A.B. 29550-1
FRANK, G. 29552
FRANK, H.A. 29124-5
FRANK, M.H. 29553
FREDERICK, S.E. 29759
FREEBERG, L.R. 29554
FREEMAN, E.A. 28711
FREEMAN, H.C. 29555
FREEMAN, T.P. 29360-1
FRENCH, C.S. 29556-8
FRENCH, S.A.W. 29559
FRENKEL, C. 28931
FREY, K.J. 28941
FREY, R. 32085
FREYSSINET, G. 29560, 30935
FRIBERG, E.E. 31278

FRICK, M. 28814-5
FRIDBERG, I. 29961
FRIDLAND, E.V. 29561
FRIDOVICH, I. 28539
FRIEDLANDER, M. 30190
FRIEDMANN, E.I. 32193
FRIEDRICH, J.W. 29562
FRIEND, D.J.C. 29563
FRIER, V. 29564
FRIESNER, R. 29310
FRISCHKNECHT, K. 31679
FRITZSCHE, I. 32203
FRÖLICH, W.G. 29565-6
FROLOV, N.S. 29422
FROSCH, S. 29567-8
FRUGE, D.R. 29521
FRY, D.J. 29569
FRY, I. 29570
FRY, S.C. 29571
FRYDRYCH, J. 30427
FUCHS, M. 29572, 30205, 30930, 31695-6,
  31936
FUCHS, M.I. 29572, 31695-6
FUJII, Y. 29573
FUJITA, Y. 30877-8, 32007
FUKAI, S. 29574-6, 31822
FUKUDA, I. 29416-7
FUKUDA, M. 30364
FULDNER, K.-H. 32417
FULLER, R.C. 28905
FULLETT, S.H. 29150
FUNNELL, G.R. 32063
FURUTA, S. 31128
FUTATSUYA, F. 30104-5
FYKSE, H. 29577

G

GÁBOR, A. 29578
GÁBORČÍK, N. 31495
GABRYŚ, H. 32552
GADAL, P. 29579, 30125, 32290-1
GAGLIANO, A.G. 29580
GALE, J. 30218
GALKIN, V.I. 29008
GALLA, H.J. 30657
GALLAGHER, J.N. 28809
GALLAHER, R.N. 29581
GALLEGOS, C.L. 29582
GALLIANO, M. 29718
GALLON, J.R. 32205
GALLOWAY, L. 29497, 29520-1
GALMICHE, J.M. 29583-4, 29667-8, 30989
GALSON, E. 32352
GALUN, M. 31798
GALUTVA, O.A. 29585
GALZIN, A.M. 29586
GAMALEĬ, Yu.V. 29587
GAMAYUNOVA, M.S. 31161, 31821
GAMBLE, J.C. 29588
GANAGO, I. 28742
GANDY, C. 30265
GANGSTAD, E.O. 29175

HINDE, R.W.   32469
HINDMAN, J.C.   29943
HINO, A.   28546, 29944
HIPKIN, C.R.   29945
HIPKINS, M.F.   29946
HIRAYAMA, O.   29947-8
HIROSAKI, S.   31130
HITAKA, N.   30351-2
HIXSON, C.S.   29672
HIYAMA, T.   29788, 29949-50
HO, L.C.   29951-4, 32339-40
HOARAU, J.   29955-6, 30854-6, 31480-1
HOBART, D.   29125
    see HOBART, D.R.
HOBART, D.R.   29126
    see HOBART, D.
HOBERG, W.   30532
HOBRO, R.   28568
HOBSON, G.E.   29957
HOCH, G.E.   29958
HOCH, J.C.   29124
HOCHMAN, A.   29959-61
HOCHSTER, R.M.   *29697*
HOCK, B.   32337-8
HODDINOTT, J.   29963-4
HODGES, C.F.   29965-6
HODGES, T.   29967, 31994
HODGES, T.K.   *29304, 31440*
HODSON, R.   32108
HOFÄCKER, W.   29968
HÖFF, A.J.   29969-73
HOFFMAN, L.R.   30758
HOFFMANN, D.   29974
HOFFMANN, F.   29975
HOFFMANN, P.   29976-9, 31042
HOFMANN, M.E.   29980-1
HÖFNER, W.   29982-3
HOFSTRA, G.   28913, 29984
HOFSTRA, J.J.   29985
HOGETSU, D.   29986
HÖHLER, T.   29987
HØJERSLEV, N.   29988
HOLADAY, A.S.   28901
HOLADAY, S.   29989
HOLDSWORTH, E.S.   29990-1
HOLE, C.C.   29992-3
HOLMES, M.G.   29994
HOLMES, N.G.   29995-7, 32251
HOLMGREN, A.   29998
HOLM-HANSEN, O.   30305, 32107-8
HOLOWKA, D.A.   29999
HOLSTEIJN, H.M.C.van
    see VAN HOLSTEIJN, H.M.C.
HOLT, S.C.   30872
HOLTEN, D.   29710
HOLTEN, J.D.   29453
HOLŮBKOVÁ, B.   29859
HOLZAPFEL, C.   30000
HOMANN, P.H.   30001-2, 31588
HONIG, B.   30055
HONSELL, E.   30003
HOOKER, T.N.   28962
HOPE, A.B.   29114, 30004
HOPE, P.   31835

HOPFIELD, J.J.   30005, 31346
HOPKINS, D.W.   29267
HOPKINS, W.G.   29875-6, 30006-7
HOPKINSON, C.S.   31217
HOPPE, H.-G.   30109
HOPPE, J.H.   30008
HOPPE, W.   *31485*
HORÁNSZKY, A.   30009
HORBACH, W.   32383
HORI, Y.   30010, 31792
HORIE, T.   30011
HORN, N.A.   28897
HORNBERGER, G.M.   29582, 32083
HORNIG, L.   32383
HORTON, M.L.   31607-8
HORTON, P.   30012-4
HORTON, R.F.   30015
HORVÁTH, G.   28792, 29196, 30016, 31049
HORVÁTH, I.   29419
HORVÁTH, M.M.   30017
HOSKINS, L.C.   30018
HOSOI, T.   30683
HOU, L.-Y.   30019
HOULIER, B.   31297
HOURSIANGOU-NEUBRUN, D.   30020
HOUSLEY, T.L.   30021-2, 31277
HOUSSIER, C.   31446
HOWARD, R.J.   29728, 30023
HOWELL, S.   29625, 30024
    see HOWELL, S.H.
HOWELL, S.H.   29624, 29626, 30025
    see HOWELL, S.
HOWES, C.D.   32271
HOXMARK, R.C.   30026
HØYER-HANSEN, G.   30027-8, 30687-8
HOZUMI, K.   30029
HOZYO, Y.   30030
HRAŠKA, Š.   30747, 31284
HRUŠKA, L.   30031
HSIAO, S.I.C.   30032
HUANG, A.H.C.   29061
HUANG KAO, M.Y.C.   32048
HUBAC, C.   30033
HUBER, D.   29218
HUBER, S.   29376
HUBER, S.C.   30034-8
HUCHZERMEYER, B.   30039
HUDÁK, J.   30040
HUDSON, J.P.   30041
HUISMAN, J.G.   30042-3
HULKOWER, S.   31667
HULL, J.C.   30044
HUMPHREY, G.F.   30045-6
HUMPHREY, J.   29622
HUMPHREYS, T.J.   30247
HUMPHRIES, E.C.   29559
HUNT, C.M.   32076
HUNT, H.W.   30048
HUNT, L.A.   28605-6, 30140-1, 30458,
    30705-6
HUNT, R.   30049
HUNT, R.D.   30151
HUNTER, C.N.   30050
HUNTER, F.   30051

MÓZSIK, L.   30953
MSHIGENI, K.E.   30954
MUALLEM, A.   30955
MUCHOW, R.C.   30956
MUELLER, P.   30957
MUGNOZZA, G.T.S.
    see SCARASCIA MUGNOZZA, G.T.
MÜHLBACH, H.   29912
MÜHLETHALER, K.   30958
MUKERJI, S.K.   30959-61
MUKHAMADIEV, B.T.   29661, 30487, 30962
MUKHERJI, S.   31234
MUKHIN, E.N.   30963-4, 31274
MUKOHATA, Y.   30785, 30965-6, 32480
MULAMBA, N.N.   29455
MULDOON, D.K.   30967, 31244
MULLER, C.H.   30044
MÜLLER, D.   28779
MÜLLER, U.   29100, 29929
MULLIGAN, H.F.   31749
MUMMERY, R.S.   32229
MUNAKATA, K.   30104-5
MUNAWAR, M.   31758
MUNDA, I.M.   30968
MURAKAMI, S.   30969, 31121
MURAMOTO, H.   31252
MURATA, N.   29526-9, 30970-4, 31141,
    31992
MURATA, T.   30975
MURATA, Y.   30077, 30097-100, 32312
MURAYAMA, S.   31077
MUREI, I.A.   30976, 31801
MURPHY, D.J.   30977-8
MURPHY, P.G.   28687
MURRAY, D.B.   30979
MURRAY, D.R.   29700, 30980
MURTY, K.S.   30131
MUSGRAVE, R.B.   32294
MUSHKETIK, L.S.   30214, 32309, 32490
MUSTÁRDY, L.A.   28792, 29458, 29578,
    30016
MUSZBEK, L.   30981
MUTO, S.   30886
MUTUSKIN, A.A.   30982-3
MUUS, L.T.   *29972*
MUZAFAROV, E.N.   *28523*, 28524, *30115*,
    30726-7, *30727*, 30984-5, *30984-5*
MVE AKAMBA, L.   30986, 31814
MYERS, G.A.   30819
MYERS, J.   31438, 32357-9
MYERS, S.P.   31955
MYHRE, D.L.   29984
MYKLESTAD, S.   30987-8
MYRON, J.C.   30337-8

N

NABEDRYK-VIALA, E.   30989
NAD', A.   30990-1 see NAGY, A.H.
NADAKAVUKAREN, M.J.   30992-3
NAGAO, N.   31139
NAGY, A.H.   29777, 30009
    see NAD', A.

NAIK, G.R.   30175
NAIR, P.K.R.   30994
NAIR, S.   30227
NAIRIZI, S.   30995
NAKAGAHRA, M.   29874
NAKAGAWA, S.   32201
NAKAJIMA, K.   29418
NAKAMURA, K.   32006
NAKANISHI, K.   31286
NAKANISHI, M.   30996
NAKANO, Y.   28599
NAKASEKO, K.   30483
NAKATA, J.   30997
NAKATANI, N.Y.   30998-9
NAKAYAMA, K.   30475, 31000
NAKAYAMA, M.   31601
NAKAZAWA, F.   31001-2
NALBANDYAN, R.M.   30746, 31099
NALBORCZYK, E.   31003
NAMBIAR, M.C.   31004
NANDI, D.L.   31005
NARUSAWA, T.   32050
NASH, G.V.   30938
NASH, T.H.III.   31006
NASITIR, M.   29840
NASRULHAQ-BOYCE, A.   31007
NASYROV, Yu.S.   *28631*, *28938*, *29008*,
    29283, *29459*, *29501-2*, *29662*, *29706*,
    30201, *30228*, *30232*, *30496*, *30613*,
    30702, *30712*, *30734*, *30896*, *30962*,
    30990, 31008-10, *31090*, *31092*, *31579*,
    31721, *31795*, *31951*, *32057*, *32142*,
    32190, *32222*, *32267*, *32317*, *32482*,
    32490
NATHANSON, B.   31011
NATO, A.   30778, 30948, 31012
NÁTR, L.   31013
NAUŠ, J.   30693, 32224
NAVARRO, S.   29048
NAZAROVA, Z.A.   32481
NEALES, T.F.   29104, 31885
NEDYALKOV, N.   31014-5
NEEMAN, E.   30319
NEGIEVICH, L.A.   31016-7
NEGISI, K.   31018-9
NEHRLICH, S.C.   28559-60
NEIFAKH, S.A.   *30488*
NEILSON, R.E.   31020
NEKRASOV, L.I.   29585, 30728-9, 31594
NEKRASOVA, G.F.   30897
NELSEN, C.E.   29823
NELSON, C.J.   30566, 31021-2, 31433,
    31530, 31870
NELSON, D.E.   31023
NELSON, L.A.   30165
NELSON, N.   28751, 29631, 30181, 31024-
    -30, 31078
NEMÉTH, J.   31031, 31305
NEMETH, J.C.   31032
NERSON, H.   31306
NETZEL, T.L.   29358, 31033
NEUBURGER, M.   31034-5
NEUGEBAUER, D.-C.   31036
NEUMANN, D.   31037, 31223

This index contains a selection of primary items chosen according to their importance in photosynthesis research and to their relevance and occurrence. The word "Photosynthesis" is not regarded as a main theme, but partial processes, photosynthetic parameters and the factors affecting photosynthesis are listed. The processes and other characteristics are summarized into several main themes when presented in combination with individual factors, *e. g.* carbon fixation pathways, electron transport chain, chlorophyll, gas exchange, ecosystem and plant productivity (*including photosynthate distribution and translocation, and canopy organization and functioning*), photorespiration, resistances to CO_2 and water vapour transfer, *etc.*
    Several items from branches related to photosynthesis research were also chosen for convenience, *e. g.* dealing with respiration, plant growth and development, water relations, anatomy, bioclimatology, *etc.* These items contain only references to papers within the scope of this bibliography.

A

Abscisic acid see Growth regulators ...

Absorbance in canopy see Canopy, radiation profile

Accumulation of dry matter see Biomass distribution ...; Dry-matter production ...;
        Ecosystem production ...

Achlorophyllous cells and organs, respiration see Respiration of achlorophyllous
        tissues in light, light inhibition of respiration

Action spectra see Irradiance, spectral composition ...

Adenosine triphosphate see ATP

Aerodynamic methods, bioclimatological methods (sampling, measurement of wind, rain,
        dew, *etc.*) 31272

Age of algae, leaf, plant see Ontogeny ...; Canopy, leaf age

Agrotechnics and ecosystem and plant productivity 28813, 29072, 29492, 29566, 30648,
        30953, 31208, 31952, 32545

Agrotechnics and gas exchange 28844, 29366-7, 29491, 30930, 31946

Agrotechnics and resistances to CO_2 and water vapour transfer 30930

Air-conditioning in photosynthesis measurement see Gasometric system, conditioning
        of air

Air-flow rate see Wind ...

Albedo, canopy see Canopy, radiation distribution

Algae and photosynthetic bacteria, cultivation (*cf.* also Algae mass cultures productivity) 29230, 29913, 30185, 30601, 31159, 31299, 31329, 31772

Algae and secondary production of reservoirs
        28985, 29311, 29426, 29588, 29658, 29739, 30407, 30605-6, 30735, 30776, 30807,
        31891, 32413

Algae, blue-green, chromatophores in see Chromatophore ...

Algae carotenoids see Xanthophylls of algae

Algae chlorophylls see Chlorophylls c, d

Algae, CO_2 and O_2 exchange see Gas exchange in algae

Algae, depth distribution in reservoirs
    28536, 28553, 28561, 28740, 28797, 28955, 28977, 28982, 28984, 29041, 29151,
    29171, 29185, 29226, 29242, 29288, 29365, 29375, 29426, 29428-9, 29511, 29588,
    29655, 29674, 29716, 29778, 29802, 29809, 29916-7, 29919, 30032, 30075, 30121,
    30147-8, 30196, 30225, 30281, 30305, 30356-7, 30397, 30407, 30645, 30724,
    30863-4, 31216, 31278, 31311-2, 31316, 31328, 31373, 31396, 31430, 31531,
    31555, 31714, 31766, 31965, 31988, 32032, 32043, 32136-8, 32174, 32308, 32368,
    32380

Algae in sewage cleaning 30925

Algae life cycles see Ontogeny of algae ...

Algae mass cultures productivity (cf. also Algae and photosynthetic bacteria, culti-
    vation) 28763, 29230, 29276, 29691, 30185, 30757, 30836, 31162, 31203, 31603

Algae photosynthesis and production
    28536, 28553, 28561, 28687, 28776, 28797, 28830, 28985, 28996, 29089, 29151,
    29242, 29653, 29809, 29821, 29825, 29842, 29914, 29917, 30072, 30075, 30147,
    30151, 30196, 30281, 30356, 30484, 30486, 30501, 30522, 30605-6, 30645, 30724,
    30732, 30757, 30843, 30873, 30906, 31183-4, 31202, 31278, 31311, 31317, 31327-
    -8, 31382, 31396, 31514, 31555, 31700, 31912, 31988, 32032, 32091, 32137-8,
    32174, 32184, 32203, 32269, 32348, 32356, 32494, 32544, 32554

Algae, primary productivity in reservoirs (cf. also Chlorophyll and pro-
    duction of algae and water reservoirs) 29346, 29448, 29469, 30585, 32136,
    32356

Algae, primary productivity, methods [cf. also O_2 determination (other than O_2 elec-
    trode); O_2 electrode]
    28568, 28615, 28687, 28983, 29242, 29511, 29588, 29716, 29802, 29915, 29919,
    29988, 30072, 30147, 30166, 30196, 30247, 30517, 30815, 31127, 31183, 31382,
    31422, 31532, 31597, 31869, 31944, 32090-1, 32137, 32174, 32184, 32274, 32396

Algae synchronous cultures see Algae and photosynthetic bacteria, cultivation;
    Ontogeny of algae ...

Altitude see Pressure, altitude ...

Amino acids see Proteins, amino acids, nucleic acids ...

δ-Aminolaevulinic acid see Chlorophyll biosynthesis ...

Amphistomatous leaf, gas exchange in (cf. also Leaf epidermis, stomata) 28623, 30721

Anaerobic atmosphere see N_2, anaerobic atmosphere ...

Antibiotics and carbon fixation pathways 28910-1, 29942, 30962, 31973, 32176, 32262,
    32291

Antibiotics and carotenoids 29266, 32014, 32018

Antibiotics and chlorophyll
    28583, 28628, 28680, 29101, 29361, 29458, 29636, 29662, 29773, 29939, 29946,
    30201, 30488, 30594, 30935, 31064, 31288, 31687-8, 31795, 32001, 32098, 32414,
    32421, 32451

Antibiotics and chloroplast (chromatophore) 28897, 28923, 29465, 29824, 29839, 30036,
    30334, 30682, 31056, 31155, 31204, 31541, 31757, 32414, 32421

Antibiotics and electron transport chain
    28592-3, 28672, 28812, 28904, 29107, 29192, 29541, 29939, 29946, 29959, 30034-
    -5, 30114, 30183, 30307, 30361, 30437, 30594, 30599, 30831, 30982-3, 31049,
    31123, 31147, 31179, 31288, 31479, 31487, 31646, 31664, 31683, 31872-4,
    32145, 32303, 32321, 32375, 32408, 32414, 32421, 32496, 32498

Antibiotics and gas exchange 29101, 29775, 29942, 30026, 30034, 30223, 30935, 30962,
    32093, 32145, 32414

Antibiotics and resistances to CO_2 and water vapour transfer   29076

Antigens see Electron transport chain, serological analysis

Antitranspirants 29072, 29212, 29486, 30507, 30930, 31379, 31416, 32352

Architecture of canopy see Canopy ...

Assimilates see Photosynthates ...

Assimilation chamber
    28502, 28540, 28571, 28728, 28939, 29065, 29372, 29400, 29491, 29675, 29722,
    29864, 30263, B30325, 30343, 30358, 30394, B30432, 30490, 30545, 30563,
    30706, 30860, 30946, 31020, 31301, 31711, 31943, 32015

ATP   28524, 28581, 28592-4, 28633, 28636, 28672-3, 28755, 28777, 28784, 28879,
    28906, 28957, 28983, 29011, 29070, 29075-6, 29101, 29151, 29304, 29314, 29353,
    29385, 29449, 29533, 29583-4, 29596, 29668, 29695, 29837, 29847, 29870, 29885,
    29942, 29961, 29974, 29977, 30048, 30062, 30126, 30136, 30166, 30197, 30215,
    30219, 30223, 30259, 30280, 30288, 30297, 30322, 30336, 30339-41, 30369,
    30420, 30429, 30470, 30528, 30567, 30572, 30690, 30744, 30805-6, 30830, 30845,
    30853, 30855, 30859, 30868, 30892, 30952, 30963-5, 30985, 31096, 31149, 31247,
    31299, 31320, 31324, 31343, 31380, 31402, 31412, 31456, 31458, 31489, 31518,
    31522-3, 31525-6, 31549, 31560, 31693, 31702, 31713, 31715-6, 31852, 31872,
    31939, 32043, 32050, 32068, 32092, 32107, 32128-9, 32134, 32217, 32223,
    32233, 32244-6, 32274, 32316, 32330, 32366, 32376, B32410, 32437, 32445,
    32489, 32496, 32507, 32514, B32515

ATP, methods 30166

ATPase, coupling factor 1
    28554, 28564, 28633, 28660, 28670-1, 28697, 28727, 28730, 28758, 28778-9,
    28789, 28802, 28817, 28906, 28908, 28992, 29044, 29076, 29105, 29114, 29216-7,
    29356, 29404, 29406, 29584, 29618, 29630-1, 29634, 29642, 29667-8, 29696,
    29700, 29719-20, 29730-1, 29749-51, 29781, 29837-8, 29844, 29856, 29934,
    29956, 29961-2, 29980-1, 29999, 30027, 30039, 30066, 30115, 30126-7, 30135,
    30153, 30178, 30192, 30292, 30340, 30409, 30411, 30498, 30548, 30726-7, 30744,
    30805-6, 30875, 30884, 30966, 30969, 30989, 31024, 31026, 31055, 31123-4,
    31147, 31234, 31295-6, 31335, 31365, 31381, 31420, 31461, 31523, 31561, 31580,
    31610, 31634, 31667, 31671-2, 31683, 31688, 31702, 31754-5, 31799, 31818,
    31871, 31915, 31924, 31930, 31942, 31955, 31985-7, 32128-9, 32157, 32233,
    32235, 32245, 32260, 32267, 32275, 32331-2, 32375-6, 32384, 32386-7, 32393,
    32438, 32445, 32480, 32507-8

ATPase, methods 29044, 29357, 29631, 29750, 29837, 30153, 30411, 30981, 31024,
    31296, 31580, 32508

Autotrophy see Carbon metabolism types ...

B

Bacteria, photosynthetic see Photosynthetic bacteria ...

Bacteriochlorophylls (*cf.* also Chlorophyll, *Chlorobium*)
      28515, 28548, 28570, 28582-3, 28644, 28672, 28700-1, 28822, 28876, 28878,
      28892, 28905, 29014, 29025-6, 29035, 29074, 29080, 29098, 29124, 29126,
      29134, 29149, 29174, 29197-200, 29263-5, 29282, 29339, 29358, 29369, 29431-2,
      29452-4, 29474, 29483-4, 29505, 29679, 29682, 29710, 29712, 29738, 29749,
      29818, 29828, 29878-9, 29943, 29947, 29969, 29973, 29996-7, 30050, 30079,
      30134, 30138, 30179, 30227, 30248, 30328, 30385-6, 30418, 30420, 30625-6,
      30671-2, 30696-7, 30718, 30742, 30780, 30789, 30915, 30934, 30939, 31033,
      31046, 31060, 31073-4, 31094, 31137-8, 31219, 31225, 31288, 31299, 31308,
      31354-5, 31369-71, 31415, 31534, 31548, 31556, 31741, 31756, 31805, 31845,
      31892-3, 31900, 31902, 31929, 32045, 32048, 32114, 32123-4, 32127, 32149,
      32168-9, 32208, 32251, 32280-2, 32367, 32409, 32549

Bacteriorhodopsin see *Halobacterium* photosynthesis

Bibliographies of photosynthesis, biographies B28522

Biliproteins see also Phycocyanins; Phycoerythrins

Biliproteins absorption spectra *in vitro* 28960, 29152, 29672, 30373, 30588, 31011,
      31423, 32071

Biliproteins absorption spectra *in vivo* 28598, 28737, 29378, 29525, 29670, 30552,
      30678, 31206, 31505

Biliproteins and production of water reservoirs 29498, 30226, 32368

Biliproteins biosynthesis, precursors 28719, 29866, 30311, 30531, 32102, 32173

Biliproteins chemical structure 28959, 29670, 32173

Biliproteins complexes *in vivo* 29590, 29670, 30372, 30552, 30877, 30940, 31901,
      32540

Biliproteins degradation 30299, 30531

Biliproteins determination, column chromatography 28960, 29672

Biliproteins determination, electrophoresis and other methods 29152, 29591, 29782,
      30373, 30588

Biliproteins determination, paper chromatography, thin-layer chromatography 30475

Biliproteins determination, spectral methods 29152

Biliproteins energetic states *in vitro* 28675

Biliproteins energetic states *in vivo* 28968, 29590, 30525, 30970, 31108, 31709,
      32357-8, 32540

Biliproteins fluorescence *in vitro* 28504, 28675, 30373, 30588, 32071, 32368

Biliproteins fluorescence *in vivo* 28737, 29014, 29378, 29590, 29592, 29723, 29836,
      30372, 30588, 30678, 30940-1, 32357, 32540

Biliproteins in model systems 28737, 29545

Biliproteins in photosynthesis mechanism 28675, 28737, 29589-90, 29670, 29723,
      29836, 30135, 30323, 30442, 30525, 30877-8, 30970, 31108, 31485, 32071, 32357-
      -8, 32540

Biliproteins in physiology of photosynthesis 28613, 32269

Bioclimatological methods see Aerodynamic methods ...

Biological clock see Diurnal changes ...

Biomass distribution and redistribution in plant
        28516-7, 28533, 28544, 28566, 28578, 28624, 28648, 28703, 28782, 28791, 28800,
        28809, 28899, 28918, 28926, 28941, 28972-3, 29007, 29016, 29022, 29042, 29053,
        29068, 29086, 29103-4, 29136, 29153, 29157, 29164, 29193, 29203, 29221, 29254,
        29290, 29331, 29335, 29351, 29366, 29373, 29443, 29509, 29524, 29550, 29566,
        29576, 29605, 29628, 29652, 29656, 29771, 29794, 29819-20, 29835, 29883, 29923,
        29968, 29975, 29977, 30011, 30044, 30053, 30195, 30203, 30272, 30353, 30412,
        30427, 30451, 30461, 30476, 30478, 30483, 30505-6, 30532, 30571, 30581, 30642,
        30680, 30691, 30699, 30711, 30733, 30820, 30824, 30891, 30932, 30957, 30967,
        30979, 31013, 31015, 31021, 31032, 31067, 31069, 31086, 31107, 31116, 31130-1,
        31148, 31163, 31180, 31201, 31228, 31242, 31272-3, 31303, 31305, 31307, 31331,
        31348, 31376-7, 31392, 31399-400, 31408, 31439, 31463, 31492, 31535, 31584-5,
        31592, 31595, 31608, 31695, 31761, 31792, 31822-3, 31825-6, 31868, 31916,
        31922, 31933-4, 31941, 31978, 31996, B32004, 32046, 32059, 32062, 32065, 32076,
        32115, 32117, 32133, 32151, 32170-1, 32231, 32243, 32257, 32322, 32325, 32347,
        32364, 32403, 32412, 32431, 32434, 32436, 32483, 32504, B32528, 32530, 32542-3

Biopotentials see Chloroplast and chromatophore biopotentials

Biosphere production see Ecosystem production ...

Blinks effect see Emerson effect, Blinks effect

Books on photosynthesis see General aspects ...

Boundary layer of leaf see Resistance, leaf boundary layer

Bundle sheaths see Carbon metabolism types ...; Carbon fixation pathways, compari-
        son in mesophyll and bundle sheath cells

C

$^{13}C/^{12}C$ ratio, $\delta^{13}C$ 28567, 28748, 29273-4, 29488, 30332, 30745, 30822, 30952, 31003,
        31039, 31212, 31398, 31467, 31653, 31697, 31798, 32087

^{14}C, ^{11}C, ^{13}C see Carbon isotopes ...

C_3 pathway of carbon fixation
        28567, 28679, 28748, 28774, 28786-7, 28807, 28837, 28841, 28901, 28970, 29016,
        29051, 29075, 29087, 29091, 29217, 29231, 29267, 29332, 29376, 29383, 29406-7,
        29463, B29468, 29489, 29500, 29525, 29571, 29579, 29604, 29637, 29709, 29744,
        29865, 29885, 29912, 29921, 29976, 30009, 30034, 30097-8, 30223, 30235, 30273,
        30277, 30288, 30333, 30429-30, 30454, 30516, 30526, 20537, 30543, 30557,
        30640, 30702, 30745, 30750, 30765, 30951-2, 31003, 31034, 31039, 31102, 31166,
        31182, 31275, 31338-9, 31372, 31393, 31400, 31417, 31419-20, 31442, 31451,
        31471, 31620, 31636-7, 32024, 32036, 32088, 32109, 32139, 32378, B32410,
        32425, 32428, 32457, 32459-62, B32515, 32547

C_4 pathway of carbon fixation
        28567, 28625, 28748, 28786-7, 28807, 28818, 28837, 28841, 28901, 28909, 28970,
        29015-6, 29047, 29051, 29075, 29085, 29087, 29091, 29181, 29187, 29217, 29231,
        29262, 29326, 29376, 29383, 29407, 29489, 29495, 29525, 29535, 29579, 29637,
        29709, 29744, 29830, 29860, 29862-3, 29865, 29885, 29912, 29921, 29976, 29987,
        30009, 30036-7, 30097-8, 30191, 30233, 30235, 30272-4, 30279, 30333, 30435,
        30462, 30498, 30537-8, 30557, 30631, 30702, 30745, 30887, 30951-2, 31003,
        31034, 31039, 31166, 31266, 31393, 31400, 31417-21, 31442, 31448-52, 31457,
        31465-6, 31613, 31620, 31636-7, 31993, 32088, 32121, 32336, 32378, B32410,
        32428-9, 32439, 32459, 32461-2, B32515, 32547

C_3, C_4, CAM pathways, comparison see Carbon metabolism types ...

Calibration of infra-red analyser see Infra-red gas analyser ...

Caloric values   see Calorimetry ...

Calorimetry 29193, 29791, 30459, B30477, 30533, 30943, 31035, 32324

Calvin-Benson cycle see C_3 pathway of carbon fixation

CAM   28703, 28787, 29092, 29142, 29216-7, 29273-4, 29281, 29376, 29463, 29489,
        29639, 29744, 29772, 29816, 29865, 29912, 30279, 30332-3, 30516, 30664-6,
        30695, 30745, 30822-3, 30927-8, 30951-2, 31003, 31057-8, 31066, 31697, 31913,
        31993, 32024-5, 32109, 32139, 32378, 32425-6, 32458, B32515

Canopy, CO_2 profiles
        28521, 29064, 29289, 29372, 30084, 30438, 31272, 31710, 31760, 31920, B32150

Canopy density, thickness
        28503, 28552, 28689, 28820, 29007, 29045, 29190, 29368, 29492, 29496, 29566,
        29574, 29633, 30011, 30029, 30146, 30376, 30409, 30438, 30674, 30689, 30874,
        30976, 31004, 31113-5, 31180, 31535, 31584, 31777, 31822, 31870, 31934, 32504,
        32545

Canopy horizontal structure 29676, 30555

Canopy, leaf age 28516, 28527, 30555, 31695

Canopy, leaf angles
        28865, 29289, B29468, 29566, 29574, 29676, B29707, 30084, 30553, 30555, 30700,
        30945, 31113-5, B31140, 31198, 31265, 31392, 31517, 31761, 31879, 31884, B32150,
        32152

Canopy microclimate and macroclimate 29289, 29368, 30071, 30084, 30899, 31528, 31762,
        31936

Canopy photosynthesis and PhAR profile (cf. also Canopy, radiation profile)
        28505, 31546-7, 31759, B32150

Canopy photosynthesis, direct measurement
        28502, 28810, 29086-7, 29166, 29289, 29294, 29373, 29575-6, 29722, 29967, 30084,
        31102, 31252, 31527, 31530, 31584, 31586, 31696, 31760, 31777, 32165, 32294

Canopy photosynthesis, energy balance 28810, 29064, 29372-3, B29707, 31220, 31633,
        31827, 31849, 32210

Canopy photosynthesis, mass and momentum balance 28810, 29373, 29656, B29707, 31849,
        32210

Canopy photosynthesis, model see Model ...

Canopy, radiation distribution; reflection, transmission, absorption, albedo, $etc.$
        (cf. also Canopy, radiation profile)
        28820, 28864-5, 29064, 29368, 29373, 29491, 29574, B29707, 29887, 29994, 30044,
        30070, 30083, 30162, 30205, B30281, 30553, 30574, 31115, B31140, 31188-9,
        31199, 31249, 31265, 31517, 31528, 31537, 31546-7, 31760, 31762, 31817, 31884,
        B32003, 32096, B32150, 32152, 32187, 32436

Canopy, radiation profile (cf. also Canopy, radiation distribution; Canopy photosyn-
        thesis and PhAR profile)
        28502, 28505, 28544, 28622, 28806, 28809, 28865, 29064, 29368, 29372-3, 29574,
        B29707, 29967, 29994, 30060-1, 30205, 30483, 30523, 30553, 31113-5, B31140,
        31198-9, 31242, 31265, 31508, 31517, 31537, 31546-7, 31584, 31633, 31760,
        31762, 31916, 31967, 32111, 32152, 32436

Canopy, resistances for CO_2 and water vapour transfer see Resistances for CO_2 and
        water vapour transfer at canopy level

(continued)

Carbon isotopes ... (continued)
    32121-2, 32172, 32200-1, 32218, 32306, 32339-40, 32363, 32372, 32413, 32427,
    32463, 32488, 32503

Carbon metabolism types and carbon fixation pathways 29581, 30009, 30109, 30213,
    31676, 32458

Carbon metabolism types and chlorophyll 29446

Carbon metabolism types and chloroplast (chromatophore) 29581, 30433-4, 30557, 31778

Carbon metabolism types and ecosystem and plant productivity 28873, 30077, 30917,
    31159, 31636

Carbon metabolism types and electron transport 29383

Carbon metabolism types and gas exchange 28970, 29050, 29091, 29166, 29489, 29985,
    30124, 30270, B30511, 30754, 30822, 30916, 31613, 31932, 32312, 32458

Carbon metabolism types and resistances to CO_2 and water vapour transfer 29232,
    30270, 31836

Carbon metabolism types and respiration 28970, 32312

Carbonic anhydrase 28613, 28901, 29503, 29709, 29776, 29990, 30630, 30991, 31107,
    31136, 31471-2, 31474, 31679, 31720

Carbonic anhydrase, methods 30618

Carboxylation see Carbon fixation pathways ...

Carboxylation resistance see Resistance, carboxylation and excitation

Carotenes 28528, 28607, 28650-1, 28708, 28715, 28760, 28944-5, 28947-8, 28978, 29170,
    29209, 29234, 29241, 29324, 29422, 29456, 29480, 29573, 29614, 29698, 29897,
    30017-8, 30102, 30237-8, 30302, 30337, 30357, 30452, 30581, 30619, 30629,
    30730, 30760, 30802, 30851, 31076, 31189, 31349, 31405-6, 31503-4, 31559,
    31574, 31642, 31831, 31961, 32016, 32055, 32142, 32185, 32190, 32284, 32289,
    32349, 32441-2, 32481

Carotenoids absorption spectra *in vitro* 28569, 29172, 29434, 29897, 30137, 30207,
    30582, 30730, 31189-90, 31223, 31349, 31583, 31673, 31851, 32349, 32501

Carotenoids absorption spectra *in vivo* 28770, 28917, 29263-4, 29525, 29573, 29666,
    29879, 29913, 30079, 30172, 30314, 30366, 30475, 30487, 30974, 31368, 32017-8,
    32103, 32288, 32500-1

Carotenoids and production of algae and water reservoirs 28790, 29802-3, 30356-7,
    31047

Carotenoids and production of higher plants 30396, 32185-6

Carotenoids biosynthesis and precursors
    28752, 28944, 28948, 28978, 29241, B29468, 29640, 29698, 29760, 29904, 29947,
    30238, 30592, 30597, 30923, 31211, 31223, 31329, 31349-50, 31504, 31680,
    31779, 32215, 32229

Carotenoids chemical structure 28767, 29241, 29804, 29924, 30102, 30592

Carotenoids complexes *in vitro* 31740

Carotenoids complexes *in vivo* 28770, 29505, 29876, 30314, 30366, 30672, 30789,
    31385, 31625, 31812, 31901

Chlorophyll absorption spectra *in vivo*
     28520, 28526, 28550, 28598, 28631, 28653, 28690, 28692, 28737, 28752, 28792,
     28817, 28836, 28917, 28932-3, 28938, 28958, 28964, 28967-8, 28976, 29030-1,
     29073-4, 29121, 29194, 29198-9, 29240, 29263-6, 29278, 29349, 29378, 29431,
     29474, 29476, 29505, 29525, 29558, 29573, 29580, 29662, 29681, 29743, 29788,
     29818, 29878-9, 29896, 29906, 29913, 29939, 29978, 30079-80, 30172, 30193-4,
     30227, 30248, 30295, 30330-1, 30366, 30475, 30487, 30552, 30570, 30616, 30669-
     -70, 30742, 30849, 30862, 30934, 30975, 31011, 31033, 31054, 31073, 31101,
     31110, 31138, 31186, 31206, 31280, 31299, 31361, 31429, 31444, 31480, 31503,
     31505, 31534, 31575, 31594, 31643, 31707, 31713, 31726, 31735, 31785, 31787-8,
     31805, 31846, 31892, 31900,. 31929, 31980, 31984, 32005, 32021, 32081, 32084,
     32099-100, 32103, 32112, 32168, 32186, 32282, 32285, 32288, 32351, 32361,
     32408-9, B32410, 32472, 32481, 32522

Chlorophyll and its products determination see also Pigments determination, sampling
     and extraction

Chlorophyll and its products determination, column chromatography 28607, 29434,
     30596, 31257, 31622

Chlorophyll and its products determination, electrophoresis and other methods
     28576, 30247

Chlorophyll and its products determination, *in vivo* 28576, 28886, 28958, 28969,
     29194-5, 29198, 29330, 29339, 29595, 29687, 30596, 30720, 31480, 31534, 31892,
     31900, 31929, 32081, 32100, 32103, 32393

Chlorophyll and its products determination, paper chromatography, thin-layer chroma-
     tography 28607, 28712, 29434, 29525, 29803, 29947, 30091, 30137, 30193,
     30357, B30432, 30475, 30582, 30629, 30923, 31430, 31673, 31782, 31797, 32093,
     32442

Chlorophyll and its products determination, spectral methods
     28346, 28348, 28565, 28962, 29026, 29293, 29561, 29743, 29919, 30142, 30164-5,
     30169, 30348, 30440, 30755, 30835, 31225, 31257, 31577, 31622, 31682, 31881,
     32083, 32116, 32405

Chlorophyll and production of algae and water reservoirs
     28553, 28565, 28673, 28721, 28773, 28776, 28790, 28797, 28861, 28880, 28955,
     28977, 28982, 28984, 29011, 29041, 29077, 29127, 29151, 29171, 29185, 29226,
     29287, 29292-3, 29311, 29327, 29336, 29345, 29355, 29375, 29396, 29428, 29448,
     29494, 29588, 29653, 29655, 29674, 29678, 29716, 29802-3, 29809, 29842, 29916,
     29919, 30045-6, 30052, 30075, 30121, 30137, 30147-8, 30169, 30196, 30225-6,
     30231, 30247, 30281, 30305, 30315, 30339, 30356-7, 30381, 30389, 30407, 30450,
     30482, 30485, 30504, 30572, 30605-6, 30613, 30676, 30724-5, 30756, 30863-5,
     30873, 30906, 30987, 31031, 31047, 31215, 31311-2, 31316-7, 31382, 31397,
     31430, 31531, 31714, 31745, 31909, 31912, 31965, 32031-2, 32043-4, 32089,
     32107-8, 32137-8, 32166, 32273-4, 32348, 32356, 32368, 32380, 32404, 32423,
     32493, 32518, 32548

Chlorophyll and production of higher plants
     29067, 29233, 29419, 30396, 30581, 31510, 31638, 31839, 32185-6, 32259, 32300

Chlorophyll biosynthesis and precursors
     28499, 28519, 28525-6, 28532, 28585-6, 28597, 28628, 28637, 28639, 28656-7,
     28667-8, 28674, 28709, 28719, 28745, 28752, 28762, 28793, 28811, 28836, 28910,
     28912, 28956, 29012, 29030, 29043, 29058-9, 29150, 29220, 29259, 29329, 29341,
     29347-9, 29361, 29392, 29403, 29433, B29468, 29476, 29523, 29546, 29573,
     29593-4, 29610-1, 29623, 29640, 29708, 29760, 29773, 29833-4, 29859, 29903-5,
     29939, 29947, 30006, 30027-8, 30176, 30200-2, 30215-6, 30229, 30240, 30252,
     30311, 30318-20, 30330-1, 30334, 30424-5, 30514, 30537, 30540-1, 30560, 30597,
     30602-3, 30610, 30612, 30660, 30673, 30740-1, 30790-2, 30795, 30827, 30837,
     30848-50, 30867, 30893-4, 30923, 30992, 31042, 31054, 31096, 31101, 31116,
                                                              (continued)

Chlorophyll energetic states *in vivo* (continued)
    29619, 29621, 29666, 29679, 29681, 29710, 29754-5, 29779, 29818, 29878, 29896,
    29936, 29946, 29969, 29971-2, 29978, 30159, 30179, 30183, 30248, 30359, 30366,
    30385, 30424, 30479, 30539, 30625, 30742, 30780, 30794, 30813, 30939, 30970,
    31063, 31074, 31108, 31219, 31299, 31369, 31444, 31571, 31638, 31642, 31709,
    31980, 32021, 32123-4, 32162, 32251, 32357-8, B32410

Chlorophyll energetics model see Model ...

Chlorophyll, enzymes of synthesis and degradation (other than chlorophyllase)
    28628, 29059, 29708, 30740, 31005, 31054, 31330

Chlorophyll fluorescence *in vitro*
    28626, 29026, 29543-4, 29548, 29662, 29774, 29943, 30173, 30367, 30388, 30641,
    30657, 30997, 31374, 31577, 31992, 32224, 32241, 32381, 32531

Chlorophyll fluorescence *in vivo*
    28518-9, 28550-1, 28585-6, 28588-9, 28591, 28595, 28636, 28657, 28676, 28678,
    28680-3, 28698, 28724-6, 28737, 28752, 28771, 28839, 28859, 28867, 28878,
    28886, 28892, 28903, 28934-6, 28938, 28956, 28963-9, 28991, 28998-9001, 29012,
    29014, 29017, 29025-7, 29030, 29074, 29096, 29134, 29194-5, 29244, 29268,
    29276, 29282, 29341, 29348-9, 29436, 29449, 29525-6, 29528-9, 29534, 29546-7,
    29592, 29595, 29599-600, 29602, 29619-21, 29635, 29661, 29681-2, 29712-3,
    29755-7, 29807, 29818, 29831, 29836, 29842, 29872, 29878, 29906, 29928, 29943,
    29946, 29955, 29991, 29997, 30000-1, 30012, 30108, 30154-6, 30158-60, 30176,
    30183-4, 30193-4, 30225, 30228-9, 30248, 30256, 30299, 30302, 30323, 30326-8,
    30330-1, 30347, 30360, 30366, 30429, 30439, 30467, 30495, 30525, 30539, 30570,
    30587, 30590-1, 30596, 30600, 30616, 30644, 30708-10, 30718-20, 30736, 30759,
    30791-2, 30794, 30833-5, 30838-9, 30848, 30850, 30859, 30875, 30877, 30892,
    30915, 30949-50, 30970-1, 30973, 31038, 31043, 31059, 31143, 31170, 31185,
    31225, 31291, 31298, 31342, 31354, 31363, 31375, 31390, 31426-7, 31444, 31454,
    31488, 31503, 31506-7, 31511, 31549, 31556, 31582, 31588, 31597, 31621, 31626,
    31631-2, 31638, 31664, 31666, 31686, 31688-92, 31707-9, 31714, 31772, 31781,
    31820, 31835, 31837, 31897-8, 31906, 31929, 31979-82, 32000, 32021, 32068,
    32081, 32102, 32116, 32162, 32238, 32245, 32249, 32251, 32255, 32287, 32310,
    32314-5, 32358, 32407, B32410, 32416, 32420, 32432-3, 32453, 32456, 32500,
    32510, 32522, 32540

Chlorophyll forms see Chlorophyll complexes ...

Chlorophyll in model systems (*cf*. also Chlorophyll energetic states *in vitro*)
    28563, 28570, 28627, 28729, 28876, 28954, 28968, 29010, 29083, 29358, 29423,
    29497, 29548-9, 29585, 29662, 29694-5, 29901, 29925, 30117, 30142, 30189,
    30293, 30417-8, 30422, 30514, 30625, 30657, 30728, 30796, 30926, 30997, 31074,
    31571, 31594, 31713, 31740, 31742, 31794, 32074, 32131-2, 32224-5, 32234,
    32354-5, 32411, 32474, 32531

Chlorophyll in photosynthesis mechanism
    28515, 28519, 28535, 28547-8, 28550, 28575-6, 28588, 28590, 28595-6, 28644,
    28657, B28677, 28691, 28698, 28700, 28720, 28724-5, 28729, 28737, 28751,
    28822, 28824, 28834, 28859, 28877, 28884-5, 28892, 28903, 28934, 28969, 28989,
    28991, 28997, 28999, 29001, 29010, 29012, 29025, 29027, 29124, 29134, 29149,
    29161, 29174, 29199, 29244, 29263-4, 29275, 29278, 29310, 29339, 29358,
    29406, 29431-2, 29440, 29445, 29452-4, B29468, 29518-21, 29525, 29528, 29580,
    29614, 29620-1, 29666, 29679, 29714, 29738, 29754, 29756-7, 29784-5, 29818,
    29836, 29876, 29879, 29896, 29922, 29971-2, 29978, 29997, 30085, 30135, 30138,
    30155, 30158, 30172, 30179, 30181, 30183, 30189, 30198, 30248, 30256, 30320,
    30322-3, 30326-8, 30347, 30359, 30366, 30418, 30422-3, 30436, 30442, 30479,
    30523, 30525, 30539-40, 30594, 30625-6, 30669, 30708, 30718, 30742, 30759,
    30779-80, 30782, 30795-6, 30812-3, 30831, 30870, 30877-8, 30881, 30915, 30926,
    30939, 30970, 31009, 31027, 31030, 31033, 31046, 31060, 31104, 31108, 31138,
    31143, 31185, 31194, 31219, 31288, 31297, 31354, 31362, 31364, 31369, 31371,
    31390-1, 31429, 31444, 31487-9, 31505, 31507, 31556, 31575, 31588, 31642,
    31666, 31713, 31740-2, 31788, 31794, 31805, 31893, 31906, 31955, 31971, 31979,

(continued)

Chlorophyll in photosynthesis mechanism (continued)
    31981-2, 32007, 32049, 32099, 32101, 32112-3, 32123, 32127, 32130, 32132,
    32148, 32162, 32168-9, 32216, 32223, 32238-9, 32246, 32249, 32280-1, 32357-8,
    32379, 32393, B32410, 32432, 32441, 32444

Chlorophyll in physiology of photosynthesis
    28656, 28776, 28797, 28835-6, 28843, 28921, 28940, 29004, 29040, 29163, 29191,
    29564, 29593-4, 29826, 30057, 30131, 30198, 30263, 30281, 30320, 30575, 30613,
    30643, 30722, 30896, 30935, 31118-9, 31230, 31281, 31340, 31510, 31550, 31587,
    31627, 31801, 32121, 32187-9, 32230, 32289, 32300, 32311, 32417

Chlorophyll in seeds and fruits
    28613, 29414, 29702, 29920, 30441, 30514, 30537, 31361, 31833

Chlorophyll luminescence see Chlorophyll delayed light emission ...

Chlorophyll, methods see Chlorophyll and its products determination ...

Chlorophyll number see Chlorophyll in physiology of photosynthesis

Chlorophyll precursors see Chlorophyll biosynthesis ...

Chlorophyll unit see Photosynthetic (chlorophyll) unit

Chlorophyllase 30186, 31221, 31385, 31605, 31782, 31852, 31855, 31991, 32077-8

Chlorophyllase and other enzymes of chlorophyll synthesis and degradation, methods
    31775

Chlorophylls a,b content and their ratio
    28518, 28529, 28532, 28550, 28568-9, 28584-7, 28589, 28595, 28607, 28625,
    28650-1, 28655-6, 28667, 28690-1, 28715, 28718, 28752, 28760, 28790-1, 28801,
    28827, 28836, 28880, 28896, 28962, 28990-1, 29004, 29039, 29067, 29104, 29150,
    29155, 29170, 29191, 29210, 29218, 29220, 29240, 29251, 29253, 29259, 29262,
    29268, 29323, 29368, 29384, 29386, 29396, 29410, 29422, 29438, 29443, 29448,
    29450, 29461, 29473, 29480, 29501, 29534, 29538, 29562, 29598, 29603, 29614,
    29661, 29669, 29677, 29682, 29687, 29704, 29706, 29711, 29717, 29753-4, 29760,
    B29763, 29765, 29777, 29803, 29807, 29833, 29835, 29876, 29898, 29939, 29943,
    29979, 29994, 30006, 30017, 30067, 30092, 30101, 30123, 30186, 30188, 30193,
    30200, 30207, 30230, 30253, 30274, 30298, 30302, 30334, 30348, 30363, 30421,
    30441, 30452, 30481, 30494, 30496, 30536, 30548, 30550, 30570, 30576,
    30579-81,   30594-7, 30611-2, 30623, 30659, 30683, 30687, 30721, 30759-60,
    30800, 30802, 30828, 30837, 30851, 30866, 30881, 30911-2, 30946, 30951, 30962,
    30993, 31012, 31063, 31109, 31116, 31119-20, 31128, 31144, 31155, 31221,
    31230, 31232, 31236, 31243, 31251, 31276, 31281, 31324, 31329-30, 31336,
    31340, 31361, 31384-5, 31395, 31405-6, 31426, 31480, 31489, 31502-3, 31540,
    31551, 31559, 31575, 31579, 31587-8, 31621, 31625, 31673, 31735, 31737, 31784,
    31795-6, 31839, 31877-8, 31885, 31930, 31971, 31979, 31991-2, 32005, 32008-9,
    32055, 32061, 32065, 32086, 32093, 32101, 32113, 32116, 32133, 32142-3, 32175,
    32185-8, 32190, 32192, 32212, 32216, 32230-1, 32237, 32254, 32258-9, 32261,
    32266, 32284-6, 32289, 32298-300, 32379, 32385, 32401, 32421, 32424, 32432,
    32441, 32481, 32488, 32501, 32526

Chlorophylls c,d
    28698-9, 28790, 29396, 29677, 29803, 29991, 30137, 30582, 30592, 30616, 30629,
    31257, 31430, 32123, 32216, 32258, 32288, 32442, 32472

Chlorophylls *Chlorobium*
    30424, 30650, 31308, 31386, 31415

Chloroplast see also Thylakoid; Stroma of chloroplast; Pyrenoid; Ribosome of chlo-
    roplast; Phycobilisome

Chloroplast and cell counting methods 32196

Chloroplast and chromatophore biopotentials
28633, 28678, 28681-2, 29265, 29583, 29668, 29956, 30036, 30126, 30156, 30178- -9, 30521, 30567, 30644, 30679, 31061, 31123, 31149, 31401, 31485, 31549, 31560, 31567, 31645-6, 31693, 31780, 32132, 32319, 32321, 32344-5, 32443, 32486

Chloroplast and chromatophore chemical composition (*cf.* also Lipids, fatty acids, and chloroplast...; Proteins, amino acids, nucleic acids, and chloroplast..., *etc.*)
28554, 28574, 28650, 28709, 28723, 28733, 28804, 28812, 28836, 28840, 28910, 28923, 28974, 29210, 29218, 29299-300, 29393, 29431, 29477, 29571, 29627, 29698, 29741, 29745, 29758, 29765, 29808, 29813, 29830, 29932, 29938, 29948, 29998, 30036-7, 30068, 30089, 30123, 30145, 30149-50, 30334, 30341-2, 30428, 30472, 30534, 30549, 30551, 30562, 30646, 30731, 30801, 30816, 30852, 30854, 30900-1, 30933, 30958, 31082, 31104, 31109, 31120, 31137, 31294, 31323, 31380, 31424, 31434-5, 31538, 31557, 31590, 31593, 31616, 31639, 31644, 31662, 31815, 31892, 31937, 32079, 32086, 32132, 32149, 32291, 32336, 32389

Chloroplast and chromatophore dimensions
28649, 28684, 28883, 28909, 29110, 29112, 29121, 29260, 29470, 29510, 29704, 29871, 30020, 30199, 30298, 30329, 30667, 30753, 30826, 30879, 30896-7, 31106, 31530, 31568-9, 31644, 31801, 32085, 32125-6, 32222, 32420, 32551-2

Chloroplast and chromatophore distribution in cell see Chloroplast and chromatophore number and distribution

Chloroplast and chromatophore fragments
28514, 28549-50, 28554, 28562, 28574, 28582, 28584, 28586-7, 28590, 28698, 28701, 28720, 28751, 28771, 28784, 28878, 28886, 28997, 29037, 29161, 29210, 29244, 29384, 29406, 29438, 29441-2, 29449, 29502, 29546-7, 29580, 29600, 29602, 29672, 29682, 29687, 29698, 29738, 29768, 29872, 29932, 29949, 29955, 30013, 30050, 30123, 30135, 30181, 30188, 30207, 30245, 30254-5, 30321, 30326-7, 30334, 30359-60, 30425, 30539, 30559, 30717, 30780, 30789, 31011, 31030, 31078-80, 31137, 31161, 31206, 31295, 31298, 31323, 31375, 31391, 31426, 31434, 31479-80, 31534, 31549, 31564, 31588, 31598, 31612, 31621, 31625, 31629-32, 31642-3, 31707, 31725-6, 31735, 31796, 31857, 31930, 31955, 31992, 32005, 32007, 32100, 32114, 32178, 32266, 32392-3

Chloroplast and chromatophore number and distribution
28649, 28655, 28684, 28883, 28909, 28925, 29110, 29112, 29128, 29360, B29468, 29470, 29704, 29871, 30199, 30298, 30329, 30643, 30660, 30896-7, 31039, 31155, 31204, 31604, 31768, 31781, 31801, 32061, 32222

Chloroplast and chromatophore replication, ontogeny
28585, 28590, 28651, 28736, 28805, 28836, 28883, 28896, 28909, 28947, 29055-6, 29128, 29150, 29260-1, 29318, 29329, 29360, 29437, 29465, 29517, 29813, 29922, 30003, 30012, 30020, 30027, 30054, 30082, 30216, 30242, 30287, 30312, 30467, 30557, 30560, 30668, 30673, 30690, 30747, 30753, 30790, 30814, 30826, 30893-4, 30900-1, 30992, 31103-4, 31133, 31222, 31318, 31443, 31468, 31504, 31644, 31680, 31719, 31743-4, 31757, 31778-9, 31807, 31830-1, 31854, 31876, 31878, 31880, 31973, 31979, 32029, 32061, 32084, 32302, 32365-6, 32381, 32388, 32397, 32399-400, 32421, 32452, 32473

Chloroplast and chromatophore volume changes
28981, 29305, 29352, 29458, 29596, 29932, 30036, 30110-1, 30199, 30889, 30910, 30984, 31207, 31210, 31530, 31568-9, 31814, 32445

Chloroplast immobilization see Photosystem stabilization ...

Chloroplast, isolated, carbon fixation in see Carbon fixation in isolated chloroplasts ...

Chloroplast, isolated, gas exchange by
28556, 28562, 28592, 28599, 28653, 28665, 28789, 28886, 29076, 29081, 29141, 29382, 29533, 29615, 29877, 29891-2, 29942, 30291, 30430, 30518, 30526, 30561,

(continued)

CO_2 and chlorophyll   28947, 29259, 29433, 29680, 30872

CO_2 and chloroplast   30872, 31950

CO_2 and ecosystem and plant productivity
        28571, 28648, 28857, 28890, 29159, 29236, 29656, 29822, 29832, 29951, 30077,
        B30405, 30890, 30957, 31394, 31400, 31509, 32516

CO_2 and electron transport chain   29383, 30291, 31320, 31410, 31948-9

CO_2 and gas exchange (*cf.* also CO_2 and gas exchange, analysis of CO_2 curves)
        28527, 28646-8, 28763, 28795, 28842, 28856, 28890, 28898, 28901, 28930, B29023,
        29091, 29115, 29156, 29159-60, 29179, 29236, 29328, 29424-5, B29468, 29503-4,
        29514, 29629, 29693, 29732, 29848, 29890, 29931, 29951, 29986, 29993, 30019,
        30084, 30097, 30220, 30224, 30289, 30381, 30453-7, 30469, 30499, B30511,
        30515, 30561, 30622-4, 30653, 30656, 30706, 30757, 30919, 30921, 30927, 30951,
        31012, 31066-7, 31090, 31135, 31154, 31314, 31400, 31470, 31521, 31524, 31592,
        31609, 31613, 31701, 31732, 31734, 31802, 31866, 31885, B32004, 32022, 32121,
        32178, 32327, 32344-5, 32541

CO_2 and gas exchange, analysis of CO_2 curves
        28571, 28782, 29015, 29090, 29156, 29656, 29693, 29986, 30097, 30139, 30456-7,
        30499, 31372, 31411, 31613, 320,73

CO_2 and photorespiration   29328, 29514, 29848, 30097, 30457, 30468, 30637, 31119,
        31731, 31928, B32004, 32430, B32515

CO_2 and resistances to CO_2 and water vapour transfer   28717, 29020, 29130, 29425,
        29503, 30289, 32073

CO_2 and respiration   28647-8, 29848, 29993, 30218, 30804, B32004

CO_2 compensation concentration
        28625, 28646, 28703, 28717, 28774, 28782-3, 28808, 28901, 28930, 29050, 29053-
        -4, 29090, 29108, 29155-6, 29178-9, 29187, 29236, 29326, 29328, 29336, 29486,
        29503, 29535-6, 29693, 29795, 29911, 29989, 30078, 30097, 30099, 30272-3,
        30294, 30453, 30499, 30542, 30596,   30622-4,   30640, 30699, 30751, 31119,
        31164-5, 31251, 31256, 31293, 31314, 31400, 31456-7, 31529, 31599, 31613,
        31636-7, 31733-4, 31802, 31836, 31861-2, 31864, 31867, 31885, 32047, 32073,
        32329, 32428, 32430, 32439

CO_2 exchange   see   Gas exchange ...

CO_2 fixation, dark   see   Dark CO_2 fixation

CO_2 measurement, infra-red gas analyser   see   Infra-red gas analyser for CO_2

CO_2 measurement (other than with infra-red gas analyser)   B30432, 30891

CO_2 transfer across membranes   29076, 30651-2, 30654-5, 30761, 31067, 31679

CO_2 transfer, theory   B29023, 29255, 29776, 30651-2, 30654, 31432

Cold (hardiness)   see   Temperature, low ...

Combustion heat   see   Calorimetry

Compensation irradiance
        28687, 28856, 29053, 29155, 29191, 29242, 29337, 29563, 29675, 30099, 31067,
        31515, 31885, 32181

Compensation point, CO_2 see CO_2 compensation concentration

Compensation point, light   see   Compensation irradiance

Competition in ecosystem 28689, 29334, 29496, 30029, 30044, 30463, 31004, 32328

Conductance for transfer of gases see Resistance ...

"Contribution" of individual organs see Biomass distribution and redistribution;
        Photosynthate translocation ...

Correlations within plant 30409, 30680, 31933-4, 32436

Cosmic radiation see Ionizing radiation ...

Coupling factor 1 see ATPase ...

Cover, vegetative see Canopy ...; Ecosystem ...

Crassulacean Acid Metabolism see CAM

Cultivar differences, carbon fixation pathways 29633, 30098, 30474

Cultivar differences, carotenoids 29414, 29762, 30396, 30481, 30802, 31144, 31384-5,
        32261, 32284

Cultivar differences, chlorophyll 28799, 29004, 29238, 29251, 29410, 29414, 29457,
        29470, 29711, 29762, 29979, 30031, 30103, 30131, 30396, 30441, 30481, 30712,
        30734, 30802, 31144, 31213, 31281, 31384-5, 31550, 31581, 32057, 32261, 32284,
        32300, 32307

Cultivar differences, chloroplast 28586, 28656, 28668, 28840, 29470, 30933

Cultivar differences, ecosystem and plant productivity
        28552, 28620, 28624, 28785, 28941, 29060, 29213, 29410, 29427, 29475, 29492,
        29633, 29992, 30031, 30113, 30162, 30221, 30461, 30555, 30648, 30663, 30677,
        30712, 30734, 30802, 31013, 31021, 31180, 31213, 31273, 31281, 31359, 31367,
        31463, 31591, 31933-4, 32057, 32059, 32062, 32171, 32243, 32261, 32431, 32434,
        32505, 32530, 32545

Cultivar differences, electron transport chain 29711, 30103, 30964, 31489

Cultivar differences, gas exchange
        28629, 28632, 28785, 29004-5, 29251, 29470, 29711, 29874, 30047, 30103, 30131,
        30221, 30257, 30663, 30705, 30712, 30743, 31281, 31550, 31596, 31978, 32057,
        32163, 32252, 32300, 32545

Cultivar differences, photorespiration 28785

Cultivar differences, resistances to CO_2 and water vapour transfer 28785, 30162,
        30221, 30931, 31596, 32252, 32485

Cultivar differences, respiration 29992, 30804, 31281, 32163

Cultivation of algae and photosynthetic bacteria see Algae and photosynthetic bac-
        teria, cultivation; Algae mass cultures productivity

Cuticular resistance see Resistance, cuticular

Cytochromes
        28525, 28580, 28590, 28599, 28641, 28643, 28668, 28670-1, 28720, 28746, 28751,
        28778, 28812, 28814, 28816, 28836, 28845, 28851, 28867, 28893-5, 28904, 28910,
        28967-8, 29074, 29078, 29180, 29182-3, 29192, 29238, 29245-6, 29263, 29358-9,
        29525-6, 29528-9, 29602, 29640, 29677, 29687, 29785-6, 29807, 29818, 29831,
        29840, 29928, 29942, 29959-60, 30012, 30014, 30050, 30118, 30123, 30134,
        30145, 30208-9, 30245-6, 30260-1, 30303, 30336-7, 30340, 30386, 30398, 30428,
        30443, 30491-3, 30594, 30780, 30839, 30845, 30859, 30878, 30983, 31000, 31007,
        31009, 31025, 31027, 31035, 31053-4, 31109, 31137, 31288, 31365, 31368, 31389,
                                                                              (continued)

Cytochromes (continued)
31455, 31462, 31478, 31511, 31522, 31534, 31556, 31626, 31665, 31736, 31785-6,
31907, 31960, 32005, 32039, 32045, 32072, 32114, 32127, 32168-9, 32250, 32266,
32301, 32315, 32393, 32407, B32410, 32417, 32419, 32421, 32453, 32464-5, 32481,
32537

Cytochromes, methods 29840, 30492

D

Dark CO_2 fixation
29040, 29092-3, 29115, 29151, 29274, 29363, 29495, 29849, 30435, 30695, 30822,
30887, 30896, 31038, 31057, 31421, 31442, 31676, 32032, 32440

Decapitation see Defoliation, decapitation ...

Decapitation, defoliation, ear and root removal, effect on carbon fixation pathways
30088

Decapitation, defoliation, ear and root removal, effect on chlorophyll 29104, 29523,
30253, 31627, 31737

Decapitation, defoliation, ear and root removal, effect on chloroplast 28649

Decapitation, defoliation, ear and root removal, effect on ecosystem and plant productivity
28704, 28873, 29046, 29104, 29381, 29475, 29559, 29771, 29794, 29895, 30513,
30554, 31305, 31399, 31606, 31748, 31763, 31792, 31926, 31934, 32431

Decapitation, defoliation, ear and root removal, effect on electron transport chain
31737

Decapitation, defoliation, ear and root removal, effect on gas exchange
28605, 28649, 29046, 29104, 29133, 30088, 30253, 30516, 31606, 31627, 32201

Decapitation, defoliation, ear and root removal, effect on resistances to CO_2 and
water vapour transfer 31627, 32194

Decapitation, defoliation, ear and root removal, effect on respiration 29104, 30516

Desiccation of tissue see Water saturation deficit

Deuterium oxide, tritium oxide 30649

Development, leaf, plant see Leaf (and plant) development and ageing

Dew see Precipitation, dew ...

Dew measurement see Aerodynamic methods ...

Dew point see Humidity of air ...

Dichroisms determination (methods and results)
28892, 28932-3, 28943, 28960, 28968, 29099, 29227, 29544, 29580, 29666, 29738,
29799, 29872, 29924-5, 29929, 29999, 30181, 30366, 30514, 30632, 30678, 30989,
31011, 31076, 31186, 31445-6, 31622, 31642, 31713, 31805, 31901, 32254, 32282

Differentiation of tissues see Leaf (and plant) development and ageing; Ontogeny...

Diffusion, diffusion coefficient see CO_2 transfer, theory

Diffusion (diffusive) conductance see Resistance ...

Diffusion (diffusive) resistance see Resistance ...

Diurnal changes in algae productivity
     29151, 29226, 29426, 29511, 29582, 29825, 30151, 30676, 30906, 31911, 32091,
     32137

Diurnal changes in biliproteins 31142

Diurnal changes in carbon fixation pathways
     29204, 29816, 30332, 30630, 30665, 30695, 30928, 31062, 32139, 32364

Diurnal changes in carotenoids 31142

Diurnal changes in chlorophyll 29623, 29674, 30381, 31142, 31362-3, 32241

Diurnal changes in ecosystem and plant productivity
     29086, 29289, 30010, 30061, 30084, 30438, 31168, 31198, 31517, 31762, 32182

Diurnal changes in electron transport chain 30630, 32242

Diurnal changes in gas exchange
     28527, 28614, 28732, 28785, 28844, 28971, 29016, 29066, 29102, 29131, 29206,
     29249, 29281, 29306, 29333, 29372-3, B29468, 29605, 29676, 29732, 29816,
     29817, 29820, 29931, 30084, 30249, 30381, 30476, B30511, 30515-6, 30630,
     30695, 30706, 30808, 30822, 30898-9, 30904-5, 30927, 30930, 30943, 30946-7,
     31062, 31068, 31142, 31258, 31261, 31278, 31321, 31362-3, 31515, 31585, 31613-
     -4, 31701, 31925, 31994, 32023, 32057, 32139, 32242, B32318, 32364, B32515

Diurnal changes in resistances to CO_2 and water vapour transfer
     28500, 28732, 28810, 29020, 29162, 29179, 29204, 29232, 29373, 29394-5, 29551,
     29816, 30212, 30249-50, 30392, B30511, 30547, 30583, 30630, 30738,,30818,
     30905, 30930, 31068, 31268, 31321, 31512, 31516, 31536, 31613, 31762, 31888,
     31994, 32051, 32110, 32120, 32139, 32364, 32369

Diurnal changes in respiration
     28527, 29065, 29820, 30476, 30516, 30904, 30947, 31077, 31362, 32023, 32339,
     32364

Drought and chlorophyll 29410

Drought and chloroplast 28529, 30879

Drought and ecosystem and plant productivity 29103, 29410, 30565, 30979, 31015,
     31868, 32010, 32200

Drought and electron transport chain 29537

Drought and gas exchange 29415, 30448, 31258, 31536

Drought and resistances to CO_2 and water vapour transfer 29467, 31536

Drought and respiration 29537, 30448

Dry-matter production, gravimetric determination 28632, 30427, 30663, 30748

E

Ear removal see Defoliation, decapitation, ear and root removal ...

Ecosystem production, primary productivity (terrestrial)(*cf*. also Biomass ...)
    28503, 28516, 28521, 28533, 28544, 28552, 28571, 28624, 28632, 28648, 28689,
    28703-4, 28711, 28788, 28800, 28809, 28813, 28857, 28864, 28899, 28937,
    28941, 28972, 29007, 29009, 29053, 29060, 29067, 29086, 29116, 29123, 29129-
    -30, 29158, 29163, 29166, 29168, 29201, 29213, 29221-2, 29236, 29254, 29290,
    29334, 29362, 29373, 29377, 29380, 29413, 29419, 29448, B29468, 29492, 29496,
    29501, 29506, 29515, 29524, 29550, 29559, 29565-6, 29575-6, 29633, 29656,
    29685, 29722, 29789-90, 29797, 29806, 29811, 29883, 29887, 29967, 29975-6,
    29984, 30029, 30044, 30049, 30070-1, 30096, 30113, 30120, 30162, 30195,
    30290, 30376, 30387, 30400, B30405, 30412, 30427, 30438, B30477, 30483,
    30489, 30505-6, 30524, 30532-3, 30554, 30565, 30573, 30674, 30689, 30691,
    30734, 30861, 30874, 30916, 30929, 30945, 30947, 30953, 30967, 30979, 30994,
    31001, 31021, 31032, 31131, 31156, 31180, 31201, 31208, 31249-50, 31259,
    31265, 31272-3, 31304, 31306-7, 31332, 31367, 31463, 31496, 31500, 31586,
    31595, 31608, 31647, 31670, 31748, 31753, 31777, 31822, 31826, 31840-2, 31848,
    31868, 31875, 31916, 31922, 31926, 31935, 31952, 31996, 32010, 32012, 32033,
    32046, 32052, 32059, 32133, 32151, 32186-7, 32210-1, 32243, 32296, 32322,
    32325, 32394, 32403, 32415, 32431, 32434, 32436, 32504-5, 32545

Ecotypes, geographical types, and carotenoids 30465, 30800, 32271

Ecotypes, geographical types, and chlorophyll 28498, 30465, 30800, 32271

Ecotypes, geographical types, and chloroplast 28498

Ecotypes, geographical types, and ecosystem and plant productivity 30044

Ecotypes, geographical types, and gas exchange 29207, 29409, 29615, 29721, 30271,
    30556, 30684-6, 31861, 31863-6, 31946, 32197, B32515

Ecotypes, geographical types, and photorespiration 30271, 30684, 30686

Ecotypes, geographical types, and resistances to CO_2 and water vapour transfer
    31861, 31864

Ecotypes, geographical types, and respiration 29409, 30271

Efficiency, photochemical (*cf*. also Irradiance and gas exchange, analysis of light
    curves)
    28783, 29333, 29675, 29790, 29975, 30047, 30722, 30916, 31359, 31470, 32175

Electron paramagnetic resonance see EPR, NMR

Electron spin resonance see EPR, NMR

Electron transport chain activity
    28511, 28589, 28695, 28990, 30415, 30630, 30765, 31051, 31338, 31521, 31738,
    31918-9, 32097, 32161, 32315

Electron transport chain components see Cytochromes; Ferredoxin ...; Ferredoxin-
    NADP reductase; NADP ...; O_2 evolution ...; Photosystems; Plastocyanin;
    Quinones

Electron transport chain, general aspects see General aspects on carbon fixation...

Electron transport chain localization in thylakoid
    28574, 28581, 28587, 28590, 28744, 28751, 29001, 29114, 29182-3, 29309-10,
    29352, 29438, 29738, 29837, 29870, 29932, 29958, 30000, 30125-6, 30156,
    30178-9, 30245, 30361, 30411, 30525, 30771, 30788, 30958, 30969, 31027,
    31030, 31097, 31172, 31186, 31287, 31335, 31410-1, 31413-4, 31426, 31483,
    31485-6, 31549, 31566, 31610, 31642, 31665, 31718, 31740, 31742, 31810,
    31883, 31901, 32077, 32157-8, 32160-1, 32204, 32250, 32260, 32266, 32282,
    32285, 32386, 32443-4, 32471, 32487, 32550

Electron transport chain model   see   Model ...

Electron transport chain, serological analysis
        28778, 28851-2, 28886, 29044, 29145, 29374, 29474, 29626, 30042, 30361-2,
        31414, 31590, 31643, 31665, 31754, 32069

Emerson effect, Blinks effect 29340, 30892, 32533-4

Energy utilization, plant and ecosystem
        28544, 28837-8, 28858, 29018, 29057, 29193, 29332, 29373, 29530, 29797, 29812,
        30113, 30375, 30378, 30380, 30505-6, 30512, 30524, 30750, 30917, 30921, 30946,
        30976, 31040, 31130-1, 31283, 31358-9, 31411, 31586, 31817, 31849, 32006,
        32175

Enzymes and carotenoids 30073, 32037-8

Enzymes and chlorophyll 28619, 29477, 32009

Enzymes and chloroplast (chromatophore)
        28619, 28727, 29393, 29430, 29634-5, 29745, 29808, 29908, 30267-8, 31050,
        31294, 31380, 31639

Enzymes and electron transport chain 29417, 30535, 30859, 31007, 31389, 31483,
        31942

Enzymes and gas exchange 31730

Enzymes of carbon fixation pathways other than RuBPC, PEPC and malic enzyme
        28555, 28557-8, 28560, 28609, 28613, 28662-3, 28775, 28818, 28853-4, 28909,
        28911, 28989, 29061-2, 29075, 29085, 29098, 29117, 29132, 29215, 29231, 29267,
        29271, 29385, 29478, 29534, 29579, 29669, 29772, 29777, 29792, 29807-8, 29833,
        29860-1, 29863, 29912, 29945, 29990, 30125, 30190-1, 30204, 30232, 30234-5,
        30266-7, 30276, 30350, 30538, 30630, 30707, 30893-4, 30897, 30936, 30951-2,
        31042, 31062, 31089, 31164, 31227, 31250-1, 31300, 31337, 31343, 31433, 31450-
        -2, 31466, 31472, 31519, 31530, 31576, 31698-9, 31973, 31977, 32176, 32223,
        32290-1, 32335, 32337-8, 32366, 32429, 32440, 32454-5

Enzymes of carbon fixation pathways other than RuBPC, PEPC, malic enzyme, malate de-
        hydrogenase, methods 28557, 28663, 29117, 29215, 29863, 30268

Enzymes of chlorophyll synthesis and degradation see Chlorophyll, enzymes ...;
        Chlorophyllase

Enzymes of electron transport chain, methods 31730

Enzymes of glycollate cycle, methods 32547

Enzymes of photorespiration see Photorespiration enzymes; Enzymes of glycollate
        cycle, methods

Epidermis see Leaf epidermis ...

EPR, NMR (methods and results)
        28570, 28572, 28667-8, 28720, 28751, 28769, 28823, 28832-3, 28935-6, 28946,
        28956, 29038, 29079-80, 29124, 29143, 29310, 29358, 29438-42, 29452-4, 29555,
        29689, 29715, 29740-1, 29768, 29779, 29784, 29796, 29799, 29830, 29880,
        29969-72, 29991, 30138, 30188, 30252, 30254-5, 30260-1, 30302, 30386, 30436,
        30444, 30657, 30717, 30812-3, 30926, 31030, 31033, 31048, 31063, 31097,
        31099, 31122, 31133, 31253, 31297, 31369-71, 31390, 31401, 31410, 31446,
        31511, 31565-6, 31605, 31635, 31638, 31649, 31713, 31785-6, 31809, 31857-8,
        31902, 31976, 32072, 32114, 32123, 32127, 32134, 32148, 32208, 32214, 32228,
        32301, 32370, 32468, 32474, 32476, 32481, 32495

Ethylene see Gases, organic ...

Evolution   see   Phylogeny ...

Excitation resistance   see   Resistance, carboxylation and excitation

Exhaust gases   see   Pollution of air ...

Exposure chamber   see   Assimilation chamber

Extension growth, leaf dimensions
          28688, 28986, 29007, 29042, 29072, 29103, 29136, 29204, 29236, 29351, 29559,
          29794, 29822, 29826, 29871, 29954, 29968, 29984, 30257, 30283, 30371, 30427,
          30460, 30480, 30699, 30711, 30739, 30764, 30844, 30861, 30916, 30979, 31004,
          31013, 31021, 31041, 31114, 31164, 31178, 31261, 31284, 31496, 31563, 31608,
          31706, 31761, 31827, 31840-1, 31853, 31870, 31887, 31925, 31947, 32052, 32170,
          32175, 32219, 32341, 32412

Extraction of pigments   see   Pigments ...

Exudation of photosynthates   see   Photosynthate translocation ...

F

Fatty acids   see   Lipids, fatty acids ...

Ferredoxin, ferredoxin-NADP reductase, methods   28543, 29271, 29718, 29872, 30042,
          30304, 31100, 31783, 32069, 32333

Ferredoxin, flavoproteins, rubredoxin
          28539, 28543, 28592-4, 28599, 28667-8, 28670, 28720, 28750, 28765, 28778,
          28816, 28832, 28851-2, 28859, 28868, 28888, 29021, 29037-8, 29070, 29143,
          29180, 29245-6, 29271, 29310, 29412, 29435,   29438-9,   29442, 29525,
          29531, 29570, 29686-9, 29796, 29798-9, 29851-4, 29880, 29998, 30034, 30042-3,
          30059, 30118, 30254, 30284, 30304, 30307-9, 30336, 30419, 30443, 30491,
          30493, 30509, 30716, 30746, 30774, 30777, 30853, 30963, 31025, 31100, 31151-3,
          31196, 31253, 31274, 31337, 31418, 31437, 31462, 31473, 31698-9, 31783, 31786,
          31844, 31874, 31918, 31976, 32039, 32069, 32072, 32333, 32454-5, 32464, 32495

Ferredoxin-NADP reductase, pteridines
          28778, 28816, 28851-2, 28859, 29037, 29043, 29143, 29246, 29397, 29531, 29718,
          30303, 30491, 30535, 30777, 30983, 31003, 31152-3, 31874, 31964, 32072

Flashes of light   see   Irradiation, flash

Flavoproteins   see   Ferredoxin ...

Flooding and chlorophyll   29136, 29338, 30857

Flooding and ecosystem and plant productivity
          28873, 29136, 30351-2, 30857, 31234, 31270, 32010, 32313, 32504

Flooding and electron transport chain   31234

Flooding and gas exchange   30351-2

Flooding and resistances to CO_2 and water vapour transfer   31270

Flooding and respiration   31234

Fluorescence, methods   29544, 30719, 31691, 31906

Fluorine   see   Pollution of air ...

Foliage   see   Canopy ...

Fraction I protein see Ribulose 1,5-bisphosphate carboxylase

Frost (hardiness) see Temperature, low ...

Fungus diseases see Phytopathological effects ...

Fusicoccin see Growth regulators ...

G

Gas exchange, general aspects see General aspects on CO_2 exchange ...

Gas exchange in algae
        28507, 28541, 28573, 28579, 28687, 28728, 28737, 28742, 28763-4, 28795, 28815,
        28846, 28879-80, 28889, 28900, 28921, 28939-40, 28979, 28981, 29118, 29140,
        29160, 29167, 29177, 29191, 29196, 29239, 29242, 29252-3, 29303, 29337, 29341,
        29378, 29416, 29429, 29451, 29500, 29503-4, 29527, 29582, 29601, 29629, 29649,
        29653, 29661, 29677, 29734, 29775, 29809, 29821, 29842, 29873, 29893, 29910,
        29928, 29945, 30023, 30026, 30124, 30137, 30147, 30151, 30172, 30185, 30193,
        30220, 30223, 30323, 30381, 30401, 30434-5, 30495, 30525, 30536, 30575, 30587,
        30622, 30624, 30630, 30757, 30768, 30845, 30848, 30944, 30962, 30968, 31051,
        31062, 31084, 31136, 31142, 31145, 31206, 31218, 31278, 31282, 31326-8, 31362-
        -4, 31396, 31409, 31429-30, 31438, 31456, 31460, 31507, 31514, 31521, 31597,
        31651, 31676, 31679, 31681, 31700, 31704, 31723, 31757, 31772, 31835, 31904,
        31960, 31964, 31969, 32031-2, 32071, 32136, 32166, 32180, 32206-7, 32242,
        32344-5, 32356, 32441, 32448, 32470, 32517

Gas exchange in isolated chloroplasts see Chloroplast, isolated, gas exchange by

Gas exchange in photosynthetic bacteria see Photosynthetic bacteria, gas exchange in

Gas exchange, model see Model ...

Gas exchange of organs other than leaf 29509, 29993, 30537, 31592, 32027, 32057

Gases, organic, and chlorophyll 29219, 29370, 29433, 31781, 31852

Gases, organic, and electron transport chain 31852

Gases, organic, and gas exchange 29219, 29239, 30502

Gasometric methods, generally 28632, B30432

Gasometric system, closed and semiclosed 28571, 31515, 31600, 31968, 32383

Gasometric system, conditioning of air 28629, 31968

Gasometric system, open 28614, 28843, 29065, 29190, 29575, B30432, 30545, 30723,
        31864, 31946, 31968, 32383

General aspects on carbon fixation pathways and electron transport chain; books
        B28677, B29006, B29767, B32159

General aspects on CO_2 exchange, photorespiration and productivity; books
        B28531, B28545, B29023, B29063, B29468, B29707, B29724, B29763, B29767, B30198,
        B30281, B30325, B30405, B30432, B30477, B 30511, B31140, B31990, B32003-4,
        B32056, B32150, B32209, B32318, B32515, B32528

Genetics cf. also Mutagens ...; Mutants ...

Genetics and ecosystem and plant productivity
        28632, 29003, 29067, 29190, 29501, 29652, 30466, 30700, 31004, 31040, 31180,
        31252, 31265, 31356, 31579, 31752, 32170, 32261, 32292, 32546

Genetics of carbon fixation pathways 29147, 29624-5, 29826, 30024, 30472-3, 30566, 30990, 31008, 31433, 31476, 32397

Genetics of carotenoids 29422, 29501, 29898, 31579, 32190, 32261, 32267, 32284

Genetics of chlorophyll 28532, 28606, 29003, 29422, 29501, 29898, 30346, 30472, 30990, 31436, 31579, 31587, 31654, 32190, 32261, 32267, 32284, 32397

Genetics of chloroplast (chromatophore) 28497, 28632, 28849, 28910, 29624, 29839, 30345, 30472, 30673, 30682, 31055, 31975, 32002, 32221, 32336, 32397, 32473

Genetics of electron transport chain 28632, 28758, 28814, 29356, 30042-3, 30415, 30472, 31055, 31433, 31937, 31957, 32267

Genetics of gas exchange 28606, 28632, 29003, 29008, 29190, 29501, 29826, 29848, 30047, 30090, 30466, 30566, 31433, 31587, 31752, 32183, 32292, 32294

Genetics of photorespiration 31008, 32529

Genetics of resistances to CO_2 and water vapour transfer 28606, 30547, 31512

Genetics of respiration 29008, 31587

Glycollate metabolism see Photorespiration ...

Glyoxysome see Peroxisome ...

Granum see Thylakoid ...

Gravimetric determination of photosynthesis see Dry-matter production ...

Gross photosynthetic rate
28613, 28617, 28687, 28972, 29054, 29311, 29336, 29340, 29564, 29613, 29820, 29842, 29848, 29887, 29967, 30011, 30081, 30381, 30476, 30500, 30622, 30643, 30891, 31430, 31592, 31633, 31759, 31941, 32046, 32104, 32118, 32165, 32415

Growth analysis, methods
28534, 29206, 29213, 29811, 30053, 30071, B30432, 30689, 30847, 30876, 31070, 31224, 31500, 31947, 32191, 32263

Growth analysis, net assimilation rate, leaf area ratio, relative growth rate
28510, 28516, 28544, 28703, 28757, 28809, 28838, 28914, 28919, 28926, 28941, 29009, 29036, 29046, 29060, 29103, 29120, 29131, 29136, 29158, 29164, 29212, 29221-2, 29233, 29372, 29380, 29427, 29550, 29559, 29652, 29794, 29823, 29975, 29984-5, 30011, 30029, 30031, 30049, 30053, 30058, 30074, 30077, 30167, 30195, 30203, 30221, 30257, 30353, 30376, 30400, 30438, 30445, 30461, 30463, 30466, 30483, 30523, 30565, 30674, 30677, 30699, 30734, 30820, 30824, 30891, 30904, 30932, 30945, 30948, 30957, 30976, 31019, 31069, 31156, 31178, 31180, 31201, 31224, 31233, 31244, 31265, 31279, 31348, 31356-9, 31392, 31400, 31416, 31439, 31494-5, 31500, 31509, 31586, 31591, 31670, 31750, 31825, 31842, 31895, 32006, 32027, 32047, 32115, 32118-9, 32170-1, 32182, 32322, 32341, 32403, 32436, 32538, 32542, 32546

Growth analysis, specific leaf area, leaf area index, leaf area duration
28502, 28505, 28533, 28552, 28566, 28622, 28685, 28757, 28785, 28788, 28809, 28864, 28899, 28941, 28989, 29003, 29007, 29009, 29022, 29060, 29064, 29068, 29086, 29103, 29120, 29162, 29164, 29190, 29221, 29233, 29269, 29294, 29335, 29362, 29372, 29427, 29443, 29475, 29492, 29496, 29550, 29559, 29565, 29574-6, 29652, 29656, B29707, 29794, 29881, 29886-7, 29923, 29967, 29975, 29984-5, 30011, 30031, 30057-8, 30070-1, 30131, 30163, 30167, 30195, 30203, 30221, 30290, 30353, 30376, 30438, 30466, 30483, 30523, 30554, 30565, 30581, 30628, 30633, 30663, 30674, 30677, 30699-700, 30711, 30733, 30802, 30945-6, 31019, 31114, 31156, 31178, 31180, 31197, 31199, 31201, 31224, 31230, 31233, 31242, 31250-2, 31265, 31348, 31356-8, 31392, 31463, 31494-5, 31500, 31509, 31517,

(continued)

Growth analysis, specific leaf area, leaf area index, leaf area duration (continued)
    31527, 31539, 31586, 31595, 31633, 31670, 31706, 31760-1, 31804, 31822, 31825,
    31840, 31875, 31895, 31925, 32006, 32012, 32046, 32096, 32111, 32115, 32120,
    32165, 32182, 32191, 32257, 32322, 32325, 32391, 32436, 32504, 32538, 32542

Growth regulators and algae productivity 28537

Growth regulators and carbon fixation pathways 28662, 28981, 29092, 29567, 29800,
    29850, 30817, 31096, 31977, 32179

Growth regulators and carotenoids 28752, 28995, 29370, 30295-6, 30595, 31014, 31144,
    31574

Growth regulators and chlorophyll
    28662, 28752, 28881, 28995, 29370, 29433, 29450, 29523, 29982, 30015, 30056,
    30173, 30240, 30296, 30431, 30992, 31014, 31096, 31144, 31601, 31855, 32001,
    32106, 32202, 32313, 32351, 32451

Growth regulators and chloroplast (chromatophore)
    28974, 28981, 29305, 29360, 29761, 30295, 30300, 30467, 30595, 30646, 30992,
    31424-5, 31541, 31768, 32240

Growth regulators and ecosystem and plant productivity
    28618, 28662, 28873, 29046, 29848, 30427, 30920, 31014-5, 31234, 31399, 31841,
    31940-1, 32201, 32218, 32313, 32516

Growth regulators and electron transport chain
    28730, 28887, 28995, 29107, 29305, 29360, 29382, 30103, 30297, 30577, 30594,
    31087, 31379, 32179, 32309, 32483-4

Growth regulators and gas exchange
    28981, 29046, 29092-3, 29296, 29350, 29363, 29486, 29800, 29849, 30103, 30427,
    30761, 30817, 30920, 30992, 31014, 31022, 31038, 31087, 31940-1, 32049, 32201,
    32483-4

Growth regulators and photorespiration 31088, 31685, 31768, 31941

Growth regulators and resistances to CO_2 and water vapour transfer 28509, 28821,
    29486, 30507-8, 30818, 31087, 32352

Growth regulators and respiration 30296, 31768

"Growth" respiration see Respiration, "growth"

H

H_2 evolution, photoreduction
    28506, 28750, 28783, 28814-5, 28889, 29135, 29211, 29444, 29570, 29683-4,
    29734-6, 29796, 29940-1, 29974, 30265, 30308-9, 30399, 30416-7, 30422-3,
    30470, 30502, 30617, 30762, 30938, 31176-7, 31274, 31415, 31437, 31473, 31724,
    31730, 32082, 32479

H_2 isotopes see Deuterium ...

H^+ transport in chloroplast see Chemiosmotic hypothesis

Halobacterium photosynthesis
    28722, 28826, 28831, 28848, 28943, 29028-9, 29034, 29084, 29095, 29099-100,
    29225, 29227-8, 29387-9, 29420, 29431, 29507, 29606-8, 29632, 29659-60, 29692,
    29742, 29797, 29847, 29854, 29900, 29926, 29929, 29935, 30055, 30063-5, 30192,
                                                                            (continued)

Halobacterium photosynthesis (continued)
    30239, 30258, 30284, 30383-4, 30390-1, 30408, 30471, 30519-21, 30614-5, 30632,
    30649, 30679, 30785, 31026, 31029, 31036, 31110-1, 31150, 31167, 31173-5,
    31286, 31309-10, 31458, 31491, 31497, 31773, 31838, 31860, 31997, 32167,
    32330, 32438, 32507

Hatch-Slack cycle see C₄ pathway ...

Herbicides see Pesticides, herbicides ...

Heterogeneity of leaf blade (organ) and carbon fixation pathways 28957, 29669, 29865,
    30233, 31266, 31578, 32462

Heterogeneity of leaf blade (organ) and carotenoids 30313

Heterogeneity of leaf blade (organ) and chlorophyll
    28529, 28655-6, 28939, 29534, 29594, 29669, 29833, 29876, 29903, 30562, 31243,
    31578, 31735

Heterogeneity of leaf blade (organ) and chloroplast
    28655, 28909, 29876, 30016, 30313, 30557, 30562, 30690, 30747, 30826, 30871,
    31243, 31325, 31568, 32061

Heterogeneity of leaf blade (organ) and electron transport chain 28529, 28655, 29534,
    30558, 30866, 31418, 31551

Heterogeneity of leaf blade (organ) and gas exchange
    28656, 28939-40, 29421, 29594, 29775, 30232, 30562, 30748, 31266, 31449, 31451-
    -2, 31734, 32226

Heterogeneity of leaf blade (organ) and photorespiration 31266, 31451

Heterogeneity of leaf blade (organ) and resistances to CO_2 and water vapour transfer
    30991, 32226

Heterogeneity of leaf blade (organ) and respiration 29421

Heterotrophy see Carbon metabolism types ...

Hill reaction see Photosystem 2 activity ...

Hill reaction, methods see Photosystem 2 activity, methods

Humidity of air and chlorophyll 29391, 31991, 32232

Humidity of air and ecosystem and plant productivity 30820, 30917, 31496, 32033

Humidita of air and gas exchange 29020, 29102, 29294, 29409, 29572, 29793, 30898,
    31585

Humidity of air and resistances to CO_2 and water vapour transfer
    28732, 29020, 29130, 29319, 29394-5, 29467, 29496, 29793, 29881, 30069, 30371,
    30387, 30392, B30477, B30511, 30643, 30738, 31268, 31762, 31769-71, 31867,
    31888, 31956, 32051, 32110, 32369

Humidity of air and respiration B30477

Humidity of air, methods (*cf.* also infra-red gas analyser for water vapour)
    B30325, B30432, 30860, 31968, 32015

Hydration level of leaf and biliproteins 30481

Hydration level of leaf and carbon fixation pathways 28717-8, 29204, 29817, 30174,
    30272, 30332, 30543-4, 30665, 30823, 31165, 31798, 32378

Hydration level of leaf and carotenoids 31851

Hydration level of leaf and chlorophyll 28529, 28718, 28789, 29554, 29727, 30272, 30481, 30825, 31052, 31581, 31851, 31903

Hydration level of leaf and chloroplast 29360, 30111, 30879, 31072

Hydration level of leaf and ecosystem and plant productivity 28566, 28809, 28986, 30272, 30310, 30524, 30739, 30808, 30820-1, 30903

Hydration level of leaf and electron transport chain 28529, 28717-8, 28789, 29360, 30111, 30386, 30892, 30919, 31581

Hydration level of leaf and gas exchange 28500, 28504, 28688, 28717-8, 28731, 28809, 28919, 29020, 29071, 29122, 29156, 29243, 29294-5, 29367, B29468, 29721, 29766, 29769, 29843, 29931, 30168, 30249, 30270, 30272, 30285-6, 30310, B30511, 30515, 30523, 30542-3, 30608, 30633, 30723, 30808, 30821, 30825, 30861, 30892, 30903, 30919, 30944, 31018, 31065, 31090, 31165, 31258, 31293, 31321, 31536, 31899, 31983, 32197, 32252, 32378, 32541

Hydration level of leaf and photorespiration 30270, 30542-4, 31018, 31293

Hydration level of leaf and resistances to CO_2 and water vapour transfer 28500-1, 28504, 28688, 28717-8, 28731, 28987-8, 29020, 29162, 29203, 29394-5, 29467, 29725, 29766, 29881, 30069, 30163, 30249, 30270, B30511, 30738-9, 30903, 30919, 30931, 31165, 31770, 31899, 32051, 32110-1, 32252, 32352, 32369, 32485, 32539

Hydration level of leaf and respiration 29843, 30270, 30515, 30808, 30825, 31018, 31065, 31293, 31882

Hydrogen see H_2

Hydrogenase see O_2 evolution mechanism and kinetics; H_2 evolution ...

Hygrometer see Humidity of air, determination; Infra-red gas analyser for water vapour

I

Ideotype see Model ...

Immobilization of chloroplasts and photosynthetic systems see Photosystems stabilization ...

Induction phenomena see Transient phenomena ...

Infra-red analyser for CO_2 28614, 29967, 29993, B30432, 30447, 30723, 30763, 30860, 31020, 31711, 31968, 32015, 32383.

Infra-red radiation, effect on photosynthetic parameters see Irradiance, spectral composition ...; Temperature, high ...

Inhibitors of electron transport chain (cf. also Pesticides ...; Antibiotics ...) 28519, 28524-5, 28541, 28560, 28562, 28571, 28588, 28592, 28594, 28604, 28633, 28678, 28681-2, 28694-6, 28705, 28738-9, 28742, 28745, 28766, 28769, 28772, 28793, 28812, 28814-5, 28827, 28847, 28867, 28875, 28879, 28884, 28886, 28893, 28903, 28935-6, 29017, 29032, 29059, 29097, 29101, 29114, 29143, 29183, 29189, 29191, 29258-9, 29261, 29270, 29304-5, 29309, 29328, 29343, 29353,
(continued)

Inhibitors of electron transport chain (continued)
        29356, 29418, 29435-6, 29471-2, 29504, 29517, 29525, 29528, 29531, 29533,
        29537, 29601, 29634-5, 29637, 29640, 29647, 29649, 29662, 29665, 29684, 29697,
        29715, 29731, 29735, 29737, 29750, 29773, 29784-6, 29831, 29856, 29889, 29904,
        29906, 29939, 30001-2, 30012, 30035, 30108, 30112, 30115, 30118-9, 30127,
        30136, 30154-5, 30158, 30160, 30183, 30188, 30215, 30219-20, 30220, 30223,
        30243, 30245, 30291, 30294, 30324, 30355, 30365, 30403, 30430, 30437, 30468,
        30470, 30502, 30536, 30539, 30589, 30594, 30596, 30599-600, 30635, 30681,
        30701, 30710, 30736, 30749, 30771, 30773, 30781-2, 30803-6, 30834, 30837,
        30841, 30853, 30878, 30970, 30983-5, 30993, 31007, 31042, 31050, 31098, 31109,
        31146, 31169, 31195, 31207, 31211, 31246, 31274, 31291, 31342, 31344, 31409,
        31410, 31416, 31420, 31447, 31449-51, 31456, 31461, 31465-6, 31483, 31503,
        31518, 31521-2, 31524, 31526, 31541, 31551, 31567, 31582, 31597, 31612, 31626,
        31660-1, 31663-4, 31667-8, 31686, 31692, 31717, 31723, 31731, 31733, 31743,
        31746-7, 31772, 31802, 31809, 31831, 31872-3, 31877-8, 31882, 31897-8, 31918,
        31942, 31950, 31957, 31959, 31969, 31998, 32036, 32039-40, 32042, 32082,
        32092, 32135, 32141, 32161, 32178, 32207, 32233-5, 32247, 32250, 32256, 32262,
        32316, 32357-8, 32384, 32414, 32418, 32432, 32451, 32497-9, 32512, 32536

Insertion   see   Ontogeny ...

Intercellular spaces, CO_2 concentration inside
        28614, 29054, 29130, 29486, 30139, 30455, 30499, 30643, 30699, 31067, 31214,
        31232, 31256, 31470

Intermediates of carbon fixation pathways, methods
        29113, 30564, 31313, 31452, 32075

Intracellular resistance   see   Resistance, intracellular (mesophyll)

Ionizing radiation (gamma, X, cosmic, *etc.*) and chlorophyll   29499, 30404, 30548

Ionizing radiation (gamma, X, cosmic, *etc.*) and chloroplast (chromatophore)   31333

Ionizing radiation (gamma, X, cosmic, *etc.*) and ecosystem and plant productivity
        29175

Ionizing radiation (gamma, X, cosmic, *etc.*) and electron transport chain   30548

Ionizing radiation (gamma, X, cosmic, *etc.*) and gas exchange   28702, 29175, 32219

Irradiance, compensation   see   Compensation irradiance

Irradiance (PhAR) and algae productivity
        28553, 28740, 28797, 29242, 29418, 29582, 29778, 29809, 30052, 30075, 30147-8,
        30151, 30281, 30356, 30605-6, 30676, 30724-5, 31260, 31316, 31428, 31714,
        32091, 32273, 32356, 32494

Irradiance (PhAR) and biliproteins   29378, 30531, 31499, 31584, 31757

Irradiance (PhAR) and carbon fixation pathways
        28775, 28835, 29015, 29267, 29313, 29316, 29462, 29533, 29772, 29808, 29889,
        29922, 29951, 30333, 30887, 31089, 31109, 31620, 31910

Irradiance (PhAR) and carotenoids
        28713, 28716, 28947, 29121, 29456, 29573, 30230, 30313, 30395, 30581, 31188,
        31236, 31276, 31330, 31349, 31430, 31673, 31757, 31810, 31831, 31851, 32142-3,
        32424

Irradiance (PhAR) and chlorophyll
        28637-8, 28713, 28719, 28752, 28775, 28817, 28828, 28835, 28866, 28896, 28902,
        28925, 28947, 28989, 28990, 29067, 29077, 29107, 29121, 29139, 29155, 29163,
                                                                        (continued)

                                                                  (continued)

Irradiance (PhAR) and resistances ... (continued)
     29373, 29394-5, 29425, 29793, 29795, 30078, 30212, 30249, 30457-8, B30511,
     30643, 30738, 30808, 30903, 30991, 31193, 31268-9, 31627, 31885, 31888, 31956,
     32051, 32194, 32369

Irradiance (PhAR) and respiration 28685, 31077, 31470, 32115, 32175

Irradiance (PhAR, total) measurement
     28735, 28806, 30071, 30129, 30315, B30325, 30407, B30432, 30553, 30634, 30882,
     30956, 31501, 31547, 31886, 32096, 32371

Irradiance, flash, and carotenoids 28651, 29266, 29760, 31219

Irradiance, flash, and chlorophyll
     28651, 28883, 29073, 29266, 29278, 29713, 29760, 29997, 30154, 30157, 30159,
     30182-3, 30229, 30240, 30718, 30740, 30779, 30848-9, 30915, 31043, 31219,
     31488, 31646, 31776, 31857, 31981, 32132, 32238, 32456

Irradiance, flash, and chloroplast (chromatophore) 28590, 28651, 28883, 29765,
     30198

Irradiance, flash, and electron transport chain
     28621, 28672, 28798, 28833, 28884-5, 28893, 28895, 28935-6, 29266, 29275,
     29278-9, 29307, 29309, 29314, 29359, 29436, 29539-41, 29583, 29668, 29686,
     29713-5, 29719, 29733, 29736, 29760, 29784, 29786, 29838, 29995, 30004, 30155,
     30180, 30246, 30338, 30355, 30365, 30411, 30708-9, 30718, 30779, 30782, 30848,
     31049, 31123, 31287, 31342, 31410, 31412, 31483, 31484-5, 31487, 31551, 31617,
     31629, 31646, 31776, 31780, 31818, 31858, 31982, 32239, 32270, 32279, 32299,
     32321, 32331-2, 32444, 32453, 32456, 32467, 32475-6, 32487

Irradiance, flash, and gas exchange 28550, 28814, 29675, 29734-6, 30848, 30892;
     32166

Irradiance, spectral composition and algae productivity 30226, 32446

Irradiance, spectral composition and biliproteins 29866, 30552, 32053, 32288

Irradiance, spectral composition and carbon fixation pathways
     28883, 29567-8, 29603, 30893-5, 31417, 31620, 32236, 32317, B32515

Irradiance, spectral composition and carotenoids
     28713, 29208, 29516, 30298, 30446, 30595, 31937, 32288

Irradiance, spectral composition and chlorophyll
     28589, 28713, 28883, 28933, 28942, 28958, 29329, 29386, 29392, 29603, 29623,
     29743, 29894, 30080, 30101, 30137, 30199, 30201, 30298, 30318-9, 30330, 30389,
     30552, 30862, 30893-4, 31044, 31064, 31254, 31324, 31429, 31719, 31744, 31829,
     31856, 31937, 32001, 32053, 32084, 32086, 32155, 32237, 32288, 32298-9, 32385,
     32433

Irradiance, spectral composition and chloroplast (chromatophore)
     28828, 28883, 28918, 28942, 29260, 29458, 29553, 29813, 30089, 30137, 30199,
     30298, 30329, 30595, 30673, 30826, 30893-5, 31045, 31106, 31284, 31335, 32288,
     32447, 32551, 32553

Irradiance, spectral composition and ecosystem and plant productivity
     28713, 28918, 29049, 29994, 30862, 31001-2, 31158-9, 31417, 31547, 31647,
     31936, 32187, 32218

Irradiance, spectral composition and electron transport chain
     28525, 28593, 28742, 28877, 28918, 28922, 29107, 29521, 29556-8, 29603,
     29805, 29955, 30089, 30179, 30324, 30329, 30401-2, 30551, 30586, 30595,
     30826, 30893, 31045, 31146, 31298, 31324, 31335, 31412, 31485, 31505, 31562,
     31564, 31957, 32084, 32086, 32134, 32236, 32246, 32248-9, 32305, 32315, 32359,
     32385, 32498, 32514

Irradiance, spectral composition and gas exchange
    28737, 28742, 28844, 28918, 28922, 29340, 29378, B29468, 29601, 29661, 29913,
    30137, 30323, 30525, 30862, 31001-2, 31084, 31157-8, 31280, 31620, 31712,
    31853, 31873, 31984, 32049, 32063, 32086, 32155, 32207, 32236, 32317, 32385,
    B32410, 32432

Irradiance, spectral composition and photorespiration 31904

Irradiation, illumination equipment and systems 29400, 29613, 30129, 32133

Irrigation and carbon fixation pathways 29816

Irrigation and chlorophyll 29443, 30217, 31855, 32008-9

Irrigation and ecosystem and plant productivity
    28864, 29203, 29223, 29269, 29334, 29411, 29566, 29883, 29965, 30041, 30071,
    30195, 30217, 30377-8, 30628, 30648, 30699, 30917, 30967, 31244, 31273,
    31321, 31496, 31539, 31558, 31585, 31706, 31710, 31826, 31843, 31868, 31925,
    31934, 32009-10, 32322, 32538

Irrigation and gas exchange 29306, 29443, 29770, 29816, 30377, 30703, 31321, 31585

Irrigation and resistances to CO_2 and water vapour transfer 29551, 29816, 31221

K

Kok effect 29733, 30099, 31520

L

Laboratory for photosynthesis studies, mobile (field laboratory) 31968

Leaf anatomy (*cf.* also Leaf thickness)
    28622, 28625, 28703, 28733, 28748, 28989, 29047, 29052, 29068, 29110-1,
    29181, 29395, 29407, 29470, 29525, 29565, 29572, 29704, 29865, 30078, 30152,
    30273-4, 30283, 30721, 30733, 30764, 30895, 30946, 31066-7, 31168, 31230,
    31232, 31239, 31243, 31252, 31394, 31568, 31610, 31627, 31769, 31801, 31836,
    31887, 32065, 32126, 32188, 32428, 32439

Leaf and plant development and ageing, morphology (*cf.* also Ontogeny ...)
    28791, 29254, 29461, 29795, 30916, 30929, 30979, 31041, 31563, 31840-1

Leaf area duration see Growth analysis, specific leaf area ...

Leaf area index see Growth analysis, specific leaf area ...

Leaf area measurement
    29320, 29408, 29455, 30379, B30432, 30692, 31267, 31729, 31905

Leaf area ratio see Growth analysis, net assimilation rate ...

Leaf chamber see Assimilation chamber

Leaf dimensions see Extension growth, leaf dimensions

Leaf epidermis, anatomy
  28501, 28622, 28703, 28986, 29069, 29232, 29247, 29399, 29402, 29616, 29737,
  30163, 30221, 30507-8, 30675, 30721, 30733, 30931, 31057, 31087, 31193, 31270,
  31442, 31867, 31887, 31899

Leaf epidermis, stomata (*cf.* also Amphistomatous leaf, gas exchange in) 29247,
  29373, 30721, 31793

Leaf life span, plastochron index 29042, 29116, 29443, 29984, 30480, 31750, 31761,
  31934

Leaf morphology
  28718, 28896, 28927, 29007, B29023, 29104, 29251, 29335, 29401, B29468, 29470,
  29496, 29559, 29675, 29794, 29881, 29951, 29984, 30058, 30095, 30168, 30253,
  30283, 30400, 30460, 30480, 30579, 30643, 30692, 30734, 30883, 30896, 30932,
  30946, 31022, 31086, 31114, 31156, 31164, 31178, 31213, 31233, 31251-2, 31267,
  31383, 31439, 31591, 31606, 31737, 31853, 31916, 32059, 32062, 32065, 32126,
  32133, 32175, 32194, 32450

Leaf movements 32329

Leaf optical properties (*cf.* also Carotenoids absorption spectra *in vivo*; Chlorophyll
  absorption spectra *in vivo*)
  28616, 28866, 28976, 29373, 29398-402, 29574, 29616-7, 30722, 30881, B31140,
  31197, 31547, 31801, 32103, 32175, 32186-9, 32552

Leaf resistance see Resistance for water vapour ...; Resistance, stomatal ...

Leaf, sun- and shade leaf see Leaf anatomy

Leaf temperature (methods and results)
  28976, B29023, 29722, 30070, 30343, B30432, 30455, 30457, 30563, 30583, 30643,
  30775, 30943, 31994, 32094, 32120, 32296

Leaf temperature measurement see Leaf temperature (methods and results)

Leaf thickness
  28546, 28622, 28785, 28866, 28989, 29068, 29190, 30187, 30283, 30643, 30733,
  30764, 30861, 30881, 30932, 31067, 31178, 31230, 31251, 31627, 31839, 31904,
  32126, 32188, 32196, 32364

Leaf volume, thickness and internal area measurement 29493

Light see Irradiance ...; Canopy, radiation ...

Lighting system see Irradiation, illumination equipment and systems

Linear dichroism see Dichroisms ...

Lipids, fatty acids, and carbon fixation pathways 29888-9, 30817, 30978, 31676

Lipids, fatty acids, and chlorophyll 28953

Lipids, fatty acids, and chloroplast (chromatophore)
  28650, 28723, 28828, 28840, 29218, 29300, 29318, 29527, 29529, 29627, 29698,
  29765, 29948, 30150, 30177, 30182, 30449, 30562, 30801, 30814, 30826, 30859,
  30900, 30971-2, 30977, 30986, 31028-9, 31109, 31120-1, 31366, 31413-4, 31557,
  31719, 31779, 31814, 31893, 31937, 32029, 32084-6, 32204, 32267

Lipids, fatty acids, and electron transport chain 28693, 30986, 31120, 31813, 31816

Lipids, fatty acids, and photorespiration 30801

Lutein see Carotenoids; Xanthophylls

M

"Maintenance" respiration   see   Respiration, "growth" and "maintenance"

Malate dehydrogenase, methods   see   Malic enzyme, malate dehydrogenase, methods

Malic enzyme
      28613, 28786-7, 28989, 29085, 29098, 29104, 29312-3, 29462, 29772, 29833,
      29846, 29850, 29860, 29862, 30232, 30234, 30332, 30431, 30537-8, 30783-4,
      30823, 30936, 30951, 31034, 31440-1, 31449, 31451-2, 31611, 32364, 32435,
      32477

Malic enzyme, malate dehydrogenase, methods 29846, 32338

Mass culture of algae see Algae mass cultures ...

Maximum photosynthetic rate see Potential photosynthetic rate ...

Mehler reaction see Photosystem 1 activity ...

Membrane transport of CO_2 see CO_2 transfer across membranes

Mesophyll resistance see Resistance, intracellular (mesophyll)

Microbody see Peroxisome

Microelements see Mineral elements (other than N,P,K) ...

Mineral elements (N,P,K) and algae productivity
      28536-7, 28830, 28996, 29226, 29375, 29466, 29653, 29825, 29917, 29952, 30281,
      30572, 31382, 31513, 32089, 32268, 32308, 32404, 32493, 32518

Mineral elements (N,P,K) and biliproteins 30531, 31423, 32102

Mineral elements (N,P,K) and carbon fixation pathways
      29153, 29581, 29891, 29945, 29983, 30037, 30584, 30844, 30952, 30988, 31315,
      32426

Mineral elements (N,P,K) and carotenoids
      29456, 29762, 30400, 30452, 30802, 31276, 31384, 31406, 31927, 32016, 32103,
      32212

Mineral elements (N,P,K) and chlorophyll
      28661, 28762, 29127, 29155, 29163, 29237, 29338, 29457, 29762, 29945, 30350,
      30400, 30452, 30620, 30693, 30802, 30933, 31083, 31139, 31276, 31406, 31495,
      31533, 31602, 31927, B31990, 32103, 32212, 32493

Mineral elements (N,P,K) and chloroplast (chromatophore) 28896, 29110, 29218, 29581,
      29654, 30933, 31056, 31533, 31678, 32102

Mineral elements (N,P,K) and ecosystem and plant productivity
      28521, 28544, 28704, 28711, 28864, 29022, 29053, 29102, 29120, 29153, 29157,
      29163, 29237, 29269, 29325, 29331, 29362, 29380, 29448, 29566, 29790, 29883,
      30074, 30203, 30206, 30290, 30377, 30380, 30387, 30400, 30530, 30565, 30573,
      30642, 30674, 30802, 30840, 30869, 30945, 30957, 30967, 30979, 31041, 31156,
      31180, 31213, 31242, 31249, 31279, 31367, 31439, 31492, 31494-6, 31527, 31535,
      31706, 31748, 31753, 31826, 31875, 31916, 31934, 31996, 32010, 32059, 32076,
      32104, 32115, 32170-1, 32322, 32505, 32530, 32542, 32545

Mineral elements (N,P,K) and electron transport chain
      28896, 30528, 30693, 30909-10, 31456, 31469, 31533, 31564, 31919, 32072,
      32102

Mineral elements (N,P,K) and gas exchange
    28527, 28556, 28783, 28899, 29102, 29155, 29331, B29468, 29726, 29945, 29975,
    30047, 30131, 30243, 30377, 30523, 30642, 30706, 30844, 31156, 31218, 31249,
    31283, 31292, 31320, 31492-4, 31525, 31527, 31570, 31939, 31978, 32104, 32115,
    32177, 32206, 32343, 32545

Mineral elements (N,P,K) and photorespiration
    28899, 31678, 32088, 32104

Mineral elements (N,P,K) and resistances to CO$_2$ and water vapour transfer 28899,
    28914, 29053, 29331, 31156, 31212

Mineral elements (other than N,P,K) and algae productivity 28536-7, 28829, 29937,
    30501, 30827, 30843, 32108, 32268

Mineral elements (other than N,P,K) and biliproteins 29498, 31423

Mineral elements (other than N,P,K) and carbon fixation pathways
    28652, 29863, 29892, 30707, 30856, 30952, 30960, 31264, 31343, 31806, 32244,
    32334, 32426, 32536

Mineral elements (other than N,P,K) and carotenoids 29498, 29947, 30494, 30549,
    31171, 31276, 31403, 32016

Mineral elements (other than N,P,K) and chlorophyll
    28585-6, 28595, 28676, 28678, 28771, 28886, 29127, 29137, 29163, 29233, 29268,
    29498, 29542, 29562, 29635, 29755, 29906, 29947, 30013, 30354, 30494, 30518,
    30541, 30549, 30569, 30576, 30610-2, 30620, 30644, 30827, 30875, 30950, 30971,
    31043, 31116, 31171, 31187, 31209, 31213, 31264, 31276, 31403, 31544, 31657,
    31692, 32072, 32108, 32302, 32380

Mineral elements (other than N,P,K) and chloroplast (chromatophore)
    28518, 28584-6, 29578, 29758, 29814, 29830, 29932, 30428, 30856, 31056, 31209,
    31264, 31375, 31697, 32042, 32079, 32319, 32535

Mineral elements (other than N,P,K) and ecosystem and plant productivity
    28711, 29042, 29157, 29163, 29233, 29448, 29705, 29874, 29921, 30095, 31107,
    31116, 31209, 31535, 31940, 32079

Mineral elements (other than N,P,K) and electron transport chain
    28564, 28676, 28682, 28753, 28766, 28771, 28822, 28884-6, 28934-5, 28991,
    29138, 29244, 29309, 29439, 29578, 29600, 29634, 29713, 29715, 29730-1,
    29755, 29755, 29906, 29932, 29962, 30013-4, 30059, 30085, 30089, 30116,
    30119, 30184, 30498, 30518, 30726-7, 30875, 30909, 30965, 31116, 31124,
    31169, 31264, 31374, 31410, 31564, 31635, 31645, 31776, 31808, 32068, 32079-
    -80, 32244, 32293, 32456, 32475-6, 32535

Mineral elements (other than N,P,K) and gas exchange
    28579, 29042, 29138, 29303, 29429, B29468, 29793, 29874, 30089, 30124, 30130,
    30243, 30490, 30518, 31107, 31116, 31264, 31292, 31940, 32063, 32079-80,
    32177, 32517

Mineral elements (other than N,P,K) and resistances to CO$_2$ and water vapour transfer
    29793, 30654, 31107, 31264

Mixotrophy see Carbon metabolism types ...

Model of aquatic community production
    29311, 29365, 29426, 29582, 29809, 29882, 31316, 31467, 31700, 31749, 32095

Model of canopy photosynthesis, prediction model
    28689, 28706, 28711, 28865, 28899, 28972, 29045, 29086-8, 29103, 29123, 29129,
    29168, 29222, 29290, 29515, 29574, 29605, 29663, B29707, 29886-7, 29967,
    29975, 30011, 30029, 30048-9, 30084, B30281, 30349, 30387, B30511, 30515,
                                                                    (continued)

Mutants, chlorophyll in
28597, 28639, 28816, 28863, 28892, 28942, 28946, 29012, 29025, 29040, 29067,
29094, 29098, 29118, 29240, 29265, 29268, 29283, 29447, 29459, 29538, 29602,
29661, 29706, 29807, 29818, 29875, 29878, 30006-7, 30027, 30050, 30228, 30302,
30318, 30487, 30495-7, 30569-70, 30687-8, 30730, 30859, 30935, 30948, 30962,
31044, 31071, 31118-9, 31139, 31223, 31298, 31349, 31355, 31431, 31575, 31588,
31619, 31623, 31640, 31655, 31666, 31722, 31756, 31772, 31878, 31917, 32034,
32144, 32190, 32222, 32228, 32272, 32326, 32360-1, 32419, 32432, 32481-2

Mutants, chloroplast (chromatophore) in
28554, 28759, 28804-5, 28816, 28850, 28863, 28897, 28909, 28942, 28946, 29538,
29706, 29777, 29807, 29824, 29839, 30006, 30027, 30345, 30569, 30673, 30688,
30730, 31104-5, 31139, 31223, 31336, 31588, 31619, 31721-2, 31820, 31832,
31834, 31876-8, 31917, 31975, 32030, 32221-2, 32276, 32350, 32397

Mutants, ecosystem and plant productivity of 29067, 29565

Mutants, electron transport chain in
28641, 28759, 28815-6, 28904, 29012, 29025, 29192, 29264, 29272, 29282, 29602,
29664, 29807, 29818, 29840, 29960, 29995, 30006, 30027, 30569-70, 30599-600,
30762, 30803, 30859, 30962, 31044, 31118-9, 31298, 31478, 31489, 31564, 31575,
31588, 31665, 31668, 31722, 31877-8, 31907, 32029, 32097, 32228, 32417, 32419,
32464, 32481-2

Mutants, gas exchange in
29040, 29091, 29118, 29268, 29629, 29661, 29706, 29807, 29848, 30228, 30495,
B30511, 30859, 30935-6, 30948, 30962, 31118-9, 31619, 31772, 32221, 32228

Mutants, photorespiration in 30886, 31118-9

Mutants, photosynthetic, isolation and selection 31668

Mutants, resistances to CO_2 and water vapour transfer in 30936, 30991

N

N_2, anaerobic atmosphere, and electron transport chain 31919, 32050

N_2, anaerobic atmosphere, and gas exchange 28994

NAD see NADP, NAD

NADP, NAD 28524-5, 28593, 28739, 28751, 28846, 28851, 28862, 29037, 29051, 29062,
29070, 29397, 29435, 29639, 29792, 29885, 29949-50, 30034, 30093, 30118,
30340, 30370, 30415, 30417, 30442, 30493, 30549, 30551, 30777, 30841, 30845,
30868, 30914, 30963-4, 30984-5, 31016-7, 31025, 31027, 31151, 31352, 31389,
31485, 31503, 31519, 31610, 31872, 31874, 31907, 31918, 32242, 32278, 32290-1,
32315-6, 32316, 32329, 32333

NADP, NAD, methods 30551, 30914, 32278

Net assimilation rate see Growth analysis, net assimilation rate ...

Net photosynthetic rate see Gas exchange ...

Nitrogen see N_2 ...; Mineral elements (N,P,K) ...

NMR see EPR, NMR

Nuclear magnetic resonance see EPR, NMR

Ontogeny of leaf, insertion level, and respiration
        28685, 29563, 29587, 29675, 29700, 30691, 31018, 31896, 32020, 32183

Ontogeny of plant   see also   Seasonal changes ...

Ontogeny of plant and carbon fixation pathways 28929, 28931, 29430, 29792, 29816,
        30537, 30907, 31164, 31227, 31250

Ontogeny of plant and carotenoids 30067, 30302, 30851, 31927

Ontogeny of plant and chlorophyll
        28606, 28674, 28713, 29250, 29501, 30067, 30253, 30302, 30660, 30694, 30712,
        30850-1, 31587, 31927, 32028, 32232

Ontogeny of plant and chloroplast 28840, 29069, 29871, 30110

Ontogeny of plant and ecosystem and plant productivity
        28578, 28713, 28873, 28899, 29022, 29223, 29334, 29338, 29427, 29475, 29509,
        29811, 29819, 29848, 29883, 29953, 29977, 30195, 30445, 30478, 30565, 30627,
        30712, 30733-4, 30932, 30995, 31013, 31233, 31250, 31262, 31279, 31394,
        31539, 31941, 32076, 32117

Ontogeny of plant and electron transport chain 28593, 29070, 29385, 29792

Ontogeny of plant and gas exchange
        28527, 28606, 28674, 28687, 28930, 29223, 29385, 29430, 29509, 29572, 29816,
        29931, 29975, 29993, 30058, 30253, 30537, 30556, 30694, 30712, 30912, 30932,
        30936, 31086, 31156, 31164, 31200, 31227, 31587, 31607, 31695, 31823, 31866,
        32023, 32163, 32226, 32294, 32449, 32545

Ontogeny of plant and photorespiration 28929, 31164

Ontogeny of plant and resistance to CO_2 and water vapour transfer
        28606, 28630, 29247, 29816, 30936, 31156, 31164, 31250, 31607, 31759, 31823

Ontogeny of plant and respiration 28527, 29223, 29509, 29819, 29993, 31587, 32023,
        32163

Optical properties, leaf see Leaf optical properties

Oscillations, short-term fluctuations, steady and non-steady state, in electron trans-
        port chain 30845, 32476

Oscillations, short-term fluctuations, steady and non-steady state, in gas exchange
        28930, 29613, 29766, 30596, 30708, 30723, 30845

Oscillations, short-term fluctuations, steady and non-steady state, in resistances to
        CO_2 and water vapour transfer 28930, 29766, 30738

Osmotically active substances and carbon fixation pathways 28841, 29092, 29817,
        30550, 31319, 32477

Osmotically active substances and chlorophyll 28584, 30215, 30550, 30693

Osmotically active substances and chloroplast (chromatophore) 28584, 29596, 30885,
        31207, 31569, 32553

Osmotically active substances and ecosystem and plant productivity 28510, 29823

Osmotically active substances and electron transport chain 28747, 29730, 30094,
        30693, 30773, 30884, 31293, 31651

Osmotically active substances and gas exchange
        28538, 29092, 30139, 30220, 30797, 30968, 31293, 31651, 31732, 32206

Osmotically active substances and photorespiration   30797, 31293

Osmotically active substances and respiration   31651

Oxygen   see   O$_2$

Ozone   see   Pollution of air ...

P

P680   28547, 28550, 28744, 28823, 28884, 28968, 28998, 29341, 29689, 29713, 29867,
         30184, 30337-8, 30782, 30812, 31390, 31410, 31486-7, 31511, 31666, 32239,
         32249, 32301, 32370, 32393, 32456

P700, P750, P890, etc.
         28515, 28528-9, 28535, 28547, 28550, 28596, 28667-8, 28698-701, 28720, 28744,
         28746, 28751, 28769, 28798, 28814, 28816, 28822, 28824, 28833, 28836, 28851,
         28859, 28878, 28893-4, 28932, 28964-9, 28995, 29073, 29079, 29124, 29126,
         29134, 29148-9, 29161, 29173, 29192, 29263, 29266, 29268, 29272, 29275, 29284,
         29310, 29406, 29438, 29440-2, 29453, 29474, 29497, 29505, 29518-20, 29525,
         29534, 29540, 29666, 29686-9, 29738, 29754-5, 29784-6, 29807, 29818, 29831,
         29867, 29872, 29878-80, 29939, 29949-50, 29955-6, 29958, 29970-1, 30050-1,
         30123, 30135, 30138, 30188-9, 30246, 30254-5, 30321, 30323, 30328, 30385-6,
         30436, 30586-7, 30594, 30626, 30718, 30742, 30780, 30799, 30915, 31009, 31025,
         31027, 31030, 31033, 31046, 31073-4, 31078, 31128, 31138, 31143, 31219, 31225,
         31230, 31297-9, 31370-1, 31374, 31391, 31410, 31426, 31478-81, 31485, 31503,
         31534, 31564-5, 31567, 31598, 31621, 31629-30, 31642, 31725, 31785-6, 31893,
         31900, 31979, 32005, 32045, 32072, 32112-4, 32114, 32127, 32134, 32168-9,
         32198-9, 32225, 32239, 32279-80, 32282-3, 32370, 32417, 32419, 32432, 32441,
         32467-8, 32550

Paramagnetic oxygen analyser   see   O$_2$ determination ...

Paramagnetic resonance   see   EPR, NMR ...

PEP carboxylase (PEPC) see Phosphoenolpyruvate carboxylase

Peroxisome, glyoxysome, microbody
         28706, 28882, 29002, 29111, 29181, 29746-7, 29759, 30374, 30526, 30559, 30801,
         31677-8, 31685, 31768, 32337-8, 32547

Pesticides   see also   Inhibitors of electron transport chain

Pesticides, herbicides and algae productivity   28996

Pesticides, herbicides and carbon fixation pathways
         29321, 29471, 29669, 29888, 29910, 30695, 30841, 30887, 30962, 31275

Pesticides, herbicides and carotenoids
         29573, 30230, 30594, 30597, 31503-4, 32215, 32424

Pesticides, herbicides and chlorophyll
         28653, 28709, 28846, 29321, 29323, 29449, 29473, 29573, 29669, 29946, 30104-5,
         30222, 30230, 30500, 30594, 30597, 30740, 30794, 30867, 31015, 31285, 31395,
         31503, 31582, 31621, 31648, 31855, 32000, 32255, 32323, 32424, 32433, 32451,
         32482

Pesticides, herbicides and chloroplast (chromatophore)   28709, 29024, 31504, 31621,
         32535

Pesticides, herbicides and ecosystem and plant productivity 28780, 29323, 29334,
        29577, 30132, 31015, 31750, 32306

Pesticides, herbicides and electron transport chain
        28743, 28777, 28846, 29143, 29279, 29449, 29634, 29751, 29810, 29946, 29978,
        30001, 30118, 30132, 30136, 30246, 30403, 30528, 30597, 30607, 30681, 30709,
        30717, 30842, 30848, 30859, 30955, 30962, 31049, 31051, 31194, 31297, 31390,
        31522, 31562, 31621, 31629, 31683, 31688, 31715, 31808, 31828, 31835, 31948,
        31995, 31998, 32029, 32140, 32217, 32239, 32255, 32293, 32407, 32418, 32476,
        32496

Pesticides, herbicides and gas exchange
        28527, 28604, 28653, 28777, 28780, 28846, 29280, 29343, 29463, 29673, 29873,
        30104-5, 30140-1, 30204, 30223, 30413, 30500, 30841-2, 30867, 30962, 31015,
        31285, 31340, 31347, 31541, 31621, 31701, 31750-1, 32063, 32470

Pesticides, herbicides and resistance to CO_2 and water vapour transfer 29709

Pesticides, herbicides and respiration 28527, 28780, 29673, 31285, 31835

Petiole see Stem, petiole, morphology, structure and physiological activity in

pH, effect on carbon fixation pathways
        28841, 28869, 28931, 29061, 29165, 29841, 29862, 30143, 30170, 30191, 30406,
        30653, 30664, 30907, 31248, 31520, 31734, 31977, 32036, 32366

pH, effect on carotenoids 31810

pH, effect on chlorophyll 28582, 28737, 28967, 29599-600, 30494, 30620, 30827,
        31110, 31129, 31254, 31323, 31488, 31491, 31782

pH, effect on chloroplast (chromatophore) 28936, 29932, 30268, 30856, 31056, 31356,
        31491

pH, effect on electron transport chain
        28562, 28581, 28727, 28744, 28747, 28753-4, 28766, 28903, 28920, 28935-6,
        28967, 29107, 29114, 29192, 29216-7, 29245, 29263, 29416-7, 29539-41, 29583,
        29599-600, 29607, 29696, 29740, 29749, 29831, 29844, 29849, 29933, 29961,
        29973, 30002, 30108, 30114, 30126, 30180, 30245, 30291, 30335, 30370, 30402,
        30561, 30653, 30718, 30772-4, 30782, 30832, 30909, 31124, 31169, 31179, 31191,
        31291, 31459, 31462, 31469, 31483, 31487, 31520, 31549, 31560, 31562, 31617,
        31630, 31664, 31683-4, 31693, 31716, 31746-7, 31808, 31813-4, 31816, 31950,
        32042, 32135, 32245, 32332, 32362, 32480, 32513

pH, effect on gas exchange
        28579, 28728, 28900, 29570, 29629, 29877, 30381, 30561, 31136, 31145, 31488,
        31514, 31520, 31732, 31734

pH, effect on photorespiration 30406, 31451, 31734

pH-stat, buffers 32013

PhAR see Irradiance ...; Canopy, radiation ...

Phosphoenolpyruvate carboxylase
        28501, 28613, 28652, 28748, 28818, 28841, 28869, 28901, 28928, 29115, 29142,
        29165, 29231, 29262, 29273, 29363, 29463, 29534-5, 29669, 29709, 29777, 29850,
        29860, 29863, 29910, 29990, 29992, 30009, 30037, 30098, 30128, 30174, 30232,
        30234, 30332-3, 30435, 30537-8, 30695, 30749, 30754, 30778, 30784, 30823,
        30886, 30897, 30907, 30928, 30951-2, 30959-61, 30990, 31166, 31223, 31300,
        31419, 31441, 31447-9, 31465, 31472, 31474, 31576, 31611, 31637, 31977,
        32036, 32262, 32317, 32425-6, 32428-9, 32440, 32457, 32502

Phosphoenolpyruvate carboxylase, methods 29165, 29990, 30749, 30784, 30907, 30959,
        31419

Phosphorus   see   Mineral elements (N,P,K) ...

Photoperiod and carbon fixation pathways   29273, 29641

Photoperiod and carotenoids   31237

Photoperiod and chlorophyll   29150, 29354, 31237, 31454

Photoperiod and chloroplast (chromatophore)   28871, 28896, 29765, 29871, 30896

Photoperiod and ecosystem and plant productivity   30680, 30896, 31237, 32436, 32504

Photoperiod and electron transport chain   31727

Photoperiod and gas exchange   28900, 29093, 29820, 30905, 30947, 32449

Photoperiod and resistances to CO_2 and water vapour transfer   30903, 30905

Photophosphorylation, cyclic
    28541,   28592-5,   28640, 28708, 28742, 28789, 28798, 28851, 28859, 28887,
    28889, 28894, 29051, 29070, 29180, 29282, 29353, 29376, 29449, 29519, 29525,
    29596-7, 29603, 29664-5, 29714, 29731, 29810, 29844, 39885, 29942, 29959,
    30035, 30050, 30103, 30110, 30114-5, 30135, 30223, 30336, 30361, 30436, 30449,
    30558, 30607, 30659, 30661, 30839, 30845, 30859, 30867, 30892, 31045, 31096,
    31291, 31379, 31416, 31420, 31455, 31458, 31461, 31503, 31522, 31533, 31551,
    31665, 31715, 31727-8, 31736, 31828, 31872, 31998, 32157, 32161, 32178,
    32198, 32204, 32216-7, 32235-6, 32247-8, 32384-5, B32410, 32496-9, 32521

Photophosphorylation in photosynthetic bacteria   see   Photosynthetic bacteria, pho-
    tophosphorylation

Photophosphorylation mechanism   see   Chemiosmotic hypothesis ...

Photophosphorylation, methods   29665, 30658, 31818, 32245

Photophosphorylation, model   see   Model ...

Photophosphorylation, non-cyclic
    28523-4, 28592-3, 28679, 28739, 28755-6, 28766, 28777, 28779, 28784, 28789,
    28798, 28833, 28839, 28859, 28879, 28887, 28896, 28906-7, 29044, 29184,
    29285, 29353, 29376, 29382, 29449, 29464, 29525, 29599, 29639, 29665, 29695,
    29730-1, 29826, 29838, 29847, 29885, 29932, 29959, 30034, 30062, 30093,
    30103, 30110-1, 30115, 30119, 30126-7, 30132, 30136, 30178, 30197, 30211,
    30291, 30297, 30329, 30361, 30429, 30442, 30528, 30540, 30548, 30558, 30661,
    30726, 30805, 30859, 30867, 30892-3, 30909, 30962-4, 30966, 30969, 30985-6,
    31045, 31082, 31096, 31119, 31128, 31191, 31238, 31275, 31322, 31356, 31379,
    31418, 31456, 31458, 31461, 31522-3, 31541, 31554, 31626, 31664-5, 31667,
    31703, 31715, 31717, 31727-8, 31736-7, 31754-5, 31818, 31970, 31987, 31998,
    32072, 32135, 32145, 32161, 32178, 32198, 32204, 32216, 32223, 32233, 32246-7,
    32309, 32311, 32343, 32385, B32410, 32437-8, 32444, 32480, 32484, 32490,
    B32515, 32536

Photophosphorylation, pseudo-cyclic   28742, 28879, 29376, 29665, 30292, 30765,
    32216-7

Photoreduction   see H_2 evolution ...

Photorespiration enzymes
    28513, 28928-9, 28931, 29109, 29231, 29512, 29536, 29604, 29637, 29700,
    29747, 29841, 30122, 30374, 30568, 30640, 30886, 30951, 31008, 31088, 31164,
    31182, 31251, 31300, 31315, 31433, 31457, 31914, 31928, 32422, 32461-2, 32529,
    32547

Photorespiration, metabolic cycles
        28600, 28772, 28774, 28928, 29002, 29032, 29109, 29159, 29176, 29286, 29294,
        29328, 29412, 29504, 29571, 29586, 29637, 29746-7, 29759, 30098, 30106, 30122,
        30202, 30288, 30294, 30369, 30374, 30429-30, 30468-9, 30499, 30543, 30546,
        30631, 30635, 30637-40, 30686, 30798, 30951, 31034, 31134-5, 31164, 31227,
        31313-4, 31352, 31451, 31520, 31677, 31731, 31733, 31802, 31904, 31908, 31928,
        32036, 32337-8, B32410, 32460-2, 32529

Photorespiration metabolic cycles enzymes, methods see Enzymes of glycollate cycle,
        methods

Photorespiration rate
        28772, 28785, 28898, 28901, 28993-4, 29043, 29054, 29087, 29090, 29108-9,
        29156, 29169, 29178, 29191, 29328, 29514, 29604, 29795, 29989, 30047, 30457,
        30542, 30546, 30622-4, 30640, 30886, 31102, 31118, 31134, 31266, 31293, 31470,
        31636-7, 31836, 31913, 32073, 32104, 32219, 32428, 32461, B32528

Photosynthate translocation and distribution
        28613, 28618, 28620, 28624, 28749, 28761, 28796, 28855, 28872-4, 28913, 28926,
        28957, 28975, 29009, 29046, 29052-3, 29102, 29119, 29132, 29157, 29221-2,
        29277, 29301-2, 29313, 29325, 29331, 29338, 29351, 29372, 29381, 29443, 29446,
        29460, 29482, 29485, 29487, 29492, 29500, 29506, 29509, 29577, 29642-6, 29648,
        29699, 29705, 29726, 29771, 29783, 29822, 29832, 29848, 29860, 29895, 29902,
        29921, 29923, 29951-4, 29963-4, 29966, 29992, 30010-1, 30021-2, 30206, 30236,
        30351-3, 30523, 30530, 30532, 30543-4, 30627, 30712-3, 30734, 30754, 30798,
        30811, 30820-1, 30840, 30858, 30896, 30920, 30929, 30937, 31229, 31245, 31262,
        31277, 31310, 31376-8, 31399, 31408, 31417, 31477, 31494, 31641, 31675, 31732,
        31759, 31763-4, 31790, 31792, 31823, 31849, 31882, 31889, 31894, 31916, 31940-
        -1, 31947, 31953-4, 31974, 31978, B32004, 32041, B32056, 32057, 32066, 32119,
        32122, 32171-2, 32182, 32200-1, 32218, 32306, 32339, 32343, 32347, 32363,
        32372, 32389, 32415, 32434, 32436, 32491, 32503, 32509, B32515, 32519-20,
        32542

Photosynthates and intermediates of carbon fixation pathways
        28567, 28841, 28855, 28940, 28957, 28975, 28981, 29019, 29040, 29098, 29142,
        29191, 29214, 29253, 29270, 29321, 29460, B29468, 29495, 29534, 29699, 29807,
        29814, 29825, 29891, 29911, 29983, 30022-3, 30038, 30088, 30098, 30109,
        30232-4, 30236, 30275, 30277, 30279, 30433, 30435, 30544, 30564, 30604,
        30810, 30817, 30841, 30887, 30897, 30936, 30968, 30988, 31040, 31048, 31109,
        31246, 31260, 31277, 31314-5, 31420, 31448, 31451, 31520, 31525, 31554,
        31599, 31628, 31676, 31765, 31889, 31908, B31990, 32036, 32075, 32092, 32121,
        32179, 32205, 32297, 32343, 32425, 32430, 32439-40, 32529

Photosynthates and intermediates of carbon fixation pathways and gas exchange
        28975, 29133, 29202, 29363, 29564, 30904-5, 31134

Photosynthetic bacteria carbon fixation pathways 29098, 30382

Photosynthetic bacteria carotenoids see Xanthophylls of photosynthetic bacteria

Photosynthetic bacteria chlorophylls see Bacteriochlorophyll; Chlorophylls, *Chloro-
        bium*

Photosynthetic bacteria chromatophores see Chloroplast and chromatophore ...; Chro-
        matophores ...

Photosynthetic bacteria electron transport chain
        28530, 28548, 28551, 28559, 28580, 28610, 28641, 28671, 28692, 28701, 28769,
        28822, 28824, 28832, 28904, 29038, 29074, 29078, 29192, 29263-4, 29266, 29272,
        29359, 29439, 29452, 29664, 29679, 29748-9, 29818, 29827, 29870, 29959-61,
        29969-70, 30050, 30116, 30265, 30340, 30398-9, 30535, 30625-6, 30718, 30780,
        30832, 30939, 31059-60, 31288, 31370, 31455, 31479, 31534, 31598, 31624,
        32014, 32169, 32250, 32279, 32281, 32370, 32467-8, 32550

Photosynthetic bacteria gas exchange 28600, 28750, 29229, 29684, 29913, 29940-1,
        30227, 30698, 31274, 31730

Photosynthetic bacteria photophosphorylation
    28640-3, 28671-2, 28992, 29630-1, 29748-50, 29959-61, 29980-1, 30050, 30280,
    30340, 30386, 30658, 30831-2, 31179, 31191, 31287, 31295, 31299, 31549, 31871,
    32375-6, 32438, 32459

Photosynthetic bacteria reaction centres see *P700* ...

Photosynthetic (chlorophyll) unit
    28518-9, 28528-9, 28589-90, 28595, 28655-6, 28724, 28726, 28779, 28997, 29001,
    29012, 29025, 29040, 29083, 29161, 29268, 29341, 29505, 29525, 29609, 29735-6,
    29785, 29830, 29978, 29991, 30034, 30179, 30181, 30719, 30759, 30831, 30969,
    30993, 31030, 31044, 31078, 31105, 31118-9, 31232, 31354, 31485, 31489, 31692,
    31713, 31725, 32113, 32124, 32407, B32410, 32432, 32441

Photosystem 1
    28514, 28528, 28539, 28547-8, 28551, 28554, 28575, 28581, 28584-5, 28587-8,
    28590, 28595, 28599, 28655, 28667-8, 28676, 28690-1, 28693-4, 28696, 28699,
    28708, 28717-8, 28720, 28739, 28744-5, 28751, 28758, 28766, 28789, 28798,
    28815, 28833, 28836, 28845-6, 28859, 28878, 28884, 28886, 28918, 28920, 28932,
    28934, 28965-6, 28968-9, 28997-9001, 29079, 29081, 29107, 29114, 29138, 29143,
    29161, 29173, 29211, 29231, 29245, 29275, 29278, 29284, 29300, 29305, 29310,
    29353, 29360, 29384, 29406, 29412, 29416, 29438-42, 29449, 29458, 29518-20,
    29525, 29527, 29533, 29540-1, 29557, 29570, 29601, 29619, 29621, 29670, 29682,
    29684, 29686, 29688, 29698, 29719, 29731, 29734, 29754-5, 29768, 29773, 29807,
    29831, 29836, 29857, 29867, 29872, 29880, 29932-3, 29939, 29948-50, 29971,
    29973-4, 30002, 30006, 30089, 30093-4, 30112, 30114, 30123, 30156-7, 30172,
    30179, 30181, 30246, 30251, 30254-5, 30291, 30320-3, 30336, 30359-62, 30402,
    30415, 30436-7, 30442-3, 30449, 30472, 30493, 30518, 30525, 30570, 30579-80,
    30586-7, 30589, 30607, 30630, 30716-8, 30771, 30777, 30780, 30799, 30839,
    30859, 30878, 30892, 30955, 30982-3, 31007, 31009, 31025, 31027, 31030, 31053,
    31078-80, 31087, 31091, 31098, 31108, 31128, 31143, 31145-6, 31155, 31161,
    31169, 31194, 31205-7, 31264, 31291, 31298, 31322, 31336, 31374-5, 31426,
    31485, 31503, 31505-6, 31521-2, 31551, 31567, 31581, 31588, 31610, 31612,
    31625-6, 31629-30, 31666, 31668, 31707, 31713, 31722, 31724-6, 31736-7,
    31785-6, 31809, 31813-4, 31816, 31828, 31835, 31877, 31918-9, 31931, 31979-82,
    32005, 32029, 32071-2, 32130, 32157, 32161-2, 32199, 32216, 32239, 32242,
    32246-8, 32256, 32283, 32303, 32315, 32331-2, 32359, 32370, 32392, 32407,
    32419, 32432, 32471, 32496, 32498-9, 32534

Photosystem 1 activity measurement
    28524, 28539, 28595, 28655, 28690, 28693, 28708, 28738-9, 28771, 28789,
    28845-7, 28887, 28918, 28995, 29138, 29143, 29384, 29533, 29557, 29637,
    29639, 29688, 29730, 29807, 29831, 29867, 29885, 29974, 30034, 30107, 30323,
    30630, 30881, 30955, 30984-5, 31087, 31091, 31206-7, 31522, 31643, 31663,
    31666-7, 31725-6, 31737, 31813, 31816, 31919, 31959, 32072, 32256, 32305,
    32315, 32418, 32499, 32512

Photosystem 1 activity measurement, methods 28739, 28852, 29768, 29933, 32013

Photosystem 1, primary acceptor
    28548, 28551, 28588, 28599, 28720, 28859, 29284, 29439, 29518, 29688, 29818,
    30716, 31027, 31030, 31629, 32432

Photosystem 2
    28514, 28518-9, 28539, 28547, 28549-51, 28554, 28575, 28581, 28584-5, 28587-8,
    28590, 28595, 28599, 28621, 28633, 28651, 28655-7, 28676, 28690-1, 28693-6,
    28699, 28718, 28738-9, 28741, 28744-5, 28754, 28759, 28766, 28777-8, 28789,
    28814-5, 28823, 28833, 28846, 28867, 28875, 28884-6, 28903, 28918, 28934-6,
    28968, 28991, 28997-9, 29001, 29081, 29096-7, 29107, 29114, 29138, 29143,
    29161, 29183-4, 29186, 29211, 29231, 29244, 29275, 29279, 29285, 29300, 29304-
    -5, 29307-9, 29341, 29353, 29360, 29384, 29406, 29416, 29436, 29439, 29442,
    29458, 29518-9, 29525, 29527, 29532, 29539-41, 29557, 29570, 29578, 29599-600,
    29602, 29619, 29634, 29670, 29684, 29688-9, 29698, 29713, 29731, 29740, 29754,
    29773, 29784-5, 29807, 29830-1, 29836, 29867, 29872, 29875, 29906, 29928,
                                                                          (continued)

Proteins, amino acids, nucleic acids, and chloroplast ... (continued)
    30569, 30589, 30682, 30688, 30752, 30816, 30872, 31027-9, 31080, 31104-5,
    31137, 31143, 31155, 31323, 31345, 31365, 31414, 31434, 31545, 31575, 31590,
    31643, 31656, 31662, 31664, 31666, 31674, 31705, 31743-4, 31754, 31815, 31877-
    -8, 31892, 31900, 31923, 31937, 31973, 32048, 32060, 32125, 32149, 32276,
    32283, 32392, 32397, 32420, 32452

Proteins, amino acids, nucleic acids, and electron transport chain 31025, 31027,
    32260

Proteins, amino acids, nucleic acids, and gas exchange 31519

Proteins, amino acids, nucleic acids, and photorespiration 30033

Protochlorophyll(ide) see Chlorophyll biosynthesis

Proton transport in chloroplast see Chemiosmotic hypothesis ...

Pteridines see Ferredoxin-NADP reductase

Pyrenoid 29510, 29728, 30758, 31589

Q

Quantum yield and requirement
    28593, 28745, 28753, 28833, 28858, 28860, 28877, 28879, 28884, 28886, 28922,
    29000, 29012, 29018, 29097, 29244, 29383, B29468, 29518-9, 29530, 29539, 29614,
    29621, 29714, 29733, 29735, 29754-5, 29797, 29805, 29847, 29885, 29936, 30062,
    30080, 30220, 30347, 30454, 30580, 30591, 30607, 30637, 30640, 30656, 30742,
    30892, 30919, 31146, 31342, 31363, 31412, 31485, 31488-9, 31504, 31564, 31571,
    31690-1, 31709, 31980, 32031, 32071, 32162, 32251, 32359, 32433, 32534

Quantum yield and requirement, methods see Quantum yield and requirement

Quinones in photosynthesis
    28535, 28549-50, 28641, 28643, 28667, 28692, 28694-6, 28744, 28746, 28769,
    28778, 28794, 28814, 28822, 28833, 28904, 28995, 29192, 29263, 29266, 29304,
    29308, 29359, 29525, 29541, 29602, 29698, 29713, 29719, 29760, 29784-5, 29818,
    29868-70, 29948, 29969, 29973, 30000, 30118, 30182, 30245, 30291, 30337-8,
    30385, 30402, 30594-7, 30625-6, 30716, 30718, 30780-1, 30812, 30859, 30939,
    31009, 31056, 31061, 31095, 31288, 31291, 31368, 31371, 31410, 31414, 31479,
    31486, 31521, 31534, 31573, 31598, 31612, 31617, 31736, 31746-7, 31785-6,
    31819-20, 31959, 31970-1, 31998, 32072, 32127, 32130, 32148, 32249, 32255,
    32279, 32281, 32295, 32303, 32408, 32467-8, 32489

Quinones, methods 30596, 30598, 30939

R

Radiation in canopy see Canopy, radiation ....

Radiation, light see Irradiance ...

Rain, precipitation, methods see Aerodynamic methods, bioclimatological methods ...

Reaction centres see P680; P700 ...

Recycling of CO_2 inside cell and leaf   29992, 30499, 31352, 31451, 32073, 32327

Relative growth rate   see   Growth analysis, net assimilation rate ...

Relative water content   see   Water saturation deficit

Resistance, carboxylation and excitation
     28785, 28896, 29090, 29156, 29319, 30221, B30281, 30456, 31067, 31232, 31251,
     31372, 31596, 31627, 31866, 32073, 32126

Resistance, cuticular   29053, 32020, 32485

Resistance, intracellular (mesophyll)
     28504, 28606, 28688, 28717-8, 28732, 28785, 28835, 28898, 28899, 28987, 29053-
     -4, 29090, 29156, 29319, 29331, 29486, 29676, 29795, 30078, 30221, 30270-1,
     B30281, 30457-8, 30608, 30633, 30904-5, 31066-7, 31164-5, 31214, 31232, 31251,
     31256, 31264, 31372, 31464, 31596, 31613, 31627, 31836, 31866-7, 32020,
     32126, 32252, 32450

Resistance, leaf boundary layer
     28602, 28785, 28898, B29023, 29054, 29090, 29130, 29179, 29855, 30633, 31164,
     31613, 31793, 31803, 32120, 32277, 32296, 32485

Resistance, stomatal (and intercellular) (*cf*. also Resistances for water vapour ...)
     28504, 28509, 28606, 28688, 28718, 28731-2, 28785, 28803, 28810, 28835, 28896,
     28898-9, 28925, 28927, 28987-8, 29015, 29020, B29023, 29053-4, 29122, 29130,
     29162, 29179, 29232, 29243, 29247, 29255, 29331, 29373, 29395, 29425, 29467,
     29486, 29551, 29676, B29724, 29766, 29793, 29816, 30076, 30212, 30221, 30249,
     30271, 30455, 30457-8, 30508, B30511, 30547, 30633, 30743, 30818, 30861,
     30903-5, 30918-9, 30931, 31087, 31164, 31214, 31231, 31250-1, 31269, 31303,
     31321, 31432, 31470, 31516, 31536, 31596, 31613-4, 31627, 31633, 31759, 31762,
     31769, 31823, 31866-7, 31956, 31994, B32004, 32020, 32051, 32064-5, 32110,
     32226, 32252, 32296, 32352, 32369, 32450, 32458, 32485

Resistances to CO_2 and water vapour transfer at canopy level   29722, B30281, 30393,
     30918, 31633, 31803-4, 31859

Resistances to CO_2 transfer
     28504, 28785, 28918, 29016, B29023, 29054, 29156, 29179, 29425, 29795, 29848,
     30221, B30325, 30633, 30643, 30654, 30757, 31067, 31164, 31214, 31230, 31255,
     31432, 31861-2, 31864-6, 31885, 32020, 32073, 32079, 32126

Resistances to water vapour transfer, "leaf resistance"
     28500-1, 28504, 28718, 28821, 28914, 28919, 28930, 29016, B29023, 29122,
     29204, 29247, 29319, 29373, 29394, 29486, B29707, 29793, 29881, 30069, 30078,
     30139, 30162-3, 30250, 30270, B30325, 30392-3, 30507, 30523, 30583, 30643,
     30737-9, 30818, 30883, 30930, 31067, 31069, 31165, 31193, 31255, 31264, 31268-
     -70, 31432, 31464, 31512, 31607, 31659, 31770, 31887-8, 31899, 31956, 31967,
     32020, 32064-5, 32079, 32111, 32139, 32220, 32277, 32296, 32364, 32458, 32485,
     32506

Respiration and photosynthesis
     28541, 28604, 28648, 28687, 28703, 28706, 28773, 28780, 28795, 28855, 28880,
     28896, 28901-2, 28918, 28979, 29008, 29101, 29191, 29201-2, 29206, 29231,
     29242, 29252-3, 29286, 29294, 29296-8, 29303, 29336-7, 29343, 29350, 29379,
     29451, 29495, 29514, 29563, 29700, 29721, 29819-20, 29842, 29893-4, 29914,
     29957, 30026, 30047, 30124, 30151, 30167, 30285, 30322, 30381, 30575, 30608,
     30622, 30624, 30750-1, 30762, 30768, 30873, 30904, 30922, 30936, 30947, 30968,
     31003, 31018, 31040, 31065, 31077, 31142, 31218, 31220, 31272, 31281, 31285,
     31327-8, 31346, 31362, 31402, 31429-30, 31460, 31468, 31493-4, 31527, 31555,
     31570, 31586-7, 31592, 31651, 31700, 31802, 31835, 31849, 31882, 31890, 31896,
     31937, 32032, 32046, 32055, 32079, 32088, 32093, 32115, 32118, 32121, 32137,
     B32150, 32174-5, 32181-2, 32312, 32327, 32329, B32410, 32432, 32484, B32515,
     32525

Respiration, dark CO_2 efflux
  28527, 28647, 28685, 28809, 28898, 28970-2, 28988-9, 28994, 29016, 29053,
  29065, 29068, 29104, 29109, 29207, 29212, 29222, 29319, 29409, 29421, 29425,
  29443, 29462, 29509, 29517, 29569, 29576, 29605, 29795, 29819, 29864, 29887,
  29923, 29952, 29967, 29992-3, 30011, 30139, 30218, 30271, 30448, 30476, 30513,
  30515-6, 30532, 30537, 30571, 30622-3, 30642-3, 30691, 30804, 30883, 30891,
  30947, 31077, 31107, 31200, 31232, 31251, 31293, 31492, 31592, 31633, 31759,
  31836, 31885, 31941, 32020, 32023, 32118, 32163, 32175, 32219, 32312, B32318,
  32339, 32341, 32364, 32412, 32491

Respiration, "growth" and "maintenance"
  29103, 29820, 29967, 30011, 30891, 30904, 31262, 32118, 32341

Respiration of achlorophyllous tissues in light, light inhibition of respiration
  29016, 31077

Ribosome of chloroplast
  28542, 28574, 28736, 28849-50, 28897, 28912, 28915-6, 28980, 29033, 29371,
  29405, 29478, 29560, 29824, 29839, 29845, 30345, 30534, 30597, 30752, 30889,
  31092, 31126, 31139, 31553, 31705, 31874, 31917, 31973, 32390, 32397

Ribulose 1,5-bisphosphate carboxylase
  28501, 28513, 28613, 28645, 28652, 28658-9, 28697, 28717-8, 28748, 28775,
  28818-9, 28835, 28883, 28891, 28901, 28910-1, 28915-6, 28928-9, 28931, 29030,
  29098, 29107-8,   29132, 29144-7, 29231, 29262, 29273, 29315-7, 29322,
  29374, 29376, 29405-6, 29463, 29478, 29503, 29534, 29567-8, 29603-4, 29612,
  29624-6, 29633, 29641, 29650-1, 29700, 29709, 29728, 29777, 29781, 29795,
  29800-1, 29807, 29826, 29841, 29865, 29910, 29922, 29945, 30009, 30024-5,
  30038, 30097-8, 30128, 30143-4, 30170, 30174, 30232, 30234, 30332-3, 30435,
  30472-4, 30537-8, 30566, 30639-40, 30647, 30695, 30714, 30754, 30769, 30778,
  30783-4, 30809-10, 30823, 30844, 30893-5, 30897, 30912, 30928, 30933, 30969,
  30980, 30990, 31008, 31062, 31096, 31102, 31117, 31164-6, 31182, 31223, 31226-
  -7, 31250-1, 31256, 31263-4, 31300, 31319, 31419-20, 31433, 31433, 31447,
  31457, 31472, 31476, 31552-3, 31576, 31578, 31589, 31602, 31660-1, 31679,
  31734, 31806, 31844, 31923-4, 31930, 31963-4, 32035-6, 32097, 32125, 32236,
  32244, 32262, 32264, 32317, 32334-5, 32373, 32397, 32406, 32421-2, 32428-9,
  32440,   32457, 32459, 32488, 32502, 32529

Ribulose 1,5-bisphosphate carboxylase, methods
  29144-5, 29374, 29626, 29650-1, 29841, 29865, 30025, 30275, 30279, 30636,
  30647, 30784, 31419, 31553, 32035, 32264, 32334

Ribulose 1,5-bisphosphate oxygenase see Ribulose 1,5-bisphosphate carboxylase ...;
  Photorespiration enzymes

Root removal see Defoliation ...

Root, underground part, and carbon fixation pathways 28855, 29016

Root, underground part, and chloroplast 32377

Root, underground part, and ecosystem and plant productivity 28521, 28937, 29016,
  29022, 29492, 29726, 30317, 30463, 30674, 31086, 31748, 32403

Root, underground part, and resistances to CO_2 and water vapour transfer 30249

Root, underground part, and respiration 30891, 32363

Rooted leaves, chlorophyll in 28791, 30683

Rooted leaves, gas exchange in 30683

Rooted leaves, productivity of 28791

RuBP carboxylase, RuBPC see Ribulose 1,5-bisphosphate carboxylase

Rubredoxin see Ferredoxin ...

S

Saccharides and carbon fixation pathways
        28855, 29253, 29430, 29571, 29629, 29699, 29891, 30526, 30686, 30809, 30817,
        30936, 30962, 31168, 31247-8, 31314-5, 31360, 31520, 31806, 31938, B31990,
        32297, B32515

Saccharides and chlorophyll 28762, 29220

Saccharides and chloroplast (chromatophore) 28896, 29728, 29871, 30016, 31616, 31677

Saccharides and ecosystem and plant productivity 31914, 31916, 32010, 32172, 32542

Saccharides and electron transport chain 29270, 30551, 31651

Saccharides and gas exchange 29133, 29220, 29500, 29921, 30546

Saccharides and photorespiration 29220

Salinity of soil and algae productivity 30175

Salinity of soil and carbon fixation pathways 29061, 29495, 30174, 30664, 31319,
        32139, 32378, 32425, 32477

Salinity of soil and carotenoids 30067

Salinity of soil and chlorophyll 29410, 30067, 30175, 30633, 31855, 32259

Salinity of soil and ecosystem and plant productivity 29331, 29410, 30633, 31041,
        31086, 31217, 32200

Salinity of soil and gas exchange 28507, 29093, 29252-3, 29331, 29495, 30139, 30633,
        30968, 31086, 31217, 31681, 32139, 32378

Salinity of soil and photorespiration 29495, 30272, 31086

Salinity of soil and resistances to CO_2 and water vapour transfer 30139, 32139

Salinity of soil and respiration 29495, 30139, 30968, 31086, 31217, 31681

Samples for pigment determination see Pigments ...

Seasonal changes see also Ontogeny of plant ...

Seasonal changes in algae productivity
        28536, 28687, 28721, 28740, 28773, 28776, 28829-30, 28861, 28880, 28955,
        28982, 29151, 29171, 29185, 29276, 29311, 29345, 29375, 29396, 29428, 29466,
        29469, 29508, 29511, 29655, 29678, 29778, 29802-3, 29809, 29821, 29842, 29937,
        30032, 30052, 30075, 30147-8, 30151, 30196, 30231, 30281, 30305, 30397, 30482,
        30572, 30605-6, 30645, 30724-5, 30732, 30735, 30756, 30776, 30863-5, 30906,
        30954, 31031, 31202, 31215-6, 31260, 31327-8, 31373, 31382, 31396, 31555,
        31745, 31749, 31758, 31909, 31989, 32032, 32044, 32138, 32227, 32269, 32273,
        32404, 32423, 32446, 32554

Seasonal changes in biliproteins 31499, 32368

Seasonal changes in carbon fixation pathways 28853, 28928, 29641, 29983, 31220

Seasonal changes in carotenoids
        28790, 29053, 29414, 29422, 29456, 29803, 30017, 30357, 30400, 30464-5, 30581,
        30760, 30800, 30802, 31047, 31385, 31403, 31579, 31961, 32055, 32185, 32284

Seasonal changes in chlorophyll
        28790, 28880, 28939-40, 29250, 29321, 29414, 29419, 29422, 29456-7, 29802-3,
        29809, 30017, 30057, 30186, 30295, 30357, 30400, 30464-5, 30572, 30581,
        30605-6, 30724, 30750-1, 30760, 30800, 30802, 30863, 31047, 31069, 31083,
        31143, 31213, 31385, 31403, 31499, 31533, 31559, 31579, 31727, 31855, 32009,
        32055, 32058, 32081, 32185, 32284, 32300, 32307, 32526

Seasonal changes in chloroplast (chromatophore) 29238, 29569, 30295, 31143, 31728

Seasonal changes in ecosystem and plant productivity
        28516, 28533, 28624, 28711, 28800, 28813, 28864, 28913, 29053, 29057, 29116,
        29164, 29254, 29334, 29373, 29377, 29411, 29469, 29551, 29577, 29656, 29685,
        29703, 29783, 29791, 29794, 29967, 29975, 30011, 30060, 30083, 30095, 30195,
        30317, 30371, 30400, B30477, 30483, 30533, 30573, 30581, 30691, 30711, 30745,
        30802, 30824, 30869, 30874, 30891, 30917, 30967, 31019, 31070, B31140, 31199,
        31220, 31259, 31383, 31508, 31539, 31558, 31586, 31633, 31647, 31706, 31748,
        31817, 31822, 31826, 31843, 31875, 31921-2, 31952, 31954, 32027, 32151, 32197,
        32243, 32265, 32306, 32363, 32372, 32403, 32415, 32436, 32543

Seasonal changes in electron transport chain 29238, 30295, 30415, 31143, 31533,
        31727-8, 32311

Seasonal changes in gas exchange
        28702, 28809, 28880, 28939-40, 28972, 29016, 29053, 29090, 29179, 29207,
        29303, 29306, 29332-3, 29344, 29372-3, 29415, 29443, 29569, 29572, 29605,
        29676, 29703, 29721, 29944, 29975, 30057, 30263-4, 30285, 30476, 30608,
        30642, 30691, 30750-1, 30919, 30946, 31019, 31220, 31261, 31396, 31515,
        31695-6, 31727-8, 31750-1, 31759, 31866, 31887, 31896, 31925, 31932, 32011,
        32022-3, 32026, 32027, 32181-3, 32300, 32311, B32318, 32353, 32415

Seasonal changes in photorespiration 30751

Seasonal changes in resistances to CO_2 and water vapour transfer
        29053, 29162, 29373, 29394-5, 29551, 30162, 30583, 30919, 31069, 31866, 32051,
        32194, 32369

Seasonal changes in respiration
        28972, 29016, 29443, 29569, 30011, 30317, B30477, 30608, 30642, 30691, 30751,
        30891, 31019, 31896, 32182

Simulation see Model ...

Sink and source of photosynthates, CO_2, etc.
        28578, 28624, 28872, 28913, 29202, 29381, 29482, 29506, 29645, 29789-90,
        29822, 29921, 29952, 29954, 29992, 30010, 30021-2, 30030, 30902, 30912, 30920,
        31086, 31214, 31277, 31281, 31306, 31399, 31790, 31792, 31823, 31879, 31978,
        32122, 32218, 32427, 32503, 32516

Soil moisture and carbon fixation pathways 28501, 32139

Soil moisture and carotenoids 31014

Soil moisture and chlorophyll 30198, 30620, 31014, 31581

Soil moisture and chloroplast (chromatophore) 31593

Soil moisture and ecosystem and plant productivity
        28566, 28788, 28873, 28899, 29009, 29022, 29120, 29221, 29496, 29550, 29790,
                                                              (continued)

T

Taxons, algae productivity 30585

Taxons, biliproteins in 30552, 32102, 32258

Taxons, carbon fixation pathways in 29322, 30098, 30273, 30474, 30538, 30980, 31910,
     32025, 32373

Taxons, carotenoids in 28528, 28715, 29208, 29256, 29342, 29780, 30017, 30592,
     31406, 32142, 32258, 32289, 32501

Taxons, chlorophyll in
     28528, 28715, 29342, 29457, 29780, 30017, 30354, 30552, 30596, 30750, 30822,
     31406, 32058, 32142, 32258, 32289

Taxons, chloroplast (chromatophore) in 29033, 29780, 32102, 32221-2

Taxons, ecosystem and plant productivity of
     29057, 29254, 30483, 30750, 31759, 31761-2, 32151, 32325

Taxons, electron transport chain in 28528, 29021, 29857, 30855, 31915, 32362

Taxons, gas exchange in 28546, 28632, 28988, 29601, 29944, 30449, 30705, 30750,
     30768, 31090, 31411, 31759

Taxons, photorespiration in 30750, 32529

Taxons, resistances to CO_2 and water vapour transfer in
     28988, 30392, 31269-70, 31759, 31762, 31887

Taxons, respiration in 30448

Temperature, high, and biliproteins 30299

Temperature, high, and carbon fixation pathways 28645, 28652, 28818, 29478, 30088,
     30236, 30749, 31611, 32502

Temperature, high, and carotenoids 29478

Temperature, high, and chlorophyll
     28588-9, 29410, 29433, 29476, 29478, 29526, 29529, 29743, 29978, 30299, 30709,
     30792, 31686, 31690-1, 31782, 31876, 32287, 32502, 32522

Temperature, high, and chloroplast (chromatophore) 29478, 31876

Temperature, high, and electron transport chain
     28508, 28588-9, 28596, 29081-2, 29416, 29529, 29639, 29933, 29978, 30708-9,
     31030, 31075, 31191, 31274, 31876, 32523

Temperature, high, and gas exchange
     28818, 29527, 30088, 30299, 30523, 30919, 31241, 31274, 31282, 31611, 31690-1,
     31701, 32287, B32318, 32502, 32523

Temperature, high, and photorespiration 29747

Temperature, high, and resistances to CO_2 and water vapour transfer 30919

Temperature, leaf see Leaf temperature ...

Temperature, low, and algae productivity 28561, 29842

Temperature, low, and biliproteins 28917

Temperature, low, and carbon fixation pathways   30236, 31530, 31908, 32297

Temperature, low, and carotenoids   28917, 30974, 31139, 31559, 31623

Temperature, low, and chlorophyll
      28526, 28917, 28963, 29177, 29291, 29337, 29354-5, 29596, 29621, 30360, 30833-
      -4, 30850, 30971, 30973, 31139, 31143, 31438, 31559, 31623, 31876

Temperature, low, and chloroplast (chromatophore)   28840, 29299, 29596, 31120, 31139,
      31143, 31243, 31530, 31616, 31727, 31876, 31955

Temperature, low, and ecosystem and plant productivity   29201, 29496, 31530, 31711

Temperature, low, and electron transport chain   28547, 29107, 29237, 30182, 30297,
      31075, 31143, 31727-8, 31876, 31955, 32486-7

Temperature, low, and gas exchange
      29337, 29726, 30172, 30529, 30631, 30684, 30743, 31258, 31711, 31727-8, 32297,
      32353

Temperature, low, and resistances to CO_2 and water vapour transfer   30392, 30743

Temperature, low, and respiration   29537, 29957, 31711

Temperature, physiological, and algae productivity   B28531, 28740, 29655, 30397,
      30724, 32273

Temperature, physiological, and carbon fixation pathways   28645, 28652, 28931, 29641,
      30097, 30684, 30686, 30907, 30959, 31166, 31251, B32003, B32515

Temperature, physiological, and chlorophyll
      29067, 29139, 29177, 29461, 29727, 30007, 30354, 30866, 31110, 31141, 31251,
      31878, 32065, 32287

Temperature, physiological, and chloroplast (chromatophore)   28924, 30668, 30866,
      31878

Temperature, physiological, and ecosystem and plant productivity
      28703, 28788, 28809, 28873, 29009, 29221, 29223, 29377, 29496, 29515, 29524,
      29575, 29663, 29790, 29895, 29984, 30010, 30096, 30203, 30249, 30371, 30387,
      30461, 30556, 30573, 30890, 30916, 30945, 30979, 31041, 31234, 31245, 31251,
      31348, 31439, 31586, 31711, 31841, 31916, 31934, B32003, 32006, 32033, 32065,
      32119, 32364, 32412, 32427, 32434, 32436, 32491, 32504-5

Temperature, physiological, and electron transport chain
      28511, 28589, 28936, 29082, 29238, 29383, 29416-7, 29583, 29607, 29630, 29639,
      30107, 30182, 30866, 30964, 31059, 31095, 31205-6, 31251, 31780, 31878

Temperature, physiological, and gas exchange
      B28531, 28614, 28629, 28688, 28702-3, 28764, 28807, 28856, 28900, 28970-1,
      29008, 29015, B29023, 29068, 29076, 29154, 29167, 29178-9, 29223, 29252-3,
      29319, 29415, 29425, B29468, 29514, 29527, 29572, 29575, 29638, 29675, 29693,
      B29707, 29721-2, 29726, 29732, 29764, 29766, 29848, 29894-5, 29911, 29931,
      30011, 30078, 30097, 30172, 30271, 30285-6, 30289, 30394, 30453-8, 30490,
      30515, 30523, 30556, 30563, 30571, 30575, 30656, 30684-5, 30699, 30706, 30751,
      30822, 30898, 30919, 30927, 30943-4, 30946, 31019, 31067-8, 31090, 31240-1,
      31245, 31256, 31258-9, 31460, 31470, 31514, 31585-6, 31613-4, 31711-2, 31732,
      31861-6, 31887, 31916, 31946, B32003-4, 32063, 32065, 32093, 32181, 32287,
      32312, B32318, 32327, 32364, 32412, 32428, B32515

Temperature, physiological, and photorespiration
      29169, 29514, 29693, 29848, 30097, 30271, 30288-9, 30457, 30684, 30686, 31251,
      31256

Temperature, physiological, and resistances to CO_2 and water vapour transfer
28732, 29015, 29020, 29319, 29394, 29766, 30078, 30271, 30289, 30455, 30457,
30547, 31251, 31256, 31268-9, 31613, 31861-2, 31864-5, 32065, 32364, 32369,
32485

Temperature, physiological, and respiration
28527, 28703, 28970, 29065, 29068, 29154, 29223, 29319, 29425, 29443, 29605,
29848, 30011, 30218, 30271, B30477, 30515, 30804, 31240-1, 31470, 31711,
B32003, 32312, 32412, 32491

Thylakoid, granum (cf. also Chloroplast ultrastructure; Electron transport chain localization in thylakoid)
28498, 28518, 28554, 28584-6, 28590, 28650-1, 28683, 28753, 28759, 28778-9,
28836, 28863, 28896, 28910, 28912, 28918, 28925, 28942, 28953, 28991, 29012,
29056, 29110-2, 29143, 29260, 29352, 29360, 29473, 29532, 29578, 29609, 29616,
29647, 29698, 29728, 29765, 29777, 29871, 29888, 29932, 29978, 30006, 30020,
30027-8, 30054, 30171, 30345, 30562, 30595, 30621, 30644, 30659, 30747, 30752-
-3, 30766, 30837, 30856, 30871, 30879, 30889, 30892, 30901, 30933, 30958,
30969, 30973, 31009, 31011, 31035, 31103-5, 31132, 31161, 31211, 31284, 31333-
-4, 31336, 31345, 31356, 31368, 31375, 31414, 31504, 31549, 31590, 31604,
31610, 31621, 31645, 31692, 31709, 31722, 31757, 31779, 31791, 31811, 31821,
31830, 31854, 31856, 31923-4, 31930, 31950, 31955, 32029, 32060, 32068, 32084,
32102, 32116, 32131, 32192-3, 32222, 32231, 32240, 32286, 32288, 32384-5,
32387, 32397, 32401, 32414, 32437, 32486, 32535, 32551

Tissue cultures, carbon fixation pathways in 30233-4, 30274, 30778, 31008, 31235,
31913

Tissue cultures, carotenoids in 29898, 30234, 30594, 30948, 32153, 32365

Tissue cultures, chlorophyll in 28686, 28736, 29219, 29283, 29450, 29898-9, 30056,
30234, 30241, 30363, 30467, 30594, 30948, 31012, 31038, 31112, 31116, 31640,
31655, 31744, 32326

Tissue cultures, chloroplast in 28736, 30068, 30467, 31012, 31743-4, 32365, 32400

Tissue cultures, electron transport chain 30594, 31012, 31038, 31116

Tissue cultures, gas exchange in 28772, 29219, 30056, 30234, 30778, 30948, 31012,
31038, 31116, 31235, 31453, 31913

Tissue cultures, growth of 30056, 31112, 31116

Tissue cultures, photorespiration in 28772, 31913, 32529

Transient phenomena in carbon fixation pathways 28989

Transient phenomena in electron transport chain activities 28784, 30799, 32524

Transient phenomena in gas exchange 28745, 28989, 30000, 31280, 31944, 32207, 32523-
-4, 32532

Transient phenomena in photorespiration 29169, 32524

Transpiration and photosynthesis
28546, 28732, 28768, 28843, 29103, 29179, 29212, 29409-10, 29816, 29944,
30273, 30457, 30628, 30695, 30703, 30705, 30733, 31068, 31228, 31252, 31255,
31303, 31464, 31887, 32023, 32220, 32226, 32252, 32541

Tritium oxide see Deuterium oxide, tritium oxide ...

X

Xanthophylls   28569, 28607, 28651, 28708, 28713-6, 28770, 29172, 29240-1, 29324,
        29396, 29422, 29480, 29614, 30017, 30102, 30207, 30237, 30302, 30356-7, 30452,
        30594, 30619, 30730, 30760, 30802, 30851, 30911, 30923, 31144, 31361, 31403,
        31406, 31414, 31503-4, 31559, 31574, 31583, 31673, 31694, 31742, 31811, 31831,
        31961, 32016, 32038, 32055, 32133, 32142, 32185, 32190, 32284, 32379, 32424,
        32441, 32481

Xanthophylls of algae
        28569, 28698, 28760, 28767, 28770, 29209, 29256, 29268, 29525, 29677, 29802-3,
        29924, 30137, 30193-4, 30238, 30356-7, 30366, 30475, 30582, 30592, 30629,
        31188-90, 31203, 31405, 31430, 31583, 31650, 31810, 31812, 32016, 32037, 32142,
        32190, 32258, 32441-2, 32500-1

Xanthophylls of photosynthetic bacteria
        28551, 28876, 28892, 28947-9, 29025, 29149, 29666, 29828, 29913, 29947, 29995,
        30079, 30134, 30269, 30672, 30788-9, 31349, 31490, 31534, 31741, 32017, 32149,
        32169

Xerophytes   see Drought...; Temperature, high ...

X-rays see Ionizing radiation ...

Y

Yield formation see Biomass distribution ...; Photosynthate translocation ...

Z

Zeaxanthin see Carotenoids ; Xanthophylls

This index contains a selection of plant genera and types interesting as experimental material for physiological, ecological and agricultural studies. Latin scientific names of plant genera and English names of plants groups and types are the main items which present the reference numbers.

A

Abies   29069, 29071, 29415, 29561, 29569, 30262, 30264, 31053, 31090, 31236

Acacia   28987, 29489, 31697, 31769, 31839, 32324, 32394

Acer   28987, 28990, 29206-7, 29212, 29243, 29254, 29320, 29394-5, 29467, 29480, 29614, 29812, 30263, 30460, 30721-2, 30764, 31081, 31090, 31269, 31411, 32111, 32175, 32185, 32188, 32194

Acetabularia   28574-7, 28579, 28609-10, 29649, 30020, 30667, 30701, 30923-4, 31847, 32240-2

Aesculus   29249, 29614, 29698, 31238, 32175

Agave   29616, 31066, 31069

Alder   see   *Alnus*

Alfalfa   see   *Medicago*

Algae (*cf.* also *Acetabularia,* Algae blue-green, Algae brown, Algae green, Algae red, *Anabaena, Anacystis, Ankistrodesmus, Chlamydomonas, Chlorella, Chrysophyta,* Diatoms, *Dinoflagellates, Dunaliella, Euglena, Nostoc, Porphyridium, Scenedesmus, Ulva*)
   28511, 28537, 28553, 28561, 28565, 28568, 28573, 28607, 28664, 28673, 28687, 28719, 28721, 28740, 28776, 28790, 28797, 28830, 28845-7, 28861, 28880, 28900, 28939, 28955, 28967, 28979, 28982, 28984-5, 28993-4, 28996, 29011, 29014, 29018, 29041, 29089, 29124, 29139, 29154, 29171, 29185, 29191, 29242, 29287-8, 29293, 29303, 29310-1, 29327, 29345, 29364-5, 29375, 29396, 29416-7, 29426, 29429, 29444, 29466, 29469, 29511, 29527, 29554, 29582, 29588, 29655, 29658, 29671, 29716, 29796, 29798, 29802-3, 29825, 29914, 29917, 29924, 29932, 29937, 30032, 30045-6, 30072, 30075, 30109, 30124, 30147-8, 30166, 30169, 30225-6, 30231, 30259, 30282, 30315, 30356, 30381, 30389, 30397, 30407, 30417, 30433, 30485-6, 30492-3, 30501, 30575, 30585, 30592, 30605-6, 30651-5, 30670, 30676, 30678, 30724-5, 30732, 30756, 30776, 30807, 30815, 30863-5, 30868, 30906, 30925, 30942, 31049, 31162, 31183, 31202, 31216, 31291, 31307, 31311, 31316-7, 31327-8, 31364, 31396-7, 31422, 31428, 31502, 31513, 31531, 31555, 31650, 31704, 31714, 31749, 31758, 31776, 31797, 31810, 31869, 31890, 31891, 31909, 31965, 31969, 31989, 32000, 32031-2, 32043-4, 32063, 32069, 32078, 32090, 32095, 32107, 32116, 32136-7, 32174, 32180, 32184, 32203, 32216, 32227, 32258, 32268, 32273-4, 32288, 32308, 32344-5, 32380, 32401, 32404, 32413, 32423, 32493, 32548, 32554

Algae, blue-green (*cf.* also *Anabaena, Anacystis, Nostoc*)
   28513, 28535, 28568-9, 28592, 28631, 28720, 28737, 28750, 28760, 28795, 28853, 28888, 28936, 28951, 28960, 28966-7, 28979, 28982, 29014, 29021, 29127, 29143-6, 29151, 29152, 29160, 29347, 29371, 29374, 29378, 29438, 29448, 29498, 29525-6, 29528-9, 29552, 29556, 29589, 29601, 29610, 29670-2, 29683, 29691, 29729, 29782, 29802-3, 29812, 29818, 29836, 29853, 29866, 29873, 29918, 29937-8, 29950, 29956, 30032, 30051, 30059, 30075, 30169, 30185, 30196, 30231, 30248, 30311, 30365, 30381, 30401, 30407, 30421, 30442-3, 30470, 30472, 30500, 30502, 30539, 30552, 30592, 30606, 30622, 30624, 30696, 30718, 30746, 30768,
(continued)

Algae, blue-green (continued)
 30772-3, 30799, 30813, 30863-5, 30872, 30971, 30974, 30983, 31011, 31031,
 31092, 31100, 31136, 31141, 31145-6, 31177, 31184, 31203, 31212, 31215, 31278,
 31327-8, 31360, 31391, 31402, 31412, 31415, 31437, 31481, 31521, 31603, 31638,
 31648, 31650, 31663, 31758, 31772, 31909, 31912, 31937, 31960, 31963-5, 31988,
 31992, 32007, 32021, 32053-4, 32069, 32071, 32102, 32113, 32157, 32205, 32305,
 32356, 32368, 32401, 32409-10, 32441, 32448, 32454, 32464, 32470

Algae, brown 28601, 28698-9, 28771, 28853, 28940, 28981, 29178, 29336-7, 29460,
 29525, 29655, 29775, 29825, 29864, 29893, 29938, 30151, 30173, 30314, 30435,
 30592, 30616, 30627, 30968, 31047, 31062, 31084, 31132-3, 31289, 31430, 31675,
 31810, 32173, 32348, 32356, 32472, 32517, 32544

Algae, green (*cf.* also *Acetabularia, Ankistrodesmus, Chlymadomonas, Chlorella, Dunaliella, Scenedesmus, Ulva*)
 28511, 28568, 28760, 28770, 28814-5, 28825, 28853, 28889, 28966, 28979, 28982,
 29014, 29118, 29127, 29140-1, 29160, 29178, 29185, 29256-7, 29337, 29403,
 29448, 29451, 29466, 29479, 29510, 29512, 29525, 29553, 29728, 29759, 29761,
 29802-3, 29815, 29873, 29908-10, 29930, 29938, 30023, 30075, 30169, 30193-4,
 30196, 30209, 30231, 30381, 30407, 30423, 30434-5, 30475, 30487-9, 30592,
 30606, 30624, 30667, 30669, 30718, 30753, 30758, 30770, 30839, 30863-4, 31011,
 31031, 31047, 31106, 31109, 31125, 31136, 31212, 31276, 31327, 31360, 31429,
 31456-7, 31460, 31589, 31597, 31626, 31650, 31758, 31810, 31837, 31909,
 31911, 31928, 31965, 31988, 32093, 32101, 32142, 32193, 32269, 32286, 32288,
 32348, 32356, 32401, 32423, 32441, 32446, 32448, 32501, 32509, 32517, 32532,
 32544

Algae, red (*cf.* also *Porphyridium*)
 28549, 28598, 28603, 28675, 28760, 28853, 28959-60, 28997, 29014, 29021,
 29152, 29178, 29182, 29252-3, 29337, 29340, 29525, 29545, 29589, 29591-2,
 29670-2, 29821, 29893, 29938, 29956, 30003, 30372-3, 30434-5, 30525, 30539,
 30586-8, 30592, 30624, 30718, 30799, 30940-1, 30954, 30972, 30974, 31047,
 31142, 31481, 31499, 31505-7, 31529, 31638, 31650, 31757-8, 32069, 32173,
 32348, 32356, 32517, 32544

Allium 28743, 30267, 30995, 31568-9, 31854, 32029, 32521

Alnus 28987-8, 29467, 30263

Aloe 29217, 30174

Alpine plants 28714-6, B29063, 29064-5, 29685, 29721-2, 30392-4, 30523-4, 30943,
 31392, 31670, 32415

Amaranthus 28875, 29085, 29231-2, 29489, 29681-2, 29862, 29987, 30078, 30498, 30951,
 31348, 31393, 31417-21, 32324, 32547

Anabaena 28528, 28541, 28631, 28889, 28960, 29014, 29145, 29211, 29239, 29371,
 29579, 29672, 29683-4, 29729, 29873, 30135, 30321, 30401, 30444, 30877-8,
 30970, 31092, 31108, 31136, 31176, 31188-90, 31328, 31423, 31590, 32035,
 32054, 32070, 32072, 32082, 32283, 32448, 32470

Anacardium 31004

Anacystis 28535, 28631, 28879, 28917, 29026, 29194-6, 29529, 29672, 29680, 29683,
 30299, 30322-4, 30531, 30539, 30564, 30845, 30926, 30970, 31051, 31092,
 31205-6, 31275, 31437-8, 31679, 31828, 31904, 32357-9, 32362, 32464-5

Ananas 28703,. 29215-7, 29489 30927-8, 30994, 31077

Ankistrodesmus 28764, 28982, 29185, 29945, 30864, 31031, 31184, 31218, 31909, 32206,
 32215, 32448

Antirrhinum 29186, 30361-2, 31413, 31662, 31664-5, 32211

Apium 31895

Apple see *Malus*

Apricot see *Armeniaca*

Aquatic macrophytes (*cf.* also *Elodea, Phragmites, Typha*)
       28567, 28687, 28711, 28748, 28837, 28901, 28975, 28982, 29018-9, 29049, 29175,
       29193, 29277, 29362, 29469, 29605, 29675, 29764, 29890, 29989, 30339, 30490,
       30623, 30645, 30746, 30843, 31039, B31140, 31239, 31259-60, 31301-2, 31307,
       31328, 31609, 31766, 32383, 32403, 32518

Arabidopsis 29283, 32221-2, 32481

Arachis 28872, 29334, 29603, 31038, 31193, 31262, 32034, 32236, 32253

Arbor vitae see *Thuja*

Armeniaca 28796, 29443, 32526, 32539

Artichoke see *Cynara*

Ash see *Fraxinus*

Asparagus 31833

Aspen see *Populus*

Atriplex 28645, 28774, 28818, 29015-6, 29085, 29330, 29383, 29787-8, 29862, 29865,
       30951, 31070, 31240-1, 31464, 31690-1, 31769-70, 31993, 32462, 32477

Avena 28709, 28743, 28836, 28941, 28973, 29107, 29113, 29120, 29123, 29289, 29363,
       29461, 29471, 29478, 29813, 29849-50, 30077, 30140, 30267, 30668, 30690,
       31277, 31323, 31393, 31607-8, 31669, 31682, 31768-9, 31952, 31977, 32060-1,
       32092, 32387-90, 32402, 32440

Avocado see *Persea*

B

Bacteria, photosynthetic (*cf.* also *Chlorobium, Chromatium, Halobacterium, Rhodopseudo-*
      monas, Rhodospirillum)
       28506, 28642-3, 28700, 28719, 28729, 28750, 28760, 28824, 28876, 28888, 28948,
       28951-2, 29014, 29025-6, 29035, 29073, 29098, 29124, 29126, 29134, 29149,
       29174, 29188, 29199, 29230, 29265-6, 29358, 29371, 29439, 29452, 29483, 29525,
       29618, 29683, 29748-9, 29779, 29796, 29798, 29818, 29896, 29973, 30066, 30079,
       30116, 30208, 30227, 30248, 30382, 30399, 30535, 30625, 30635, 30658, 30696-8,
       30702, 30718, 30762, 30780, 30934, 31072-4, 31138, 31191, 31219, 31253,
       31274, 31290, 31308, 31360, 31369, 31386-7, 31398, 31415, 31437, 31455,
       31458, 31624, 31730, 31741, 31789, 31810, 31929, 32114, 32251, 32282, 32409

Banana see *Musa*

Barley see *Hordeum*

Bean see *Phaseolus*

Beech see *Fagus*

Bermuda grass see *Cynodon*

Carpinus 29206-7, 29320, 29394, 30096, 30460, 31163

Carrot see *Daucus*

Carthamus 30995, 31729

Carya 29254, 29320, 29467

Cashew see *Anacardium*

Cassava see *Manihot*

Castanea 28546, 29249, 31508

Castor bean see *Ricinus*

Cat's tail see *Typha*

Cattail flag see *Typha*

Cauliflower see *Brassica*

Cedar see *Cedrus*

Cedrus B31990

Celery see *Apium*

Cerasus 29846

Cereals see *Avena, Hordeum, Oryza, Panicum, Secale, Sorgum, Triticum, Zea*

Chenopodium 28591, 28875, 29267, 29548, 29681-2, 29788, 29891, 30056, 30173, 30681,
          30746, 30947, 31092, 31099, 31129, 32008, 32214, 32329, 32547

Cherry see *Cerasus*

Chestnut see *Castanea*

Chick pea see *Cicer*

Chicory see *Cichorium*

Chinese cabbage see *Brassica*

Chlamydomonas 28513, 28554, 28646-7, 28690-1, 28723, 28758-9, 28775, 28795, 28808,
          28849-50, 28910, 28912, 29012, 29118, 29177, 29185, 29208, 29275, 29384,
          29437, 29459, 29466, 29560, 29602, 29624-6, 29661, 29673, 29683, 29736, 29759,
          29824, 29839, 29918, 30024-5, 30042, 30085, 30308-9, 30487-9, 30495-7, 30539,
          30599-601, 30622, 30624, 30718, 30752, 30803, 30836, 30944, 31031, 31063,
          31085, 31104-5, 31170, 31298, 31336, 31405-6, 31643, 31668, 31674, 31721-3,
          31726, 31844, 31907, 31914, 31917, 31930, 31975, 32029-30, 32113, 32190,
          32285, 32360-1, 32406, 32432, 32448, 32464-5, 32473

Chlorella 28550, 28726, 28742, 28762-3, 28795, 28808, 28814, 28867, 28888, 28893-5,
          28921-2, 28934, 28961, 29026, 29096, 29159-60, 29167, 29185, 29188, 29208,
          29278-9, 29308-9, 29347, 29382, 29418, 29436, 29466, 29500, 29610, 29637,
          29662, 29683, 29733-6, 29873, 29912, 29918, 29986, 30136, 30156-8, 30223-4,
          30347, 30365, 30401, 30416-7, 30452, 30539, 30613, 30622, 30624, 30635,
          30669, 30708-10, 30718-9, 20757, 30782, 30799, 30827, 30886, 30892, 30926,
          30949, 30962, 30983, 31011, 31104, 31141, 31170, 31212, 31280, 31282-3,
          31285, 31329, 31342, 31412, 31472, 31532, 31544, 31582, 31631-2, 31720,
          31723, 31726, 31738, 31778, 31782, 31835, 31957, 32050, 32075, 32101, 32125,
          32162, 32166, 32238-9, 32293, 32357, 32361, 32396, 32410, 32416-7, 32419,
          32432-3, 32453, B32515, 32524, 32534

Chlorobium   28878, 28905, 28992, 29484, 29913, 30138, 30398, 31137, 31845, 31900,
        32036

Chromatium   28506, 28582-3, 28592, 28600, 28832, 29074, 29230, 29431-2, 29454, 29479,
        29630-1, 29851, 29913, 29997, 30386, 30626, 30774, 31033, 31059, 31092,
        31094, 31157-9, 31437, 31534, 31576, 32036, 32127, 32250

Chrysanthemum   29086, 29858, 30683, 31208

Chrysophyta   28568, 28767, 29127, 29658, 29677, 29928, 29937, 30046, 30582, 30906,
        31031, 31184, 31328, 31758, 31788, 31918, 31965, 31988, 32101, 32448

Cicer   31395, 31568

Cichorium   28743, 29829

Citrullus   31769, 32337-8

Citrus   28546, 28930, 29020, 29048, 29370, 29414, 29944, 30080, 30187, 30205, 30250,
        30571, 30737-9, 30996, 31181, 31496, 31692, 31769, 31781-2, 31961, 32491,
        32506

Clover   see *Trifolium*

Cocksfoot   see *Dactylis*

Cocoa   see *Theobroma*

Coconut palm   see *Cocos*

Cocos   30858, 30979, 30994

Coffea   29563, 30699

Coffee tree   see *Coffea*

Coniferous plants (*cf*. also *Abies, Cedrus, Juniperus, Larix, Picea, Pinus, Pseudotsuga,
        Thuja, Tsuga*)
        28719, 28838, 28863, 28987, 29071, 30787, 31019-20, 31053, 31154

Corchorus   29240

Corn   see *Zea*

Cornelian cherry   see *Cornus*

Cornus   29243, 29254, 29467, 29480, 31269, 31411, 32185, 32191

Corylus   29207

Cotton   see *Gossypium*

Cottonwood   see *Populus*

Cowberry   see *Vaccinium*

Cowpea   see *Vigna*

Crabgrass   see *Digitaria*

Crataegus   30911

Cucumber   see *Cucumis*

Cucumis 28571, 28597, 28653-4, 28732, 29058-9, 29150, 29258, 29329, 29392, 29450,
      29555, 29596-7, 29711, 30235-6, 30244, 30298, 30374, 30396, 30529, 30790-2,
      31001, 31099, 31160, 31586, 31602, 31876, 31880-1, 32041, 32073, 32103, 32154,
      32382, 32547

Cucurbita 29218, 29902, 30173, 30300, 30431, B30477, 30514, 30621, 31122, 31160,
      31411, 31446, 32202

Currant see *Ribes*

Cyanobacteria see Algae, blue-green

Cynara 29538

Cynodon 31022, 31530

Cyperus 29047, 29865, 30951

D

Dactylis 28973, 30140, 30829, 31208, 31495, 31510, 31760

Dallis grass see *Paspalum*

Daucus 28666, 29280, 29555, 30276-7, 30467, 31038, 31855, 31933-4, 32153

Deciduous trees and shrubs (*cf*. also *Acer, Aesculus, Alnus, Armeniaca, Betula,*
      Carpinus, Carya, Castanea, Cerasus, Citrus, Cornus, Corylus, Crataegus,
      Eucalyptus, Fagus, Fraxinus, Hevea, Hibiscus, Juglans, Malus, Mangifera,
      Morus, Persea, Persica, Pirus, Platanus, Populus, Prunus, Quercus, Ribes,
      Robinia, Rubus, Salix, Sambucus, Sorbus, Tilia, Ulmus, Vitis)
      28801, 28838, 28987, 29121, 29202, 29209, 29212, 29250, 29254, 29301, 29320,
      29398, 29467, 29480, 29587, 29614, 29721-2, 30060-1, 30465, 30683, 30695,
      30764, 30787, 30994, 31090, 31470, 31517, 31659, 31839, 31854, 31887-8,
      32058, 32185, 32263, 32325

Desert plants and ecosystems
      28498, B28545, 28588-9, 29116, 29281, 29529, 30505-6, 30515, 30675, 30695,
      30919, 30974, 31066, 31068-70, 31148, 31535, 31887-8, 32020, 32026-7, B32318,
      32347, B32515

Dewberry see *Rubus*

Diatoms 28507, 28511, 28536, 28760, 28977, 28982, 29151, 29176-7, 29185, 29276,
      29354-5, 29428, 29434, 29466, 29479, 29588, 29655, 29677, 29691, 29802-3,
      29842, 29937, 29990-1, 30032, 30072, 30075, 30137, 30172, 30196, 30231, 30305,
      30381, 30407, 30450, 30592, 30606, 30616, 30624, 30865, 30906, 30987-8, 31031,
      31126, 31184, 31215, 31317, 31327-8, 31382, 31650, 31681, 31714, 31745, 31758,
      31810, 31889-90, 31909, 31965, 31988, 32077, 32107, 32138, 32274, 32288,
      32396, 32423, 32472

Digitaria 29376, 29489, 29771, 29865, 30035-7, 30083, 30140-1, 30951, 31393, 31447-8,
      31450-2, 32047, 32152

Dinoflagellatae, Dinophyceae
      29077, 29428, 29588, 29990, 30366, 30484, 30629, 31326, 31362-3, 31676, 31714,
      31758, 31812, 31901, 31965, 31988, 32195, 32442

Dogwood see *Cornus*

Douglas fir see *Pseudotsuga*

Dunaliella 29177, 29210, 30018, 30207, 30219-20, 31171, 32093, 32184, 32379

E

Egg plant see *Solanum*

Elaeis 28838, 29496

Elder see *Sambucus*

Elm see *Ulmus*

Elodea 29262, B29468, 29469, 29616, 29778, 30316, 30490, 30767, 31327-8, 31541, 31766, 32518

Endive see *Cichorium*

Equisetum 29021, 29205, 29852-3, 32069

Eucalyptus 28838, 29018, 29119, 29136, 29162, 30494, 31270, 31769, 31861-6, 32394

Euglena 28760, 28804-5, 28828, 28923-4, 28966, 28980, 28982, 29014, 29017, 29183, 29318, 29341, 29364, 29371, 29374, 29412, 29465-6, 29479, 29516-7, 29560, 29637, 29773, 29801, 29803, 29815, 29840, 29918, 30169, 30208, 30260-1, 30307, 30309, 30439-40, 30536, 30592, 30630, 30673, 30848, 30926, 30935, 31011, 31031, 31044, 31064, 31104-5, 31170, 31211, 31222-3, 31538, 31612, 31638, 31644, 31650, 31656, 31674, 31705, 31767, 31778-9, 31810, 31930, 31965, 32069, 32077-8, 32101, 32260, 32276, 32286, 32333, 32349-50, 32421-2, 32448, 32452, 32464-5

Euphorbia 28799, 29018-9, 29057, 29232, 30750, 31697, 32324

Evergreen plants see Sempervirent plants

F

Fagus 28622, 29344, 29398, 29467, 29489, 29614, 30263, 30476, 30594, 30881, 31696, 32197, 32257

Ferns 29835, 32263, 32377, 32394, 32458

Fescue see *Festuca*

Festuca 28973, 29064, 29217, 29581, 29725, 30393, 30483, 30566, 30829, 31021-2, 31208, 31307, 31383, 31433, 31510, 31759, 31761-2, 31870, 32097, 32144

Ficus 28546, 29155, 29163, 30833, 31839

Fig see *Ficus*

Fir see *Abies*

Flax see *Linum*

Forage crops (*cf.* also *Brassica,* Grasses, Leguminous plants, *Lupinus, Medicago, Trifolium, Vicia, Vigna, etc.*)
   28516-7, 28973, 29120, 30378, 30380, 31130-1, 31244, 31498, 31759-62

Forest (including undergrowth) plants and   ecosystems (*cf.* also Coniferous plants,
        Deciduous trees and shrubs, Ferns, *Fragaria*, Grasses, Lichens, Liverworts,
        Medicinal plants, Mosses, *Sphagnum*, *Vaccinium*, etc.)
        28801, 28857, 28890, 29071, 29561, 29614, 29628, 29703-4, 29791, 30060-1,
        30096, 30120, 30459, 30523, 30691, 31018, 31695, 31879, 32051, 32175, 32325,
        32394

Fountain-grass see *Pennisetum*

Foxtail millet see *Setaria*

Fragaria 29068, 29614, 29663, 31208, 32200

Fraxinus 28987-8, 29243, 29249, 29320, 29467, 29614, 30764, 31081, 31090, 31238,
        31269-70, 31692, 32175

Fruit plants and trees (*cf.* also *Ananas*, *Armeniaca*, *Cerasus*, *Citrullus*, *Citrus*, *Cocos*,
        Cucumis, *Cucurbita*, *Ficus*, *Fragaria*, *Malus*, *Mangifera*, *Musa*, *Persea*,
        Persica, *Pirus*, *Prunus*, *Ribes*, *Rubus*, *Sorbus*, *Vaccinium*, *Vitis*)
        28546, 30017, 32011

Fungi (parasitic) 28623, 28705, 28898, 29711, 29859, 29965-6, 30008, 30445, 31181,
        31229, 31557, 31784, 32064, 32402

G

Garlic see *Allium*

Gherkin see *Cucumis*

Glycine 28505, 28595, 28689, 28785, 28788, 28838, 28986, 28989, 29003-4, 29104,
        29113, 29376, 29381, 29475, 29482, 29705, 29868-9, 29881, 29887, 29898-9,
        29984, 30021-2, 30073, 30077, 30080, 30221, 30276-8, 30455, 30530, 30833,
        30912, 30952, 30980, 30994-5, 31071, 31107, 31340, 31348, 31377, 31393,
        31439, 31464, 31494, 31570, 31592, 31606, 31731-4, 31823, 31849, 31953-4,
        32062, 32073, 32218-9, 32304, 32312-3, 32364, 32374, 32466, 32488

Gossypium 28500, 28505, 29652, 28749, 28882, 29007, 29164, 29203-4, 29334, 29536,
        29662, 29984, 30033, 30067, 30095, 30287, 30301-2, 30415, 30445, 30513,
        30734, 30818, 30995, 31008, 31230-1, 31233, 31252, 31293-4, 31340, 31348,
        31439, 31512, 31537, 31557, 31593, 31769, 31843, 31849, 32024, 32103, 32287,
        32459, 32462, 32482

Gourd see *Cucurbita*

Gram chick pea see *Vigna*

Grape fruit see *Citrus*

Grape vine see *Vitis*

Grasses (*cf.* also *Avena*, *Bromus*, *Carex*, *Cynodon*, *Cyperus*, *Dactylis*, *Digitaria*, *Festu-*
        ca, *Hordeum*, *Lolium*, *Oryza*, *Panicum*, *Paspalum*, *Pennisetum*, *Phalaris*, *Phleum*,
        Poa, *Saccharum*, *Secale*, *Setaria*, *Sorgum*, *Triticum*, *Zea*)
        28838, 28875, 28970-3, 29024, 29057, B29063, 29064-5, 29129, 29181, 29187,
        29232, 29332, 29376, 29407, 29409, 29419, 29456-7, 29480, 29515, 29542,
        29573, 29581, 29638, 29663, 29669, 29676, 29685, 29704, 29744, 29863, 29865,
        29911, 29985, 30009, 30044, 30048, 30077, 30083, 30098, 30104-5, 30140-1,
        30270, 30273, 30354, 30379, 30393, 30459, 30483, 30533, 30573, 30633, 30763,
        30847, 30945, 30951, 31070, 31131, 31217, 31383, 31392, 31411, 31464, 31476,
                                                                    (continued)

Grasses (continued)
       31510, 31530, 31633, 31697, 31748, 31759-62, 31766, 31849, 31920, 31993,
       32047, 32098, 32117, 32144, 32151-2, 32186-7, 32189, 32259, 32324, 32363,
       32477, 32542-3, 32547

Groundnut see *Arachis*

H

Halobacterium 28722, 28826, 28831, 28848, 28943, 29028-9, 29034, 29084, 29095, 29099-
       -100, 29225, 29227-9, 29387-9, 29420, 29507, 29606-8, 29632, 29659-60, 29692,
       29742, 29847, 29854, 29900, 29926, 29929, 29935, 30064-5, 30192, 30258, 30269,
       30284, 30384, 30391, 30519-21, 30614-5, 30632, 30649, 30679, 30785, 31029,
       31036, 31110-1, 31150, 31167, 31173-5, 31286, 31309-10, 31458, 31491, 31497,
       31548, 31773, 31838, 31997, 32167, 32330, 32507

Halophilous plants (*cf.* also Salt marsh and strand plants)
       28728, 29061, 29489, 29495, 29865, 31070, 32373, 32477

Hawthorn see *Crataegus*

Hazel see *Corylus*

Hedera 29398, 29489, 30173, 30276-7, 30730, 31471, 31619, 31692, 31769

Helianthus 28505, 28774, 29054, 29328, 29372-3, 29485, 29514, 29523, 29793, 29858,
       29912, 29932, 29982, 30011, 30074, 30106, 30132, 30146, 30199, 30253, 30353,
       B30511, 30542-4, 30622, 30808, 30852, 30892, 31168, 31257, 31267, 31411,
       31464, 31494, 31537, 31604, 31620, 31633, 31790, 31801, 31859, 31926, 31932,
       32073, 32111, 32220, 32304, 32310, 32343, 32462, 32469

Hemlock see *Tsuga*

Hemp see *Cannabis*

Hevea   29018-9, 30929, 31596

Hibiscus   28838

Hickory see *Carya*

Holy see *Ilex*

Hordeum 28496, 28521, 28578, 28595, 28618, 28623, 28628, 28639, 28667-8, 28689,
       28745, 28755, 28777, 28789, 28809-10, 28817, 28836, 28973, 29008, 29040,
       29113, 29123, 29221-3, 29270, 29289, 29338, 29350, 29405, 29410, 29456, 29478,
       29489, 29546-7, 29555, 29593, 29610-1, 29699, 29708, 29760, 29765, 29823,
       29859, 29875, 29903-5, 29932, 29939, 30027-8, 30034, 30077, 30081, 30097-8,
       30132, 30163, 30201-2, 30215, 30312, 30334, 30345, 30404, 30478, 30557, 30568,
       30594, 30597, 30602-3, 30610-2, 30669, 30687-8, 30693-4, 30718, 30740-1,
       30793, 30801, 30826, 30844, 30847, 30857, 30891, 30917, 30920, 30995, 31007,
       31013, 31054, 31075, 31160, 31169, 31192, 31209, 31226-7, 31279, 31381, 31431,
       31444, 31463-4, 31494, 31504, 31533, 31547, 31572, 31587, 31715, 31742, 31752,
       31795, 31829, 31830, 31842, 31849, 31876-8, 31930, 31952, 31978, 32046, 32084,
       32086, 32118, 32176, 32232, 32266, 32298-9, 32312, 32361, 32387, 32390, 32397-
       8, 32431, 32477, 32483, 32530

Hornbeam see *Carpinus*

Horse chestnut see *Aesculus*

Horsetail see *Equisetum*

I

Ilex   30787

Ipomoea 29427, 29789-90, 30030

Ivy see *Hedera*

J

Jerusalem artichoke see *Helianthus*

Jointgrass see *Paspalum*

Juglans 29243, 31515

Juniper see *Juniperus*

Juniperus 28987, 29250

Jute see *Corchorus*

K

Kalanchoë 29092-3, 29142, 29217, 29273, 29376, 29401, 29772, 29865, 30665-6, 30937,
        30952, 31999, 32109, 32458

Kale see *Brassica*

Kenaf see *Hibiscus*

Kohlrabi see *Brassica*

L

Lactuca 28509, 28682, 28689, 28754, 28844, 28934, 28936, 29105, 29271, 29529, 29858,
        29907, 30182-3, 30291, 30718, 30801, 30950, 30971-2, 30974, 31028-9, 31684,
        31726, 31746-7, 31799, 31809-10, 31849, 31858, 31942, 32103, 32122, 32252,
        32412

Larch see *Larix*

Larix 28793, 29569, 30040, 30263, 30736, 30889, 31053

Lathyrus 29094, 31208, 31569

Leguminous plants (*cf.* also *Arachis, Cajanus, Cicer, Glycine, Lathyrus, Lens, Lupinus,
        Medicago, Phaseolus, Pisum, Trifolium, Vicia, Vigna*)
        28782, 29380, 29832, 29923, 30376, 31228, 31771

Lemna   see *Lemnaceae*

Lemnaceae   29385-6, 29616, 30350, 30533, 30549-51, 30817, 30914, 30957, 31344, 31545,
     31810, 32278, 32447

Lemon   see *Citrus*

Lens   31842

Lentil   see *Lens*

Lettuce   see *Lactuca*

Lichens   28607, 28702, 28961-2, 29421, 29435, 29806, 30204, 30285-6, 30392, 30523-4,
     31006, 31052, 31388, 31392, 31500, 31540, 31798, 31882, 32146

Linden see *Tilia*

Linseed   see *Linum*

Linum   28527, 28743, 30068, 30400, 31194, 31237, 31361

Liverworts   29446, 29769, 30596, 31112

Locust   see *Robinia*

Lolium   28521, 28527, 28602, 28689, 28704, 28743, 28757, 28896, 28899, 28925, 28973,
     29122, 29604, 29819-20, 29883, 30140, 30167-8, 30257, 30310, 30463-4, 30483,
     30891, 31220, 31261, 31527, 31759-62, 31910, 32170-1, 32449-50

Lucerne see *Medicago*

Lupine   see *Lupinus*

Lupinus   28614, 28727, 29115, 29323, 29758, 30507-8, 31379, 31657

Lycopersicon   28705, 28732, 28928-9, 28931, 28978, 29036, 29050, 29424, 29462-3,
     29586, 29603, 29951, 29954, 30010, 30080, 30090, 30199, 30294, 30377, 30427,
     B30477, 30537, 30656, 30797, 30833-4, 30883, 30920, 30972, 30974, 30976,
     31122, 31187, 31229, 31303, 31399, 31558, 31657, 31692, 31792, 31855, 31983,
     32339-41, 32373

M

Macereed   see *Typha*

Maize   see *Zea*

Malus   28629-30, 28661, 28685, 29005, 29066, 29251, 29324, 29368, 29702, 29822, 30017,
     30081, B30511, 30620, 30748, 30917, 31187, 31197-9, 31591, 31711-2, 31750-1,
     31782, 32011, 32120

Mangifera   31841

Mango   see *Mangifera*

Mangold   see *Beta*

Mangrove communities   29116, 29327

Manihot   28606, 28838, 30705-6

Manioc   see   *Manihot*

Maple   see   *Acer*

Marrow   see   *Cucurbita*

Meadowgrass   see   *Poa*

Medicago   28813, 28827, 28838, 29018, 29021, 29237, B29468, 29663, 29853, 30019,
          30371, 30455, 30458, 30619, 30707, 30824, 30847, 30932, 30995, 31264, 31313-5,
          31408, 31784, 31849, 31859, 32067, 32069, 32147, 32547

Medicinal plants (*cf.* also *Cynodon, Hibiscus, Papaver, Ricinus, etc.*)
          29055-6, 29342, 30584, 31416, 32324

Melon   see   *Cucumis*

Millet   see   *Panicum*

Morus   29320, 31627-8

Mosses (*cf.* also *Sphagnum*)
          28521, 28596, 29101, 29727, 29770, 29931, 30120, 30237-8, 30242, 30448, 30523,
          30596, 31065, 31067, 31204, 31378, 31392, 31694, 32012, 32230-1, 32551-2

Mulberry   see   *Morus*

Mung bean   see   *Vigna*

Musa   30994, 32033

Musk-melon   see   *Cucumis*

Mustard   see   *Sinapis*

N

Napier grass   see   *Pennisetum*

Nicotiana   28652, 28659, 28686, 28736, 28766, 28772, 28897, 28910, 29033, 29050,
          29108, 29217, 29268, 29322, 29356, 29376, 29464, 29560, 29586, 29683, 39698,
          29762, 29891, 29895, 29956, 30042-3, 30054, 30057-8, 30241, 30276-7, 30361,
          30363-4, 30472-4, 30498, B30511, 30546, 30569-70, 30647, 30662, 30714, 30778,
          30948, 30980, 31012, 31037. 31117-9, 31134-5, 31160, 31274, 31411, 31437,
          31439, 31480-1, 31489, 31588, 31640, 31663-6, 31743-4, 31769, 31782, 31785,
          31802, 31850-2, 31856, 31991, 32002, 32103, 32113, 32326, 32365-6, 32397,
          32420, 32462, 32516, 32529

Nostoc   28528, 29037, 29421, 29557, 29590, 29752, 29853, 30336, 31151-3, 31176, 31389,
          31529, 31783, 31882, 31919, 32070, 32470, 32540

O

Oak   see   *Quercus*

Oat   see   *Avena*

Oil palm   see   *Elaeis*

Onion see *Allium*

Orange see *Citrus*

Orchardgrass see *Dactylis*

Ornamental plants (*cf.* also *Agave, Antirrhinum, Asparagus,* Coniferous plants, *Cyperus,*
    Deciduous trees and shrubs, *Eucalyptus, Euphorbia, Ficus, Hedera, Hibiscus,*
    Ilex, Lathyrus, Lupinus, Pelargonium, Perilla, Rosa, Tradescantia, Tulipa,
    etc.)
    28502, 28620, 28853, 28870, 29039, 29232, 29250, 29313, 29373, 29425, 29612,
    29614, 29616-7, 29780, 29807, 29817, 29858, 29953, 29963, 30002, 30042,
    30346, 30392-3, 30665, 30683, 30771, 30787, 31231, 31233, 31348, 31434-5,
    31471, 31550, 31563, 31854, 32005, 32101, 32185, 32229, 32319-20, 32324,
    32397

Oryza 28787, 28809, 29110-2, 29233, 29865, 29874, 30047, 30077, 30080, 30084, 30097-
    -9, 30105, 30131, 30351-2, 30565, 30648, 30933, 31139, 31234, 31419, 31509,
    31578, 31601-2, 31623, 31777, 32006, 32047, 32312, 32504-5

P

Paddy see *Oryza*

Palms see *Cocos, Elaeis*

Panicum 28838, 29008, 29085, 29187, 29217, 29407, 29534-5, 29860, 29862-3, 29865,
    30099, 30140-1, 30233, 30235, 30951, 31400, 31447-8, 31451-2, 31465-6, 31633,
    31993, 32047, 32259, 32312, 32324, 32335-6, 32462

Papaver 31235

Paprika see *Capsicum*

Para-rubber tree see *Hevea*

Parasitic plants 29487, 29614, 30704, 31739, 31927

Parsley see *Petroselinum*

Parsnip see *Pastinaca*

Paspalum 30140, 31613, 32047

Pastinaca 29280, 29456

Pasture plants see Forage plants

Pea see *Pisum*

Peach see *Persica*

Peanut see *Arachis*

Pear see *Pirus*

Peavine see *Lathyrus*

Pecan see *Carya*

Pelargonium 29343, 29487, 29894, 31471, 31769

Pennisetum   29217, 29231, 29862, 30140, 30213, 30538, 30967, 31243-5, 31419, 32259

Pepper see *Capsicum, Piper*

Perilla 30080

Persea 30801, 30835, 31956

Persica 28546, 29622, 29944, 31885, 32011, 32539

Petroselinum 29182-3, 30244, 32069

Phalaris 28973, 29676, 30483

Phaseolus 28518-20, 28585, 28604, 28634, 28657, 28662-3, 28697, 28782, 28856, 28874,
            28883, 28887, 28909-12, 28956, 29030-1, 29046, 29054, 29138, 29214, 29247,
            29259-61, 29269, 29296-8, 29301,   29347-9,   29351, 29391, 29555, 29571,
            29579, 29603, 29610, 29648, 29841, 29922, 29932, 29964, 30042, 30088-9, 30139,
            30182-4, 30229, 30244, 30295, 30348, 30422-3, 30425, 30441, 30455, 30548,
            30643, 30797, 30814, 30837, 30849, 30862, 30897, 30901, 30903-5, 30974, 30994-
            -5, 31002, 31014, 31034, 31087-8, 31164-5, 31211, 31250-1, 31262, 31300,
            31324, 31352, 31366, 31411, 31427, 31543, 31547, 31568, 31574, 31620-1, 31692,
            31737, 31874, 31899, 31940, 31979, 31982-3, 32066, 32099-101, 32126, 32266,
            32275, 32290-1, 32297, 32314, 32352, 32385, 32399-400, 32414, 32502, 32547

Phleum 29663, 30483, 30829, 31760

Photosynthetic bacteria see Bacteria, photosynthetic

Phragmites 29362, 29675, 31307, 31817, 31965, 32047, 32518

Picea   28717-8, 28793, 28914, 28927, 29071-2, 29250, 29569, 29572, 29704, 30040,
          30085, 30262-4, 30283, 30476, B30511, 30563, 30574, 30579-81, 30736, 30800,
          30876, 31128, 31695-6, 31727-8, 31946-7

Pigeon pea see *Cajanus*

Pine see *Pinus*

Pineapple see *Ananas*

Pinus   28616-7, 28803, 28806, 28838, 28987-8, 28990, 29071-2, 29119, 29137, 29250,
          29299, 29325, 29641, 29693, 29717, 29732, 29792, 29806, 30029, 30040, 30120,
          30249-50, 30262-4, 30556, 30563, 30608, 30642, 30691, 30715, 30787, 30850,
          30869, 30898-9, 31019, 31032, 31045, 31053, 31143, 31258, 31268, 31477, 31516,
          31536, 31634, 31946-7, 31962, 32023, 32198-9, 32271-2, 32369, 32525

Piper 28976, 30880, 30994

Pirus 29737, 30481, 32485

Pisum 28508, 28521, 28524-5, 28540, 28542, 28555, 28557-60, 28585-7, 28590, 28595,
        29613-4, 28680, 28738-9, 28743, 28755-6, 28781-3, 28886, 28918, 28934, 28936,
        28991, 29021, 29081, 29106, 29114, 29157, 29165, 29238, 29285, 29323, 29404-6,
        29447, 29449, 29477-8, 29499, 29509, 29603, 29646, 29700, 29706, 29709, 29848,
        29875, 29906, 29934, 29939, 29946, 29992-3, 30013-4, 30034, 30038, 30066,
        30103, 30110-1, 30114-5, 30150, 30182-4, 30190, 30200, 30214, 30235, 30251,
        30267, 30278, 30303-4, 30309, 30327, 30329, 30360, 30369, 30376, 30395, 30436-
        -7, 30548, 30727, 30733, 30749, 30761, 30769, 30794, 30797, 30801, 30816,
        30867, 30937, 30950, 30963-4, 30982-5, 30998-9, 31043, 31075, 31092, 31161,
        31191, 31195, 31210, 31238, 31274, 31281, 31322, 31333-5, 31356, 31380-1,
        31424-6, 31454, 31503, 31523-6, 31542, 31552-3, 31568, 31708, 31742, 31807,
        31858, 31930, 31938-9, 31959, B32004, 32029, 32049, 32228, 32236, 32284,
        32309, 32312, 32315-6, 32418, 32427, 32456, 32489-90, 32512-3, B32515, 32547,
        32551, 32553

Plane tree   see   *Platanus*

Platanus   29042

Plum   see   *Prunus*

Poa   28725, 29057, 29663, 29865, 29965-6, 30140, 31208, 31383, 32185, 32547

Poplar   see   *Populus*

Poppy   see   *Papaver*

Populus   28616, 29071, 29209, 29212, 29301-2, 29306, 29400, 29467, 29555, 29753,
      30087, 30130, 30563, 30787, 31090, 31238, 31270, 31403, 31599-600, 31769,
      31793, 32087, 32175, 32196, 32297, 32541

Porphyridium   28960, 29590, 29672, 29955-6, 30877, 30892, 30926, 31709, 32448

Portulaca   29865, 29912, 30174, 30272, 30274-5, 30498, 31993, 32430, 32547

Potato   see   *Solanum*

Prune   see   *Prunus*

Prunus (*cf.* also *Armeniaca, Cerasus*)   29179, 29443, 29467, 29614, 31769, 31801, 31947,
      32526

Pseudotsuga   28821, 29335, 29717, 30262-3, 30476, B30511, 31394, 32051, 32372

Pumpkin   see   *Cucurbita*

Purslane   see   *Portulaca*

Q

Quercus   28622, 28988, 29071, 29121, 29169, 29206-7, 29209, 29249, 29254, 29320,
      29344, 29372-3, 29394, 29467, 29480, 29489, 29628, 29703, 30096, 30263, 30460,
      B30511, 30787, 31081, 31083, 31090, 31163, 31208, 31270, 31411, 31508, 32064,
      32110-1, 32185, 32194, 32257

R

Radish   see   *Raphanus*

Rape   see   *Brassica*

Raphanus   28995, 29182-3, 30173, 30199, 30594, 30597, 30992-3, 31291, 31775, 31895,
      31941, 32122, 32133

Raspberry   see   *Rubus*

Reed   see   *Phragmites*

Rhodopseudomonas   28530, 28551, 28580, 28640-1, 28692, 28822, 28904, 28949, 29098,
      29148, 29192, 29263-6, 29266, 29339, 29347, 29359, 29431-2, 29474, 29505,
      29610, 29627, 29650-1, 29683, 29712, 29878-9, 29940-1, 29947, 29959-61,
      29969-70, 29995-7, 30050, 30134, 30153, 30265, 30280, 30328, 30340-2, 30382,
      30385, 30421, 30626, 30671-2, 30788-9, 30831, 30915, 30926, 31005, 31033,
                                                                      (continued)

Rhodopseudomonas (continued)
       31046, 31059, 31092, 31179, 31225, 31287-8, 31354-5, 31360, 31368, 31371,
       31478, 31490, 31805, 32014, 32017-9, 32045, 32048, 32114, 32123, 32168-9,
       32279-81, 32346, 32367, 32467-8, 32492, 32511

Rhodospirillum
       28644, 28672, 28701, 28769, 28892, 29038, 29078, 29197-8, 29200, 29272,
       29282, 29371, 29453-4, 29479, 29664, 29666, 29683, 29750-1, 29827-8, 29980-1,
       30567, 30626, 30742, 30926, 30939, 31056, 31093, 31095, 31147, 31295-6,
       31299, 31360, 31370, 31479, 31598, 31660, 31756, 31871, 31892-3, 31958,
       32123, 32149, 32375-6, 32406, 32495, 32549-50

Ribes 29008

Rice see *Oryza*

Ricinus 29316-7, 30801, 30840, 31584-5, 32457

Robinia 28919, 28963, 30760, 30764, 30833-4

Rosa 29952, 31321

Rose see *Rosa*

Rubber tree see *Hevea*

Rubus 29467, 29489, 29920, 30760, 31692

Rye see *Secale*

Ryegrass see *Lolium*

S

Saccharum 28787, 28838, 29018-9, 29050, 29060, 29489, 29812, 29912, 30175, 30677,
       30811, 30951, 31041, 31411

Safflower see *Carthamus*

Salix 29064-5, 29209, 29467, 30393, 31270, 31595, 32415

Salt marsh and strand plants (*cf.* also Halophilous plants)
       29061, 29332-3, 29489, 29865, 30631, 30735, 30906, 31217, 32544

Sambucus 28543, 29212, 29467, 29614, 32069

Sandal see *Santalum*

Santalum 31221

Scenedesmus
       28814-6, 28845, 28847, 28942, 28944-7, 29014, 29021, 29118, 29127, 29185,
       29208, 29448, 29503-4, 29629, 29653, 29673, 29735, 29802, 29853, 30085,
       30231, 30318, 30381, 30416, 30528, 30594, 30607, 30622, 30624, 30718, 30798,
       30812-3, 30864, 30873, 31031, 31089, 31280, 31349-51, 31409, 31411, 31532,
       31564, 31720, 31723-6, 31738, 31767, 31909, 31998, 32016, 32069, 32089,
       32145, 32207, 32333, 32370, 32448, 32522-4, 32533

Secale 29476-7, 30077, 30202, 30568, 30713, 30731, 31055, 31144, 31782, 32266-7

Sedge see *Carex*

Sempervirent plants (*cf.* also *Coffea*, Coniferous plants, *Hedera*, *Ilex*, etc.)
       29250, 29587, 30465, 31659, 31947, 32324

Service-tree see *Sorbus*

Tsuga   28963, 29300, 29569, 30833-4

Tulip   see *Tulipa*

Tulipa   29953, 29957, 31442

Tundra plants and ecosystems   29290, 31758

Turnip   see *Brassica*

Typha   29362, 29605, B31140, 31239, 31307, 32518

U

Ulmus   29207, 29243, 29320, 29467, 31090, 31269, 31270

Ulva   28940, 28950, 28979, 29178, 30026, 31429, 32142-3, 32500

V

Vaccinium   29704, 30120, 30523, 31670

Vegetables (*cf.* also *Allium, Asparagus, Beta, Brassica, Capsicum, Cichorium, Cucumis,*
      Cucurbita, Cynara, Daucus, Lactuca, Lycopersicon, Pastinaca, Petroselinum,
      Phaseolus, Pisum, Portulaca, Raphanus, Solanum, Spinacia)
      28751, 29131-3, 29329, 29603, 29743, 29747, 29979, 30078, B30477, 30718,
      31399, 31547, 31586, ·31605, 31616, 31657, 31833, 31895, 32069, 32103, 32106,
      32122, 32211

Vetch   see *Vicia*

Vicia   28833, 28875, 29323, 29477, 29616, 29663, 29689, 29740-1, 29758, 29932,
      30188, 30212, 30369-70, 30376, B30477, 30659-61, 30801, 30828, 30901, 30909,
      30920, 30982-3, 31082, 31155, 31180, 31292, 31494, 31564-9, 31586, 31769,
      31826, 31921-2, 31983, 32134, 32147, 32317, 32322, 32462

Vigna   28866, 29640, 29923, 30173, 30980, 31075, 31570, 31996, 32059

Vine   see *Vitis*

Vitis   28546, 28552, 28761, 29331, 29367, 29490-2, 29944, 29968, 30480, 30703,
      30745, 31187, 31200, 31547, 31827, 31855, 31974, 32011, 32055, 32201

W

Walnut   see *Juglans*

Watermelon   see *Citrullus*

Weeds (*cf.* also *Amaranthus, Atriplex, Avena, Bromus, Chenopodium, Digitaria, Setaria,*
      etc.)
      28498, 28855, 28875, 28881, 29024, 29433, 30104, 30230, 30270, 31166, 31201,
      31232, 31347-8, 31633, 31884, 32047, 32165, 32519-20

Wheat   see *Triticum*

Whortleberry see *Vaccinium*

Willow see *Salix*

Wolffia see *Lemnaceae*

Z

Zea   28505, 28527, 28529, 28532, 28599, 28652, 28655-6, 28688, 28713, 28730,
          28743, 28786, 28792, 28807, 28838, 28865, 28896, 28909, 29018, 29021, 29045,
          29050-1, 29054, 29090-1, 29103, 29106, 29147, 29153, 29165-6, 29187, 29190,
          29262, 29376, 29405-6, 29408, 29430, 29455, 29458, B29468, 29478, 29489,
          29501, 29562, 29565-6, 29578-9, 29594, 29603, 29637, 29709, 29826, 29830,
          29833-4, 29861, 29865, 29875-6, 29885, 29912, 29984, 30006-7, 30016, 30049,
          30069, 30077, 30080, 30083, 30087, 30097-9, 30140-1, 30150, 30214-5, 30228,
          30232-3, 30235, 30243, 30252, 30267, 30272, 30275, 30288, 30296, 30313,
          30319-20, 30371, 30438, 30446, 30461-2, B30477, 30498, B30511, 30540-2,
          30557-8, 30576, 30643, 30669, 30681, 30705, 30754, 30783-4, 30801, 30819-21,
          30859, 30866, 30871, 30887, 30900-1, 30907, 30936, 30951-2, 30959-61,
          30980, 30991, 30995, 31015, 31034, 31037, 31053-4, 31075, 31244, 31261, 31265-
          -6, 31272-4, 31304-5, 31325, 31337, 31348, 31357, 31359, 31407, 31411, 31443,
          31474, 31492, 31494, 31504, 31539, 31551, 31567-8, 31579, 31586, 31611,
          31620, 31633, 31674, 31769, 31802, 31821, 31849, 31859, 31903, 31932, 31966-7,
          31977, 31984, 31993, 32047, 32065, 32076, 32084, 32096, 32103, 32118, 32147,
          32163, 32172, 32192, 32220, 32223, 32226, 32261-2, 32292, 32294, 32300,
          32304, 32312, 32317, 32322-3, 32382, 32391, 32399, 32429-30, 32462, 32484,
          32489, 32503, 32545

Zebrina see *Tradescantia*